Turbomachine Blade Vibration

Turbomachine Blade Vibration

J.S. RAO

Department of Mechanical Engineering
Indian Institute of Technology
New Delhi, India

JOHN WILEY & SONS
NEW YORK CHICHESTER BRISBANE TORONTO SINGAPORE

First published in 1991 by
WILEY EASTERN LIMITED
4835/24, Ansari Road, Daryaganj
New Delhi 110 002, India

Distributors:

Australia and New Zealand
JACARANDA WILEY LTD.,
P.O. Box 1226, Milton Old 4064
Australia

Canada
JOHN WILEY & SONS CANADA LIMITED
22 Worcester Road, Kexdale, Ontario, Canada

Europe and Africa
JOHN WILEY & SONS LIMITED
Baffins Lane, Chichester, West Sussex, England

South East Asia
JOHN WILEY & SONS (PTE) LIMITED
05-04 Block B, Union Industrial Building
37 Jalan Pemimpin, Singapore 2057

Africa and South Asia
WILEY EASTERN LIMITED
4835/24 Ansari Road, Daryaganj
New Delhi 110 002, India

North and South America and rest of the World
JOHN WILEY & SONS, INC.
605 Third Avenue, New York, NY 10158, USA

Library of Congress Cataloging-in-Publication Data

Rao, J.S.
Turbomachine blade vibration/J.S. Rao

p. cm.
Includes bibliographical references and index.

1. Turbomachines—Blades—Vibration I. Title
TJ267.5.B5R36 1991 91–554
621.406–dc20 CIP

ISBN 0-470-21764-2 JOHN WILEY & SONS, INC.
ISBN 81-224-0304-2 WILEY EASTERN LIMITED

Printed at Rajkamal Electric Press, G T Karnal Road, Delhi 110 033

To

my wife
Indira Rao

daughter
Shailaja

grand daughter
Niha

and

daughter-in-law
Kelly

Preface

I was fascinated by Machine Tools, which Indian Engineering Institutions used to flood their curriculum in 1950s and 1960s, as Workshop Theory and Practice. I always wanted to do my graduate research work in Machine Tool vibration, but my supervisor, Professor B.M. Belgaumkar, would not let me do so. Like arranged matches in India, I was asked to work on Vibration Characteristics of Tapered Beams, which I felt was a boring topic of solving some differential equations. When I wanted to change this subject for my doctoral work, again the supervisor prevailed and tied me to Tapered Beam Vibrations. I thought it is a good chance to get away from the tapered beams, for my post-doctoral work, which I chose to do in England under the sponsorship of British Council. Even the British Council found a place for me to continue work in Beam Vibrations, this time with more application to Turbine Blades in the group of Professor William D. Carnegie at the University of Surrey, and since then I was stuck in this field, and began to develop more liking to this topic. One thing led to another, and I got introduced to another colleague and friend, Professor Neville F. Rieger at Rochester Institute of Technology, with whom the ideas of Blade Excitation and Damping Mechanisms grew into the puzzle of Blade Vibration problems. Fortunately, we developed a good group at the Indian Institutes of Technology in Kharagpur and New Delhi. As a result of all the above, a large amount of information was generated over last 25 years and this book is an outcome of this effort.

With ever increase in demand for larger capacity steam turbines, high speed gas turbines and more efficient rotating compressors, blade vibration and the resulting fatigue became more and more an important subject in the mechanical engineering design of turbomachinery. Fatigue failure of turbine blades is one of the most vexing problems of turbine manufacturers, ever since the steam turbine became the mainstay for power generating equipment and gas turbines are increasingly used in the air transport. The problem is an extremely good example of demonstrating a mechanical vibration exercise encompassing the excitation due to aerodynamic stage interaction; damping due to material deformation, friction at slip surfaces and aerodynamic damping; vibration of a complex structure with an asymmetric airfoil cross-section tapered along its length and mounted at a stagger angle on a rotating rigid or flexible disk, sometimes several of such blades tied together with shrouds or lacing wires to form packets.

One can use classical exotic energy principles, or adopt discretized models with good applications of the finite element method. The problem can be described by linear theories under simple conditions or by nonlinear theories when large amplitudes are considered, particularly when damping and aeroelastic conditions are included. The problem is also governed by heat transfer analysis and the thermal stresses play an important role along with the dynamic stresses that arise out of resonances under steady or transient conditions of operation. There one can find all the romance in the Theory and Practice of Mechanical Vibrations.

This book assumes a background of at least one course each on the theory of vibration, unsteady aerodynamics, applied elasticity, numerical methods and finite elements. The book addresses itself to the application of various principles in these subjects, all the time leading to the analysis of turbine blade vibrations that may lead to the studies of its fatigue. The approach has been as simple as possible that is consistent with the demands of this subject and that would lead to the development of a computer code. The solution of technical problems is illustrated by various examples.

I owe a very large debt of gratitude to the management of Indian Institutes of Technology, at New Delhi and Kharagpur, Aeronautical Research and Development Board of Ministry of Defence in India, University of Surrey, Guildford, England, Rochester Institute of Technology, Rochester, New York and to the Stress Technology Inc., Rochester, New York. I am also deeply indebted to the Department of Science and Technology in the Government of India and to the Embassy of India, Washington, D.C., for their support to my academic activities during 1984–88 when I was the Science Counselor consolidating and developing further Indo-US collaboration in Science and Technology. I am particularly thankful to my various colleagues in India at the Indian Institutes of Technology, New Delhi and Kanpur, Bharat Heavy Electricals Ltd., Hyderabad and Gas Turbine Research Establishment, Bangalore for constant support in this field of work. I am also thankful to M/s Wiley Eastern for their care in publication of this book.

Over the years, our two children graduated and became independent, with a vacuum at home. Despite busy hours of work and travel demanded by the diplomatic job in the years 1984–88, thanks to the support from my wife, Indira, time could be found in the evenings and weekends to be devoted to the academic work. No words of appreciation can be found to the type of support and understanding given by her, without which, nothing of this would have been possible. I sincerely thank her for this support.

Washington DC **J.S. Rao**
March 1991

Contents

Preface *vii*

1. Introduction **1**

1.1 Importance of Blade Vibrations *1*
1.2 The Problem *4*
1.3 Scope and Purpose *4*

2. Single Blade Vibrations **14**

2.1 Introduction *14*
2.2 Formulation of the Problem for Beam Type of Blades *18*
2.3 Effect of Taper *29*
2.4 Effect of Pre-Twist on Bending Modes *34*
2.5 Effect of Asymmetry of Cross-section *42*
2.6 Effect of Centrifugal Forces *46*
2.7 Effect of Disc Radius *47*
2.8 Effect of Setting Angle *49*
2.9 General Blade Problem *54*
2.10 Effect of Support Stiffness *57*

3. Discrete Analysis of Blades **67**

3.1 Introduction *67*
3.2 Myklestad-Prohl Method *69*
3.3 Extended Holzer Method *78*
3.4 Finite Element Method *83*

4. Small Aspect Ratio Blades **98**

4.1 Introduction *98*
4.2 Energy Expressions and Method of Solution *100*
4.3 Bending Modes *115*
4.4 Torsion Modes *123*
4.5 Plate Modes *127*
4.6 Finite Element Method for Shells *131*

5. Blade Group Frequencies and Mode Shapes **148**

5.1 Introduction *148*
5.2 Two Dimensional Continuous System Analysis *151*
5.3 Prohl's Analysis *156*
5.4 Complex Group with Lacing Wires — Deak and Baird Analysis *167*

6. Excitation **181**

6.1 Introduction *181*
6.2 Basic Isolated Airfoil Theories *190*
6.3 Interference Effects in a Turbomachine Stage *200*
6.4 Effect of Vortex Wakes Shed by Stator *214*
6.5 Effect of Viscous Wake Interaction *216*
6.6 Some Results *221*
6.7 Hydraulic Analogy *232*

7. Damping **248**

7.1 Introduction *248*
7.2 Damping Mechanisms *252*
7.3 Modal Damping Envelopes *276*

8. Forced Vibrations **289**

8.1 Introduction *289*
8.2 Modal Analysis of Continuous Systems *291*
8.3 Prohl Method *302*
8.4 Reissner's Variational Principle *305*
8.5 Influence of Nonlinear Damping *318*

9. Transient Vibrations **324**

9.1 Introduction *324*
9.2 Transient Response of a Cantilever *325*
9.3 Response of Accelerating Turbine Blade *329*

10. Coupled Blade-Disc Vibrations **334**

10.1 Introduction *334*
10.2 Disc Receptances *338*
10.3 Blade Receptances *343*
10.4 Armstrong's Analysis for Tuned Systems *347*
10.5 Ewins' Analysis *350*
10.6 Multi Stage Disc-Blade-Shroud Analysis *357*
10.7 Finite Element Method *375*

11. Some Thoughts on Fatigue Life Estimation **386**

11.1 Fatigue Data *387*

11.2 Miner's Cumulative Rule *390*

11.3 Endurance Limit Modifying Factors *390*

11.4 Mean Stress Due to Blade Rotation *392*

11.5 Illustrations *392*

12. Some Reflections **400**

Appendix 1: Heat Transfer and Thermal Stress of Turbine Blades **403**

Appendix 2: SI Units **421**

Appendix 3: Nomenclature **426**

Index **441**

Chapter 1

Introduction

1.1 IMPORTANCE OF BLADE VIBRATIONS

Since the advent of steam turbines and their application in various sectors of industry, it is a common experience that blade failures is a major cause of breakdown in these machines. Blade failures due to fatigue are predominantly vibration related. The dynamic loads on the blading can arise from many sources, the predominant being the source of the operation principles on which the machine is designed. When a rotor blade passes across the nozzles of the stator, it experiences fluctuating lift and moment forces repeatedly at a frequency given by the number of nozzles multiplied by the speed of the machine. The blades are very flexible structural members, in the sense that a significant number of their natural frequencies can be in the region of possible nozzle excitation frequencies. Though a machine can be normally designed to avoid resonance at its steady operating speed, it experiences resonances several times during the starting and shutting of the machine, i.e. whenever the instantaneous speed of the machine gives rise to a nozzle excitation coinciding with the blade frequencies. Thus it is not infrequent to find major shut-downs of these machines arising because of blade failures. Typical examples can be seen from recently reported failures [1] and [2].

Fleeting and Coats [1] reported their experiences on the blade failures that occurred in the HP turbines of RMS "Queen Elizabeth 2" in 1968. This ship left the manufacturers on 19 November 1968 and failures occurred quite early on 24 December 1968 during the ship's maiden voyage from Tail O' the Bank, resulting in a complete damage to the 9th stage starboard HP turbine rotor and partial damage to the 10th stage starboard HP turbine rotor and the 9th stage port HP turbine rotor. This failure is attributed, by them, to resonances of the blade packet in the respective stages with nozzle excitation.

Frank [2] reported considerable damage that occurred when a 600 MW turbo-set was restarted after a general inspection. The major overhaul was carried out after 25000 hours of service. Before pressing the machine into service, the electrical overspeed protection device was checked to respond at the preset 108% of the nominal speed. Then the speed was reduced to check the mechanical overspeed protection device set to switch the turbo-set off at about 110% of the rated speed. Just before attaining this mark, the machine exploded. The generator shaft at the turbine end ruptured over half its length and broke in the middle. Two fractures had occurred at the exciter end of the shaft. The shafts of three low pressure sections were fractured both at the turbine and the generator ends. The medium pressure section was also broken at the generator end. The high pressure section of the turbine was not affected. The incident has been attributed to a sudden instability in the train of shafts, which may have been caused by a blade fracture in the last low-pressure stage or the breakage of a bearing. The turbo-set damage alone is estimated to be $40 million. A study conducted by Dewey and Rieger [3] reveals that high cycle fatigue alone is responsible for at least 40% failures in high pressure stages of steam turbines.

Steam turbine capacity continually increased till recently, to meet the rising demand for power generation. Typically, most electric utilities use 500 to 1000 MW range machines. Machines upto 1500 MW capacity have been developed and peaked at this level due to a stagnant growth in the power industry in recent times. The blade loadings have continually risen typically in increased hostile environment, temperature, wetness, speed and corrosion. Typically, the blade materials used are high chromium steel and titanium. Small turbines continue to have stainless steel blades.

Recent technological developments made it possible to raise the permissible operating temperatures for nickel and cobalt base superalloys in gas turbines. These alloys allow temperatures above 750 deg C and are used routinely in jet engines. The higher operating temperatures around 950 deg C are achieved by designing microstructures that give the alloys unusual resistance to deformation and fracture, resulting in higher efficiencies of jet engines.

Unique high temperature strength and performance in the nickel base superalloys is achieved by directional solidification, which yields a structure that is strongly anisotropic. The blades can therefore be designed so that they are strongest where they must bear the heaviest load. Also, the use of controlled solidification to produce single crystal turbine blades has significantly increased the prospects of further raise in temperatures for application in jet engines. The absence of grain boundaries in these materials frees them of certain detrimental effects of grain boundaries on high temperature behaviour. Coatings with unusual resistance to oxidation have been developed to protect the high performance superalloys in the hot environment of turbines. These are some of the key factors in the outstanding experience and long periods between the maintenance of jet aircraft engines.

The above developments demand (a) greater output per blade in MW generated or flow/hour (b) smaller frame size or greater output per unit volume (c) higher operating speeds (d) higher process temperatures. Above all, the reliability is the most demanding requirement. In general, simultaneous satisfaction of these demands has called for an increased attention to the structural problems of high cycle and low cycle fatigue of the blading.

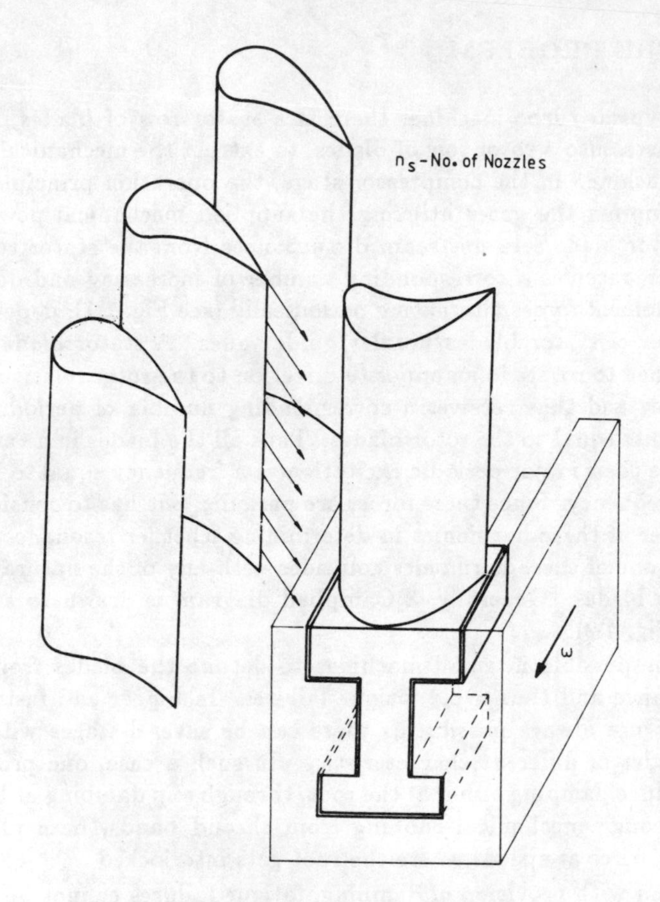

Fig. 1.1 Nozzle passing excitation

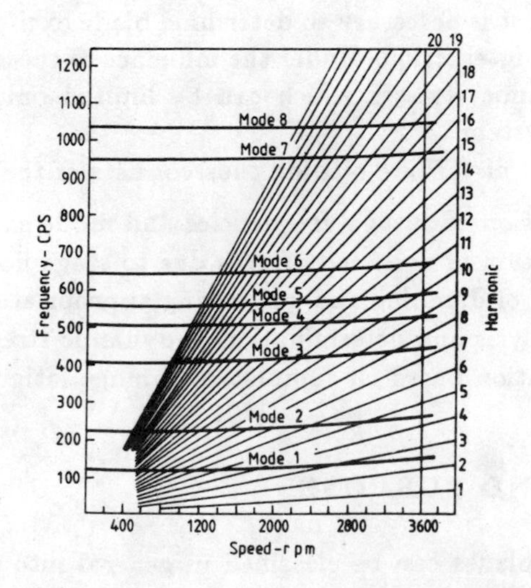

Fig. 1.2 Typical Campbell diagram

1.2 THE PROBLEM

In a typical turbo-machine, there is a stator row of blades, which guide the gases onto a rotor row of blades, to extract the mechanical power from the machine. In the compressor stage, the operation principle is reversed to compress the gases utilizing the supplied mechanical power. A typical rotor blade sees upstream disturbances from the stator row and as it rotates, receives a corresponding number of increasing and decreasing lift and moment forces alternating periodically (see Fig. 1.1), depending on the number of stator blades/nozzles/guide vanes. A stator blade can also be imagined to rotate in an opposite direction to the rotor relative to the moving row and thus receives a corresponding number of periodic forces and moments equal to the rotor blades. Thus all the blades in a turbo-machine receive their major periodic excitation at a frequency equal to Nozzle Passing Frequency. Since these forces are periodic, one has to consider a several number of these harmonics in determining whether resonance takes place, when one of these harmonics coincides with any of the natural frequencies of the blades. Generally, a Campbell diagram is drawn to ascertain this (see Fig. 1.2).

It is possible in small machines to detune the blades from a possible resonance and thus avoid fatigue failures. In bigger and faster machines, this is not always possible as there can be several stages with thousands of blades of different characteristics. In such a case, one procedure is to introduce damping either at the root, through slip damping at lower speeds, or through mechanical rubbing from shroud bands, base platforms and lacing wires at speeds where the root gets interlocked.

Even with provision of damping, fatigue failures cannot be avoided due to resonances taking place under transient conditions while the machine is started or shut down at the critical speeds defined by the Campbell diagram. Hence, it is necessary to determine blade excitation forces arising out of stage flow interaction. Under the influence of these forces, the blades experience dynamic stresses which can be limited only by the damping present in the system.

A good design of turbine blading thus consists of the following steps:

- Determination of natural frequencies and mode shapes.
- Determination of nonsteady forces due to stage flow interaction.
- Evaluation of damping and generating appropriate models.
- Modal analysis and determination of dynamic stresses.
- Life estimation based on cumulative damage fatigue theories.

1.3 SCOPE AND PURPOSE

Turbo-machine blades can be classified in general into two categories depending on their manner of operation as either impulse blades or reaction blades. Impulse blades function by redirecting the passing steam or gas flow, through a specified angle. A work producing force is developed by the resulting change of momentum of the passing fluid flow. The tangential force thus applied to each blade drives the turbine. Reaction blades function as airfoils by developing a gas dynamic lift from the pressure difference which the airfoil causes between the blade's upper and lower surfaces. High pressure stages are generally impulse stages and the low pressure stages are reaction stages. Most reaction stages develop some impulse effect in the lower vane section. Thus, a single free-standing blade can be considered as a pretwisted cantilever beam with an asymmetric aerofoil cross-section mounted at a stagger angle on a rotating disc (see Fig. 1.3). Nomenclature typically used to describe the blades is given in Fig. 1.4. Vibration characteristics of such a blade are always coupled between the two bending modes in the flapwise and chordwise directions and the torsion mode. Centrifugal loads have predominant stiffening effect, particularly for the gas turbine rotors which operate at high speeds. The problem is also complicated by several second order effects such as shear defections, rotary inertia, fiber bending in torsion, warping of the cross-section, root fixing and Coriolis accelerations. Various researchers at different stages have derived state equations of motion for such rotating blades. Here is a comprehensive set of them by Rao & Rao [4].

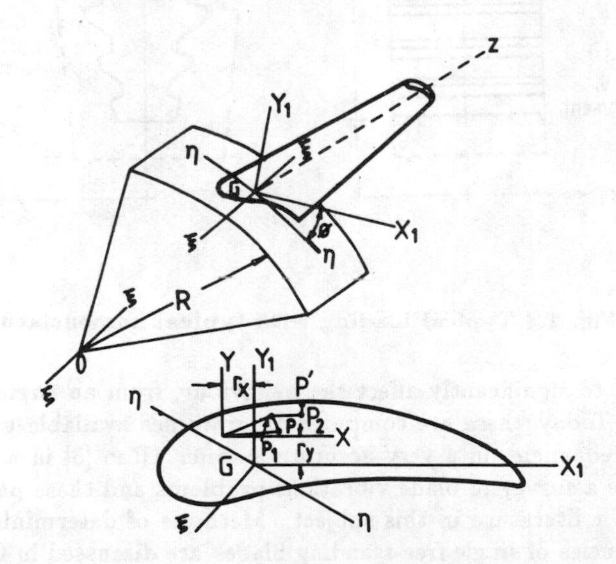

Fig. 1.3 Rotating blade with a typical cross-section

Though it is theoretically an uphill task to determine the free vibration characteristics of turbine blades taking into account all the above mentioned influencing factors, fortunately not all the second order parameters

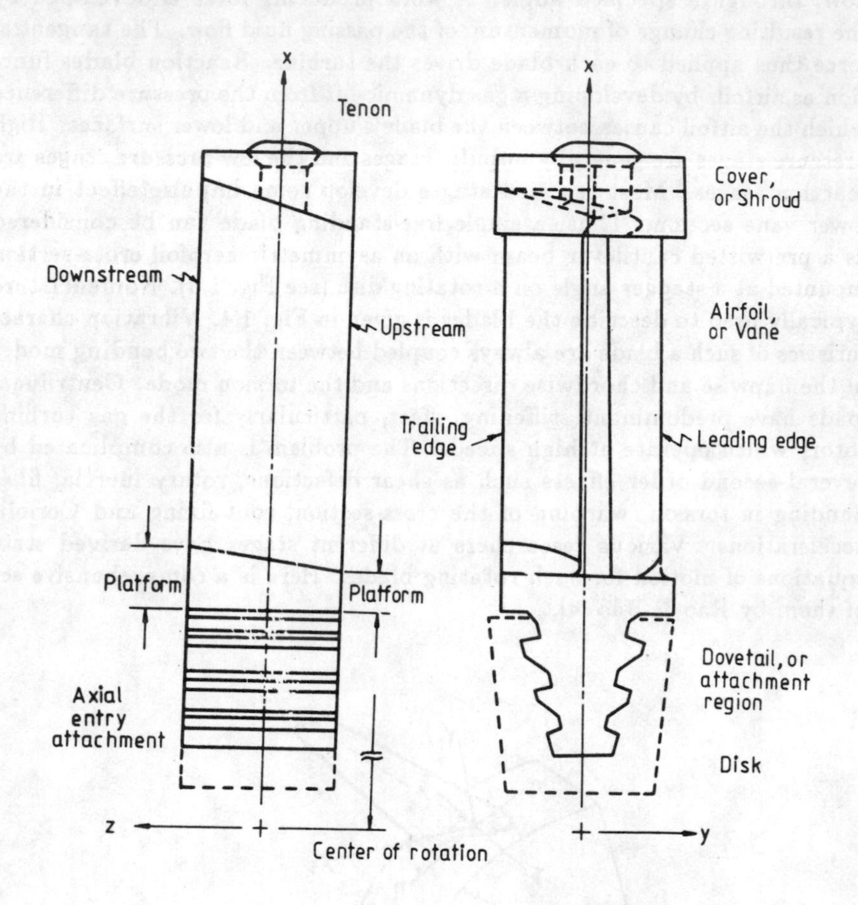

Fig. 1.4 Typical blading with typical nomenclature

are found to significantly affect the behaviour, from an engineering point of view. Today there are computer programmes available to predict the natural frequencies in a very accurate manner. Rao [5] in a series of papers made a survey of blade vibrations problems and these papers may be referred for literature in this subject. Methods of determining the natural frequencies of single free-standing blades are discussed in Chapter 2 by continuous system modelling and energy methods.

Discrete methods of analysis can also be used to determine the free vibration characteristics of blades in bending and torsion vibrations. Chapter 3 describes the discrete system modelling techniques, viz., Myklestad-Prohl, and extended Holzer methods, to determine the free vibration characteristics of free-standing blades. Finite element method analysis is also described here, using 20-noded curved serendipity isoparametric elements [6], to determine the natural frequencies of blades with complex geometry.

Certain class of turbine and compressor blades cannot be adequately dealt with by beam theories. It is essential to use plate and shell theories to accurately model such pretwisted rotating shell like structures. Finite element methods offer relatively easier approach to this problem compared to classical theory of elasticity methods. However, for pretwisted shells, finite element methods have not given very accurate results. Leissa et al [7] considered 20 different twisted plates and 19 different theoretical methods and found considerable disagreement among them. Rao and Gupta [8] set up the differential geometry of the blade in curvilinear coordinates and established the strain displacement relations. The strain and kinetic energies of the rotating and vibrating blade can then be determined and Lagrangian function set up. Following Ritz procedure, the natural frequencies and mode shapes can be determined. The methods of determining the natural frequencies of pretwisted-rotating-plate type of blades are discussed in Chapter 4. This chapter also describes a finite element method using Ahmad's 8 noded superparametric thick shell element [9], to determine the natural frequencies and mode shapes of small aspect ratio blades.

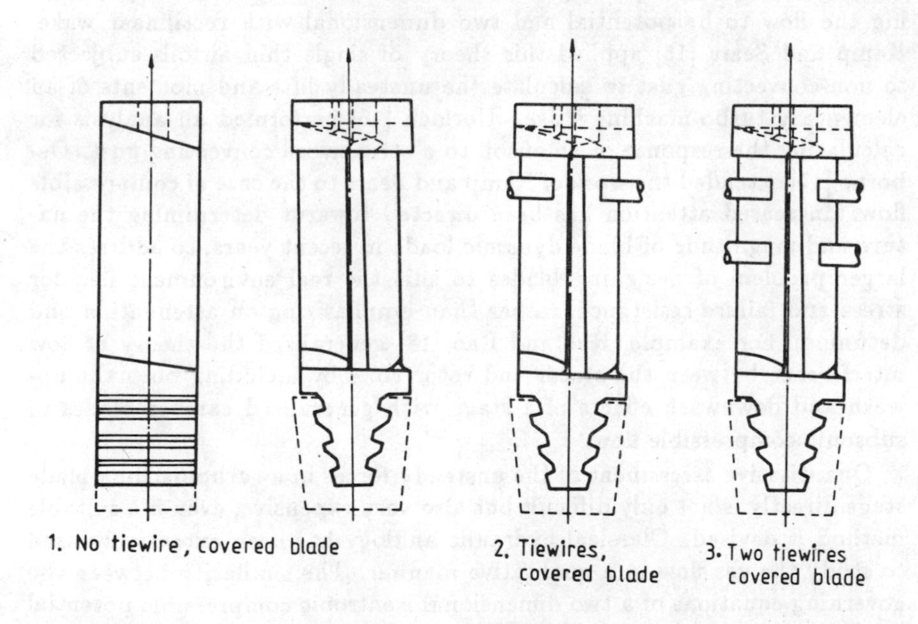

Fig. 1.5 Schematic diagram of blade with various tiewire and cover combinations

Turbine blades are often grouped together with a cover to form packets with a shroud at the tip of the blades. They are also tied together by tie or lacing wires at one or two locations as shown in Fig. 1.5. This arrangement changes the vibration characteristics of the blading. Smith [10] analyzed such blade packets. Rao [11] used variational principles to derive the equations of motion of uniform and straight blades in a packet and determined the natural frequencies. Discrete methods of analysis become handy when the blades are not of uniform cross-section. Prohl [12] gave a general method for calculating vibration frequency and stresses of a banded group of turbine buckets. Deak and Baird [13] gave a general procedure for laced blades with tie wires. These methods are discussed in Chapter 5.

Most of the dynamic loads acting on the blades result from distortions of the process gas stream. They arise from the presence of flow guides, nozzles, diaphragms, structural components, moisture separators, etc., within the gas flow path. Flow distortions may also arise from structural components plus atmospheric turbulence and other items in gas turbines. Blade dynamic loads may also arise from vibrations transmitted from the rotor, or elsewhere in the machine structure.

In normal operating conditions of a turbo-machine, the major source of excitation arises out of the interaction between the moving blade rows and stationary blade rows. Basic work relating to this field comes from Karman and Sears [14] who presented an analysis for airfoil theory for non uniform motion, by replacing the airfoil by a vortex sheet and by assuming the flow to be potential and two dimensional with rectilinear wake. Kemp and Sears [15] applied this theory of single thin airfoils subjected to non-convecting gust to calculate the unsteady lifts and moments of an elementary turbo-machine stage. Horlock [16] performed an analysis for calculating the response of an airfoil to a streamwise convecting gust. Osborne [17] extended the work of Kemp and Sears to the case of compressible flow. Increased attention has been directed towards determining the nature and magnitude of blade dynamic loads in recent years, to address the larger problem of designing blades to suit the real environment i.e., for stress and failure resistance, rather than emphasizing on attenuation and detuning. For example, Rao and Rao [18] generalized the theory of flow interference between the stator and rotor rows by including both the upwash and downwash effects of a stage with generalized camber blades in subsonic compressible flow.

Quantitative assessment of the unsteady forces in a turbomachine blade stage directly is not only difficult but also very expensive, even if a suitable method is devised. Classical hydraulic analogy has been extensively used to study the gas flow in a qualitative manner. The similarity between the governing equations of a two dimensional isentropic compressible potential gas flow and a free surface incompressible water flow makes it possible to simulate complex gas flow with ease on a water table and the essential gas dynamic quantities can be obtained from the water table results. However, the classical analogy has its own inherent limitations because the analogy is valid only for a hypothetical gas of specific heat ratio equal to 2, which no real gas possesses. This is a serious limitation which has to be taken into account before co-relating the simulated flow results with those obtained from gas dynamic solution. Various attempts have been made before, but they were without success. Rao in a series of reports [19] made extensive studies and established a modified analogy for gases with any specific heat ratio. The modified analogy uses correction factors derived for Mach numbers, pressure ratios and temperature ratios, based on the fact that the ratio of areas at any cross-section to that where the Mach number becomes unity is same for all gases with any specific heat ratio. This analogy was used by Rao et al [20] to build a rotating water table to determine the non-steady forces in a turbine stage, using the modified hydraulic analogy. Chapter 6 discusses various aspects of determining the non-steady forces by analytical methods and by the modified analogy.

Dissipation of vibratory energy in a turbine blade, is the most important part of the design of a blade, to limit the dynamic stresses that occur during transient or steady resonant conditions of operation. The damping in a turbine blade comes essentially from three different sources, Material damping, Friction damping and Gas damping. Material damping is complex in its behaviour and understanding, but a fair amount of knowledgehas been developed by Lazan [21]. Since this damping source from the material alone is not sufficient to contain the resonant amplitudes under transient conditions during start-up and shut-down operations, friction damping has been practiced for a long time in the turbine industry. The friction source is provided at the blade root or attachment region itself in the form of a fir tree, T root or any other form suitable to provide maximum slip action to dissipate the energy, where other necessary forms of friction surfaces are also provided as shroud bands, platforms and lacing wires, as discussed earlier.

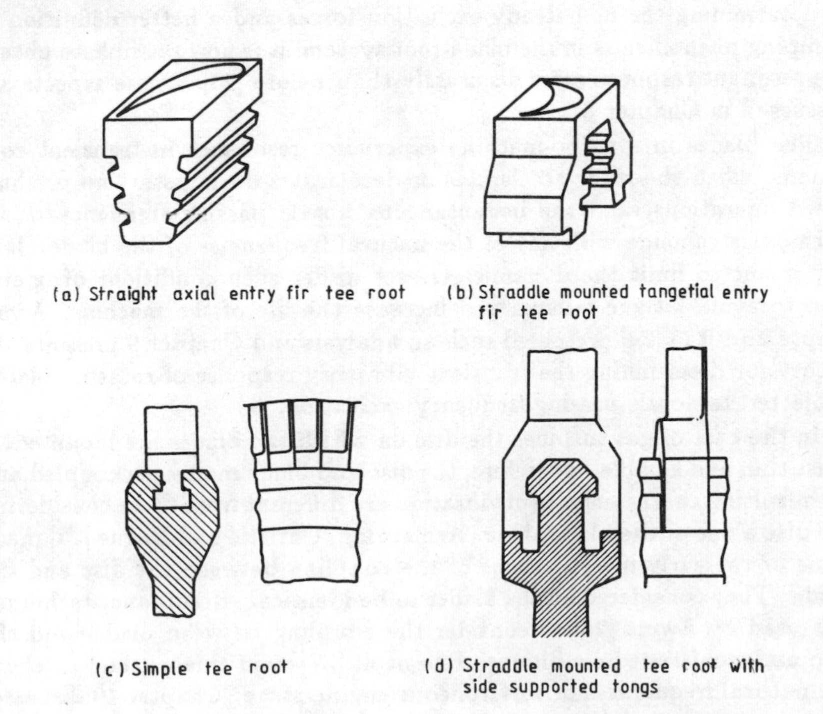

Fig. 1.6 Type of blade attachments

Most attachment regions can be classified as tangential entry or axial entry attachments as shown in Fig. 1.6. The tangential attachments are generally straddle mounted as shown. Both the fir tree type and T type are shown in this figure. It is practically very difficult to determine the friction damping arising out of such complicated blade root attachments. Traditionally, the turbine industry depended on experimental tests to determine the damping coefficients for blade root attachments. The damping in a turbine blade is a complicated problem, as it is a combination of three different complex phenomena. Slip or friction damping occurs at relatively low speeds until the root gets firmly locked due to the centrifugal loads. At high speeds the dissipation of energy is entirely due to material only, unless other friction dampers are provided like platforms, lacing wires, and shrouds. Gas dynamic damping also adds to a certain amount of dissipation of the vibratory energy. Chapter 7 discusses the latest trends, see Rao Gupta and Vyas [22], in determining the non-linear damping of blade root system, as a function of rotational speed and tip displacement for different vibratory modes.

First attempts to determine forced vibration response were made by Prohl [10] who assumed 100% steady load as the dynamic load for the purpose of calculating resonant response factors for blade packets. He used the basic principle that by resonance, all the input energy is dissipated by damping in the blade, while arriving at these factors. With improvements in determining the non-steady excitation forces and a better definition of damping phenomenon in the blade-root system, it is now possible to obtain the resonant response more accurately than before [23]. These aspects are discussed in Chapter 8.

The blades in a turbo-machine experience resonance in transient conditions, when the rotor accelerates or decelerates during start-up or shut-down operations, and the instantaneous nozzle passing frequency or its harmonics coincide with any of the natural frequencies of the blade. It is important to limit the dynamic stresses under such conditions of operation to avoid fatigue failures and increase the life of the machine. Vyas, Gupta and Rao [24] presented such an analysis and Chapter 9 presents the theory for determining the transient vibratory response of rotating blades subjected to nozzle passing frequency excitation.

In the case of gas turbines the disc on which the blades are mounted, is quite thin and flexible. Therefore, the disc and blade modes get coupled and the resulting characteristics of vibration are different from those considering the disc alone or the blade alone. Armstrong, Christie and Hague [25] made some of the early investigations of the coupling between the disc and the blade. They considered all the blades to be identical. Receptance technique was used by Ewins [26] to consider the coupling between blades and the disc under mistuned conditions. Rao et al [27] used this method to check the natural frequencies of an Orpheous engine stage. Chapter 10 discusses some of these principles to determine the natural frequencies of misturned-bladed- disc assemblies. The application of NASTRAN finite element code [28] to determine the natural frequencies and mode shapes of bladed-disc

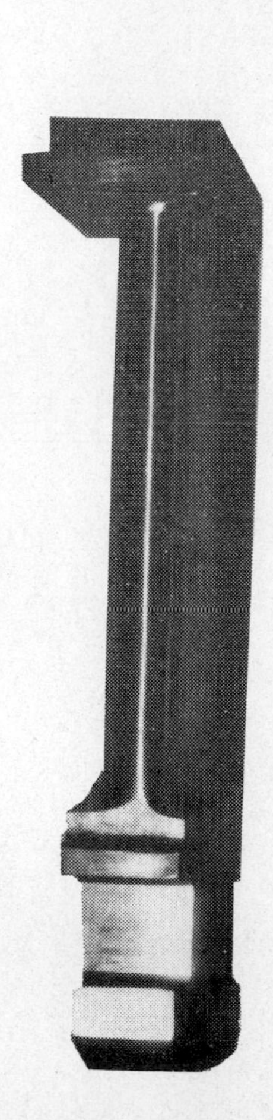

Cylindrical T-root blade with integral shroud.

T-root twisted blade with a hole for damping wire.

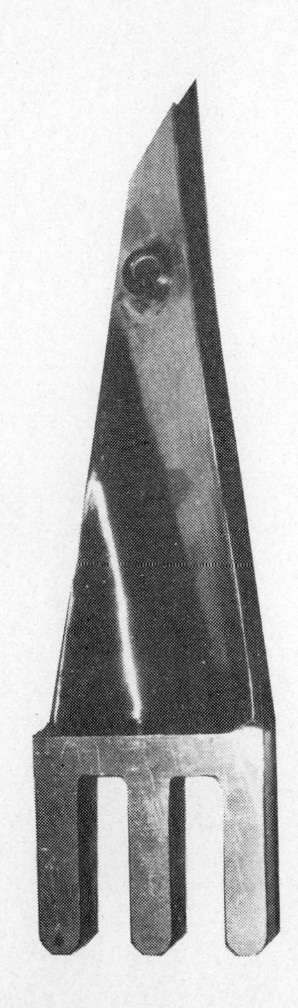

Fork root twisted blade with a boss for damping wire.

Courtesy: BHEL, Hyderabad

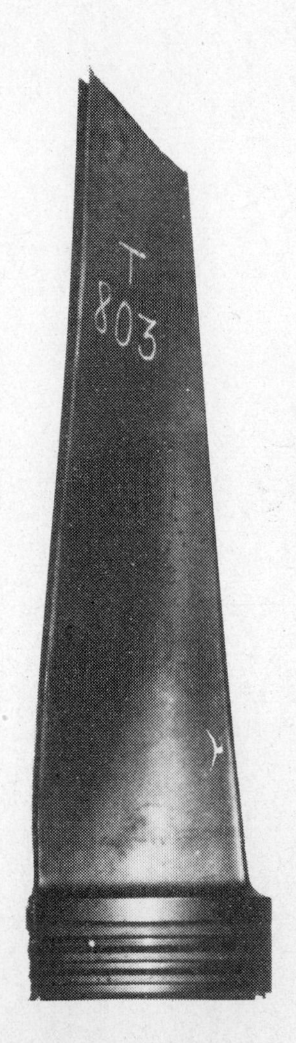

Twisted blade with a Fir-tree root

Twisted blade with a Fir-tree root

Courtesy: BHEL, Hyderabad

Cylindrical blade with integral shrouds assembled on a rotor.
Twisted blades with damping wire can be seen in the row next to the cylindrical blades.

Twisted blades with Fir-tree root and damping pins assembled on a rotor.
Note axial entry of blades into the rotor at the root.

Courtesy: BHEL, Hyderabad

Twisted blades with damping wires assembled on the rotor.
The last stage twisted blades have fork-root, and are attached to the rotor by taper pins. The ends of taper pins are visible in this photograph.
The last-but-one stage has twisted blades with T-root.
Both the rows of twisted blades have damping wires, divided into regiments.
The row on the extreme left is of cylindrical blades with T-root and integral shroud.

Twisted blade with Fir-tree root assembled on a rotor.
The last-but-one stage has a damping wire, whereas the last stage has damping pins.
Also note the axial entry of the blades into the rotor at the root.

Courtesy: BHEL, Hyderabad

structures is also described in this chapter.

Vibration induced fatigue failure is a problem of major concern in turbomachine blading. The behaviour of most blades is strongly influenced by the mean and alternating stresses experienced during the operation, by the applied load history, by the elastic and fatigue properties of the blade material and by the operating environment. The critical aspect of the problem constitutes resonant vibration at an integral order speed when the principal component of the vibrating modes in the spinning blade matches, in both time and space, a corresponding pattern in the gas/steam stream, caused by any distortion in the flow field. This may give rise to dynamic stresses leading to fatigue failure. The practice of tuning the blade away from critical frequencies may not always be possible in an engine of advanced design with as many as thousands of blades of different characteristics. As compared to typical steam turbine power generating systems which have heavy rotors operating at low speeds, the nature of the problem becomes more apparent for high performance gas turbines operating in supersonic regimes containing several critical speeds. The blades get excessively stressed during their transit past several criticals during acceleration and deceleration operations of the machine. The dynamic stress analysis of the blading under forced conditions of vibration, assessment of damaging fatigue influences and estimation of blade life, hence, becomes imperative. Fatigue crack usually initiates in a zone of high stress at some metallurgical or structural discontinuity and if critical conditions of blade operations sustain, the crack may grow and lead to the failure.

Rao and Vyas [29] discussed a procedure to estimate the life of a blade subjected to a typical loading pattern during start up/shut down cycles. The alternating stresses on the blade consisting of several harmonics of nozzle passing frequency corresponding to the harmonics in the excitation function, were considered to act independently of each other, generating independent loading patterns, to finally have a cumulative damaging influence on the blade. They used Bagci fatigue failure surface and conventional Miner's rule, to obtain the cumulative damage due to each stress block, for an occurrence of a stress peak, above the defined failure value of the stress. A significant outcome of this analysis is, a blade that would fail in 1732 load blocks of start up and shut down operations at 800 rpm/min would operate safely, if the start up and shut down accelerations are doubled to 1600 rpm/min. In Chapter 11, some of these aspects on life estimation are discussed.

Some reflections on the state of work and development on this subject and suggestions for future work are suggested in Chapter 12. This book is intended for advanced study of those interested in the field of turbine blade dynamics and design. It may also be of interest to those studying advanced vibration theories, since the book outlines the approach to a typical complex problem considering all aspects of excitation, free vibrations, damping, forced and transient vibrations, coupled structures and life estimation.

REFERENCES

1. Fleeting, R. and Coats, R., "Blade Failures in the HP Turbines of RMS Queen Elizabeth 2 and Their Rectification", *Trans. Instn. Marine Engr.*, **82**, 1970, p. 49.
2. Frank, W., Schaden Spiegel, **25**, #1, 1982, p. 20.
3. Dewey, R.P. and Rieger, N.F. "Survey of Steam Turbine Blade Failures" *Proc. EPRI Workshop on Steam Turbine Reliability*, Boston, MA, 1982.
4. Rao, J.S. and Rao, D.K., "Equations of Motion of Rotating Pretwisted Blades in Bending-Bending-Torsion with Effects of Warping, Shear, Rotary Inertia, etc.," *Proc Silver Jubilee Conf., Aeronautical Society of India*, Bangalore, p. 4, 1974.
5. Rao, J.S., "Natural Frequencies of Turbine Blading — A Survey", **5**, p. 1, 1973; "Turbine Blading Excitation and Vibration", **9**, p. 15, 1977; "Turbomachine Blade Vibration", **12**, p. 19, 1980; "Turbomachine Blade Vibration", **15**, p. 3, 1983; "Turbomachine Blade Vibration", **19**, p. 3, 1987, *The Shock and Vibration Digest.*
6. Bahree, R., Sharan, A.M. and Rao, J.S., "The Design of Rotor Blades due to the Combined Effects of Vibratory and Thermal Loads", *Trans. ASME, J. Gas Turbines*, **111**, p. 610, 1989.
7. Leissa, A.W., MacBain, J.C. and Kielb, R.E., "Vibration of Twisted Cantilever Plates — Summary of Previous and Current Studies", *J. Sound and Vib.*, **96**, p. 159, 1984.
8. Rao, J.S. and Gupta, K., "Free Vibrations of Rotating Small Aspects Ratio Pretwisted Blades", *Mechanism and Machine Theory*, **22**, v. 2. p. 159, 1987.
9. Gupta, D.K., Ramakrishnan, C.V. and Rao, J.S., "Fluid Structure Interaction Problems in Turbine Blade Vibration", *Advances in Fluid-Structure Interaction — 1984*, ASME PVP-**78**, AMD-**64**, p. 89, 1984.
10. Smith, D.M., "Vibration of Turbine Blades in Packets", *Proc 7th Conf. for Applied Machanics*, **3**, p. 178, 1949.
11. Rao, J.S., "Application of Hamilton's Principle to Shrouded Turbine Blades", *Proc. 19th Cong. Ind. Soc. Theo. and Appld. Mechanics*, p. 93, 1974.
12. Prohl, M.A., "A Method for Calculating Vibration Frequency and Stresses of A Banded Group of Turbine Buckets", *Trans. ASME*, **80**, p. 169, 1958.
13. Deak, A.L. and Baird, R.D., "A Procedure for Calculating the Frequencies of Steam Turbine Exhaust Blades", *Trans. ASME*, **85**, p. 324, 1963.
14. Karman, Th. and Sears, W.R., "Airfoil Theory for Non-uniform Motion", *J. Aero Sci.*, **5**, p. 371, 1938.
15. Kemp, N.H. and Sears, W.R., "Aerodynamic Interference Between Moving Blade Rows", *J. Aero Sci.*, **20**, p. 585, 1953.
16. Horlock, J.H., "Fluctuating Lift on Airfoils Moving Through Transverse and Chordwise Gusts", *J. Basic Engrg., Trans. ASME*, **90**, p. 494, 1968.
17. Osborne, C., "Compressibility Effects in the Unsteady Interaction Between Blade Rows", Ph.D. Thesis, Cornell University, 1971.
18. Rao, V.V.R., and Rao, J.S., "Effect of Downwash on the Non-steady Forces in a Turbomachine Stage", *Bladed Disk Assemblies*, ASME DE-vol **6**, p. 21, 1987.
19. Rao, J.S., "Hydraulic Analogy for Compressible Gas Flow in Converging Nozzles" 80-ID002-1; "Simulation of Compressible Gas Flow in Converging-Diverging Nozzles by the Use of Hydraulic Analogy" 80-ID002-2; "An Examination of Errors in Hydraulic Analogy for Nozzle Flows with Compressible

Normal Shock", 80-ID002-3; "The Effect of Straight Oblique Shock Waves on Hydraulic Analogy", 80-ID002-4; "A Study of Nozzle Exit Flows by Hydraulic Analogy", 80-ID002-5; Stress Technology Inc., Rochester, NY, 1980-81.

20. Rao, J.S. et al., "Nonsteady Force Measurement in an Orpheus Gas Turbine Engine Using Hydraulic Analogy", *Def. Sci. Journal,* **35**, p. 391, 1985.
21. Lazan, B.J., "Damping of Materials and Members in Structural Mechanics", Pergamon Press Inc., NY, 1968.
22. Rao, J.S., Gupta, K. and Vyas, N.S., "Blade Damping Measurements in A Spin Rig with Nozzle Passing Excitation Simulated by Electromagnets", *Shock and Vib. Bull.* no. **56**, pt. 2, p. 109, 1986.
23. Rao, J.S. and Vyas, N.S., "Resonant Stress Determination of A Turbine Blade with Modal Damping as a Function of Rotor Speed and Vibrational Amplitude", ASME 89-GT-27; "Transient Stress Response of a Turbine Blade Under Nonlinear Damping Effects", ASME 90-GT-269.
24. Vyas, N.S., Gupta, K. and Rao, J.S., "Transient Response of Turbine Blade", *Proc. 7th World Cong. IFToMM,* Sevilla, Spain, p. 697, 1987.
25. Armstrong, E.K. and Christie, P.I. and Hague, W.M., "Natural Frequencies of Bladed Discs", *Proc. Instn. Mech. Engrs.,* **180**, pt. 31, p. 110, 1965.
26. Ewins, D.J., "Vibration Characteristics of Bladed Disc Assemblies", *J. Mech. Engrg. Sci.,* **15**, p. 165, 1973.
27. Rao, J.S., Shah, C.B., Ganesh, Ch.L. and Rao, Y.V.K.S., "Vibration Characteristics of Aircraft Engine-Blade Disc Assembly", *Def. Sci. Journal,* **36**, p. 9, 1986.
28. Midturi, S., Soni, M.L., Stange, W.A. and Reed, J.D., "On Model Generation and Modal Analysis of Flexible Bladed-Disk Assemblies", *Bladed Disk Assemblies,* ASME Publication, p. 49, 1987.
29. Rao, J.S. and Vyas, N.S., "On Life Estimation of Turbine Blading", *7th World Cong. IFToMM — Rotor Dynamics Committee Proc.,* Sevilla, Spain, 1987.

Chapter 2

Single Blade Vibrations

2.1 INTRODUCTION

A single free-standing blade can be considered as a pre-twisted cantilever beam with an asymmetric aerofoil cross-section mounted at a stagger angle on a rotating disc. The starting solution for a simple stationary blade is obtained from the classical Euler-Bernoulli beam [1] with cantilever boundary conditions for bending vibration and St. Venant's non-circular rod for torsional vibration [1,2]. Coupled bending-torsion vibrations occur when the centre of flexure does not coincide with the centroid as in the aerofoil blade cross-section and the vibrations are coupled between the two bending modes because of pre-twist. The problem becomes further complicated because of second order effects such as shear deflections, rotary inertia, fiber bending in torsion, warping of the cross section, root fixing and Coriolis accelerations. In general the equations of motion will be six coupled-partial-differential equations [3] coupled between the two bending deflections, the two shearing deflections, the torsional deflection and the longitudinal deflection. Further the warping function will be obtained by a modified Poisson equation, taking into account the dynamic conditions of the blade. Thus, theoretically it is an uphill task to determine the natural frequencies of an actual turbine blade with all the effects mentioned.

Thus various researchers have derived solutions to the problem by considering individual aspects such as taper, pre-twist, asymmetry of cross-section centrifugal forces and making simplified assumptions with regard to second order effects. Broadly speaking, this work was carried through two different approaches viz., continuous system and discrete system modelings of the blade. In this chapter, the continuous system modeling is used which enables a systematic parametric study for the estimation of natural frequencies of free standing turbine blades.

The classical approach of solving the differential equations of motion of a cantilever blade is possible only under very simplified conditions, typical examples are Ward [4], Wrinch [5], Meyer [6] on tapered beams and Sutherland and Goodman [7] on the effects of shear and rotary inertia. Energy methods for continuous systems are very powerful in the solution of turbine blade problems. Different versions of energy methods are Rayleigh [8], Rayleigh-Ritz[9], Galerkin [10], Lagrange [11], Reissner [12] and Nemat-

Nassar [13] based on the basic Hamilton's principle. Other methods used for solving differential equations of motion of cantilever blades are Integral equation approach [14], Perturbation [15] and Collocation [10] procedures. The literature in this area is reviewed by Rao [16, 17, 18 and 19].

Effects of Taper

Meyer [6], Ward [4], Nicholson [20], Wrinch [5], Akimasa [21], Conway [22], Watanabe [23], Nubuo [24] and Wang [25] gave solutions of lateral vibrations of cantilever beams with variable cross-section. Martin [15] derived correction factors for the effect of taper on flexural frequencies by expressing the frequency as a double power series on two taper parameters of breadth and thickness. Power series solutions of blade natural frequencies have been obtained by Taylor [26] for the case of a uniform beam and a completely tapered beam. Using the Galerkin method, Rao [27] obtained the formula for the fundamental flexural frequency of a tapered cantilever beam with rectangular corss-section. With the help of a digital computer, Rao and Carnegie [28] determined the first three lateral frequencies of tapered cantilever beams by the use of the Ritz-Galerkin process. Rao [10] applied the Collocation method to determine the first three natural frequencies of tapered cantilever beams with rectangular cross-section and showed close agreement with the results obtained by the Ritz-Galerkin process. Walker [29] gave simple formulae for the fundamental torsional mode obtained by substituting the original system by a torsional member without mass and having single moment inertia. Vet [30] gave results of torsional vibrations of rectangular cross-section cantilever beams for general ratios of length to depth and width to depth, obtained by the use of the Rayleigh-Ritz method. Rao [31] used the Collocation method to obtain a formula for the fundamental mode of a tapered cantilever beam with a triangular cross- section in torsional vibration. Rao, Belgaumkar and Carnegie [32] applied the Collocation method for a tapered cantilever blade with rectangular cross section, determined the fundamental mode in torsional vibration, which was later extended to higher modes [33]. Rayleigh introduced the concept of rotary inertia, which was later extended by Timoshenko [34] who obtained an equation allowing for both rotary inertia and shear deflection by a conventional Newtonian approach.

Sutherland and Goodman [7] solved the Timoshenko beam equation for the case of simply supported and cantilever beams. Huang [35, 36, 37] first treated the problem by the approximate methods of Ritz and Galerkin and later published useful design charts for eigen values and modifying quotients of beams which had several end conditions. The results obtained by Huang [36] for the first mode were incorrect and corrected values were later given by Huang and King [38]. Lee and Bishop [14] used integral equations for the solution of flexural vibrations of a wedge taking into account both the effects of rotary inertia and shear. Lee [39] used the minimum principle to determine the vibration characteristics of a wedge taking into account the rotary inertia and shear. Carnegie [40] studied the effects of shear deflection and rotary inertia for straight and pre- twisted uniform cantilever beams. Dawson [41] used the Rayleigh-Ritz method and showed good agreement between his theoretical results and those of Sutherland and Goodman.

Vibration of Rotating Cantilever Blades

Lo and Renbarger [42] derived the differential equation of motion of a cantilever blade mounted on a rotating disc at a stagger angle and showed that the frequencies of a bar, vibrating transverse to a plane, inclined at an angle to the plane of rotation, can be found by a simple transformation of the frequencies of a bar, vibrating perpendicular to the plane of rotation. Lo [43] in his analysis simplified the non-linear problem by assuming the blade to be rigid everywhere except at the root and presented the solution in a phase plane. Boyce, Diprima and Handleman [44] used Rayleigh-Ritz and Southwell methods, to determine the upper and lower bounds of natural frequencies of a turbine blade, vibrating perpendicular to the plane of rotation. Yntema [45] making use of the bending mode of a non-rotating beam determined the first three modes of a rotating beam by the use of a Rayleigh energy approach. Boyce [46] used Rayleigh-Ritz and Southwell methods to determine the upper and lower bounds of natural frequencies and showed that the frequencies depend almost linearly on the hub radius for various rotational speeds. Carnegie [47] derived a theoretical expression for the work done due to centrifugal effects for small vibrations of rotating cantilever beams and established an equation for the fundamental frequency of vibration by the use of Rayleigh's method. Schilhansl [48] investigated the stiffening effect of the centrifugal forces on the first mode frequency of a rotating cantilever blade by the use of successive approximation. Hirsch [49] considered the effect of non-rigid supports. Horway [50] considered hinged ends and Niordson [51] considered the effect of loose hinged supports on the natural frequencies of rotating blades. Kissel [52] has given the modified Southwell factor for the effect of rotation. Rao [53] used the Galerkin method to study the effects of rotational speed, disc radius and stagger angle of the blade and obtained a simple relation to determine the frequencies of lateral vibrations. Rao and Carnegie [54, 55] used the Ritz energy process and an averaging procedure to solve the differential equation derived by Carnegie [56] and obtained the non-linear response of a cantilever blade.

Vibration of Pre-Twisted Cantilever Blades

If the blade has pre-twist, uncoupled bending modes in flapwise and chordwise directions are not possible. Both the lateral deflections are always coupled and coupled bending-bending vibrations occur. Dunholter [57] considered the static displacements and natural frequencies of pre-twisted beams. White [58] employed Green's functions to derive the conditions of orthogonality for a uniform pre-twisted blade executing bending-bending vibrations. Diprima and Handelman [59] solved the equations of motion of a pre-twisted cantilever blade by the Rayleigh-Ritz principle. Troesch,

Anliker and Zeigler [60] gave the frequency equation for the case of an infinitely thin cross-section. Martin [61] used the mathematical theory of naturally bent and twisted rods as given in Love's theory of elasticity and applied the perturbation method to solve the pre-twisted beam problem. Schimogo [62] gave a solution under various assumptions for lateral vibrations of pre-twisted rods with variable cross-section. Sato [63] considered the position of an elastic axis and its effects on lateral vibrations of pre-twisted rods. Sylper [64] extended the known condition of orthogonality of normal modes, for a beam in simple bending in one transverse plane, to that of coupled bending-bending modes of vibrations in two planes. Anliker and Troesch [65] gave the solutions of pre-twisted rods for various boundary conditions. Dawson [66] presented the solution for pre-twisted beams executing lateral vibrations by the Ritz method. Rao [67] used Galerkin method to determine the first five frequencies of a pre-twisted tapered cantilever blade. Bogdanoff and Horner [68] gave results for the influence of rotation on the first three natural frequencies in torsion of a uniform cantilever beam with different base setting angles and constant pre-twist rates. Brody and Targoff [69] considered an additional effect due to torsion on the vibrations of pre-twisted cantilever blades. Carnegie [70] and Rao [71] considered the effects of fiber bending on the torsional vibration of pre-twisted cantilever blades. Rao and Carnegie [33] applied the Collocation method to determine the natural frequencies in torsion of pre-twisted tapered cantilever blades. Galerkin method was used by Rao [72] to study the torsional vibrations of pre-twisted tapered cantilever blades. Subrahmanyam et al used Reissner variational method for pre-twisted blades [73] and Rayleigh-Ritz method was used for pre-twisted rotating blades [74, 9], including the effects of shear deformation and rotary inertia.

Cantilever Blades with Asymmetrical Cross-Section

If the geometry of the beam is such that the centre of flexure and centroid of a cross section do not coincide, coupling between flexural and torsional modes exist. Garland [75] considered the case of such a cantilever blade and used the Rayleigh-Ritz method to determine the frequencies and amplitude ratios. Houbolt and Brooks [76] derived the equations of motion for pre-twisted cantilever beams with asymmetrical aerofoil cross-section and suggested that Rayleigh-Ritz method for their solution. Carnegie [77] described an experimental method to determine the centres of flexure and torsion of aerofoil cross-sections. Dawson [78] determined the natural frequencies of cantilever beams executing coupled bending-bending-torsion vibrations by Rayleigh-Ritz method. Rao and Carnegie [79] used Galerkin method to determine the first five coupled frequencies of a straight uniform blade with asymmetric aerofoil cross-section. Collocation method was used by Rao [80] to solve the coupled equations of a straight cantilever blade with asymmetric aerofoil cross section and showed good agreement with the theoretical results obtained by Galerkin method and the experimental results of Carnegie and Dawson [81].

Coupled bending-torsional vibrations, including shear and rotary inertia, have been studied [82]; simple relations to determine the natural frequencies were used to examine the effects of disc radius, rotational speed, asymmetry of cross-section and stagger angle. Rao [83] used the Galerkin method to study coupled bending-torsional vibrations of pre-twisted asymmetric blades mounted on a rotating disc at a stagger angle and the design charts were given. The study has been extended to the development of a general computer program for a free-standing blade having specified sectional properties [84, 11]. Zuladzinski [85] used a work-energy formulation to determine the effect of rotation on fundamental frequency. Using Lagrange's equations, Jadvani [11] made free vibration analysis of rotating pre-twisted blades of asymmetric aerofoil cross-section mounted on a disc at a stagger angle and provided several design charts to determine the coupled natural frequencies. Reissner variational method was extended to take into account asymmetry [86], rotation [87] and then to general blades [88].

In this chapter, the motion of the blade is first defined and using this response, the energy expressions are set up. Using Lagrange's equations, the eigen-value problem for a general blade is obtained. The results obtained in non- dimensional parameters of the blade are presented to aid the designer in predicting the natural frequencies. A general computer program is described which can be used to determine nine coupled frequencies of a blade, mounted on rotating disc, with the profile defined at specified stations.

2.2 FORMULATION OF PROBLEM FOR BEAM TYPE OF BLADES

The response of the blade can be described in terms of bending displacements $x(z,t)$, $y(z,t)$ of the point O, which is the centre of flexure of the cross-section of the blade shown in figure 2.1 and torsional displacement $\theta(z,t)$ with respect to the same point. A particle initially at point P with coordinates x, y in $0xy$ axis system, moves to point P_1 under the influence

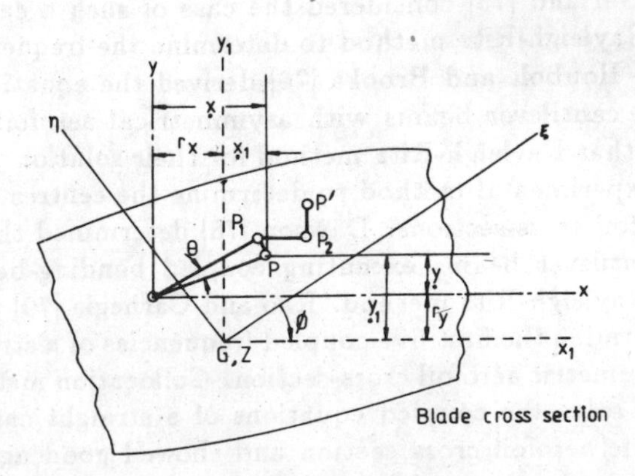

Fig. 2.1 Coordinate axis–displacement vector

of angle of rotation θ, resulting in an inward displacement $y\theta$, and outward displacement $x\theta$ in x and y directions respectively. This point P further-moves from P_1 to P_2 by $x(z,t)$ and then vertically to a position P' by $y(z,t)$ under the influence of bending displacements. Hence, the x and y displacements u_x and u_y of the particle P can be written as

$$\begin{aligned} u_x &= x - y\theta \\ u_y &= y + x\theta \end{aligned} \tag{2.1}$$

In figure 2.1, an axis system $x_1 y_1$ through the centroid G is also used to describe the bending displacements, which identifies the stagger angle ϕ with respect to the ξ axis in the plane of rotation of the disc as shown in figure 2.2. In figure 2.2 the blade is occupying a position Γ, for the instance as shown in fixed axis system CMN.

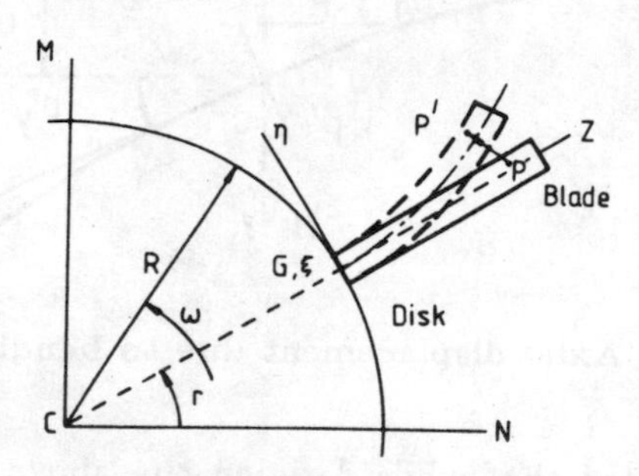

Fig. 2.2 Displacement vector in fixed axis system

xy and x_1y_1 axis in figure 2.1 are displaced by r_x in x direction and r_y in y direction respectively. The displacements in equation (2.1) can therefore be written as

$$\begin{aligned} u_x &= (x + r_y\theta) - y_1\theta = x_1 - y_1\theta \\ u_y &= (y + r_x\theta) + x_1\theta = y_1 + x_1\theta \end{aligned} \tag{2.2}$$

In the above equation x_1 and y_1 denote the bending displacements of the centroid of the cross-section and x and y denote the bending displacement of the centre of flexure O.

The point P described before, moving from P to P_1, P_1 to P_2 and P_2 to P', further moves longitudinally in z direction by $u(z,t)$ under the action of centrifugal forces. It has a further motion in the axial direction due to bending action given by $-x_1u_x' - y_1u_y'$, as given in figures 2.3 and 2.4. The displacement in axial direction z, due to warping can be neglected. Then, the total longitudinal placement u_z can be obtained as

$$\begin{aligned} u_z &= -x_1u_x' - y_1u_y' + u \\ &= -x_1(x_1' - y_1\theta') - y_1(y_1' + x_1\theta') + u \end{aligned} \tag{2.3}$$

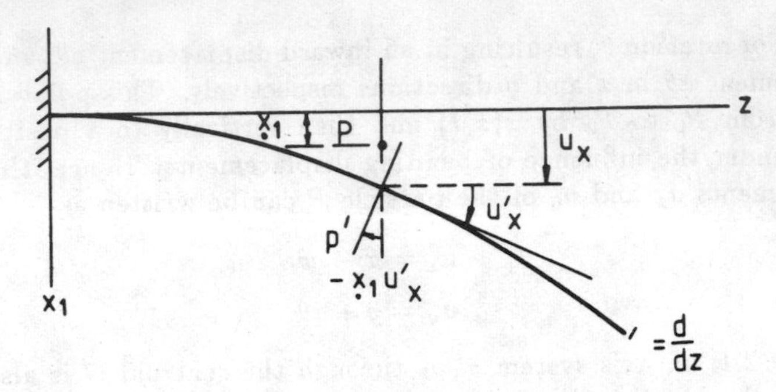

Fig. 2.3 Axial displacement due to bending in x_1z plane

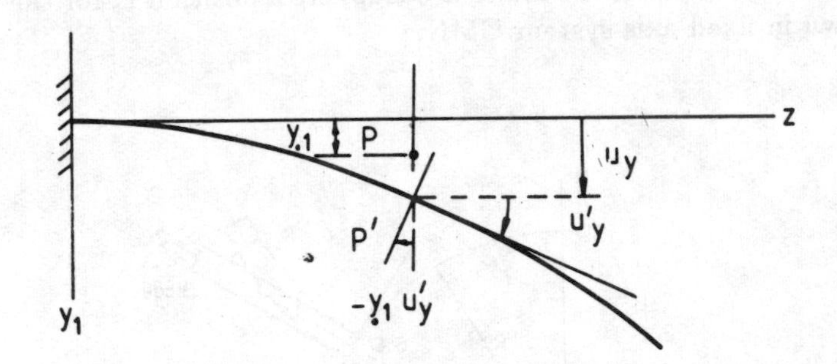

Fig. 2.4 Axial displacement due to bending in y_1z plane

It should be noted that while deriving the above equation, the effect of shear deformation is neglected. Equations (2.2) and (2.3) give the state of the blade, which are used in determining the kinetic and potential energies as follows.

Kinetic Energy of Rotating Blade

The displacement vector u of the particle P can be written as

$$\overline{u} = \overline{PP'} = u_x\overline{e}_{x1} + u_y\overline{e}_{y1} + u_z\overline{e}_{z1} \tag{2.4}$$

where $\overline{e}_{x1}$, $\overline{e}_{y1}$ and $\overline{e}_{z1}$ denote the rotating unit vectors in x_1, y_1 and z directions respectively. The unit vectors $\overline{e}_{x1}$, $\overline{e}_{y1}$ are related to the rotating unit vectors $\overline{e}_\xi$ and $\overline{e}_\eta$ in ξ and η directions by

$$\begin{aligned} \overline{e}_{x1} &= \overline{e}_\xi \cos\phi - \overline{e}_\eta \sin\phi \\ \overline{e}_{y1} &= \overline{e}_\xi \sin\phi + \overline{e}_\eta \cos\phi \end{aligned} \tag{2.5}$$

where ϕ denotes the stagger angle of the blade.

Equation (2.4) can also be written as

$$u = u_\xi\overline{e}_\xi + u_\eta\overline{e}_\eta + u_z\overline{e}_z \tag{2.6}$$

where u_ξ andu_η are components of displacement vector along ξ and η directions and are given by

$$
\begin{aligned}
u_\xi &= u_x \cos\phi + u_y \sin\phi \\
&= (x_1 - y_1\theta)\cos\phi + (y_1 + x_1\theta)\sin\phi \\
u_\eta &= -u_x \sin\phi + u_y \cos\phi \\
&= -(x_1 - y_1\theta)\sin\phi + (y_1 + x_1\theta)\cos\phi
\end{aligned} \tag{2.7}
$$

The position vector $\overline{CP'}$ of the displaced particle in figure 2.2 is

$$
\begin{aligned}
\bar{r} &= \overline{CP'} = \overline{CG} + \overline{GP} + \overline{PP'} \\
&= (R\bar{e}_z) + (\xi\,\bar{e}_\xi + \eta\bar{e}_\eta + z\bar{e}_z) + u_\xi\bar{e}_\xi + u_\eta\bar{e}_\eta + u_z\bar{e}_z
\end{aligned} \tag{2.8}
$$

The above can be rewritten as

$$
\begin{aligned}
\bar{r} &= (\xi + u_\xi)\,\bar{e}_\xi + (\eta + u_\eta)\,\bar{e}_\eta + (R + z + u_z)\,\bar{e}_z \\
&= r_\xi\bar{e}_\xi + r_\eta\bar{e}_\eta + r_z\bar{e}_z
\end{aligned} \tag{2.9}
$$

Therefore, the velocity of particle P can be obtained as

$$
\dot{\bar{r}} = \dot{r}_\xi\bar{e}_\xi + r_\xi\dot{\bar{e}}_\xi + \dot{r}_\eta\bar{e}_\eta + r_\eta\dot{\bar{e}}_\eta + \dot{r}_z\bar{e}_z + r_z\dot{\bar{e}}_z \tag{2.10}
$$

Noting that

$$
\begin{aligned}
\dot{\bar{e}}_z &= \omega\bar{e}_\eta \\
\dot{\bar{e}}_\eta &= -\omega\bar{e}_z \\
\dot{\bar{e}}_\xi &= 0
\end{aligned} \tag{2.11}
$$

equation (2.10) becomes

$$
\dot{\bar{r}} = \dot{r}_\xi\bar{e}_\xi + (\dot{r}_\eta + \omega r_z)\bar{e}_\eta + (\dot{r}_z - \omega r_\eta)\bar{e}_z \tag{2.12}
$$

From the above, the square of the velocity of the particle is obtained as

$$
\begin{aligned}
v^2 &= \dot{\bar{r}} \cdot \dot{\bar{r}} \\
&= (\dot{r}_\xi^2 + \dot{r}_\eta^2 + \dot{r}_z^2) + 2\omega(\dot{r}_\eta r_z - r_\eta\dot{r}_z) + \omega^2(r_\eta^2 + r_z^2)
\end{aligned} \tag{2.13}
$$

With the help of equations (2.9), (2.7) and (2.3), the first group of terms in equation (2.13) simplifies to

$$
\begin{aligned}
\dot{r}_\xi^2 + \dot{r}_\eta^2 + \dot{r}_z^2 &= \dot{x}_1^2 + \dot{y}_1^2 + (x_1^2 + y_1^2)\dot{\theta}^2 \\
&\quad + 2(x_1\dot{y}_1 - y_1\dot{x}_1)\dot{\theta} + \dot{u}^2 + x_1^2\dot{x}_1'^2 + y_1^2\dot{y}_1'^2 \\
&\quad + 2x_1y_1\dot{x}_1'\dot{y}_1' - 2\dot{u}(x_1\dot{x}_1' - y_1\dot{y}_1')
\end{aligned} \tag{2.14}
$$

Ignoring the axial rotary inertia terms, the above becomes

$$
\dot{r}_\xi^2 + \dot{r}_\eta^2 + \dot{r}_z^2 = \dot{x}_1^2 + \dot{y}_1^2 + (x_1^2 + y_1^2)\dot{\theta}^2 \tag{2.15}
$$

The second term in equation (2.13) represents Coriolis components and are neglected here. The third group of terms in equation (2.13) represents the kinetic energy due to rotation, which is

$$\omega^2(r_\eta^2 + r_z^2) = \omega^2\left[\{\eta + (y_1 + x_1\theta)\cos\phi - (x_1 - y_1\theta)\sin\phi\}^2 + \{(R+z) + u - x_1(x_1' - y_1\theta') - y_1(y_1' + x_1\theta')\}^2\right] \tag{2.16}$$

Accounting for centrifugal components due to the displacement of the particles on the line of the centroid only, the above equation simplifies to

$$\omega^2(r_\eta^2 + r_z^2) = \omega^2\left[2(R+z)u + y_1^2\cos^2\phi + x_1^2\sin^2\phi - x_1 y_1 \sin 2\phi\right] \tag{2.17}$$

In the above, rigid body motion terms and u^2 terms are set to zero. It will be mostly seen that u is a small quantity and its square becomes a second order term.

The total kinetic energy T of the blade is given by

$$T = \frac{1}{2}\int v^2\,dm = \frac{1}{2}\int \rho v^2\,dA\,dz \tag{2.18}$$

With the help of equations (2.15) and (2.17), we get

$$T = \int_0^L \left[\frac{m}{2}(\dot{x}_1^2 + \dot{y}_1^2) + \frac{I_{cg}}{2}\dot{\theta}^2 + \omega^2(R+z)um + \frac{\omega^2}{2}m(y_1^2\cos^2\phi + x_1^2\sin^2\phi - x_1 y_1 \sin 2\phi)\right] dz \tag{2.19}$$

In the above ρ is mass density and m is mass per unit length of blade, and

$$I_{cg} = m\int_A (x_1^2 + y_1^2)\,dA = mI_p$$

Figures 2.5 and 2.6 show the rotation of an element dz at distance z in x, z and y, z planes respectively. The inward displacement of the element

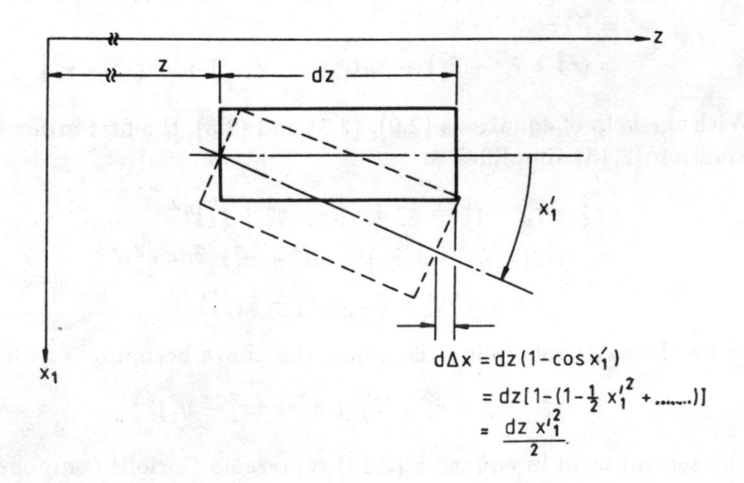

Fig. 2.5 Rotation of element in $x_1 z$ plane

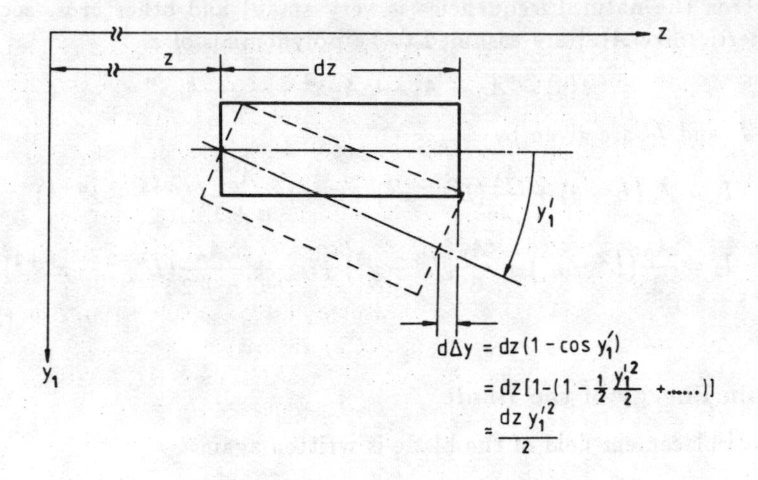

Fig. 2.6 Rotation of element in y_1z plane

is therefore

$$d\Delta = d\Delta_x \text{ and } d\Delta_y$$
$$= -\frac{1}{2}(x_1'^2 + y_1'^2)dz \tag{2.20}$$

Therefore u in equation (2.19), which is the total axial displacement at z, due to bending is given by

$$u = \int_0^z \frac{1}{2}(x_1'^2 + y_1'^2)\, dz \tag{2.21}$$

With the help of equations (2.2) and the above, the kinetic energy expression becomes

$$\begin{aligned}
T = {} & \frac{\rho}{2}\int_0^L A(z)\left[(\dot{x} + r_y(z)\dot{\theta})^2 + (\dot{y} - r_x(z)\dot{\theta})^2\right] dz \\
& + \frac{1}{2}\int_0^L I_{cg}(z)(\dot{\theta})^2\, dz - \frac{\rho\omega^2}{2}\int_0^L [(x' + r_y'(z)\theta + r_y(z)\theta')^2 \\
& + (y' - r_x'(z)\theta - r_x(z)\theta')^2]\,(RI_1 + I_2)\, dz \\
& + \frac{\rho\omega^2}{2}\sin^2\phi \int_0^L A(z)(x + r_y(z)\theta)^2\, dz \\
& + \frac{\rho\omega^2}{2}\cos^2\phi \int_0^L A(z)(y - r_x(z)\theta)^2\, dz \\
& - \frac{\rho\omega^2}{2}\sin 2\phi \int_0^L A(z)(x + r_y(z)\theta)(y - r_x(z)\theta)\, dz
\end{aligned} \tag{2.22}$$

In the above expression a negative sign for r_x is introduced, see figure 2.1. $r_x(z)$ and $r_y(z)$ are assumed to be quadratic functions of z, (since their effect on the natural frequencies is very small) and other cross-sectional properties like $A(z)$ are assumed to be polynomials of z.

$$A(z) = A_0 + A_1 z + A_2 z^2 + \cdots + A_n z^n$$

and I_1 and I_2 are given by

$$I_1 = A_0(L - z) + \frac{A_1}{2}(L^2 - z^2) + \cdots + \frac{A_n}{n+1}(L^{n+1} - z^{n+1})$$

$$I_2 = \frac{A_0}{2}(L^2 - z^2) + \frac{A_1}{3}(L^3 - z^3) + \cdots + \frac{A_n}{n+2}(L^{n+2} - z^{n+2}) \tag{2.22a}$$

Strain Energy of the Blade

The displacement field of the blade is written again

$$u_x = x_1 - y_1\theta$$

$$u_y = y_1 + x_1\theta \tag{2.2}$$

$$u_z = x_1(x_1' - y_1\theta') - y_1(y_1' + x_1\theta') + u \tag{2.3}$$

The longitudinal strain due to bending is

$$\begin{aligned} e_{zz} &= -x_1(x_1'' - y_1\theta'') - y_1(y_1'' + x_1\theta'') \\ &= -x_1 x_1'' - y_1 y_1'' \end{aligned} \tag{2.23}$$

In the above the strain due to axial displacement is not accounted. The extensional strain energy is therefore given by

$$\begin{aligned} V_1 &= \frac{1}{2}\int_V E e_{zz}^2 \, dv \\ &= \frac{1}{2}\int_0^L E \int_A (x_1^2 x_1''^2 + y_1^2 y_1''^2 + 2x_1 y_1 x_1'' y_1'') \, dA \, dZ \\ &= \frac{1}{2}\int_0^L (EI_{y1y1} x''^2 + EI_{x1x1} y''^2 + 2EI_{x1y1} x'' y'') \, dz \end{aligned} \tag{2.24}$$

In the above I_{y1y1}, I_{x1x1} and I_{x1y1} are familiar second moments of area about the centroidal axes. Also the effect of torsional displacement on extensional energy is neglected and hence x_1 and y_1 are replaced by x and y.

The two shearing strains are

$$\begin{aligned} e_{xz} &= \frac{1}{2}(x_1' - y_1\theta') + \frac{1}{2}[-(x_1' - y_1\theta') - y_1\theta'] \\ &= -\frac{1}{2} y_1\theta' \end{aligned} \tag{2.25}$$

$$\begin{aligned} e_{yz} &= \frac{1}{2}(y_1' - x_1\theta') + \frac{1}{2}[x_1\theta' - (y_1' + x_1\theta')] \\ &= \frac{1}{2} x_1\theta' \end{aligned} \tag{2.26}$$

The shearing strain energy therefore is

$$\begin{aligned} V_2 &= 2G \int_V (e_{xz}^2 + e_{yz}^2)\, dv \\ &= \frac{G}{2} \int_0^L \int_A (x_1^2 + y_1^2)\theta'^2 \, dA\, dz \\ &= \int_0^L \frac{1}{2} C\theta'^2 \, dz \end{aligned} \tag{2.27}$$

where

$$C = \int_A G(x_1^2 + y_1^2)\, dA$$

For non-circular sections, the axial displacement must include the effect of warping, $\phi_c\,(x, y)$, which is $\phi_c \theta'(2, 89)$. Then it can be shown

$$C = G \int_A \left[(\phi_{c,x} - y_1)^2 + (\phi_{c,y} + x_1)^2 \right] dA \tag{2.28}$$

Hence the total strain energy is

$$\begin{aligned} V &= V_1 + V_2 \\ &= \frac{1}{2} \left[\int_0^L EI_{x1x1}(z)(y'')^2 dz + 2 \int_0^L EI_{x1y1}(z)(y'')(x'')\, dz \right. \\ &\quad \left. + \int_0^L EI_{y1y1}(z)(x'')^2 dz + \int_0^L C(z)(\theta')^2\, dz \right] \end{aligned} \tag{2.29}$$

The differential equations of motion can be derived using the energy expressions and the Hamilton's principle. These equations can then be solved by Galerkin procedure. Alternatively Ritz method can be used by minimizing the Lagrangian expression $L = T - V$ directly to form the eigen value problem. Here the Lagrangian procedure will be illustrated to formulate the eigenvalue problem.

The following shape functions are assumed in series form for bending-bending and torsion

$$\begin{aligned} x &= \sum_{i=1}^{n} f_i(Z) q_{1i} \\ y &= \sum_{i=1}^{n} f_i(Z) q_{2i} \\ \theta &= \sum_{i=1}^{n} F_i(Z) q_{3i} \end{aligned} \tag{2.30}$$

where

$$Z = z/L$$

$$f_i(Z) = \frac{(i+2)(i+3)}{6} Z^{i+1} - \frac{i(i+3)}{3} Z^{i+2} + \frac{i(i+1)}{6} Z^{i+3}$$

$$F_i(Z) = Z^i - \frac{i}{i+1} Z^{i+1}$$

and q_{1i}, q_{2i} and q_{3i} are generalized coordinates.

The shape functions above satify the boundary conditions of a cantilever beam, viz.

$$f_i(Z) = f_i'(Z) = F_i(Z) = 0 \text{ at } Z = 0$$
$$f_i''(Z) = f_i'''(Z) = F_i'(Z) = 0 \text{ at } Z = 1$$

The expressions for potential and kinetic energies thus become

$$V = \frac{1}{2}\Bigg[\int_0^L EI_{x1x1}(z)\left(\sum f_i'' q_{2i}\right)^2 dz + 2\int_0^L EI_{x1y1}(z)\left(\sum f_i'' q_{1i}\right)\left(\sum f_i'' q_{2i}\right) dz + \int_0^L EI_{y1y1}(z)\left(\sum f_i'' q_{1i}\right)^2 dz + \int_0^L C(z)\left(\sum F_i' q_{3i}\right)^2 dz\Bigg] \tag{2.31}$$

$$\begin{aligned}
T = {} & \frac{\rho}{2}\int_0^L A(z)\left[\left(\sum f_i \dot{q}_{1i} + r_y(z)\sum F_i \dot{q}_{3i}\right)^2 + \left(\sum f_i \dot{q}_{2i} - r_x(z)\sum F_i \dot{q}_{3i}\right)^2\right] dz \\
& + \frac{1}{2}\int_0^L I_{cg}(z)\left(\sum F_i \dot{q}_{3i}\right)^2 dz \\
& - \frac{\rho\omega^2}{2}\int_0^L \left[\left(\sum f_i' q_{1i} + r_y'(z)\sum F_i q_{3i} + r_y(z)\sum F_i' q_{3i}\right)^2 \right. \\
& \left. + \left(\sum f_i' q_{2i} - r_x'(z)\sum F_i q_{3i} - r_x(Z)\sum F_i' q_{3i}\right)^2\right](RI_1 + I_2)\, dz \\
& + \frac{\rho\omega^2}{2}\sin^2\phi\int_0^L A(z)\left(\sum f_i q_{1i} + r_y(z)\sum F_i q_{3i}\right)^2 dz \\
& + \frac{\rho\omega^2}{2}\cos^2\phi\int_0^L A(z)\left(\sum f_i q_{2i} + r_x(z)\sum F_i q_{3i}\right)^2 dz
\end{aligned}$$

$$- \frac{\rho\omega^2}{2} \sin 2\phi \int_0^L A(z) \left(\sum f_i q_{1i} + r_y(z) \sum F_i q_{3i} \right)$$
$$\left(\sum f_i q_{2i} - r_x(z) \sum F_i q_{3i} \right) dz \tag{2.32}$$

Using Lagrangian equations

$$\frac{d}{dt}\left(\frac{\partial T}{\partial \dot{q}}\right) - \frac{\partial T}{\partial q} + \frac{\partial V}{\partial q} = 0$$

the equations of motion can be obtained as

$$[M]\{\ddot{q}\} + [K]\{q\} = 0 \tag{2.33}$$

where

$$[M] = \begin{bmatrix} \overline{A} & \overline{B} & \overline{C} \\ \overline{D} & \overline{E} & \overline{F} \\ \overline{G} & \overline{H} & \overline{I} \end{bmatrix} \qquad \{\ddot{q}\} = \begin{Bmatrix} \ddot{q}_{1i} \\ \ddot{q}_{2i} \\ \ddot{q}_{3i} \end{Bmatrix}$$
$$[K] = \begin{bmatrix} \overline{J} & \overline{K} & \overline{L} \\ \overline{M} & \overline{N} & \overline{O} \\ \overline{P} & \overline{Q} & \overline{R} \end{bmatrix} \qquad \{q\} = \begin{Bmatrix} q_{1i} \\ q_{2i} \\ q_{3i} \end{Bmatrix} \tag{2.34}$$

$$\overline{A}_{i,j} = \rho \int_0^L A(z) f_i f_j \, dz$$
$$\overline{B}_{i,j} = 0$$
$$\overline{C}_{i,j} = \rho \int_0^L A(z) r_y(z) F_i f_j \, dz$$
$$\overline{D}_{i,j} = 0$$
$$\overline{E}_{i,j} = \overline{A}_{i,j}$$
$$\overline{F}_{i,j} = -\rho \int_0^L A(z) r_x(z) F_i f_j \, dz$$
$$\overline{G}_{i,j} = \rho \int_0^L A(z) r_y(z) f_i F_j \, dz$$
$$\overline{H}_{i,j} = \rho \int_0^L A(z) r_x(z) f_i F_j \, dz$$
$$\overline{I}_{i,j} = \int_0^L I_{cf}(z) F_i F_j \, dz$$
$$\overline{J}_{i,j} = \int_0^L EI_{y1y1}(z) f_i'' f_j'' \, dz + \rho\omega^2 \int_0^L f_i' f_j' (RI_1 + I_2) \, dz$$
$$- \rho\omega^2 \sin^2 \phi \int_0^L A(z) f_i f_j \, dz$$
$$\overline{K}_{i,j} = \int_0^L EI_{x1y1}(z) f_i'' f_j'' \, dz + \frac{\rho\omega^2}{2} \sin 2\phi \int_0^L A(z) f_i f_j \, dz$$

$$\overline{L}_{i,j} = \rho\omega^2 \left[\int_0^L r'_y(z)(RI_1 + I_2)F_i f'_j \, dz + \int_0^L r_y(z)(RI_1 + I_2)F'_i f'_j \, dz \right.$$
$$\left. - \sin^2\phi \int_0^L A(z) r_y(z) F_i f_j \, dz - \frac{1}{2}\sin 2\phi \int_0^L A(z) r_x(z) F_i f_j \, dz \right]$$
$$\overline{M}_{i,j} = \int_0^L EI_{x1y1}(z) f''_i f''_j \, dz + \frac{\rho\omega^2}{2}\sin 2\phi \int_0^L A(z) f_i f_j \, dz$$
$$\overline{N}_{i,j} = \int_0^L EI_{x1x1}(z) f''_i f''_j \, dz + \rho\omega^2 \left[\int_0^L f'_i f'_j (RI_1 + I_2)\, dz \right.$$
$$\left. - \cos^2\phi \int_0^L A(z) f_i f_j \, dz \right]$$
$$\overline{O}_{i,j} = -\rho\omega^2 \left[\int_0^L (r'_x(z) F_i f'_j + r_x(z) F'_i f'_j)(RI_1 + I_2)\, dz \right.$$
$$\left. - \cos^2\phi \int_0^L A(z) r_x(z) F_i f_j \, dz - \frac{1}{2}\sin 2\phi \int_0^L A(z) r_y(z) F_i f_j \, dz \right]$$
$$\overline{P}_{i,j} = \rho\omega^2 \left[\int_0^L (r'_y(z) f'_i F_j + r_y(z) f'_i F'_j)(RI_1 + I_2)\, dz \right.$$
$$\left. - \sin^2\phi \int_0^L A(z) r_y(z) f_i F_j \, dz - \frac{1}{2}\sin 2\phi \int_0^L A(z) r_x(z) f_i F_j \, dz \right]$$
$$\overline{Q}_{i,j} = -\rho\omega^2 \left[\int_0^L (r'_x(z) f'_i F_j + r_x(z) f'_i F'_j)(RI_1 + I_2)\, dz \right.$$
$$\left. - \cos^2\phi \int_0^L A(z) r_x(z) f_i F_j \, dz - \frac{1}{2}\sin 2\phi \int_0^L A(z) r_y(z) f_i F_j \, dz \right]$$
$$\overline{R}_{i,j} = \int_0^L C(z) F_i F_j \, dz + \rho\omega^2 \left[\int_0^L \{r'^2_x(z) + r'^2_y(z)\} F_i F_j \right.$$
$$+ \{r_y(z) r'_y(z)(F_i F'_j + F'_i F_j)\} + \{r^2_x(z) + r^2_y(z)\} F'_i F'_j$$
$$\left. + \{r_x(z) r'_x(z)(F_i F'_j + F'_i F_j)\} \right] (RI_1 + I_2)\, dz$$
$$-\rho\omega^2 \cos^2\phi \int_0^L A(z) r^2_x(z) F_i F_j \, dz$$
$$-\rho\omega^2 \sin^2\phi \int_0^L A(z) r^2_y(z) F_i F_j \, dz$$
$$-\rho\omega^2 \sin 2\phi \int_0^L A(z) r_x(z) r_y(z) F_i F_j \, dz \tag{2.35}$$

$j = 1, n$

$i = 1, n$

The cross-sectional properties can be expressed as polynomials of a suitable degree, e.g.,

$$I_{x1x1} = I_{x0} + I_{x1}z + I_{x2}z^2 + \cdots + I_{xn}z^n$$

and the integrals in equation (2.35) can be evaluated by a suitable numerical integration procedure, thus setting up the eigenvalue problem of equation (2.33). The eigenvalues and corresponding mode shapes can be obtained by any library routine. It has been found that the lowest three eigenvalues converge if a five term solution is used in equation (2.30) for each x, y and θ. Thus with a 15×15 matrix size problem, 9 useful frequencies can be obtained. Results from such a program for specific cases are discussed in the following sections.

2.3 EFFECT OF TAPER

Blade taper has a significant influence on both bending and torsional vibrations since the mass and stiffness distributions are considerably changed along the blade length. There has been considerable interest in using rectangular cross-section to obtain correction factors for determining tapered blade frequencies from the corresponding uniform cantilever frequencies. This is because the properties of an aerofoil section blade, the area and second moment of area, can be used to determine the equivalent rectangular section for bending vibrations. The breadth and thickness of a tapered beam are expressed as

$$b = b_0(1 - \alpha Z)$$
$$t = t_0(1 - \beta Z)$$

where b_0 and t_0 are root section values and α, β are taper parameters. Using a one term approximation for uncoupled bending vibration in yz plane, one can obtain the following expression for the fundamental mode [90]:

$$p_1^2 = \frac{12EI_0}{m_0 L^4} \frac{\left(\frac{12}{5} - \frac{2\alpha}{5} - \frac{6\beta}{5} + \frac{12\alpha\beta}{35} + \frac{12\beta^2}{35} - \frac{9\alpha\beta^2}{70} - \frac{3\beta^3}{70} + \frac{2\alpha\beta^3}{105}\right)}{\left(\frac{104}{45} - \frac{584\alpha}{315} - \frac{584\beta}{315} + \frac{5353\alpha\beta}{3465}\right)}$$

The above expression gives an error of 0.4% in estimating the fundamental natural frequency of a uniform cantilever beam. Defining

$$\lambda_{si}^2 = \frac{\rho A_0 L^4}{EI_0} p_i^2$$

the frequency parameter ratios for the first three modes obtained from a five term solution are given in figures 2.7 to 2.9. Experimental results of reference [91] are also given for comparison. Important conclusions from the figures are:

- A decrease of cross-sectional dimension in breadth with constant thickness, increases the fundamental bending frequency as well as the higher mode bending frequencies.
- A decrease of cross-sectional dimension in thickness with constant breadth, increases the fundamental bending frequency, whereas decreases the higher mode bending frequencies.

Because the torsional stiffness is given by a series expression involving hyperbolic tangent terms, it is not easy to derive an equivalent rectangular cross-section for a given aerofoil sectioned blade. Even then, to understand the effect of taper on torsional vibrations, the frequency ratio as a function of thickness taper for various breadth and taper parameters is given in figures 2.10 to 2.12 for the first three modes. From these it can be observed that:

- A decrease in breadth dimension, with constant thickness increases the torsional mode frequencies.
- A decrease in thickness dimension, with constant breadth, decreases the torsional mode frequencies.

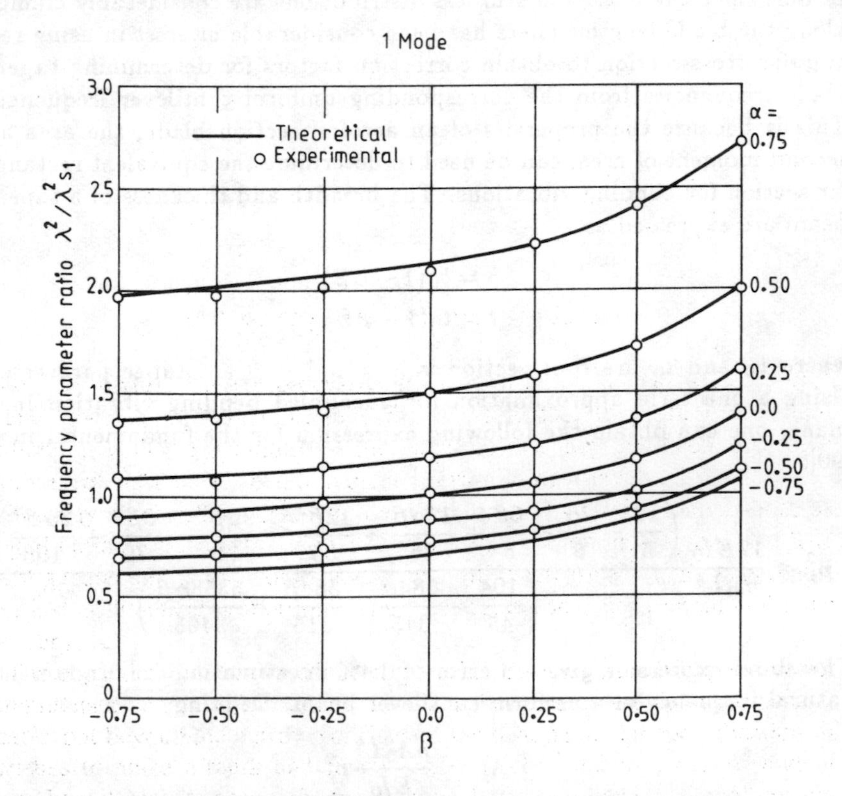

Fig. 2.7 Effect of thickness taper on frequency parameter ratio for various values of breadth taper

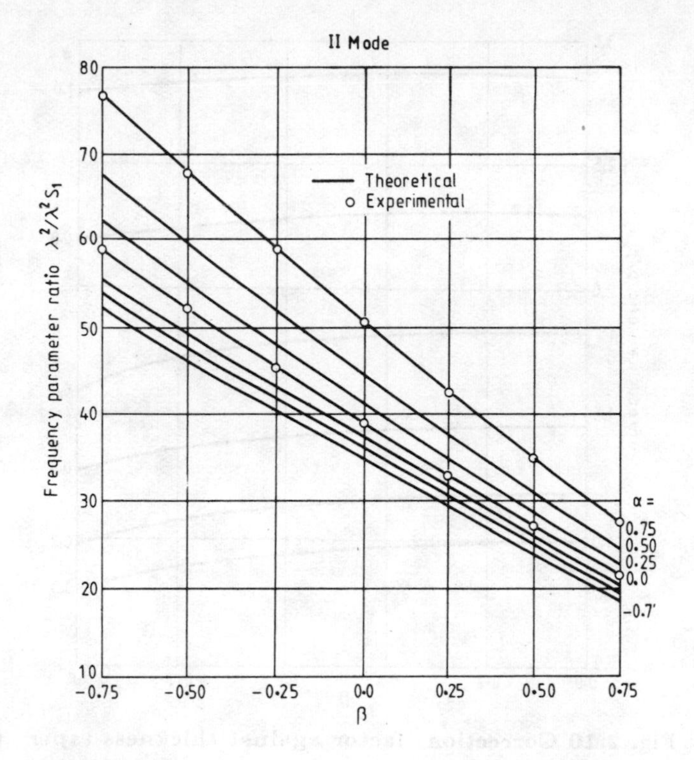

Fig. 2.8 Effect of thickness taper on frequency parameter ratio for various values of breadth taper

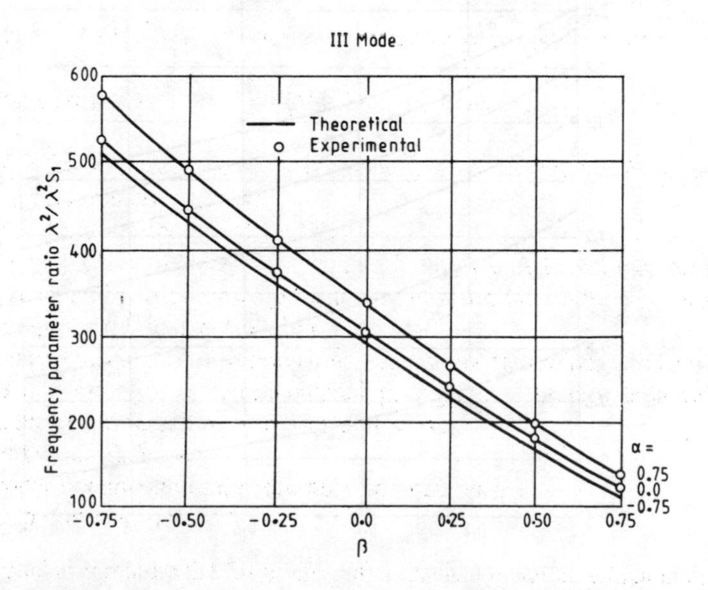

Fig. 2.9 Effect of thickness taper on frequency parameter ratio for various values of breadth taper

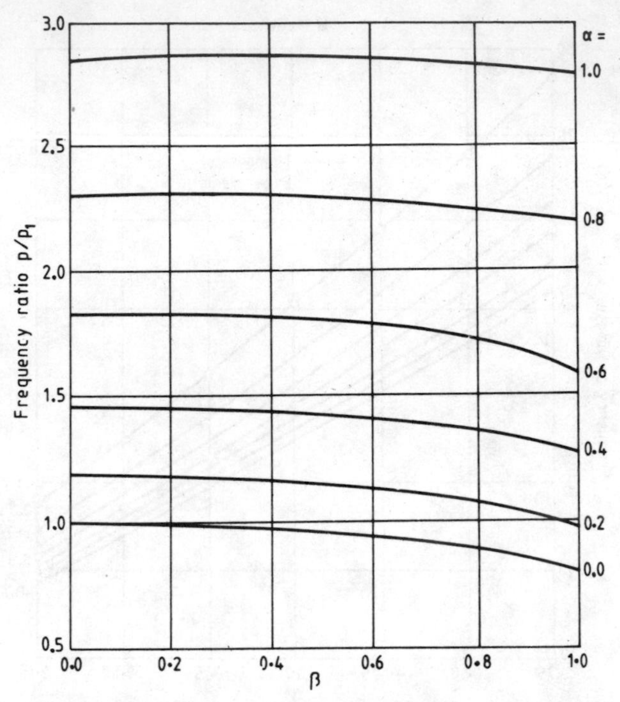

Fig. 2.10 Correction factor against thickness taper for various breadth tapers for 1st torsional mode

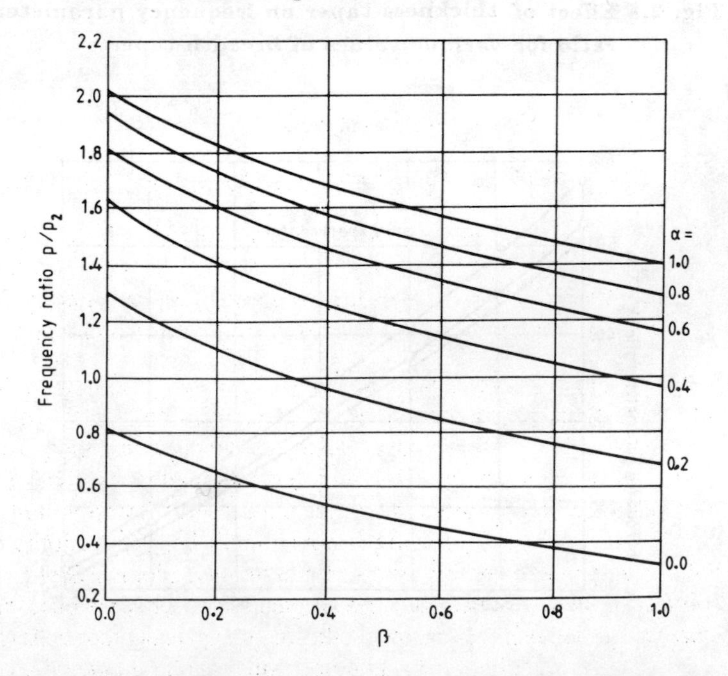

Fig. 2.11 Correction factor against thickness taper for various breadth tapers for 2nd torsional mode

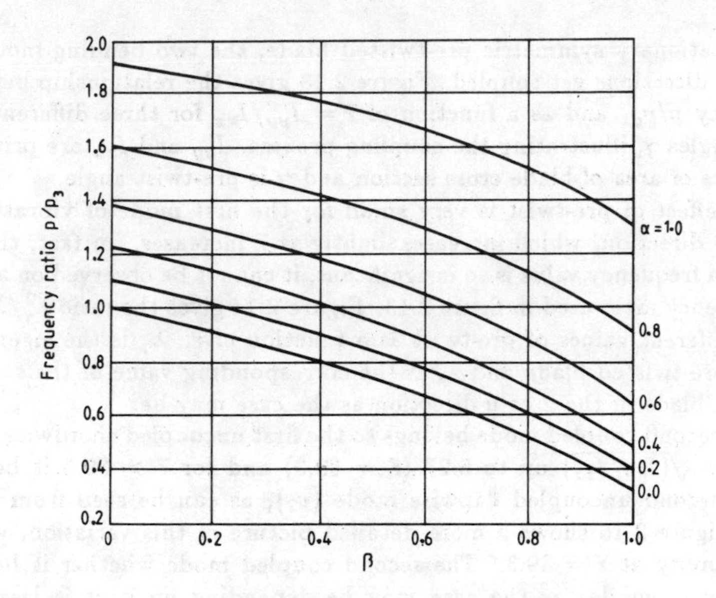

Fig. 2.12 Correction factor against thickness taper for various breadth tapers for 3rd torsional mode

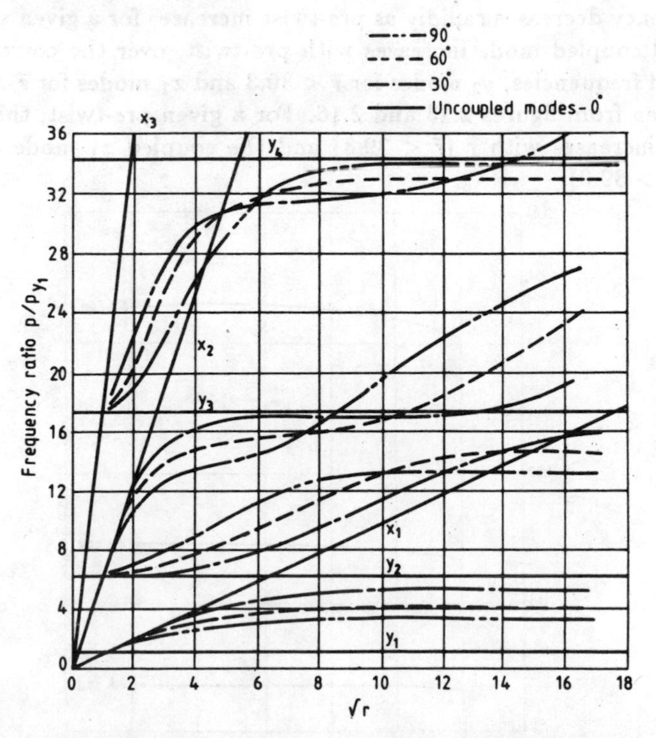

Fig. 2.13 Theoretical relation between frequency ratio and $\bar{r}$

2.4 EFFECT OF PRE-TWIST ON BENDING MODES

For a stationary symmetric pre-twisted blade, the two bending modes in y and x directions get coupled. Figure 2.13 gives the relationship between frequency p/p_{y1} and as a function of $\overline{r} = I_{yy}/I_{xx}$ for three different pre-twist angles γ, illustrating the coupling process. I_{yy} and I_{xx} are principal moments of area of blade cross section and γ is pre-twist angle.

The effect of pre-twist is very small for the first mode of vibration in flapwise direction, which increases slightly as $\overline{r}$ increases. In fact, this increase in frequency value is so insignificant, it cannot be observed on a scale of frequency ratio used in figure 2.13. Figure 2.14 gives the ratio $\lambda_\gamma^2/\lambda_s^2$, for three different values of pre-twist as a function of $\overline{r}$. λ_γ is the eigenvalue of the pre-twisted blade and λ_s is the corresponding value of the straight uniform blade in the x or y direction as the case may be.

The second coupled mode belongs to the first uncoupled chordwise mode (x_1), for $\sqrt{(I_{yy}/I_{xx})}$ up to 6.27 ($\overline{r} = 39.3$) and for $\overline{r} > 39.3$ it belongs to the second uncoupled flapwise mode (y_2), as can be seen from figure 2.13. Figure 2.15 shows a more detailed picture of this variation, with a discontinuity at $\overline{r} = 39.3$. The second coupled mode whether it belongs to x_1 or y_2 modes as the case may be depending on $\overline{r}$, it is less than the corresponding uncoupled x_1 or y_2 modes. x_1 mode decreases as $\overline{r}$ increases up to 39.3, the effect being more predominant for larger pre-twists. For $\overline{r} > 39.3$, the y_2 mode increases with $\overline{r}$ for a given pre-twist and the frequency decreases rapidly as pre-twist increases for a given value of $\overline{r}$. The third coupled mode increases with pre-twist, over the corresponding uncoupled frequencies, y_2 modes for $\overline{r} < 39.3$ and x_1 modes for $\overline{r} > 39.3$, as can be seen from figures 2.13 and 2.16. For a given pre-twist, the coupled y_2 mode increases with $\overline{r}$ ($\overline{r} < 39.3$) and the coupled x_1 mode decreases with $\overline{r}$ ($\overline{r} > 39.3$).

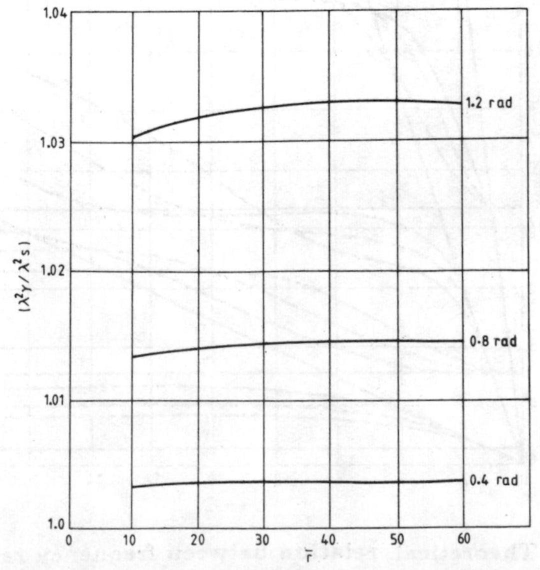

Fig. 2.14 Effect of $\overline{r}$ on frequency parameter ratio in first coupled mode for different pre-twists

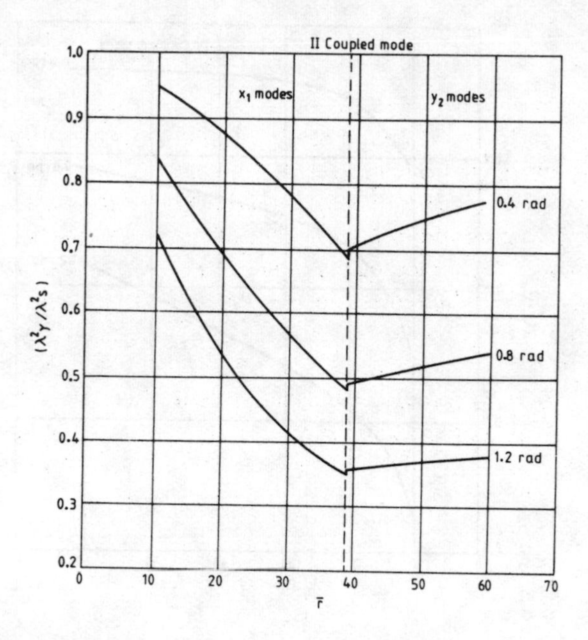

Fig. 2.15 Effect of $\bar{r}$ on frequency parameter ratio in second coupled mode for different pre-twists

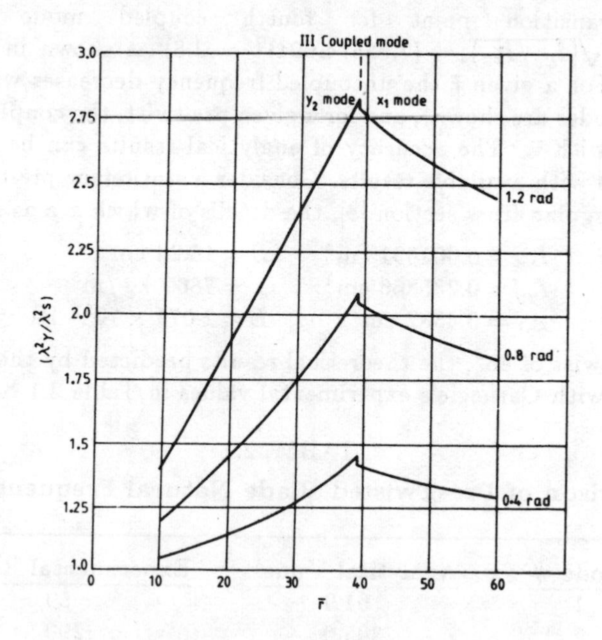

Fig. 2.16 Effect of $\bar{r}$ on frequency parameter ratio in third coupled mode for different pre-twists

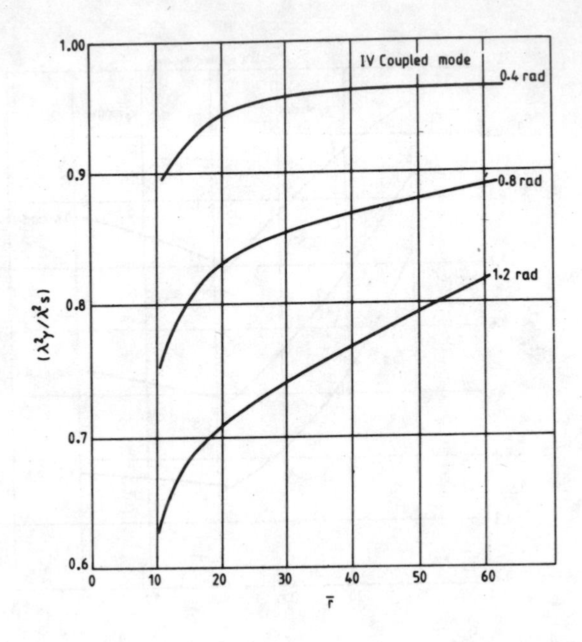

Fig. 2.17 Effect of $\overline{r}$ on frequency parameter ratio in fourth coupled mode for different pre-twists

The transition point for fourth coupled mode occurs at $\overline{r} = 7.84$ $(\sqrt{(I_{yy}/I_{xx})} = (7.855/4.694)^2 = 2.8)$ as shown in figures 2.13 and 2.17. For a given $\overline{r}$ the uncoupled frequency decreases with pre-twist (only y_3 modes are shown), and for a given pre-twist, the coupled frequency increases, with $\overline{r}$. The accuracy of analytical results can be judged by a comparison with available results. Consider a stationary pre-twisted blade with rectangular cross section [8], the details of which are as follows:

$$I_{xx} = 0.001091 \text{ cm}^4 \qquad L = 15.24 \text{ cm}$$
$$I_{yy} = 0.235866 \text{ cm}^4 \qquad \rho = 7860 \text{ kg/m}^3$$
$$A = 0.4387 \text{ cm}^2 \qquad E = 2.071 \times 10^{11} \text{ Pa}$$

For a pre-twist of 45°, the theoretical results predicted by the analysis are compared with Carnegie's experimental values in Table 2.1 below.

TABLE 2.1

Comparison of Pre-Twisted Blade Natural Frequencies — Hz

Mode #	Analytical Value	Experimental Result
1	61.9	59
2	305.0	290
3	949.0	920
4	1220.0	1110

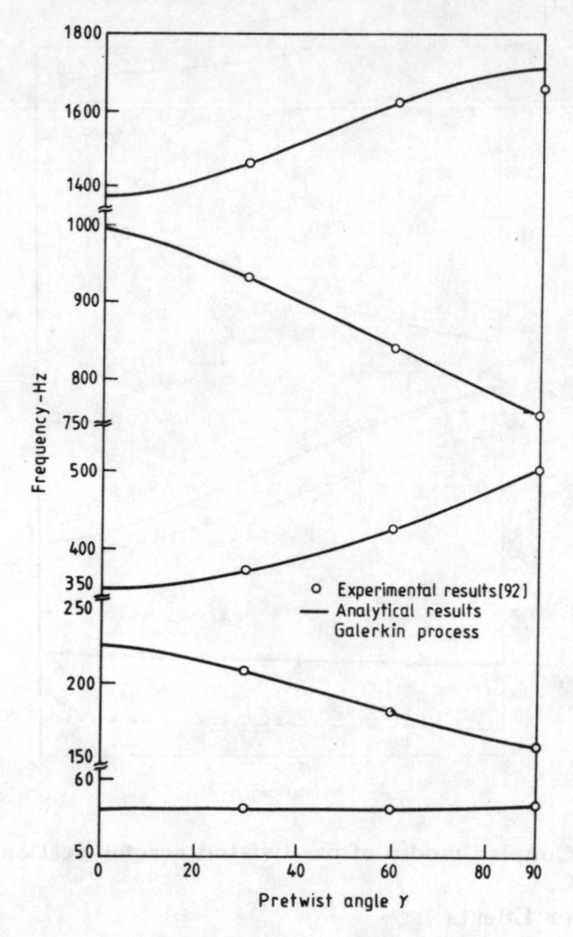

Fig. 2.18 Coupled bending-bending frequencies

It can be noted that the experimental results are lower than the theoretical values, which is due to the higher order effects neglected in the theory. For a rectangular section of 2.54 × 0.635 cm with 30.48 cm length, the coupled frequencies obtained are plotted in figure 2.18 as a function of pre-twist angle. Experimental results of Dawson and Carnegie [92] are also shown in the same figure, which agree well with the analytical prediction. For airfoil cross-section blading tested by Carnegie [8] the data is as follows:

$$I_{xx} = 0.000416 \text{ cm}^4 \qquad E = 2.14 \times 10^{11} \text{ Pa}$$
$$I_{yy} = 0.2793 \text{ cm}^4 \qquad \rho = 7860 \text{ kg/m}^3$$
$$L = 15.24 \text{ cm} \qquad A = 0.5897 \text{ cm}^2$$

The analytical results, using figures 2.14 and 2.17, are plotted in figure 2.19. The theoretical prediction for the first mode by Carnegie and his experimental results are also shown in the same figure. Results for x_1 coupled mode are not quoted by him. However, the experimental observations are consistently lower, over the theoretical predictions, but follow the same trend.

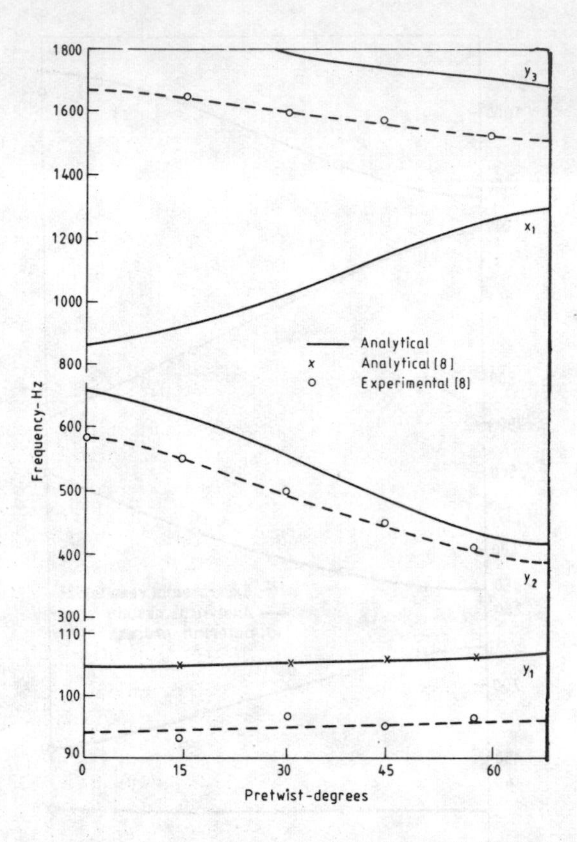

Fig. 2.19 Coupled modes of pre-twisted aerofoil section blades

Higher Order Effects

When a beam vibrates, an element not only has pure translation (lateral deflection), but also rotates, by virtue of the slope of the beam at that location. This causes an additional inertia, first called by Rayleigh as rotary inertia. In classical theory, this inertia is neglected. Also, in the derivation of Euler's equation, the deflection due to shear is neglected, so that the angle of rotation of the element is equal to the slope. The additional shear energy should be accounted in derivation of the equations of motion [3]. Sutherland and Goodman [7] gave plots of corrections to be applied for different types of beams considering the shear deformation and rotary inertia effects. For a beam, the correction is dependent on the factor $\bar{r}_g = \sqrt{(I/AL^2)}$ which is non- dimensional radius of gyration. The corrections are predominant for large values of $\bar{r}_g$ and higher modes. For normal values of the $\bar{r}_g$ encountered in practice, the corrections are not of significance. Rao [82] gave simple formulae for the correction factor, for the first three modes and $\bar{r}_g < 0.02$.

$$\lambda_{rg}^2/\lambda_s^2\}_I = 1.0 + 0.025\bar{r}_g - 18.75\bar{r}_g^2$$

$$\lambda_{rg}^2/\lambda_s^2\}_{II} = 1.0 + 0.0125\bar{r}_g - 115.625\bar{r}_g^2$$

$$\lambda_{rg}^2/\lambda_s^2\}_{III} = 1.0 + 0.084\bar{r}_g - 256.763\bar{r}_g^2$$

Experimental observations of bending frequencies of beams have always been found to be lower than frequencies determined by classical theory and the discrepancy has been conveniently attributed to higher order effects such as rotary inertia and shear. However, the effects of shear and rotary inertia do not account for all the discrepancy between the classical and modified theory. The analytical results considering the effects of shear deformation and rotary inertia [93] for the following blades are compared with the experimental values in Table 2.2.

$$\begin{aligned} I_{xx} &= 0.1002 \text{ cm}^4 \\ A &= 2.32 \text{ cm}^2 \\ L &= 20.32 \text{ cm} \\ \rho &= 7833 \text{ kg/m}^3 \\ E &= 2.071 \times 10^{11} \text{ Pa} \\ r_x &= 0.297 \text{ cm} \\ C &= 318.5 \text{ Nm}^2/\text{rad} \end{aligned}$$

TABLE 2.2

Effect of Shear and Rotary Inertia on Bending Frequencies

Mode #	Theoretical Frequency Hz		Experimental Values Hz
	Classical	Corrected	
1	144.7	144.5	142
2	904.5	900.0	891
3	2525.0	2489.0	2450

The effects of shear and rotary inertia have been considered for pre-twisted and tapered blades and reference may be made to the work of Carnegie and Thomas [94].

Torsional Mode

When a blade of thin walled section is pre-twisted, increased torsional stiffness is evident. The additional amount c_γ over and above that of St. Venant, is due to inclination of the blade longitudinal fibers and Carnegie [8] derived the following relationship:

$$c_\gamma = c_3 \left(\frac{\gamma}{L}\right)^2 \tag{2.36}$$

where the constant c_3 depends on the cross-section of the blade.

$$c_3 = E \int_A \left(r^4 - \frac{r^2 I_p}{A}\right) dA \tag{2.37}$$

In the above, r is distance of an element from centroid in a symmetric cross-section and I_p is its second polar moment of area. For non-symmetrical cross-section this is to be modified.

$$c_3 = E \int_A \left[r_1^2 r_2^2 - \frac{r_1 r_2}{A} \int r_1 r_2 dA \right] dA \qquad (2.38)$$

where r_1 and r_2 are the distances of an element from centroid and centre of torsion, respectively. For a rectangular section of breadth b and thickness t,

$$c_3 = E \frac{(b^4 + t^4)}{180} bt \qquad (2.39)$$

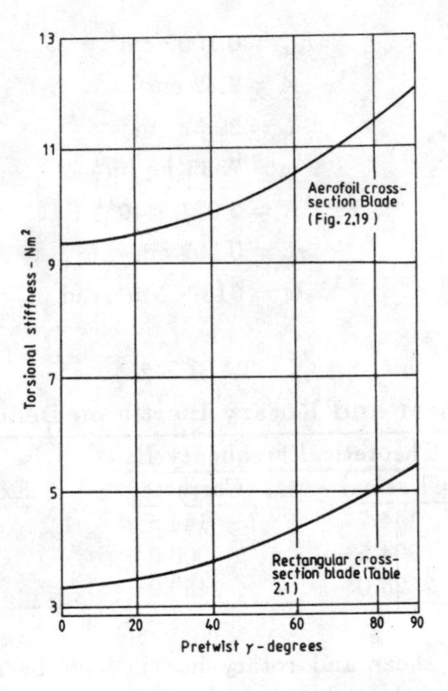

Fig. 2.20 Torsional stiffness of pre-twisted blades [8]

Figure 2.20 gives the torsional stiffness values estimated for pre-twisted blades as a function of pre-twist, obtained by adding St. Venant stiffness to the additional stiffness given by (2.39) for rectangular cross-section and evaluated by the integral in (2.38) for aerofoil cross-section blades, whose cross section is defined in earlier examples of Table 2.1 and figure 2.19. Carnegie [8] verified these results through experimental observations.

The estimation of torsional natural frequencies is now a relatively simple matter and the following formula [1] can be straightaway used.

$$p_n = \frac{n\pi}{2L} \sqrt{\left(\frac{C}{I_{cg}}\right)} \text{ rad/sec} \quad n = 1, 3, 5, \ldots \qquad (2.40)$$

where C is total torsional stiffness and I_{cg} is polar mass moment of inertia about the centroid. Figures 2.21 and 2.22 give the theoretical and experimental values of Carnegie [8] for rectangular and aerofoil cross-section blades.

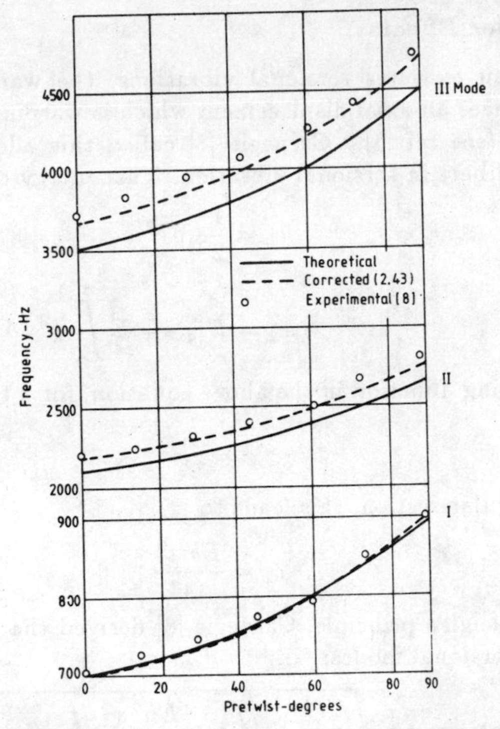

Fig. 2.21 Effect of pre-twist on torsional frequencies of rectangular cross-section blades

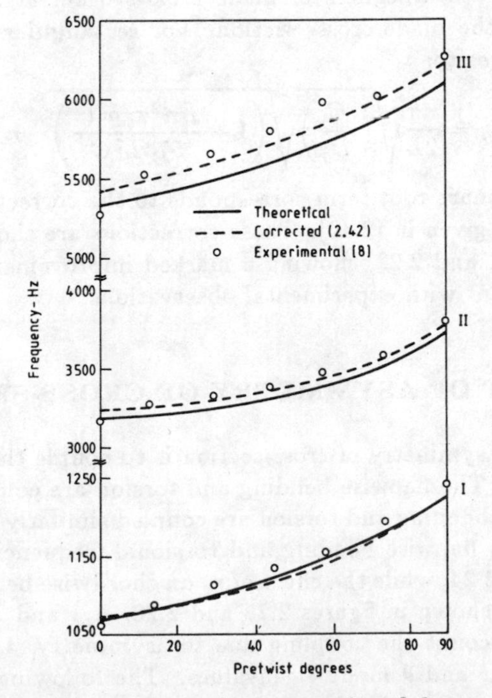

Fig. 2.22 Effect of pre-twist on torsional frequencies of aerofoil blades

Higher Order Effects

While a beam executes torsional vibrations, the warping of any cross-section produces an axial displacement which is warping function ϕ_c times the slope θ' (see ref. 3). Carnegie [8] called this additional bending of longitudinal fibers in torsional vibration. This energy can be shown to be

$$v_b = \frac{1}{2} c_1 \theta''^2$$

where

$$c_1 = EI_{\phi\phi} = E \int_A \phi^2 dA$$

The warping function in the above equation, for a thin section is given by

$$\phi = xy$$

For a rectangular section, this leads to

$$c_1 = \frac{Eb^3t^3}{144} \tag{2.41}$$

Using Rayleigh's principle, Carnegie [8] derived the additional effect of bending on torsional modes:

$$p_n = \frac{n\pi}{2L} \sqrt{\left(\frac{C}{I_{cg}}\right)} \sqrt{\left(1 + \frac{En^2\pi^2}{48L^2C} \int_A x^2 t^3 \, dx\right)} \tag{2.42}$$

In the above t is thickness of blade cross-section at a distance x along the chord of the blade cross- section. For rectangular section, the above formula reduces to:

$$p_n = \frac{n\pi}{2L} \sqrt{\left(\frac{C}{I_{cg}}\right)} \sqrt{\left(1 + \frac{En^2\pi^2 b^3 t^3}{576L^2C}\right)} \quad n \text{ is odd} \tag{2.43}$$

The second square root term corresponds to the correction to be made for the frequency given in (2.40). These corrections are shown by dotted lines in figures 2.21 and 2.22, showing a marked improvement of the theory to be in agreement with experimental observations.

2.5 EFFECT OF ASYMMETRY OF CROSS-SECTION

The effect of asymmetry of cross-section is to couple the bending and torsional modes. The flapwise bending and torsion are coupled because of r_x; the chordwise bending and torsion are coupled similarly because of r_y. The effect of r_x on flapwise bending and torsional frequencies is shown in figures 2.23 and 2.24, while the effect of r_y on chordwise bending and torsional frequencies is shown in figures 2.25 and 2.26. λ_{rx} and λ_{ry} are eigenvalues taking into account the coupling due to asymmetry, λ_{sy}, λ_{sx} and λ_θ are uncoupled y, x and θ mode eigenvalues. The following observations may be made:

- Both bending frequencies decrease with asymmetry.
- Torsional frequency increases with asymmetry.
- The asymmetry has more influence on higher modes.
- Generally, the effect of asymmetry is negligible on both torsional and bending frequencies.

The aerofoil section blade of Carnegie [8] having the following properties can be studied to find the effect of asymmetry.

$$
\begin{aligned}
I_{xx} &= 0.0035 \text{ cm}^4 & r_x &= 0.0193 \text{ cm} \\
I_{yy} &= 0.2793 \text{ cm}^4 & r_y &= 0.1194 \text{ cm} \\
L &= 15.24 \text{ cm} & E &= 2.14 \times 10^{11} \text{ Pa} \\
A &= 0.5897 \text{ cm}^2 & C &= 9.14 \text{ Nm}^2/\text{rad} \\
\rho &= 7860 \text{ Kg/m}^3 & &
\end{aligned}
$$

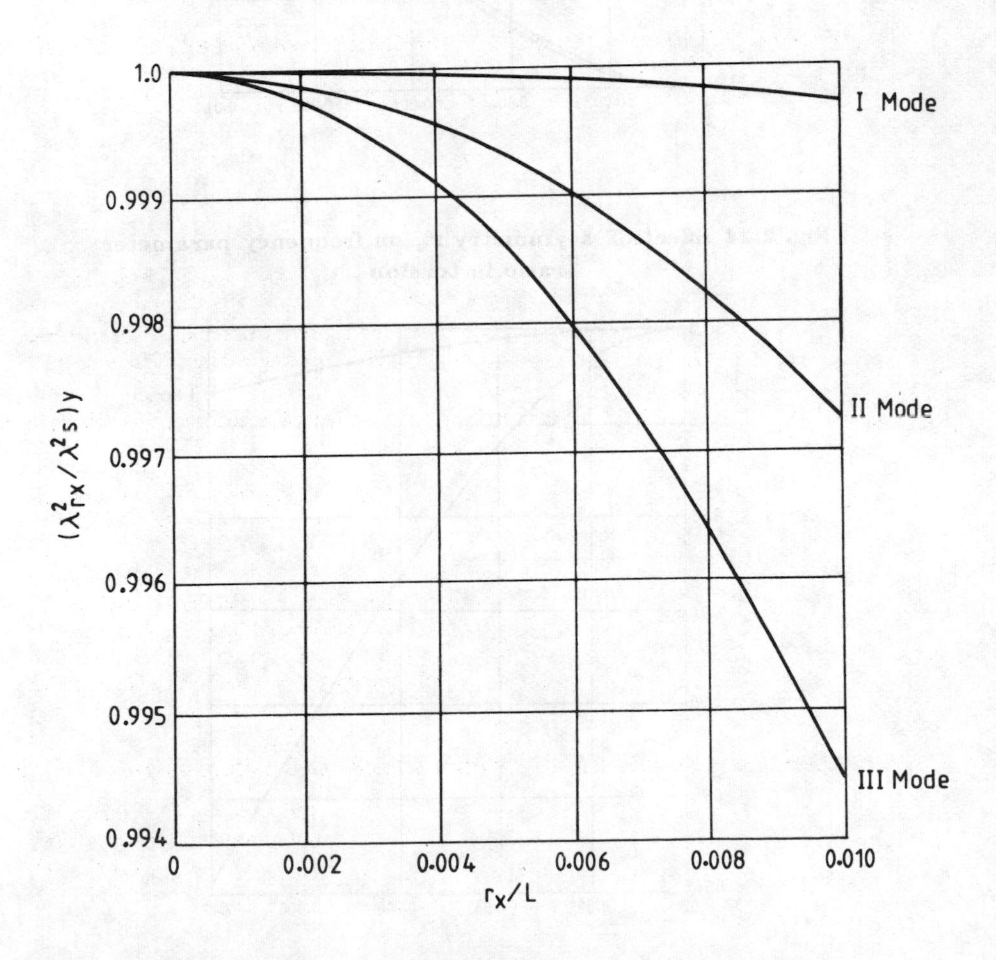

Fig. 2.23 Effect of asymmetry on frequency parameter ratio in flapwise bending

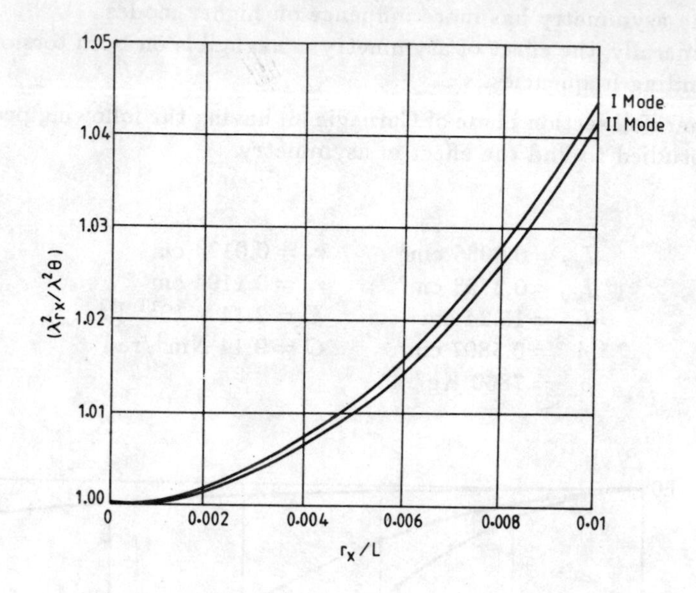

Fig. 2.24 Effect of asymmetry r_x on frequency parameter ratio in torsion

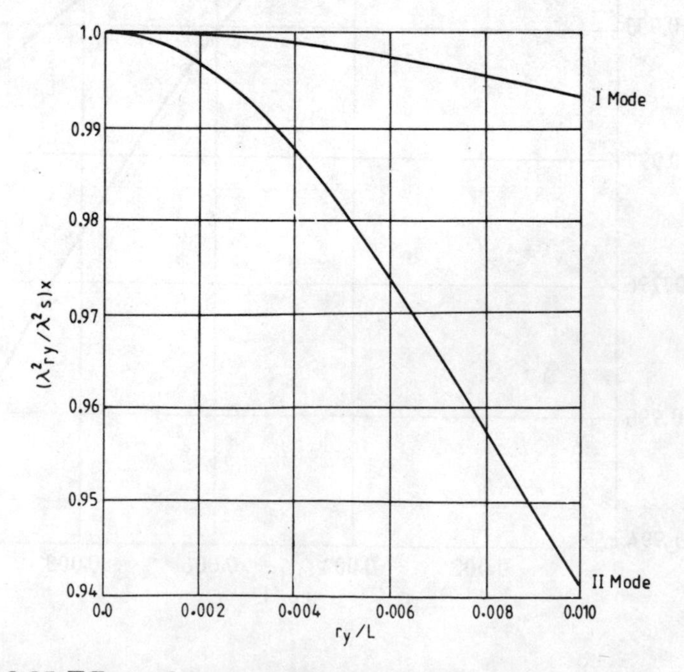

Fig. 2.25 Effect of asymmetry r_y on frequency parameter ratio in chordwise bending

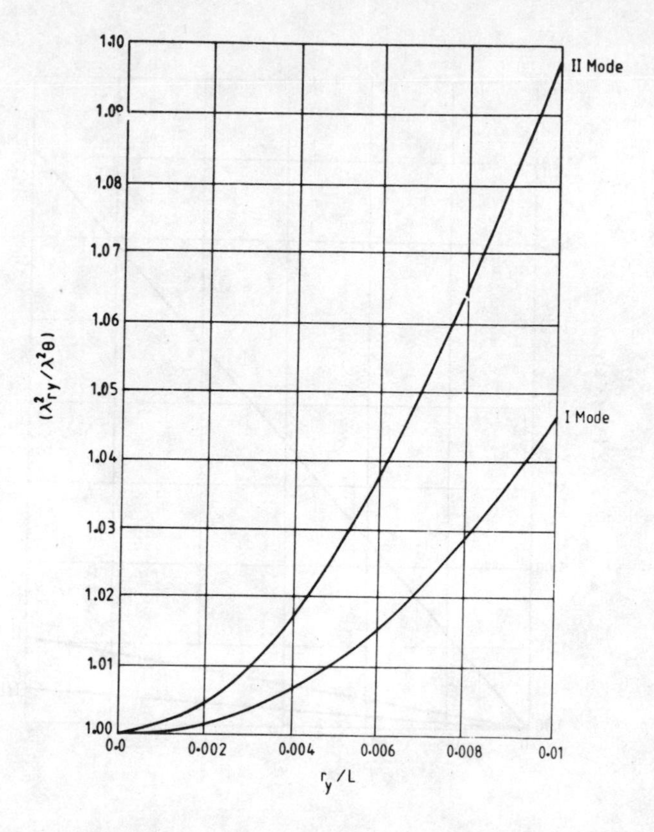

Fig. 2.26 Effect of asymmetry r_y on frequency parameter ratio in torsion

The analytical results are compared with experimental values below in Table 2.3.

TABLE 2.3

Comparison of Asymmetric Cross-Section Blade Frequencies — Hz

Mode #	Uncoupled Frequency	Analytical Values	Experimental Values
y_1	96.47	96.5	97.0
y_2	604.6	604.9	610.0
x_1	862.17	842.0	790.0
θ_1	1050.0	1089.4	1102.0
y_3	1693.1	1694.9	1693.0

There is a good agreement between analytical and experimental results and it may be safely assumed that the effect of asymmetry can be ignored in the calculation of natural frequencies, when compared with other effects such as pre-twist and rotation, in bending vibrations. It may be considered if necessary, for torsional frequencies.

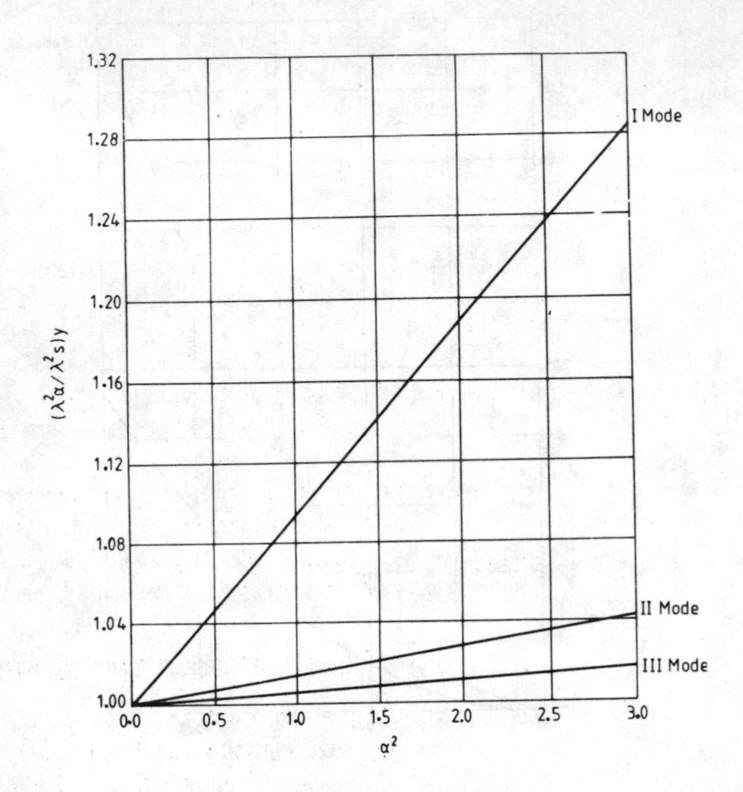

Fig. 2.27 Effect of speed of rotation on frequency parameter in flapwise bending mode at setting angle 0°

2.6 EFFECT OF CENTRIFUGAL FORCES

The blade gets stiffened in bending because of centrifugal forces and thus the bending natural frequencies increase with speed of rotation. There is no effect of centrifugal forces in the torsional motion of the blade. Denoting λ_α as the eigenvalue taking into account the effect of rotation defined by non-dimensional parameter $\alpha^2 = m\omega^2 L^4/EI_{xx'}$ the frequency parameter ratio for the first three modes in flapwise bending of the blade is given in figure 2.27 as a function of α^2 for a setting angle of $\phi' = 0$. (Stagger angle $\phi = 90^o$) and disc radius $R = 0$. From this figure it can be observed that:

- The non-dimensional frequency parameter ratio increases linearly with the rotational parameter.
- The centrifugal forces have pronounced influence on the fundamental mode and relatively negligible influence on higher modes.

The chordwise bending modes depend not only on the parameter α^2 but also on the ratio of principal second moments of area $\overline{r}$. For the first three chordwise modes, the frequency parameter ratio is given in figures 2.28 to 2.30 as a function of $\bar{r}$ for three different values of α^2. The following points may be noted:

- The centrifugal force has relatively lesser influence on chordwise modes compared to the flapwise modes.
- The centrifugal force has negligible influence on higher modes.
- The influence of centrifugal forces decreases considerably as $\bar{r}$ increases.
- For a given $\bar{r}$, the frequency parameter ratio increases linearly with α^2.

2.7 EFFECT OF DISC RADIUS

The effect of disc radius is to increase the centrifugal forces thus stiffening the blade further and increase the natural frequencies in bending modes.

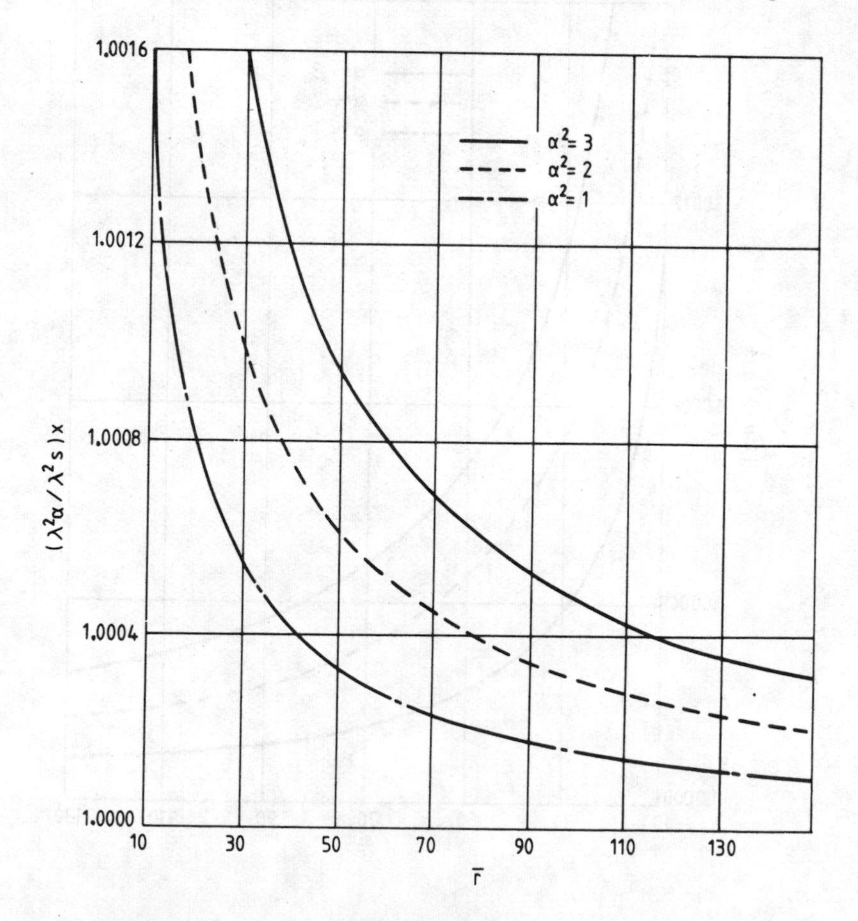

Fig. 2.28 Effect of $\bar{r}$ on frequency parameter ratio in the first chordwise bending mode at setting angle 0°

For flapwise bending, the frequency parameter ratio is given as a function of disc parameter $\overline{R} = R/L$ in figure 2.31 for three different values of α^2 in the first three modes. λ_R is the eigenvalue taking into account the parameter of rotation in the flapwise mode as in figure 2.27. It can be seen that:

- The disc radius has predominant influence on the flapwise modes, particularly in the fundamental mode.
- The frequency parameter ratio increases linearly with the disc radius.

The effect of disc radius is to stiffen the blade chordwise bending modes and the frequency parameter ratio increases linearly with $\overline{R}$ as in the case of flapwise modes. The chordwise modes are also dependent on $\overline{r}$ and hence the results are given in figures 2.32 to 2.34 as a function of $\overline{r}$ for $R/L = 7$ at setting angle $\phi' = 0$ and $\alpha^2 = 1, 2, 3$. For any other value of R/L, a proportional factor is to be taken since the variation of the frequency parameter ratio is linear with $\overline{R}$.

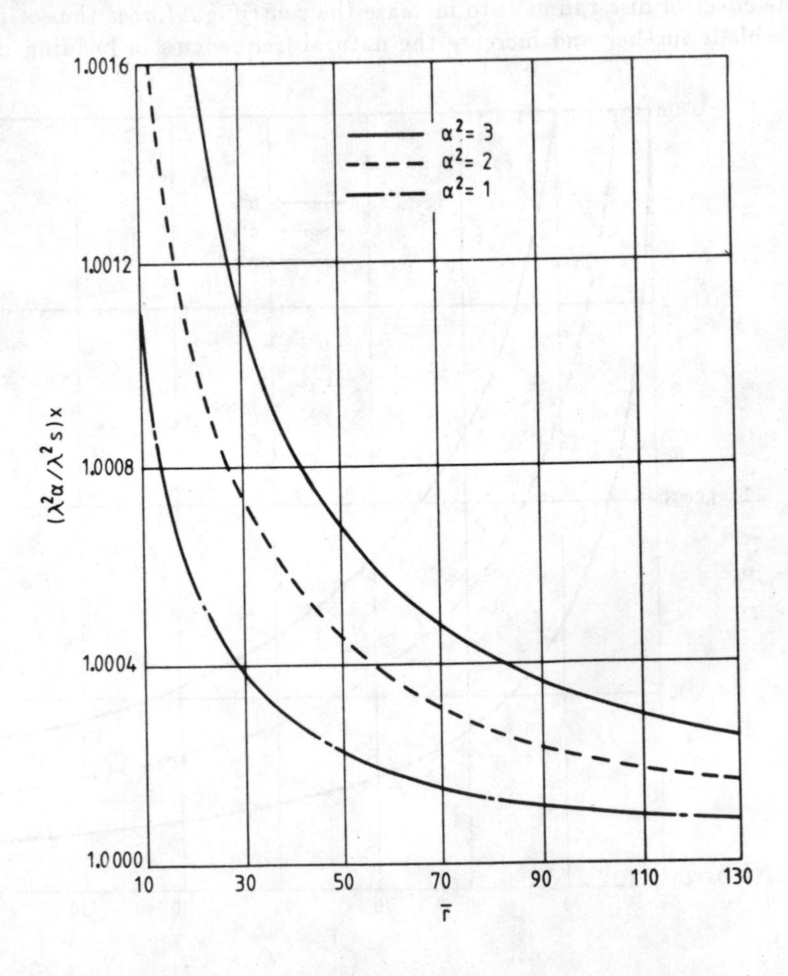

Fig. 2.29 Effect of $\overline{r}$ on frequency parameter ratio in the second chordwise bending mode at setting angle 0°

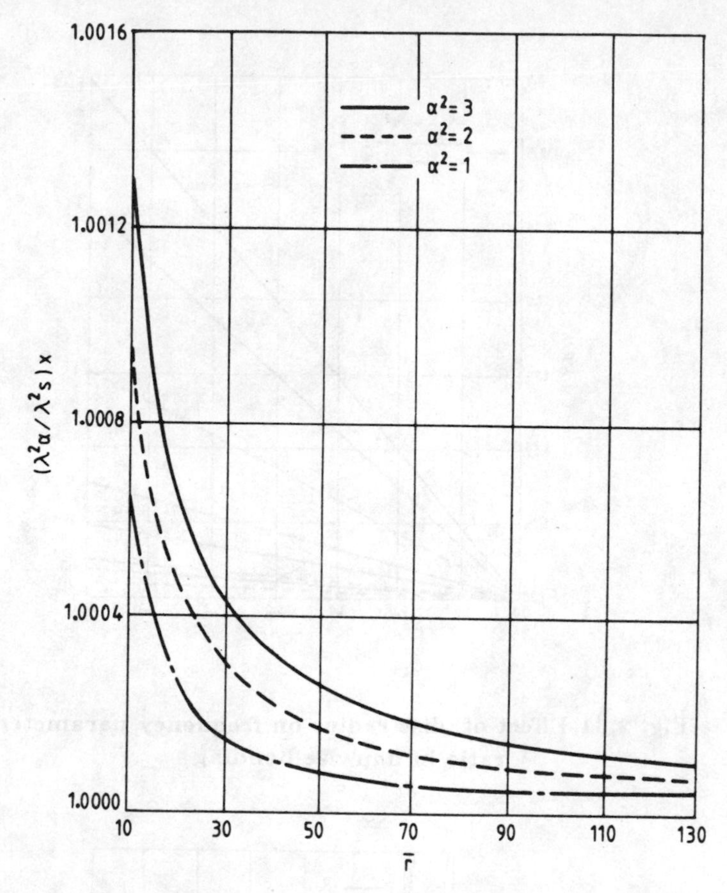

Fig. 2.30 Effect of $\bar{r}$ on frequency parameter ratio in the third chordwise bending mode at setting angle 0°

With the help of figures 2.27 and 2.31 for flapwise modes and figures 2.28 - 2.30 and 2.32 - 2.34 for chordwise modes, it is possible to estimate the natural frequencies of a blade mounted on a rotating disc at a setting of 0°, (the blade chord in the plane of the disc).

2.8 EFFECT OF SETTING ANGLE

The effect of setting angle $\phi' = (\phi - 90°)$ on the flapwise modes is given in figure 2.35 for three different values of α^2. The variation is found to be linear with $\sin^2 \phi'$. λ'_ϕ is the eigenvalue taking into account the setting angle. The following may be observed.

- The effect of setting angle is to decrease the bending mode natural frequencies linearly with $\sin^2 \phi'$.
- The effect is predominant in the fundamental mode and is insignificant for higher modes.

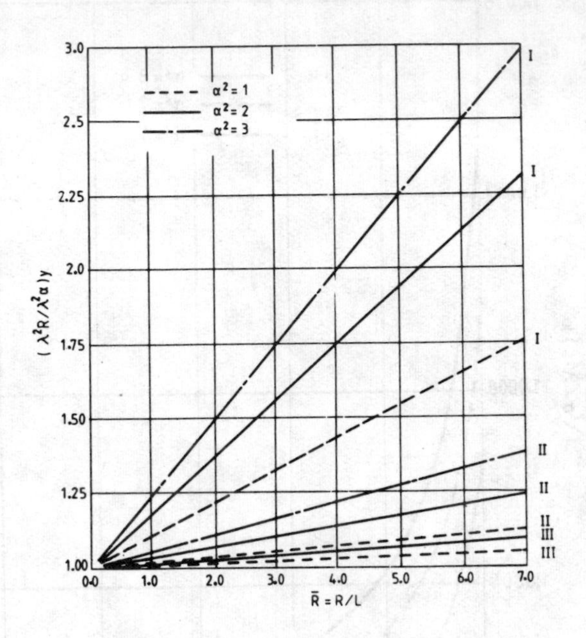

Fig. 2.31 Effect of disc radius on frequency parameter ratio in flapwise bending

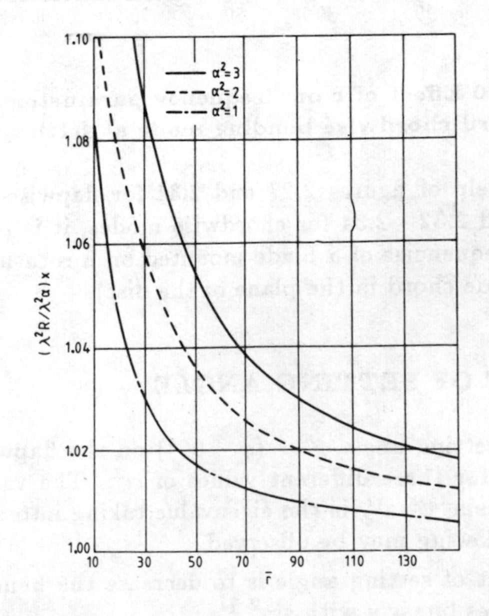

Fig. 2.32 Effect of $\overline{\mathbf{r}}$ on frequency parameter ratio in the first chordwise bending mode at $\phi' = 0$, $\overline{\mathbf{R}} = 7$

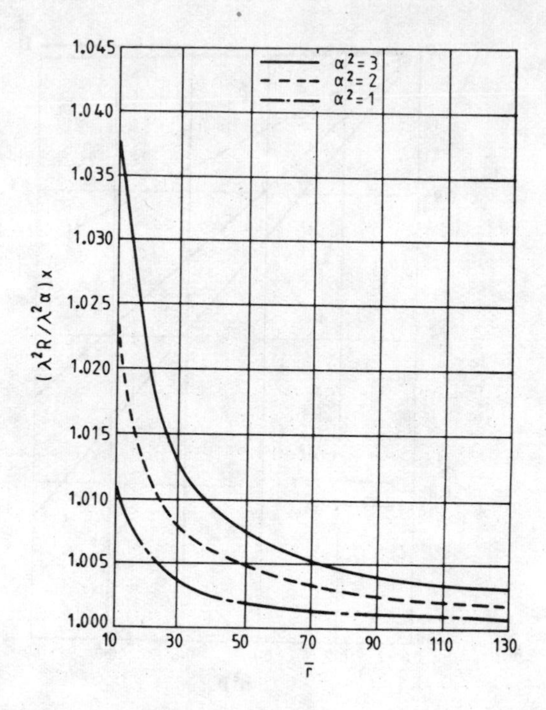

Fig. 2.33 Effect of $\overline{r}$ on frequency parameter ratio in the second chordwise bending mode at $\phi' = 0$, $\overline{R} = 7$

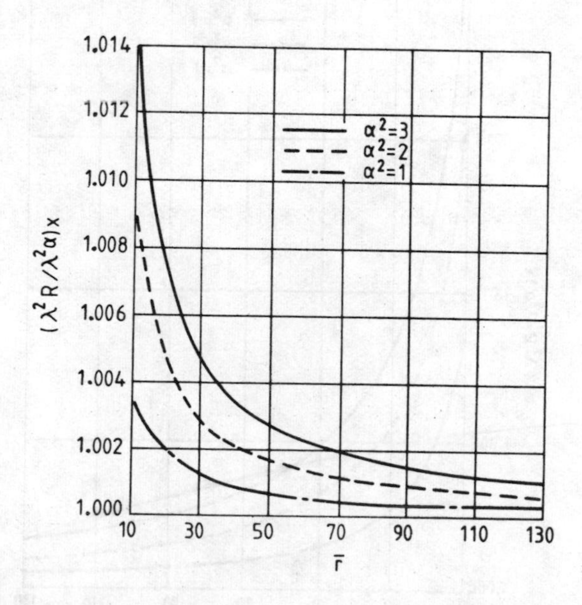

Fig. 2.34 Effect of $\overline{r}$ on frequency parameter ratio in the third chordwise bending mode at $\phi' = 0$, $\overline{R} = 7$

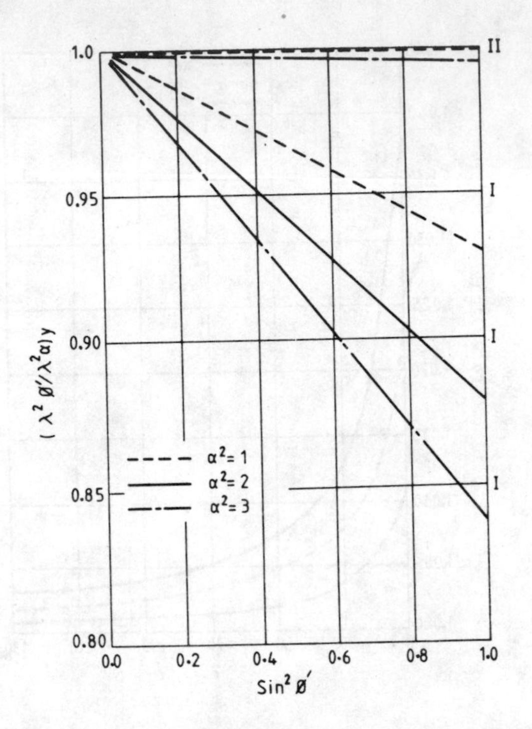

Fig. 2.35 Effect of setting angle on frequency parameter ratio in flapwise bending

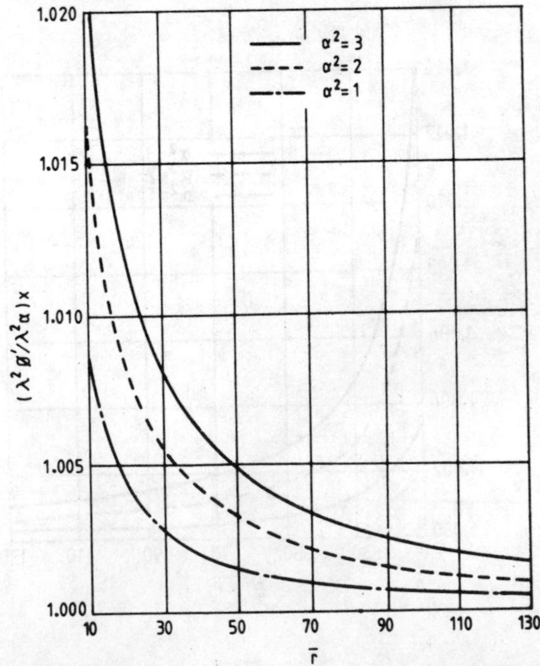

Fig. 2.36 Effect of $\overline{r}$ on frequency parameter ratio in the first chordwise bending mode at setting angle of 90°

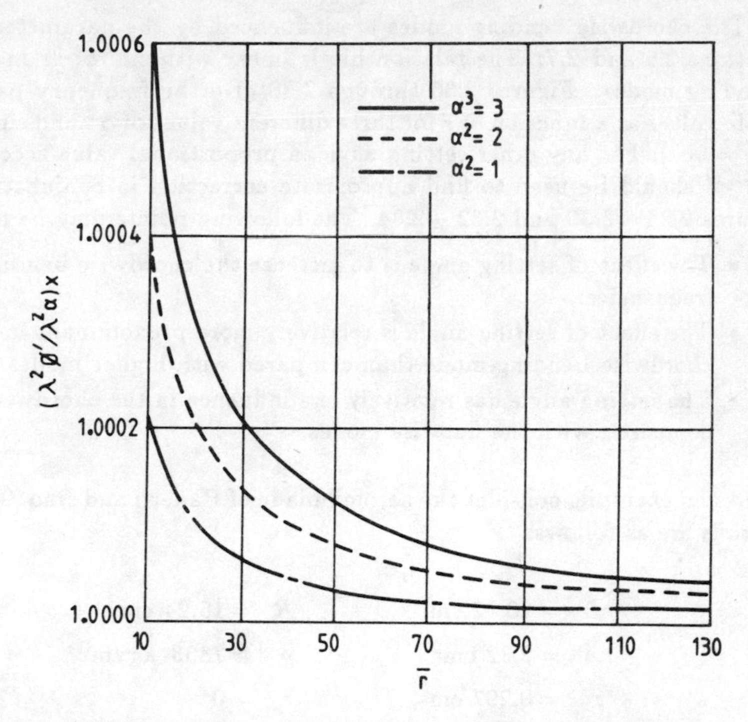

Fig. 2.37 Effect of $\bar{r}$ on frequency parameter ratio in the second chordwise bending mode at setting angle of 90°

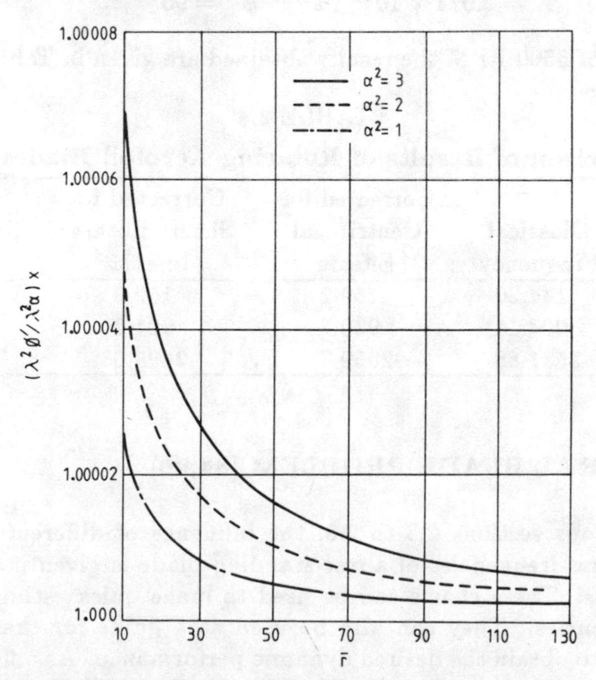

Fig. 2.38 Effect of $\bar{r}$ on frequency parameter ratio in the third chordwise bending mode at setting angle of 90°

The chordwise bending modes are influenced by the parameter $\bar{r}$ as in sections 2.6 and 2.7. The relationship is linear with $\sin^2 \phi'$ as in flapwise bending modes. Figures 2.36 through 2.38 give the frequency parameter ratio values as a function of $\bar{r}$ for three different values of α^2 and $\sin^2 \phi' = 1$ ($\phi' = 90°$). For any other setting angle, a proportional value according to $\sin^2 \phi'$ should be used to find appropriate correction in conjunction with figures 2.28 – 2.30 and 2.32 – 2.34. The following points may be noted.

- The effect of setting angle is to increase the chordwise bending mode frequencies.
- The effect of setting angle is relatively more predominant in the first chordwise bending mode than compared with higher modes.
- The setting angle has relatively less influence in the chordwise modes compared with the flapwise modes.

As an example, consider the aerofoil blade of Banerji and Rao [95] whose details are as follows:

$$
\begin{array}{llll}
L & = 20.32 \text{ cm} & R & = 15.24 \text{ cm} \\
A & = 2.32 \text{ cm}^2 & \rho & = 7833 \text{ kg/m}^3 \\
r_x & = 0.297 \text{ cm} & r_y & = 0 \\
I_{xx} & = 0.1002 \text{ cm}^4 & I_{yy} & = 2.034 \text{ cm}^4 \\
E & = 2.071 \times 10^{11} \text{ Pa} & \phi & = 90°
\end{array}
$$

For a speed of 3500 RPM the results obtained are given in Table 2.4 below:

TABLE 2.4

Comparison of Results of Rotating Aerofoil Blades — Hz

Mode #	Classical Frequency	Corrected for Centrifugal Effects	Corrected for Shear, Rotary Inertia	Experimental Results
1	144.26	169.2	169.0	165
2	904.14	926.7	921.2	914
3	2531.88	2559.7	2495.0	2475

2.9 GENERAL BLADE PROBLEM [84,96]

In the previous sections 2.3 to 2.8, the influence of different parameters on the natural frequencies of a free-standing blade is given in the form of design charts. These charts can be used to make quick estimation of the blade frequencies. They can also be used as a guide for changing blade parameters to obtain the desired dynamic performance. As a final check, it is always desirable to obtain the dynamic characteristics from a computer program. Using a continuum model, two such programs using Galerkin and Lagrangian approaches are available. The input to these programs are as follows.

- Blade profile coordinates in the disc plane and rotational axis system at different sections along the blade length.
- Blade material properties, ρ, E, G.
- Disc radius and setting angle.

The program determines sectional properties *viz.*, area, second moments of area, blade asymmetry, torsional stiffness and curve fitting program is used to obtain the best possible functions of these properties up to a sixth degree polynomial in the axial distance z. The blade length is then divided in 10 intervals and 19th degree Gaussian quadrature formula is used in integration for all the intervals for evaluating the matrix elements in the eigenvalue problem. A five term solution for each bending and torsion mode is used to obtain the first nine coupled frequencies and mode shapes with the help of a QR algorithm.

Two case studies are given here to determine the frequencies of a general blade. The first blade is from Carnegie (8) given below and in figure 2.39.

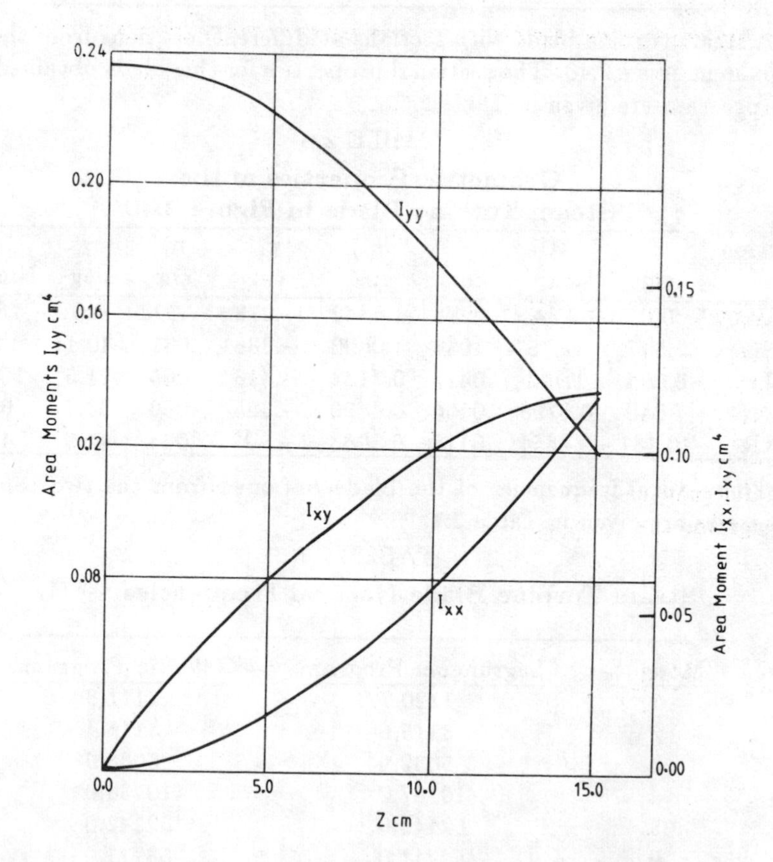

Fig. 2.39 Variation of area moments along the blade length

$E = 2.07 \times 10^{11}$ Pa $\quad G = 8.28 \times 10^{10}$ Pa
$\rho = 7860$ kg/m^3 $\quad L = 15.24$ cm
$A = 0.438709$ cm^2 $\quad I_{cf} = 0.471467 \times 10^{-3}$ kg cm^2/cm

The results obtained from Lagrangian program are compared with experimental values in Table 2.5.

TABLE 2.5

Comparison of Natural Frequencies of a Blade with Experimental Values — Hz

Mode #	Theoretical Values	Experimental Values
1	62.085	59
2	305.52	290
3	761.44	760
4	951.35	920
5	1222.90	1110
6	2284.30	2300
7	3813.20	3902

A steam turbine blade with sections at different locations from the root is given in figure 2.40. The sectional properties for this blade obtained from the program are given in Table 2.6.

TABLE 2.6

Geometric Properties of the Steam Turbine Blade in Figure 2.40

Section	Z	A	I_{xx}	I_{yy}	r_x	r_y	γ	C
	cm	cm^2	cm^4	cm^4	cm	cm	deg	Nm^2/rad
AA	0.0	2.34	.2278	1.6842	−.278	−.023	0.0	2088.7
EE	2.547	1.615	.1036	1.0890	−.388	.035	10.1	1497.6
JJ	5.094	1.102	.0418	0.7184	−.413	.065	21.5	1058.6
NN	7.640	0.6798	.0166	0.4406	−.355	.069	33.7	681.5
RR	10.187	0.4854	.0118	0.3063	−.212	.044	41.9	472.3

The natural frequencies of the blade obtained from the two computer programs are given in Table 2.7.

TABLE 2.7

Steam Turbine Blade Natural Frequencies — Hz

Mode #	Lagrangian Program	Galerkin Program
1	1120.7	1121.8
2	3115.6	3116.4
3	5082.9	5083.9
4	10137.0	10138.0
5	52142.0	52142.0
6	58711.0	58711.0

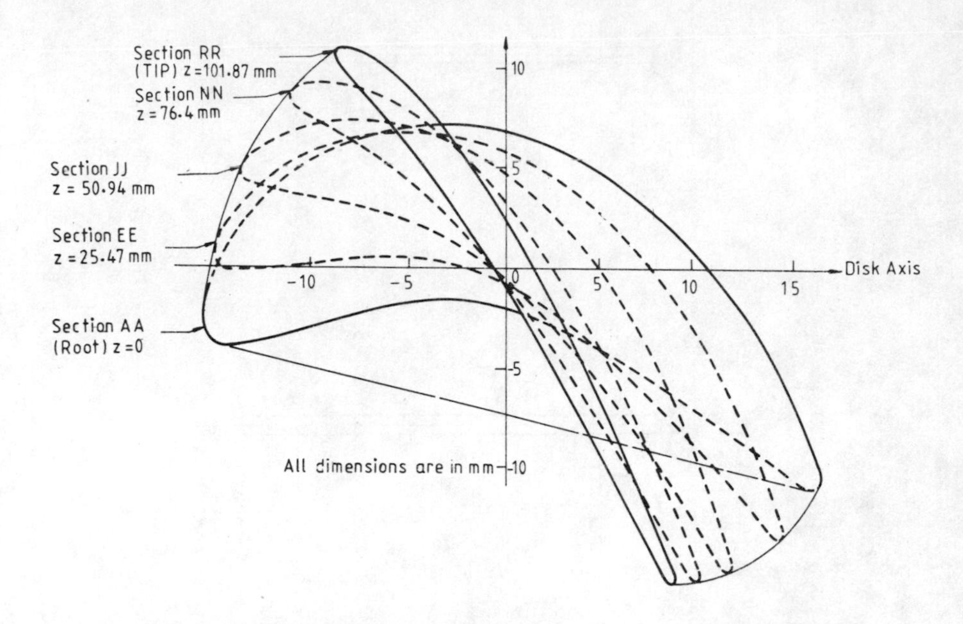

Fig. 2.40 Steam turbine blade with sections at different distances from root

2.10 EFFECT OF SUPPORT STIFFNESS

Though a turbine blade is loosely mounted on the disc, the root gets encastered due to the centrifugal force, thus almost satisfying the cantilever fixed end conditions. For all practical purposes, the cantilever conditions used in previous sections give very accurate estimation of the natural frequencies. The effect of support stiffness on the uncoupled bending and torsional frequencies is dealt with here, to illustrate how the frequencies change with loose mounting of blades of uniform cross-section. The blade is assumed to be restrained by a spring K_x for bending motion in yz plane and by a spring K_z for torsional motion as shown in figure 2.41.

2.10.1 Bending Vibration

Because of support stiffness, modeled by spring K_x the additional strain energy is

$$V_s = \frac{1}{2}K_x y_0'^2 \tag{2.44}$$

Hence according to Hamilton's principle,

$$\delta \int_{t_1}^{t_2} \left[\int_0^L (\frac{1}{2}EI_{xx}y''^2 - \frac{1}{2}m\dot{y}^2)dz + \frac{1}{2}K_x y_0'^2 \right] dt = 0 \tag{2.45}$$

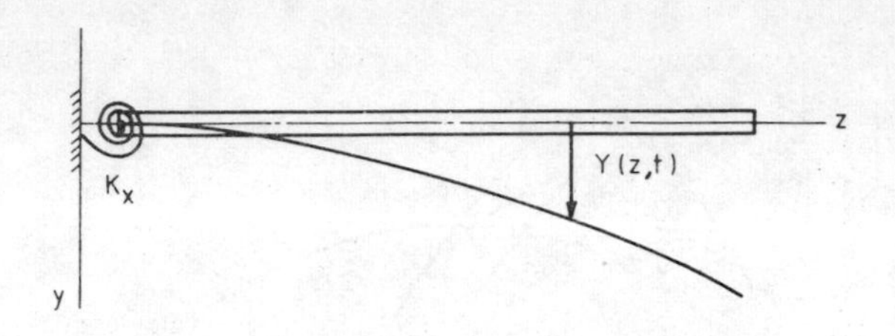

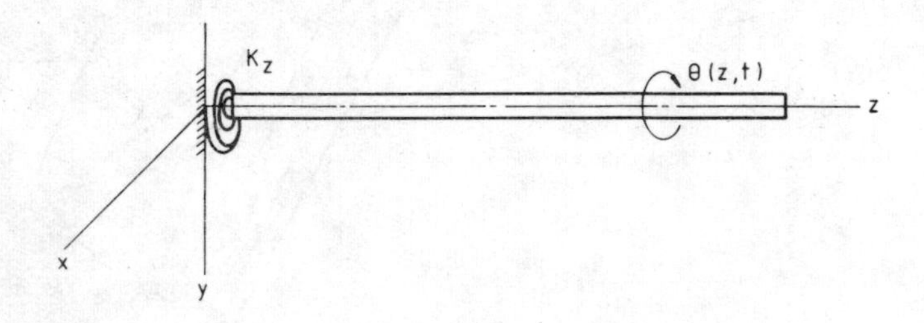

Fig. 2.41 Support stiffness in bending and torsional motions

Taking the variation above, the following is obtained

$$\int_{t_1}^{t_2} \left[\int_0^L (m\ddot{y} + EI_{xx}y'''')\delta y\, dz + EI_{xx}y''\delta y' \Big\}_0^L - EI_{xx}y'''\delta y \Big\}_0^L + K_x y_0' \delta y_0' \right] dt = 0 \tag{2.46}$$

From the above, the following differential equation and boundary conditions in nondimensional form can be obtained

$$Y'''' - \lambda^4 Y = 0 \tag{2.47}$$

$$Y = 0;\ Y'' - \beta Y' = 0;\ Z = 0$$
$$Y'' = 0;\ Y''' = 0;\ Z = 1 \tag{2.48}$$

In equation (2.47) and (2.48)

$$Y = \frac{y}{L};\ Z = \frac{z}{L}$$

$$\lambda^4 = m\frac{L^4\omega^2}{EI_{xx}}$$

$$\beta = \frac{K_x L}{EI_{xx}}$$

$$' = \frac{d}{dZ} \tag{2.49}$$

λ is nondimensional frequency parameter and β is nondimensional support stiffness parameter. Solution of (2.47) is

$$Y = A \cosh \lambda Z + B \sinh \lambda Z + C \cos \lambda Z + D \sin \lambda Z \tag{2.50}$$

Using the boundary conditions in (2.48), the above leads to

$$\begin{bmatrix} \left\{\frac{\beta}{2\lambda}(Ch + C) + Sh\right\} & \left\{\frac{\beta}{2\lambda}(Ch + C) - S\right\} \\ \left\{\frac{\beta}{2\lambda}(Sh - S) + Ch\right\} & \left\{\frac{\beta}{2\lambda}(Sh - S) - C\right\} \end{bmatrix} \begin{Bmatrix} B \\ D \end{Bmatrix} = 0 \tag{2.51}$$

where

$$\begin{aligned} C &= \cos \lambda \\ Ch &= \cosh \lambda \\ S &= \sin \lambda \\ Sh &= \sinh \lambda \end{aligned} \tag{2.52}$$

Expanding the determinant in (2.51), the following frequency equation is obtained

$$(1 + C \cdot Ch)\frac{\beta}{\lambda} + Sh \cdot C - Ch \cdot S = 0 \tag{2.53}$$

The mode shape can be shown to be

$$\begin{aligned} Y(Z) = {} & \sin \lambda Z - \frac{\beta}{2\lambda}\left(1 - \frac{b}{a}\right) \cos \lambda Z - \frac{b}{a} \sinh \lambda Z \\ & + \frac{\beta}{2\lambda}\left(1 - \frac{b}{a}\right) \cosh \lambda Z \end{aligned} \tag{2.54}$$

where

$$a = \frac{\beta}{2\lambda}(Ch + C) + Sh$$

$$b = \frac{\beta}{2\lambda}(Ch + C) - S$$

For $\beta = 0$, $K_x = 0$, the boundary conditions are for a hinged-free beam and the frequency equation is:

$$Sh \cdot C - Ch \cdot S = 0 \tag{2.55}$$

For $\beta = \infty$, $K_x = \infty$ the boundary conditions are for a cantilever beam, then

$$1 + \dot{C} \cdot Ch = 0 \tag{2.56}$$

For any other conditions, in between the hinged and fixed conditions at the root, equation (2.53) should be used. Figure (2.42) shows these frequencies as a function of nondimensional parameter β.

2.10.2 Torsional Vibration

For torsional vibration, Hamilton's principle gives:

$$\delta \int_{t_1}^{t_2} \left[\int_0^L \left(\frac{1}{2} C\theta'^2 - \frac{1}{2} I_{cg}\dot{\theta}^2 \right) dz + \frac{1}{2} K_z \theta_0^2 \right] dt = 0 \tag{2.57}$$

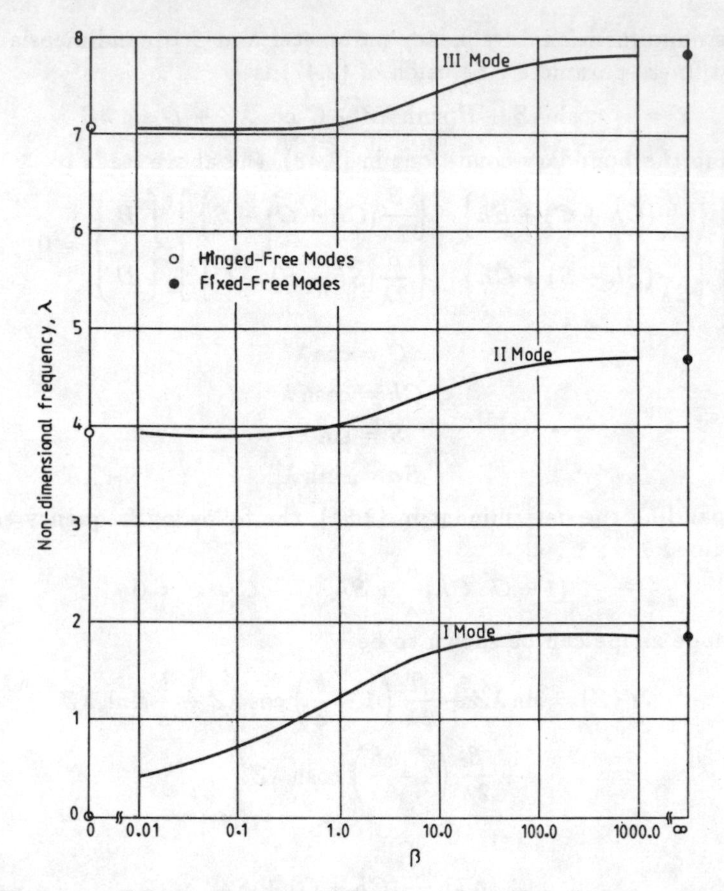

Fig. 2.42 Effect of root stiffness on bending frequencies

Taking the variation above,

$$\int_{t_1}^{t_2}\left[\int_0^L (I_{cg}\ddot{\theta} - C\theta'')\delta\theta dz + C\theta'\delta\theta\}_0^L + K_z\theta_0\delta\theta_0\right] dt = 0 \tag{2.58}$$

Nondimensionalizing,

$$\theta'' + \lambda^2\theta = 0$$

$$\alpha\theta - \theta' = 0 \qquad Z = 0 \tag{2.59}$$

$$\theta' = 0 \qquad Z = 1 \tag{2.60}$$

where,

$$\alpha = \frac{K_z L}{C}$$

$$\lambda^2 = \frac{I_{cg}L^2p^2}{C};$$

$$Z = \frac{z}{L}; \quad ' = \frac{d}{dZ} \tag{2.61}$$

Solution of (2.59) is

$$\theta = A\cos\lambda Z + B\sin\lambda Z$$

Using boundary conditions (2.60),

$$\alpha\cot\lambda - \lambda = 0 \tag{2.62}$$

$$\theta = \cot\lambda\cos\lambda Z + \sin\lambda Z \tag{2.63}$$

For $\alpha = 0$, the frequency equation is

$$\sin\lambda = 0 \tag{2.64}$$

For $\alpha = \infty$

$$\cos\lambda = 0 \tag{2.65}$$

Figure 2.43 shows the effect of root stiffness on the torsional frequencies.

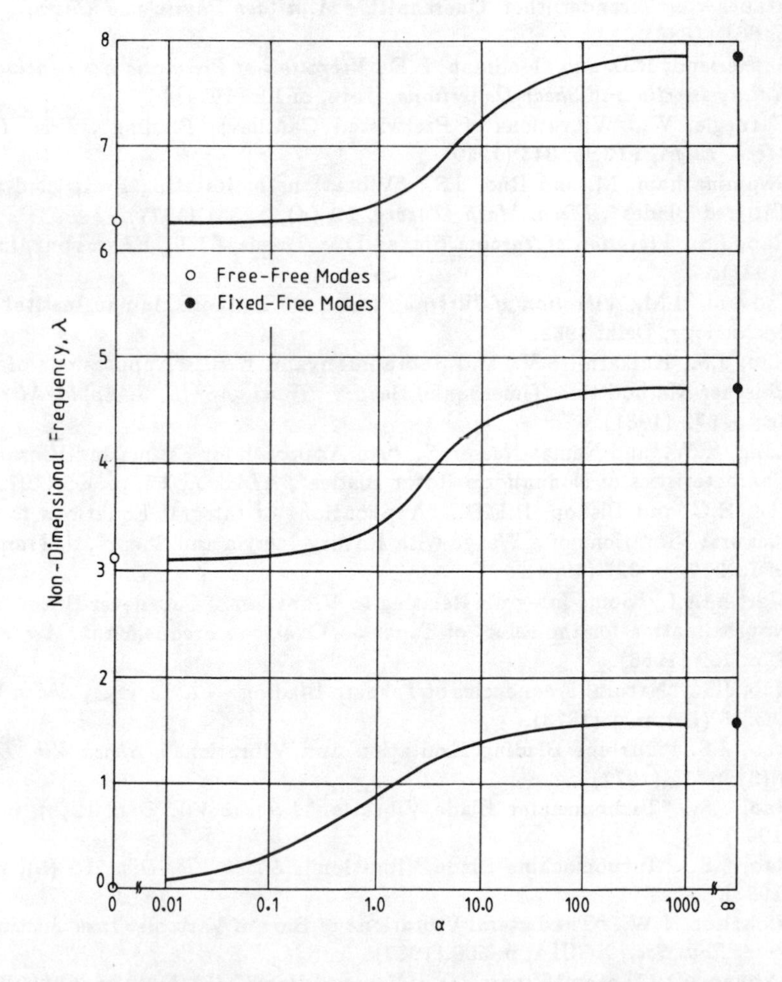

Fig. 2.43 Effect of root stiffness on torsional frequencies

REFERENCES

1. Timoshenko, S.P., *Vibration Problems in Engineering*, Van Nostrand Publications (1955).
2. Timoshenko, S.P. and Goodier, J.N., *Theory of Elasticity*, McGraw-Hill Publications (1961).
3. Rao, J.S., and Rao, D.K., "Equations of Motion of Rotating Pre-Twisted Blades in Bending-Bending-Torsion with Effects of Warping, Shear, Rotary Inertia, etc." *Proc. Silver Jubilee Conf.*, Aeronautical Soc. of India, Bangalore, p. 4-13 (1974).
4. Ward, P.F., "Transverse Vibrations of a Rod of Variable Cross Section", *Phil. Mag.*, **25**, p. 85 (1913).
5. Wrinch, D.M., "Lateral Vibrations of Bars of Conical Type," *Proc. Roy. Soc.*, **101**, p. 493 (1922).
6. Meyer, F., "Mathematische Theorie der Transversalen Schwingungen Eines Stabes von Veranderlichen Querchnitt", *Ann. der Physik und Chemie*, **33**, p. 661 (1888).
7. Sutherland, R.L. and Goodman, L.E., *Vibration of Prismatic Bars Including Rotary Inertia and Shear Deflections*, Univ. of Ill. (1954).
8. Carnegie, W., "Vibrations of Pretwisted Cantilever Blading", *Proc. Inst. Mech. Engr.*, **173**, p. 343 (1959).
9. Swaminatham, M. and Rao, J.S., "Vibrations of Rotating Pretwisted and Tapered Blades", *Mech. Mach. Theory*, **12** (4), p. 331 (1977).
10. Rao, J.S., *Vibration of Turbine Blades*, DSc Thesis, I.I.T., Kharagpur, India (1971).
11. Jadvani, H.M., *Vibration of Turbine Blades*, Ph.D. thesis, Indian Institute of Technology, Delhi 1982.
12. Rao, J.S., Kulkarni, S.V., and Subrahmanyam, K.B., "Applications of the Reissner Method to a Timeshenko Beam", *Trans. ASME, J. Appld. Mechs.*, **48**, p. 672 (1981).
13. Lang, K.W., and Nemat-Nasser, S., "An Approach for Estimating Vibration Characteristics of Nonuniform Rotor Blades", *AIAA J.*, **17**, p. 995 (1979).
14. Lee, H.C. and Bishop, R.E.D., "Applications of Integral Equations to the Flexural Vibrations of a Wedge with Rotary Inertia and Shear", *J. Franklin Inst.*, **277**, p. 327 (1964).
15. Martin, A.I., "Some Integrals Relating to Vibration of Cantilever Beams and Approximation for the Effect of Taper on Overtone Frequencies", *Aero. Q.*, **7**, p. 109 (1956).
16. Rao, J.S., "Natural Frequencies of Turbine Blading — A Survey", *Shock Vib. Dig.*, **5** (10), p. 1 (1973).
17. Rao, J.S., "Turbine Blading Excitation and Vibrations", *Shock Vib. Dig.*, **9** (3), p. 15 (1977).
18. Rao, J.S., "Turbomachine Blade Vibration", *Shock Vib. Dig.*, 12(2), p. 19 (1980).
19. Rao, J.S., "Turbomachine Blade Vibration", *Shock Vib. Dig.*, **15** (5), p. 3 (1983).
20. Nicholson, J.W., "The Lateral Vibrations of Bars of Variable Cross Section", *Proc. Roy. Soc.*, XCIIIA, p. 506 (1917).
21. Akimasa, O., "Lateral Vibrations of Tapered Bars", *JSME*, **28**, p. 429 (1925).
22. Conway, H.D., "Calculations of Frequencies of Truncated Pyramids", *Aircraft Engr.*, **18**, p. 235 (1946).

23. **Watanabe, I.**, "Natural Frequencies of Flexural Vibrations of Tapered Cantilevers with Uniform Thickness", *Proc. 1st Jap. Natl. Cong. Appl. Mech.*, p. 547 (1951).
24. Nobuo, I., "A Technique for Computing Natural Frequencies of Trapezoidal Vibrators", *J. Aero. Sci.*, **24**, p. 239 (1957).
25. Wang, H.C., *A General Treatment of the Transverse Vibration of Tapered Beams*, Ph.D. Thesis, Univ. of Ill. (1966).
26. Taylor, J.L., "Natural Vibrational Frequencies of Flexible Rotor Blades", *Aircraft Engr.*, **30**, p. 331 (1958).
27. Rao, J.S., *Effect of Taper on Uncoupled Natural Frequencies of Cantilever Beams*, Ph.D. Thesis, I.I.T., Kharagpur, India (1965).
28. Rao, J.S. and Carnegie, W., "Determination of the Frequencies of Lateral Vibrations of Tapered Cantilever Beams by the Use of Ritz-Galerkin Process", *BMEE*, **10**, p. 239 (1971).
29. Walker, P.B., *Simple Formulae for Fundamental Natural Frequencies of Cantilevers*, R&M 1831 (1938).
30. Vet, M., "Torsional Vibration of Beams Having Rectangular Cross Sections", *J. Acoust. Soc. Amer.*, **34**, p. 1570 (1962).
31. Rao, J.S., "Fundamental Torsion Vibration of a Cantilever Beam of Triangular Cross Section with Uniform Taper", *Proc. 10th Cong. ISTAM*, p. 66 (1965).
32. Rao, J.S., Belgaumkar, B.M. and Carnegie, W., "Torsional Vibration of a Cantilever Beam of Rectangular Cross Section with Uniform Taper", *BMEE*, **9**, p. 61 (1970).
33. Rao, J.S. and Carnegie W., "Torsional Vibration of Pretwisted Tapered Cantilever Beams Treated By Collocation Method", *Ind. J. Pure and Appl. Physics*, **10**, p. 566 (1972).
34. Timoshenko, S.P., "On the Correction for Shear of the Differential Equation for Transverse Vibrations of Prismatic Bars", *Phil. Mag.*, **41**, p. 566 (1921).
35. Huang, T.C., "Effect of Rotary Inertia and Shear on Vibration of Beams Treated by Approximate Methods of Ritz and Galerkin," *Proc. 3rd US Natl. Cong. Appl. Mech., ASME*, p. 189 (1958).
36. Huang, T.C., "Effect of Rotary Inertia and Shear Deformations on the Frequency and Normal Mode Equations of Uniform Beams with Simple End Conditions", *J. Appl. Mech., ASME*, **28**, p. 579 (1961).
37. Huang, T.C., *Eigenvalues and Modifying Quotients of Vibration of Beams*, Univ. Wis., Engr. Expt. Stn. Rept. No. 25 (1964).
38. Huang, T.C. and King, C.S., *New Tables and Eigenvalues Representing Normal Modes of Vibration of Timoshenko Beams*, Plenum Press (1963).
39. Lee, H.C., "A Generalized Minimum Principle and Its Application to the Vibration of a Wedge with Rotary Inertia and Shear", *J. Appl. Mech., Trans. ASME*, **30**, p. 176 (1963).
40. Carnegie, W., "Vibrations of Pretwisted Cantilever Blading Allowing for Rotary Inertia and Shear", *JMES*, **6**, p. 105 (1964).
41. Dawson, B., "Rotary Inertia and Shear Treated by Ritz Method", *J. Aero. Soc.*, (London) **72**, p. 341 (1968).
42. Lo, H. and Renbarger, J., "Bending Vibrations of a Rotating Beam", *Proc. 1st US Natl. Cong. Appl. Mech.*, ASME, p. 75 (1952).
43. Lo, H., "A Nonlinear Problem in the Bending Vibration of a Rotating Beam", *J. Appl. Mech., Trans. ASME*, **19**, p. 461 (1952).

44. Boyce, W.E., Diprima, R.C. and Handelman, G.H., "Vibrations of Rotating Beams of Constant Section", *Proc. 2nd US Natl. Cong. Appl. Mech.*, ASME, p. 165 (1954).
45. Yntema, R.T., *Simplified Procedures and Charts for the Rapid Estimation of Bending Frequencies of Rotating Beams*, NACA TN 3459 (1955).
46. Boyce, W.E., "Effect of Hub Radius on the Vibrations of a Uniform Bar", *J. Appl. Mech., Trans. ASME*, **23**, p. 287 (1956).
47. Carnegie, W., "Vibrations of Rotating Cantilever Blading: Theoretical Approaches to the Frequency Problem Based on Energy Methods", *JMES*, **1**, p. 235 (1959).
48. Schilhansl, M.J., "Bending Frequency of a Rotating Cantilever Beam", *J. Appl. Mech., ASME*, **25**, p. 28 (1958).
49. Hirsch, G., "Investigation of Vibrations in Bending of Rotating Turbine Blades on Nonrigid Support", Jahrbuck Wissenschaft., *Gessellsch Luftfahrt*, p. 174 (1958).
50. Horway, G., "Chordwise and Beamwise Bending Frequencies of Hinged Rotor Blades", *J. Aero. Sci.*, **15**, p. 497 (1948).
51. Niordson, F., "Vibration of Turbine Blades with Loose Hinge Supports", *Engr. Digest*, **15**, p. 359 (1954).
52. Kissel, W., "The Lowest Natural Bending Frequency of a Rotating Blade of Uniform Cross Section", *Escher Wyss. News*, **31**, p. 28 (1958).
53. Rao, J.S., "Flexural Vibration of Turbine Blades", *Archiwum Budowy Maszyn*, Tom XVII (3), p. 375 (1970).
54. Rao, J.S. and Carnegie, W., "Nonlinear Vibration of Rotating Cantilever Beams", *J. Roy. Aero. Soc.*, **74**, p. 161 (1970).
55. Rao, J.S. and Carnegie, W., "Nonlinear Vibrations of Rotating Cantilever Beams Treated by Ritz Averaging Procedure", *J. Roy. Aero. Soc.*, **76**, p. 566 (1972).
56. Carnegie, W., "The Application of the Variational Method to Derive the Equations of Motion of Vibrating Cantilever Blading under Rotation", *BMEE*, **6**, p. 29 (1967).
57. Dunholter, R.J., "Static Displacement and Coupled Frequencies of a Twisted Bar", *J. Aero. Sci.*, **13**, p. 214 (1946).
58. White, W.T., "An Integral Equation Approach to Problems of Vibrating Beams", *J. Franklin Inst.*, **245**, pp. 25 and 117 (1948).
59. Diprima, R.C. and Handelman, G.H., "Vibration of Twisted Beams", *Q. Appl. Math.*, **12**, p. 241 (1954).
60. Troesch, B.A., Anliker, M. and Zeigler, H., "Lateral Vibrations of Twisted Rods", *Q. Appl. Math.*, **12**, p. 163 (1954).
61. Martin, A.I., "Approximation for the Effect of Twist on Vibration of a Cantilever Blade", *Aero. Q.*, **8**, p. 291 (1957).
62. Schimogo, T., "Lateral Vibrations of Twisted Rods with Variable Cross Section", *Proc. 7th Natl. Cong. Appl. Mech.*, p. 343 (1957).
63. Sato, Y., "Vibrations of Twisted Rods and Effect of Position on Elastic Axis", *Trans. JSME*, **24**, p. 866 (1958).
64. Slyper, N.A., "Coupled Bending Vibrations of Pretwisted Cantilevers", *JMES*, **4**, p. 365 (1962).
65. Anliker, M. and Troesch, B.A., "Lateral Vibrations of Pretwisted Rods with Various Boundary Conditions", *ZAMP*, **14**, p. 218 (1963).
66. Dawson, B., "Coupled Bending Vibrations of Pretwisted Cantilever Blading Treated by Raleigh-Ritz Method", *JMES*, **10**, p. 381 (1968).

67. Rao, J.S., "Flexural Vibration of Pretwisted Tapered Cantilever Beams Treated by Galerkin Method", *J. Eng. Indus., ASME*, **94**, p. 343 (1972).
68. Bogdonoff, J.L. and Horner, J.T., "Torsional Vibrations of Rotating Twisted Bars", *J. Aero. Sci.*, **23**, p. 393 (1956).
69. Brody, W.G. and Targoff, W.P., *Uncoupled Torsional Vibrations of Rotating Twisted Bars*, WADC, TR 56-501 (1957).
70. Carnegie, W., "Vibrations of Pretwisted Cantilever Blading: An Additional Effect Due to Torsion", *Proc. Inst. Mech. Engr.*, **176**, p. 315 (1962).
71. Rao, J.S., "Torsional Vibration of Pretwisted Cantilever Beams", *J. Inst. Engr. Ind.*, CE Div. **52**, p. 211 (1972).
72. Rao, J.S., "Vibration of Pretwisted Tapered Cantilever Beams in Torsion", *Archiwum Budwy Maszyn*, Tom **18** (3), p. 443 (1971).
73. Subrahmanyam, K.B., Kulkarni, S.V. and Rao, J.S., "Coupled Bending-Bending Vibrations of Pre-Twisted Cantilever Blading Allowing for Shear Deflection and Rotary Inertia by the Reissner Method", *Int. J. Mech. Sci.*, **23** (9), p. 517, (1981).
74. Swaminadham, M., "Study of the Vibrations of Rotating Blades", *ASME* Paper No. 77-DET-147.
75. Garland, C.F., "The Normal Modes and Vibrations of Beams Having Non-collinear Elastic and Mass Axes", *J. Appl. Mech., Trans. ASME*, p. A-97 (1940).
76. Houbolt, J.C. and Brooks, G., *Differential Equations of Motion for Combined Flapwise Bending Chordwise Bending and Torsion of Twisted Nonuniform Rotor Blades*, NACA 1346 (1958).
77. Carnegie, W., "Experimental Determination of Center of Flexure and Center of Torsion Coordinates of Asymmetrical Aerofoil Cross Section", *JMES*, **1**, p. 241 (1959).
78. Dawson, B., *Vibration Characteristics of Cantilever Beams of Uniform Cross Section*, Ph.D. Thesis Univ. London (1967).
79. Rao, J.S. and Carnegie, W., "Solution of the Equations of Motion Coupled Bending-Bending Torsion Vibrations of Turbine Blades by the Method of Ritz-Galerkin", *Intl. J. Mech. Sci.*, **12**, p. 875 (1970).
80. Rao, J.S., "Coupled Bending-Bending Torsion Vibration of Cantilever Beams", *J. Aero. Soc. Ind.*, **24**, p. 265 (1972).
81. Carnegie, W. and Dawson, B., "Vibration Characteristics of Straight Blades of Asymmetrical Airfoil Cross Section", *Aero. Q.*, **20**, p. 178 (1969).
82. Rao, J.S., *Vibration of Turbine Blades*, Aero. R. and D. Board, Ministry of Defence ,India, ARDB-STR-5002 (1977).
83. Rao, J.S., "Coupled Vibrations of Turbomachine Blades", *Shock Vib. Bull.*, U.S. Naval Res. Lab., Proc. No. 47, Pt. 2, p. 107 (1977).
84. Rao, J.S., "Turbine Blade Vibration", *23rd Cong. Ind. Soc. Theort. Appl. Mech.*, p. 150 (1978).
85. Zuladzinski, C., "Effect of Centrifugal Forces on Lateral Deflection and Fundamental Frequency of Turbine Blades", *ASME*, Paper No. 77-Pet-12.
86. Subrahmanyam, K.B., Kulkarni, S.V. and Rao, J.S., "Coupled Bending-Torsion Vibrations of Rotating Blades of Asymmetric Aerofoil Cross Section with Allowance for Shear Deflection and Rotary Inertia by Use of the Reissner Method", *J. Sound Vib.*, **75** (1), p. 17 (1981).
87. Subrahmanyam, K.B., Kulkarni, S.V. and Rao, J.S., "Analysis of Lateral Vibrations of Rotating Cantilever Blades Allowing for Shear Deflection and

Rotary Inertia by Reissner and Potential Energy Methods", *Mechanism and Machine Theory*, **17** (4), p. 235 (1982).

88. Kulkarni, S.V., Subrahmanyam, K.B. and Rao, J.S., "Coupled Bending-Bending-Torsion Vibrations of Turbomachine Blading Treated by Reissner Method", *Proc. 1st Nat. Conf. Mechanisms and Machines*, Dec. 2-3, 1981, Bombay, India, p. R-19.
89. Roark, R.J., *Formulas for Stress and Strain*, McGraw Hill Publication, (1975).
90. Rao, J.S., "The Fundamental Flexural Vibration of a Cantilever Beam of Rectangular Cross Section with Uniform Taper", *Aero. Q.*, **16**, p. 139 (1965).
91. Carnegie, W., Dawson, B. and Thomas, J., "Vibration Characteristics of Cantilever Blading", *Proc. Inst. Mech. Engr.*, **180** (31), p. 71, (1965-66).
92. Dawson, B. and Carnegie, W., "Modal Curves of Pre-Twisted Beams of Rectangular Cross-Section", *JMES*, **11**, p. 1 (1969).
93. Banerji, S., Rao, J.S., "Relations for Natural Frequencies of Turbine Blading", *Proc. 22nd Congr. Ind. Soc. Theo. & Appl. Mech.*, p. 34 (1977).
94. Carnegie, W. and Thomas, J., "The Effect of Shear Deformation and Rotary Inertia on the Lateral Frequencies of Cantilever Beams in Bending", *J. Engr. Indus. Trans. ASME*, **94**, p. 267 (1972).
95. Banerji, S. and Rao, J.S., "Coupled Bending Torsion Vibrations of Rotating Blades", *ASME*, Paper No. 76-GT-43.
96. Rao, J.S. and Jadvani, H.M., "Free and Forced Vibration of Turbine Blades", *Vibration of Bladed Disc Assemblies*, ASME publication, p. 11 (1983).

Chapter 3

Discrete Analysis of Blades

3.1 INTRODUCTION

The continuum approach adopted in Chapter 2 for a free standing blade, requires a lot of analytical work before a numerical procedure can be adopted. Discretising the blade and using appropriate element relations is simpler than the analytical work that goes with continuum methods and thus the discrete methods are favoured by several workers. Other distinct advantages of discrete models are their compatibility to complicated problems like laced and packeted blades and the availability of well-tested finite element programs to model any complicated blade group. The discrete analysis techniques can be broadly classified into the following methods as applied to turbine blade problems using beam theory:

- Holzer Method
- Myklestad/Prohl Method
- Matrix Methods
- Finite Difference Method
- Finite Element Method

Holzer [1] developed the basic discrete analysis technique for torsional vibration calculations of a given system. The given system is discretised into several rigid inertias connected by massless elastic torsional shafts. The dynamic properties are transferred from station to station using the field (spring) and point (mass) dynamic properties. The boundary conditions are used to set up a criterion and determine whether an assumed frequency satisfies this criterion for a natural frequency. The method is very popular [2,3]. Rao [4] used this method to determine the correction factors for the effect of taper on the torsional natural frequencies of a cantilever beam. The trial and error procedure in Holzer's method is eliminated by deriving a polynomial frequency equation for a cantilever beam executing uncoupled torsional vibrations [5]. This method saves computer time compared with classical Holzer method.

The application of Holzer method to bending vibration problems is tedious as the bending problem involves four state quantities compared with only two state quantities in torsional problems. Myklestad [6] and Prohl [7] developed two convenient tabular procedures to determine, bending natural frequencies of beams. As in Holzer method, the beam is discretised into several masses connected by massless beam elements having the original flexural rigidity of the beam between the stations. Depending on the boundary conditions, a suitable criterion can be set up and the natural frequency determined by trial and error root search technique. Rao [8] developed a polynomial frequency equation to determine the bending natural frequencies of a cantilever blade to eliminate the trial and error root searching technique. Dawson and Davies [9] developed an algorithm to automatically search a root in the Myklestad method, to save computer time. Mahalingam [10,11] also gave improvement of Myklestad method for bending vibration of beams.

Housner and Keightley [12] determined the natural frequencies and mode shapes of a variable cross-section cantilever by means of a digital computer using Myklestad, Prohl and Stodola methods. Mabie and Rogers [13] gave the results of tapered cantilever beams with end loads. Targoff [14] combined Holzer and Myklestad methods for rotating twisted blades. Myklestad method was used by Rosard [15] to study the coupled bending-bending vibration of pre-twisted blades. Isakson and Eisley [16, 17] also used Holzer-Myklestad method to determine coupled bending torsion modes of rotating twisted blades of asymmetric cross-section. An extended Holzer method was suggested by Carnegie [18] to determine coupled bending-bending vibration frequencies of a rotating pre-twisted blade. This procedure was programmed by Rao and Carnegie [19] to determine the natural frequencies and modes of rotating pre-twisted blades. Rao extended his polynomial frequency equation approach to rotating blades [20], asymmetric cross-section blades [21] and combined these two approaches for rotating aerofoil blades [22]. Krupka and Baumanis [23] developed a lumped parameter model of Carnegie's field equations using Myklestad type approach to consider bending-bending mode of a rotating tapered twisted blade including shear and rotary inertia effects. The Myklestad method has been applied by Nagaraja et al [24] to study the effects of flexibility of the root on tapered blades.

The bending vibration problem of a beam can be set up with the help of influence coefficients; the deflections of the beam under the inertia loadings are set up with the help of influence coefficients and then the eigenvalue problem is obtained [25]. Duncan and Collar [26], Duncan [27], Thomson [28,29], Leckie and Lindberg [30,31] used matrix methods with influence coefficients. Morris and Head [32] adopted influence coefficients method for a pre-twisted blade. Mendelson and Gendler [33,34] assumed the inertia loads to be distributed in each section, instead of lumping them and developeda method of station functions for coupled bending-torsion vibration of cantilever blades. Holzer and Myklestad methods are developed in the form of transfer matrix methods by Thomson [35]. Downs [36] described a new dynamic discretization technique and considered transverse vibrations of tapered beams. Influence coefficients and matrix algebra have been used by Wilkerson [37] to develop a general computer program for a rotor blade to determine the first three bending modes and fundamental torsion mode. A point transmission matrix method has been used for a similar problem in which a mass is attached to a rotor blade by Murthy and Barna [38]. Downs [39] extended his modified discretization technique to twisted beams with asymmetric cross-section.

Thomas [40-44] used finite difference method to determine the effect of taper and twist on bending and torsional vibrations of cantilever blades. Carnegie, Stirling and Fleming [45] used finite difference method to study the effects of rotation and stagger angle on bending vibrations of cantilever blades. Carnegie et al [46,47] used finite difference method for asymmetric cross-section beams to determine coupled bending torsion modes.

The finite element method [48] has become very popular in the recent years. Dokumaci [49-51] used matrix displacement method to determine natural frequencies of pre-twisted tapered blades. Nagraj [52] extended the matrix displacement method to determine coupled bending-bending-torsion vibration characteristics of a rotating blade. Singh [53-55] used the finite element method to study the coupled bending-bending-torsional vibration modes taking into account the effect of root flexibility. Murthy [56] developed a finite element formulation of rotating pre-twisted tapered beams with five degrees of freedom for each element. Gupta and Rao [57] extended tapered twisted elements to Timoshenko beam elements; Abbas [58] developed simple finite elements for thick pre-twisted blades accounting for shear and rotary inertia; Putter and Manor [59] included the effect of tip mass and also presented a high precision rotating beam element based on the fifth degree polynomial. Singh and Rawtani [60] solved the classical wave equation for torsional vibration of a blade including root stiffness at the fixed boundary. The effect of attachment flexibility was considered by Nagaraj and Sahu [61] to determine the torsional vibrations of pre-twisted rotating blades by finite element method based on Rayleigh-Ritz and Galerkin procedures.

3.2 MYKLESTAD-PROHL METHOD

A straight airfoil blade mounted at a 90° stagger angle on a rotating disc is considered. The coupling between y and θ modes is only accounted. Figure 3.1 shows the blade with discrete masses m_0, m_1, ..., $m_{\lambda-1}$, m_λ, $m_{\lambda+1}$, ... and inertias J_0, J_1, ..., $J_{\lambda-1}$, J_λ, $J_{\lambda+1}$, ... located at stations 0, 1, ..., $(\lambda-1)$, λ, $(\lambda+1)$, ... respectively. Let the horizontal distances between the centres of flexure and the corresponding centroids at stations 0, 1, ..., $(\lambda-1)$, λ, $(\lambda+1)$, ... be s_0, s_1, ..., $s_{\lambda-1}$, s_λ, $s_{\lambda+1}$, ... respectively. Also, let the lengths between stations 0 and 1, 1 and 2, ..., $(\lambda-1)$ and λ, λ and $(\lambda+1)$, ... be L_0, L_1, ..., $L_{\lambda-1}$, L_λ, ... and the distances from the root to the stations 0, 1, ..., $(\lambda-1)$, λ, $(\lambda+1)$, ... along the axial direction be z_0, z_1, ..., $z_{\lambda-1}$, z_λ, $z_{\lambda+1}$, ... respectively.

The deflected position of the elastic axis through the centre of flexure of blade cross-section is shown in figure 3.2. Let y_0, y_1, ..., $y_{\lambda-1}$, y_λ, $y_{\lambda+1}$, ...

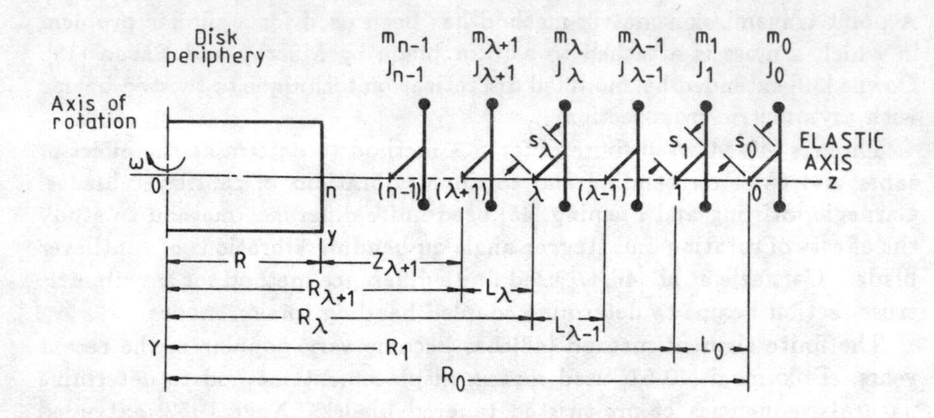

Fig. 3.1 Discretized rotating cantilever blade

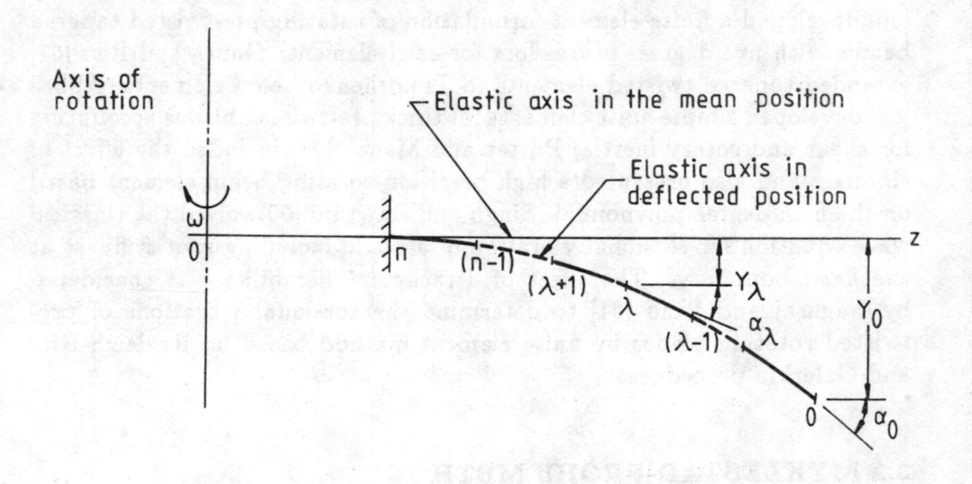

Fig. 3.2 Deflected position of the rotating blade

be the linear deflections of the elastic axis, and θ_0, θ_1, ..., $\theta_{\lambda-1}$, θ_λ, $\theta_{\lambda+1}$, ... be the angles of twist at stations 0, 1, ..., $(\lambda - 1)$, λ, $(\lambda + 1)$, ... respectively.

For a rotational speed ω, when the blade is vibrating harmonically with frequency p, the shear force S, bending moment M and the twisting moment T can be written by considering the inertia and centrifugal terms as follows:

$$S_\lambda = \sum_{i=0}^{\lambda} m_i p^2 (y_i + s_i\theta_i) - \alpha_\lambda \sum_{i=0}^{\lambda} m_i R_i \omega^2, \tag{3.1}$$

$$M_\lambda = \sum_{i=0}^{\lambda-1} m_i p^2 (y_i + s_i\theta_i)(R_i - R_\lambda)$$
$$- \sum_{i=0}^{\lambda-1} m_i R_i \omega^2 (y_i + s_i\theta_i - y_\lambda - s_\lambda\theta_\lambda) \tag{3.2}$$

and,

$$T_\lambda = \sum_{i=0}^{\lambda} (m_i s_i y_i + J_i\theta_i) p^2 - \alpha_\lambda \sum_{i=0}^{\lambda} m_i s_i \omega^2 R_i \tag{3.3}$$

where,

$$R_i = R + z_i \tag{3.4}$$

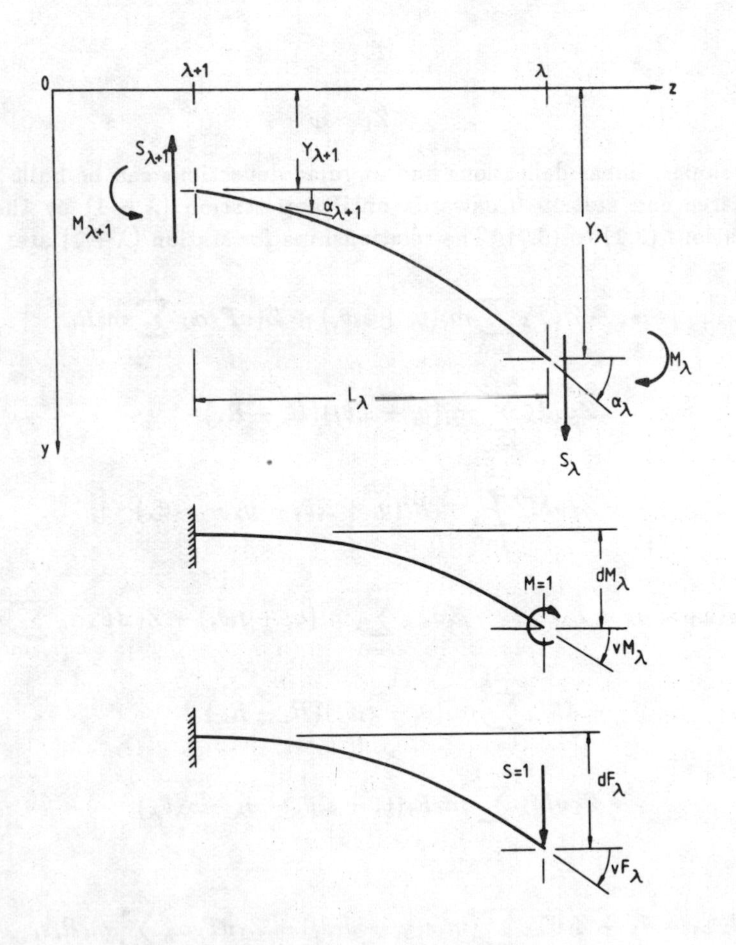

Fig. 3.3 Elastic relations for λth element

Assuming that the sections of the blade are short and the slopes α, linear deflections y and torsional deflections θ at stations $0, 1, \ldots, \lambda$ are known, the corresponding quantities at stations $(\lambda + 1)$ can be expressed as (see Figure 3.3).

$$\alpha_{\lambda+1} = \alpha_\lambda - vF_\lambda\, S_\lambda - vM_\lambda\, M_\lambda \tag{3.5}$$

$$y_{\lambda+1} = y_\lambda - L_\lambda \alpha_{\lambda+1} - dF_\lambda\, S_\lambda - dM_\lambda\, M_\lambda \tag{3.6}$$

and,

$$\theta_{\lambda+1} = \theta_\lambda - vT_\lambda\, T_\lambda \tag{3.7}$$

where dF_λ and vF_λ denote deflection and slope respectively at station λ, relative to station $(\lambda+1)$ due to a unit force applied at station λ; dM_λ and vM_λ are deflection and slope at station λ, relative to station $(\lambda+1)$ due to a unit moment applied at station λ; and, vT_λ represents torsional deflection at station λ, relative to station $(\lambda + 1)$ due to a unit torque applied at station λ. With

$$Z = p^2 \tag{3.8}$$

and

$$Z_1 = \omega^2 \tag{3.9}$$

the slopes, linear deflections and angular deflections can be built up from the free end station 0 onwards until any station $(\lambda + 1)$ by the use of equations (3.1) to (3.7). The relationships for station $(\lambda + 1)$ are:

$$\begin{aligned}\alpha_{\lambda+1} = \alpha_\lambda &- ZvF_\lambda \sum_{i=0}^{\lambda} m_i(y_i + s_i\theta_i) + Z_1 vF_\lambda \alpha_\lambda \sum_{i=0}^{\lambda} m_i R_i \\ &- ZvM_\lambda \sum_{i=0}^{\lambda-1} m_i(y_i + s_i\theta_i)(R_i - R_\lambda) \\ &+ Z_1 vM_\lambda \sum_{i=0}^{\lambda-1} m_i R_i(y_i + s_i\theta_i - y_\lambda - s_\lambda\theta_\lambda)\end{aligned} \tag{3.10}$$

$$\begin{aligned}y_{\lambda+1} = y_\lambda - L_\lambda\alpha_{\lambda+1} &- ZdF_\lambda \sum_{i=0}^{\lambda} m_i(y_i + s_i\theta_i) + Z_1\, dF_\lambda \alpha_\lambda \sum_{i=0}^{\lambda} m_i R_i \\ &- ZdM_\lambda \sum_{i=0}^{\lambda-1} m_i(y_i + s_i\theta_i)(R_i - R_\lambda) \\ &+ Z_1 vM_\lambda \sum_{i=0}^{\lambda-1} m_i R_i(y_i + s_i\theta_i - y_\lambda - s_\lambda\theta_\lambda)\end{aligned} \tag{3.11}$$

and

$$\theta_{\lambda+1} = \theta_\lambda - Z\, vT_\lambda \sum_{i=0}^{\lambda}(m_i s_i y_i + J_i\theta_i) + Z_1 vT_\lambda \alpha_\lambda \sum_{i=0}^{\lambda} m_i R_i s_i \tag{3.12}$$

For uncoupled torsional vibrations, there are only two boundary conditions relating to T and θ at the root and free end. Since at free end $T_0 = 0$, θ_0 can be set equal to 1 and for free vibrations the simplified equations (3.3) and (3.12) can be used to find T and θ at all stations λ beginning from 0 until the root station n, provided the frequency of vibration p is known. Holzer [1] suggested a simple tabular procedure [3,25] where these calculations can be carried out with an assumed value of p. If the assumed value is correct then the boundary condition $\theta_n = 0$ will be satisfied, otherwise several trials can be made until the correct frequency is found. The calculations can be carried out in transfer matrix form also [35]. The trial and error procedure is eliminated to reduce computer time, see Rao [5], by developing a polynomial frequency equation for the system.

For uncoupled bending vibrations, there are four state quantities; shear force, bending moment, slope and deflection as in equations (3.1), (3.2) and (3.10), (3.11), with torsional deflection set to zero. At the free end station 0, the shear force and bending moment are zero, but the slope α_0 and y_0 are finite. For free vibrations either α_0 or y_0 can be set to unity, as in Holzer method, however, the station to station transfer calculations become difficult because of the retention of an unknown quantity of α_0 or y_0 in addition to the frequency p. Myklestad [6] defined two sets each of four amplitude coefficients, which can be independently determined in two separate tables for all stations of the beam. These amplitude coefficients are then used to obtain a criterion for p to be the natural frequency, using the fixed end boundary conditions. Then as in Holzer method for torsional vibrations, a trial and error procedure can be used to determine the correct natural frequency.

Prohl [7] developed another procedure in which two principal solutions are used e.g., one with $\alpha_0 = 1$, $y_0 = 0$; and the other with $\alpha_0 = 0$, $y_0 = 1$. Both of these starting conditions enable station to station calculations to be made. Using the principle of linear superposition, the two principal solutions are then combined to obtain the state quantities at all stations in terms of unknown values α_0 and y_0. The boundary conditions at the root are then used to obtain a frequency criterion. The calculations in Myklestad or Prohl procedures can be carried out in transfer matrix form for straight beams [35]. For uncoupled bending vibrations Rao [8] derived a polynomial frequency equation for the system. Rao's procedure to set up the frequency equation for coupled bending-torsion modes of the rotating beam in figure 3.1 is discussed below.

Equations (3.10) and (3.12) indicate that the slope, linear deflection and angular deflection at every station depend on α_0, y_0, θ_0 and Z, in addition to the system characteristics. A close examination further reveals that the slope, linear and angular deflections involve powers of Z which increase as the station number increases. With these observations in mind, the following generalized polynomial expressions for station λ can be written as

$$\alpha_\lambda = \alpha_0 \sum_{i=0}^{\lambda-1} B_{\lambda,i} Z^i - y_0 \sum_{i=0}^{\lambda} B'_{\lambda,i} Z^i - \theta_0 \sum_{i=0}^{\lambda} C_{\lambda,i} Z^i \tag{3.13}$$

$$y_\lambda = y_0 \sum_{i=0}^{\lambda} A_{\lambda,i} Z^i - \alpha_0 \sum_{i=0}^{\lambda-1} A'_{\lambda,i} Z^i + \theta_0 \sum_{i=0}^{\lambda} C'_{\lambda,i} Z^i \tag{3.14}$$

$$\theta_\lambda = \theta_0 \sum_{i=0}^{\lambda} D_{\lambda,i} Z^i - y_0 \sum_{i=0}^{\lambda} D'_{\lambda,i} Z^i + \alpha_0 \sum_{i=0}^{\lambda-1} E_{\lambda,i} Z^i \tag{3.15}$$

The polynomial coefficients in the above equations can be obtained by equating corresponding terms in equations (3.13) to (3.15) with those in equations (3.10) to (3.12) respectively for all stations. Fortunately, these relations are recursive in nature and the following general relations are obtained.

$$\begin{aligned}
B'_{\lambda+1,j} = B'_{\lambda,j} &+ vF_\lambda \sum_{i=j-1}^{\lambda} (m_i A_{i,j-1} - m_i s_i D'_{i,j-1}) \\
&+ vM_\lambda \sum_{i=j-1}^{\lambda-1} (m_i A_{i,j-1} - m_i s_i D'_{i,j-1})(R_i - R_\lambda) \\
&+ vF_\lambda Z_1 \sum_{i=0}^{\lambda} m_i R_i B'_{\lambda,j} \\
&+ vM_\lambda Z_1 \sum_{i=0}^{\lambda-1} (m_i R_i A_{\lambda,j} - m_i R_i s_\lambda D'_{\lambda,j}) \\
&+ vM_\lambda \dot{Z}_1 \sum_{i=j}^{\lambda-1} (m_i R_i A_{i,j} - m_i R_i s_i D'_{i,j});
\end{aligned}$$

$$j = 0, 1, 2, \ldots, (\lambda + 1)$$

$$\begin{aligned}
A_{\lambda+1,j} = A_{\lambda,j} + L_\lambda B'_{\lambda+1,j} &- dF_\lambda \sum_{i=j-1}^{\lambda} (m_i A_{i,j-1} - m_i s_i D'_{i,j-1}) \\
&- dM_\lambda \sum_{i=j-1}^{\lambda-1} (m_i A_{i,j-1} - m_i s_i D'_{i,j-1})(R_i - R_\lambda) \\
&- dF_\lambda Z_1 \sum_{i=0}^{\lambda} m_i R_i B'_{\lambda,j} \\
&- dM_\lambda Z_1 \sum_{i=0}^{\lambda-1} (m_i R_i A_{\lambda,j} - m_i R_i s_\lambda D'_{i,j})
\end{aligned}$$

$$+ dM_\lambda Z_1 \sum_{i=j}^{\lambda-1} (m_i R_i A_{i,j} - m_i R_i s_i D'_{\lambda,j});$$

$$j = 0, 1, 2, \ldots, (\lambda + 1).$$

$$D_{\lambda+1,j} = D_{\lambda,j} - vT_\lambda \sum_{i=j-1}^{\lambda} (J_i D_{i,j-1} + m_i s_i C'_{i,j-1})$$

$$- vT_\lambda Z_1 \sum_{i=0}^{\lambda} m_i R_i s_i C_{\lambda,j};$$

$$j = 1, 2, \ldots, (\lambda + 1)$$

$$B_{\lambda+1,j} = B_{\lambda,j} + vF_\lambda \sum_{i=j}^{\lambda} (m_i A'_{i,j-1} - m_i s_i E_{i,j-1})$$

$$+ vM_\lambda \sum_{i=j}^{\lambda-1} (m_i A'_{i,j-1} - m_i s_i E_{i,j-1})(R_i - R_\lambda)$$

$$+ vF_\lambda Z_1 \sum_{i=0}^{\lambda} m_i R_i B_{\lambda,j}$$

$$+ vM_\lambda Z_1 \sum_{i=0}^{\lambda-1} (m_i R_i A'_{\lambda,j} - m_i R_i s_\lambda E_{\lambda,j})$$

$$- vM_\lambda Z_1 \sum_{i=j+1}^{\lambda-1} (m_i R_i A'_{i,j} - m_i R_i s_i E_{i,j});$$

$$j = 0, 1, 2, \ldots, \lambda.$$

$$A'_{\lambda+1,j} = A'_{\lambda,j} + L_\lambda B_{\lambda+1,j} - dF_\lambda \sum_{i=j}^{\lambda} (m_i A'_{i,j-1} - m_i s_i E_{i,j-1})$$

$$- dM_\lambda \sum_{i=j}^{\lambda-1} (m_i A'_{i,j-1} - m_i s_i E_{i,j-1})(R_i - R_\lambda)$$

$$- dF_\lambda Z_1 \sum_{i=0}^{\lambda} m_i R_i B_{\lambda,j}$$

$$- dM_\lambda Z_1 \sum_{i=0}^{\lambda-1} (m_i R_i A'_{\lambda,j} - m_i R_i s_\lambda E_{\lambda,j})$$

$$+ dM_\lambda Z_1 \sum_{i=j+1}^{\lambda-1} (m_i R_i A'_{i,j} - m_i R_i s_i E_{i,j});$$

$$j = 0, 1, 2, \ldots, \lambda.$$

$$D'_{\lambda+1,j} = D'_{\lambda,j} + vT_\lambda \sum_{i=j-1}^{\lambda} (m_i s_i A_{i,j-1} - J_i D_{i,j-1})$$

$$+ vT_\lambda Z_1 \sum_{i=0}^{\lambda} m_i R_i s_i B'_{\lambda,j};$$

$$j = 0, 1, 2, \ldots, (\lambda + 1)$$

$$C_{\lambda+1,j} = C_{\lambda,j} + vF_\lambda \sum_{i=j-1}^{\lambda} (m_i C'_{i,j-1} + m_i s_i D_{i,j-1})$$

$$+ vM_\lambda \sum_{i=j-1}^{\lambda-1} (m_i C'_{i,j-1} + m_i s_i D_{i,j-1})(R_i - R_\lambda)$$

$$+ vF_\lambda Z_1 \sum_{i=0}^{\lambda} m_i R_i C_{\lambda,j}$$

$$+ vM_\lambda Z_1 \sum_{i=0}^{\lambda-1} (m_i R_i C'_{\lambda,j} + m_i R_i s_\lambda D_{\lambda,j})$$

$$- vM_\lambda Z_1 \sum_{i=j}^{\lambda-1} (m_i R_i C'_{i,j} + m_i R_i s_i D_{i,j});$$

$$j = 0, 1, 2, \ldots, (\lambda + 1)$$

$$C'_{\lambda+1,j} = C'_{\lambda,j} + L_\lambda C_{\lambda+1,j} - dF_\lambda \sum_{i=j-1}^{\lambda} (m_i C'_{i,j-1} + m_i s_i D_{i,j-1})$$

$$- dM_\lambda \sum_{i=j-1}^{\lambda-1} (m_i C'_{i,j-1} + m_i s_i D_{i,j-1})(R_i - R_\lambda)$$

$$- dF_\lambda Z_1 \sum_{i=0}^{\lambda} m_i R_i C_{\lambda,j}$$

$$- dM_\lambda Z_1 \sum_{i=0}^{\lambda-1} (m_i R_i C'_{i,j} + m_i R_i s_\lambda D_{\lambda,j})$$

$$+ dM_\lambda Z_1 \sum_{i=j}^{\lambda-1} (m_i R_i C'_{i,j} + m_i R_i s_i D_{i,j});$$

$$j = 0, 1, 2, \ldots, (\lambda + 1).$$

$$E_{\lambda+1,j} = E_{\lambda,j} + vT_\lambda \sum_{i=j}^{\lambda} (m_i s_i A'_{i,j-1} - J_i E_{i,j-1})$$

$$+ vT_\lambda Z_1 \sum_{i=0}^{\lambda} m_i R_i s_i B_{\lambda,j};$$

$$j = 0, 1, 2, \ldots, \lambda. \tag{3.16}$$

The boundary conditions for the cantilever blade are:

$$\begin{aligned} \alpha_0 &= \alpha_0, \\ y_0 &= 1 \\ \theta_0 &= \theta_0; \alpha_n = y_n = \theta_n = 0 \end{aligned} \tag{3.17}$$

From equations (3.13), (3.14) and (3.17), α_0 and θ_0 can be obtained.

$$\alpha_0 = \frac{\sum_{i=0}^{n} B'_{n,i} Z^i \sum_{i=0}^{n} C_{n,i} Z^i - \sum_{i=0}^{n} C_{n,i} Z^i \sum_{i=0}^{n} A_{n,i} Z^i}{\sum_{i=0}^{n-1} B_{n,i} Z^i \sum_{i=0}^{n} C'_{n,i} Z^i - \sum_{i=0}^{n} C_{n,i} Z^i \sum_{i=0}^{n-1} A'_{n,i} Z^i} \tag{3.18}$$

$$\theta_0 = \frac{\sum_{i=0}^{n} B'_{n,i} Z^i \sum_{i=0}^{n-1} A'_{n,i} Z^i - \sum_{i=0}^{n-1} B_{n,i} Z^i \sum_{i=0}^{n} A_{n,i} Z^i}{\sum_{i=0}^{n-1} B_{n,i} Z^i \sum_{i=0}^{n} C'_{n,i} Z^i - \sum_{i=0}^{n} C_{n,i} Z^i \sum_{i=0}^{n-1} A'_{n,i} Z^i} \tag{3.19}$$

Using the above equations in equation (3.15) for the fixed end i.e., equation (3.17), the following polynomial frequency equation is obtained for the vibrating blade:

$$\begin{aligned} &\sum_{i=0}^{n} D_{n,i} Z^i \left(\sum_{i=0}^{n} B'_{n,i} Z^i \sum_{i=0}^{n-1} A'_{n,i} Z^i - \sum_{i=0}^{n} A_{n,i} Z^i \sum_{i=0}^{n-1} B'_{n,i} Z^i \right) \\ &- \sum_{i=0}^{n} D'_{n,i} Z^i \left(\sum_{i=0}^{n-1} B_{n,i} Z^i \sum_{i=0}^{n} C'_{n,i} Z^i - \sum_{i=0}^{n} C_{n,i} Z^i \sum_{i=0}^{n-1} A'_{n,i} Z^i \right) \\ &+ \sum_{i=0}^{n-1} E_{n,i} Z^i \left(\sum_{i=0}^{n} B'_{n,i} Z^i \sum_{i=0}^{n} C'_{n,i} Z^i - \sum_{i=0}^{n} C_{n,i} Z^i \sum_{i=0}^{n} A_{n,i} Z^i \right) = 0 \end{aligned} \tag{3.20}$$

The frequency equation above is of degree $(3n-1)$ in p^2. For a n stationed system, there are only $2n$ coupled bending-torsion modes. Rao [8] has shown that the additional $(n-1)$ coefficients in the polynomial frequency equation in bending motion are actually zero, thus the last $(n-1)$ terms in (3.20) can be ignored. A computer program can be developed based on the above analysis, to extract as many roots as desired from the frequency equation.

As an example consider the blade given in section 2.8. The blade is divided into 8 sections and the masses are lumped at nine stations. The first three coupled modes in bending as well as torsion modes are given in Table 3.1 along with the analytical and experimental values of Rao and Banerji [62].

TABLE 3.1

Coupled Bending-Torsion Modes of a Rotating Cantilever Blade

		Coupled Bending Mode			Coupled Torsion Mode	
RPM	Mode No.	Galerkin Method, Hz	Experiment Hz	Freq. Eq. Hz.	Galerkin Method, Hz.	Freq. Eq. Hz.
	I	158.3	155	158.0	1700	1699
2500	II	916.9	902	910.8	5123	5096
	III	2538.0	2460	2509.0	8554	8448
	I	170.2	165	170.0	1700	1699
3500	II	928.5	914	922.6	5123	5096
	III	2550.0	2475	2522.0	8554	8448

3.3 EXTENDED HOLZER METHOD

For coupled bending-bending vibrations of rotating blades, extended Holzer method is more convenient. The blade is divided into m number of elements of equal length dZ each element being regarded as massless cantilever with its mass concentrated at the root and equal to half the sum of the masses of the two adjacent elements. The blade element $(n+1)$ in both xz and yz planes is shown in figure 3.4 along with the forces and moments acting on it. For a frequency of vibration p, the conditions of equilibrium for the nth element of the rotating blade are:

$$F_{x,n+1} = F_{x,n} + m_n p^2 x_n dZ + m_n \omega^2 dZ[x_n \sin^2 \phi - y_n \sin \phi \cos \phi] \tag{3.21}$$

$$F_{y,n+1} = F_{y,n} + m_n p^2 y_n dZ + m_n \omega^2 dZ[y_n \cos^2 \phi - x_n \sin \phi \cos \phi] \tag{3.22}$$

$$\begin{aligned} M_{x,n+1} = M_{x,n} &+ F_{x,n+1} dZ + (x_{n+1} - x_n) \sum_{n=n+1}^{n=m} [m_n \omega^2 (R + Z_n) dZ] \\ &- I_{yy,n} \rho p^2 x'_n dZ - I_{xy,n} \rho p^2 y'_n dZ \end{aligned} \tag{3.23}$$

$$\begin{aligned} M_{y,n+1} = M_{y,n} &+ F_{y,n+1} dZ + (y_{n+1} - y_n) \sum_{n=n+1}^{n=m} [m_n \omega^2 (R + Z_n) dZ] \\ &- I_{xx,n} \rho p^2 y'_n dZ - I_{xy,n} \rho p^2 x'_n dZ \end{aligned} \tag{3.24}$$

Making use of the flexure equations of a short cantilever blade subjected to an end force, lateral force and bending moment, the following equations

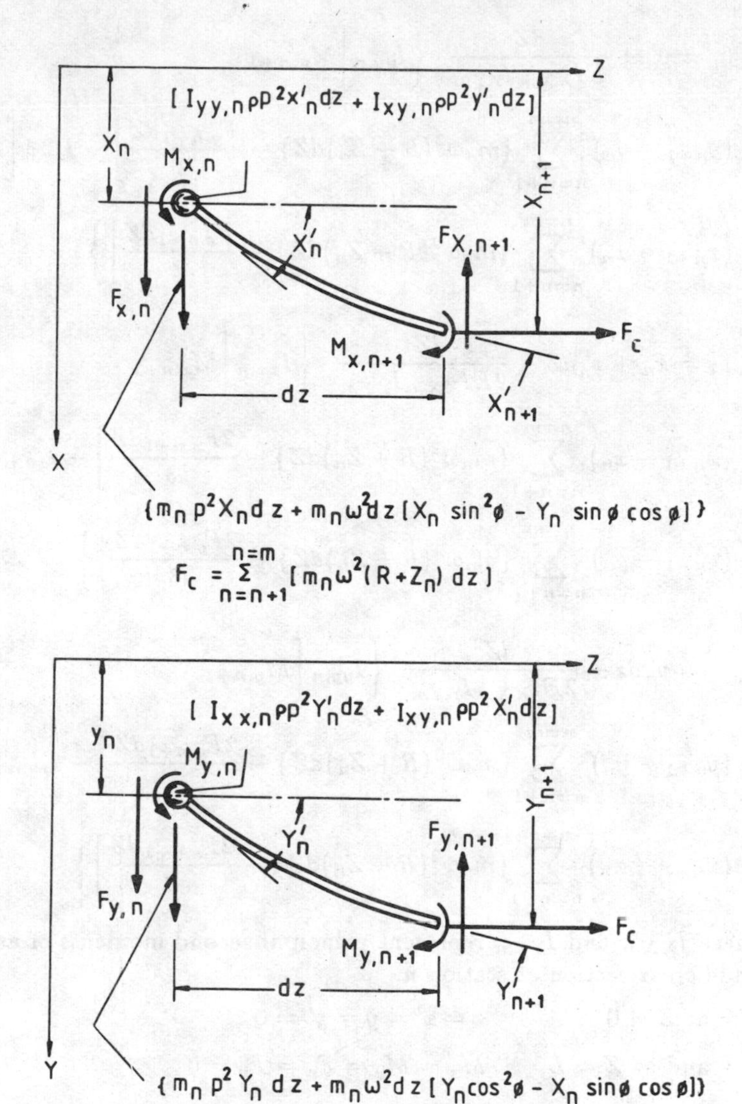

Fig. 3.4 Forces and moments acting on the blade element

can be obtained:

$$x'_{n+1} = x'_n + \frac{dZ}{EI_{XX,n}I_{YY,n}} \cdot \Bigg\{ I_{xx,n} \Bigg[M_{x,n+1}$$

$$- (x_{n+1} - x_n) \sum_{n=n+1}^{n=m} \{m_n\omega^2(R + Z_n)dZ\} - \frac{F_{x,n+1}dZ}{2}\Bigg] - I_{xy,n}\Bigg[M_{y,n+1}$$

$$- (y_{n+1} - y_n) \sum_{n=n+1}^{n=m} \{m_n\omega^2(R + Z_n)dZ\} - \frac{F_{x,n+1}dZ}{2}\Bigg]\Bigg\} \qquad (3.25)$$

$$y'_{n+1} = y'_n + \frac{dZ}{EI_{XX,n}I_{YY,n}} \cdot \Bigg\{ I_{yy,n} \Bigg[M_{y,n+1}$$

$$- (y_{n+1} - y_n) \sum_{n=n+1}^{n=m} \{m_n\omega^2(R + Z_n)dZ\} - \frac{F_{y,n+1}dZ}{2} \Bigg] - I_{xy,n} \Bigg[M_{x,n+1}$$

$$- (x_{n+1} - x_n) \sum_{n=n+1}^{n=m} \{m_n\omega^2(R + Z_n)dZ\} - \frac{F_{x,n+1}dZ}{2} \Bigg] \Bigg\} \qquad (3.26)$$

$$x_{n+1} = x_n + x'_n dZ + \frac{dZ^2}{2EI_{XX,n}I_{YY,n}} \cdot \Bigg\{ I_{xx,n} \Bigg[M_{x,n+1}$$

$$- (x_{n+1} - x_n) \sum_{n=n+1}^{n=m} \{m_n\omega^2(R + Z_n)dZ\} - \frac{2F_{x,n+1}dZ}{3} \Bigg] - I_{xy,n} \Bigg[M_{y,n+1}$$

$$- (y_{n+1} - y_n) \sum_{n=n+1}^{n=m} \{m_n\omega^2(R + Z_n)dZ\} - \frac{2F_{y,n+1}dZ}{3} \Bigg] \Bigg\} \qquad (3.27)$$

$$y_{n+1} = y'_n dz + \frac{dZ^2}{2EI_{XX,n}I_{YY,n}} \cdot \Bigg\{ I_{yy,n} \Bigg[M_{y,n+1}$$

$$- (y_{n+1} - y_n) \sum_{n=n+1}^{n=m} \{m_n\omega^2(R + Z_n)dZ\} - \frac{2F_{y,n+1}dZ}{3} \Bigg] - I_{xy,n} \Bigg[M_{x,n+1}$$

$$- (x_{n+1} - x_n) \sum_{n=n+1}^{n=m} \{m_n\omega^2(R + Z_n)dZ\} - \frac{2F_{x,n+1}dZ}{3} \Bigg] \Bigg\} \qquad (3.28)$$

where $I_{XX,n}$ and $I_{YY,n}$ represent principal second moments of area of the blade cross-section at station n.

$$\text{at } Z = 0 \qquad x = x' = y = y' = 0 \qquad (3.29)$$

$$\text{and at } Z = L \qquad M_x = M_y = F_x = F_y = 0 \qquad (3.30)$$

Determination of Natural Frequencies

To determine the natural frequencies the process consists essentially of assuming a value for the natural frequency p, and applying equations (3.21) to (3.28) for each element of the blade successively, from the root to the free end with the help of boundary conditions (3.29) and (3.30). The following factors are to be taken into account for a successful application of equations (3.21) to (3.28).

1. The magnitudes of shear forces F_x,F_y and bending moments M_x, M_y at the root are not known.

2. Equations (3.23) to (3.28) involve x_{n+1} and y_{n+1}, which are not yet determined when dealing with the element $(n + 1)$.

Since the problem is a linear eigenvalue problem the first factor above is dealt with by repeating the numerical calculations of equations (3.21) to (3.28), for unit values F_x, F_y, M_x and M_y at the root successively. Then the total shear forces F_x, F_y and bending moments M_x, M_y at $Z = L$, for given shear forces F_{x0}, F_{y0} and bending moments M_{x0}, M_{y0} at the root are

$$\begin{aligned}
F_{x,L} &= F_{x,0}F'_{x,L} + F_{y,0}F''_{x,L} + M_{x,0}F'''_{x,L} + M_{y,0}F''''_{x,L} \\
F_{y,L} &= F_{x,0}F'_{y,L} + F_{y,0}F''_{y,L} + M_{x,0}F'''_{y,L} + M_{y,0}F''''_{y,L} \\
M_{x,L} &= F_{x,0}M'_{x,L} + F_{y,0}M''_{x,L} + M_{x,0}M'''_{x,L} + M_{y,0}M''''_{x,L} \\
M_{y,L} &= F_{x,0}M'_{y,L} + F_{y,0}M''_{y,L} + M_{x,0}M'''_{y,L} + M_{y,0}M''''_{y,L}
\end{aligned} \tag{3.31}$$

where the primes $'$, $''$, $'''$, and $''''$ refer to the values determined at the free end due to the unit values of F_x, F_y, M_x and M_y at the root, respectively. Equations (3.31) can be expressed in the matrix form as follows.

$$\begin{Bmatrix} F_{x,L} \\ F_{y,L} \\ M_{x,L} \\ M_{y,L} \end{Bmatrix} = \begin{bmatrix} F'_{x,L} & F''_{x,L} & F'''_{x,L} & F''''_{x,L} \\ F'_{y,L} & F''_{y,L} & F'''_{y,L} & F''''_{y,L} \\ M'_{x,L} & M''_{x,L} & M'''_{x,L} & M''''_{x,L} \\ M'_{y,L} & M''_{y,L} & M'''_{y,L} & M''''_{y,L} \end{bmatrix} \begin{Bmatrix} F_{x,0} \\ F_{y,0} \\ M_{x,0} \\ M_{y,0} \end{Bmatrix} \tag{3.32}$$

If the assumed value of frequency p corresponds to one of the natural modes, then according to the boundary conditions (3.30) and equation (3.32), the following equation must be satisfied

$$\Delta = \begin{vmatrix} F'_{x,L} & F''_{x,L} & F'''_{x,L} & F''''_{x,L} \\ F'_{y,L} & F''_{y,L} & F'''_{y,L} & F''''_{y,L} \\ M'_{x,L} & M''_{x,L} & M'''_{x,L} & M''''_{x,L} \\ M'_{y,L} & M''_{y,L} & M'''_{y,L} & M''''_{y,L} \end{vmatrix} = 0 \tag{3.33}$$

To determine the correct natural frequency, different values for p can be tried until equation (3.33) is satisfied. Values for p are increased at regular intervals over a range until the value of the determinant changes its sign, after which a linear interpolation is applied for determining the desired value of natural frequency.

As regards the second factor mentioned earlier an iteration process is used until two successive values of x and y agree with each other. Since the blade element is very short the first approximations of x_{n+1} and y_{n+1} can be written as:

$$\begin{aligned}
x_{n+1} &= x_n + x'_n dZ \\
Y_{n+1} &= y_n + y'_n dZ
\end{aligned} \tag{3.34}$$

Equations (3.34) are used as the initial values of x_{n+1} and y_{n+1} in the iteration process for each element of the blade.

Determination of Mode Shapes

Once the natural frequency is determined the mode shapes can be obtained according to the following process.

Since the problem is an eigenvalue one either F_{x0}, F_{y0}, M_{x0}, M_{y0} can be fixed as unity, say $F_{x0} = 1$. Accordingly equations (3.31) consistent with the boundary conditions (3.30) can be written as:

$$\begin{aligned}
F_{y,0}F''_{x,L} + M_{x,0}F'''_{x,L} + M_{y,0}F''''_{x,L} &= -F'_{x,L} \\
F_{y,0}F''_{y,L} + M_{x,0}F'''_{y,L} + M_{y,0}F''''_{y,L} &= -F'_{y,L} \\
F_{y,0}M''_{x,L} + M_{x,0}M'''_{x,L} + M_{y,0}M''''_{x,L} &= -M'_{x,L} \\
F_{y,0}M''_{y,L} + M_{x,0}M'''_{y,L} + M_{y,0}M''''_{y,L} &= -M'_{y,L}
\end{aligned} \tag{3.35}$$

Using any three equations from equations (3.35), proportional values for F_{y0}, M_{x0} and M_{y0} can be determined for the assumed value of $F_{x0} = 1$.

Using the values of F_{x0}, F_{y0}, M_{x0} and M_{y0} and the natural frequency p, the total deflections at each station can be determined using equations (3.27) and (3.28).

Using the analysis presented above, the natural frequencies are determined by Rao and Carnegie [19] for a cantilever blade having the following data:

Breadth $B = 1.27$ cm
Thickness $D = 0.635$ cm
Length $L = 25.4$ cm
Density $\rho = 7833$ kg/m^3
Young's modulus $E = 2.071 \times 10^{11}$ Pa

The blade is divided into 30 elements.

The problem is analyzed for seven values of angle of pre-twist viz., $\gamma = 0$, 15, 30, 45, 60, 75 and 90 degrees and the results obtained for coupled bending-bending vibration frequencies are plotted in figure 3.5 for the first six modes of vibration. It can be seen that modes 1, 3 and 5 show an increase in frequency with increase of pre-twist angle, while modes 2, 4 and 6 show a decrease in frequency with increase of pre-twist angle.

The uncoupled natural frequencies for the first three modes of vibration in the $y-z$ plane are determined for seven values of rotational speed viz., $\omega = 0$, 50, 100, 150, 200, 250 and 300 c/s and these results are plotted in figure 3.6. The curves show an increase in frequency of all three modes of vibration with increase of rotational speed.

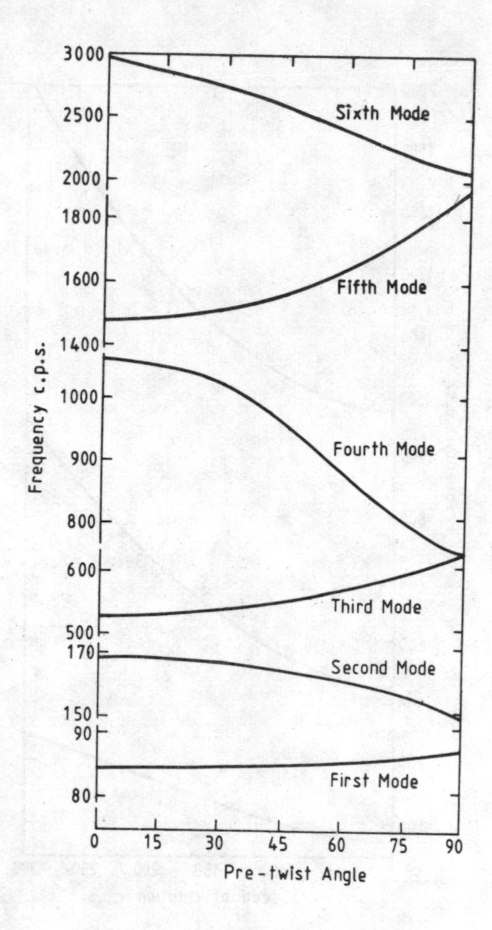

Fig. 3.5 Frequency against pre-twist angle

3.4 FINITE ELEMENT METHOD

In this section, we will discuss the finite element method for the free vibration analysis, of turbine blades by using curved, solid, C^o continuity, 'serendipity', 20-noded isoparameteric elements. This element is chosen by Bahree, Sharan and Rao [63] because of its versatility in accurately mapping the complex geometry of the turbine blade. The geometry of the blade is considered to have an airfoil cross-section, being asymmetric and pre-twisted and having taper along its length. A typical blade with the root and tip cross-sections along its length is shown in Fig. 3.7.

3.4.1 The Isoparametric Finite Element Formulation

The most apparent features of isoparametric elements are that their sides may be curved and that they make use of a special coordinate system

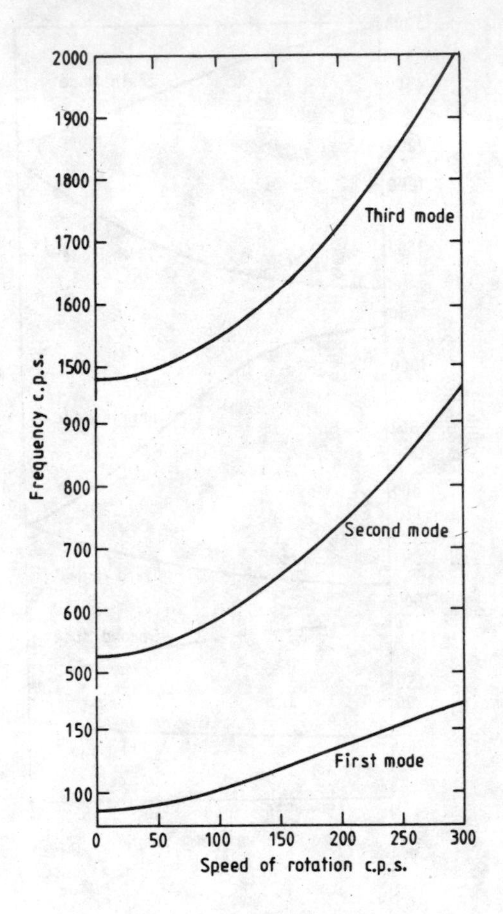

Fig. 3.6 Frequency against speed of rotation

(ξ, η, ς) as shown in Fig. 3.8. These features can be understood in the following manner.

1. Nodal degrees of freedom, $\{d\}$, dictate displacements $\{u\,v\,w\}$ of a point in the element. This can be written as

$$\begin{Bmatrix} u \\ v \\ w \end{Bmatrix} = [N]\{d\} \tag{3.36}$$

2. Nodal local coordinates, $\{c\}$, define global coordinates $\{x\,y\,z\}$ of a point in the element. This can be expressed as

$$\begin{Bmatrix} x \\ y \\ z \end{Bmatrix} = [\overline{N}]\{c\} \tag{3.37}$$

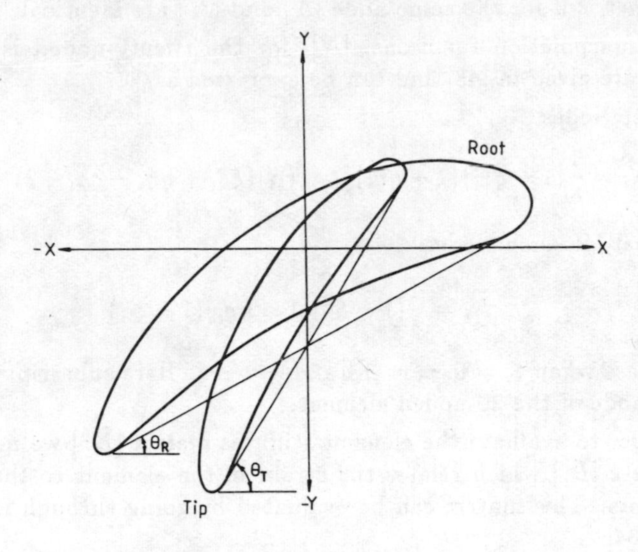

Fig. 3.7 The blade airfoil cross-sections at the root and the tip

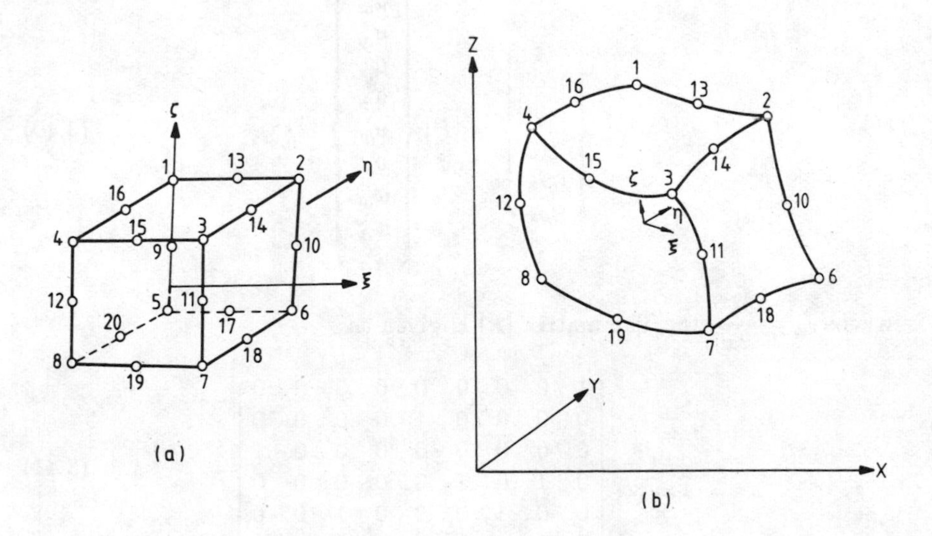

Fig. 3.8 The solid isoparametric 'serendipity' 20-noded element (a) Local coordinate system, (b) Global cartesian coordinate system

The matrices $[N]$ and $[\overline{N}]$ are functions of ξ, η and ς. An element is isoparametric if for the same node $[N]$ and $[\overline{N}]$ are identical.

The interpolation functions, $[N]$, for the twenty-noded isoparametric element are given in [48] and can be expressed as

Corner Nodes

$$N_i = \frac{1}{8}(1+\xi\xi_i)(1+\eta\eta_i)(1+\varsigma\varsigma_i)(\xi\xi_i+\eta\eta_i+\varsigma\varsigma_i-2) \tag{3.38}$$

Typical Mid-side Node

$$N_i = \frac{1}{4}(1-\xi^2)(1+\eta\eta_i)(1+\varsigma\varsigma_i) \tag{3.39}$$

for the case when $\xi_i = 0$, $\eta_i = \pm 1$ and $\varsigma_i = \pm 1$. Here subscript i represents the ith node of the 20-noded element.

In order to evaluate the element stiffness matrix $[K^e]$ we need to know the matrix $[B^e]$ which relates the strain in the element to the nodal displacements. This matrix can be evaluated by going through four steps as follows [64].

Step 1

The strain vector in the global coordinate system can be related to the derivatives of the displacement field as

$$\begin{Bmatrix} \epsilon_x \\ \epsilon_y \\ \epsilon_z \\ \gamma_{xy} \\ \gamma_{xz} \\ \gamma_{yz} \end{Bmatrix} = [P] \begin{Bmatrix} u_{,x} \\ u_{,y} \\ u_{,z} \\ v_{,x} \\ v_{,y} \\ v_{,z} \\ w_{,x} \\ w_{,y} \\ w_{,z} \end{Bmatrix} \tag{3.40}$$

where $_{,x} = \dfrac{\partial}{\partial x}$ etc. The matrix $[P]$ is given as

$$\underset{6\times 9}{[P]} = \begin{bmatrix} 1 & 0 & 0 & 0 & 0 & 0 & 0 & 0 & 0 \\ 0 & 0 & 0 & 0 & 1 & 0 & 0 & 0 & 0 \\ 0 & 0 & 0 & 0 & 0 & 0 & 0 & 0 & 1 \\ 0 & 1 & 0 & 1 & 0 & 0 & 0 & 0 & 0 \\ 0 & 0 & 1 & 0 & 0 & 0 & 1 & 0 & 0 \\ 0 & 0 & 0 & 0 & 0 & 1 & 0 & 1 & 0 \end{bmatrix} \tag{3.41}$$

Step 2

The vector containing the derivatives of the displacement field in the global coordinate system can be transformed to the vector containing derivatives of the displacement field in the local coordinate system as:

$$\begin{Bmatrix} u_{,x} \\ u_{,y} \\ u_{,z} \\ v_{,x} \\ v_{,y} \\ v_{,z} \\ w_{,x} \\ w_{,y} \\ w_{,z} \end{Bmatrix} = [Q] \begin{Bmatrix} u_{,\xi} \\ u_{,\eta} \\ u_{,\varsigma} \\ v_{,\xi} \\ v_{,\eta} \\ v_{,\varsigma} \\ w_{,\xi} \\ w_{,\eta} \\ w_{,\varsigma} \end{Bmatrix} \tag{3.42}$$

The matrix $[Q]$ contains terms of the inverse of the Jacobian matrix $[J]$ and can be shown as

$$\underset{9\times9}{[Q]} = \begin{bmatrix} [J^{-1}] & [0] & [0] \\ [0] & [J^{-1}] & [0] \\ [0] & [0] & [J^{-1}] \end{bmatrix} \tag{3.43}$$

where $[J^{-1}]$ is given by

$$\underset{3\times3}{[J^{-1}]} = \begin{bmatrix} J_{11}^{-1} & J_{12}^{-1} & J_{13}^{-1} \\ J_{21}^{-1} & J_{22}^{-1} & J_{23}^{-1} \\ J_{31}^{-1} & J_{32}^{-1} & J_{33}^{-1} \end{bmatrix} \tag{3.43a}$$

(see equations (3.52) to (3.54) for details regarding $[J]$).

Step 3

The vector containing derivatives of the displacement field in the local coordinates system can be related to the nodal displacement vector as

$$\begin{Bmatrix} u_{,\xi} \\ u_{,\eta} \\ u_{,\varsigma} \\ v_{,\xi} \\ v_{,\eta} \\ v_{,\varsigma} \\ w_{,\xi} \\ w_{,\eta} \\ w_{,\varsigma} \end{Bmatrix} = [R] \begin{Bmatrix} d_1 \\ d_2 \\ d_3 \\ \cdot \\ \cdot \\ \cdot \\ \cdot \\ \cdot \\ d_{20} \end{Bmatrix} \tag{3.44}$$

In the above $[R]$ contains derivatives of the interpolation functions with respect to the local coordinate system and is given by

$$\underset{9\times60}{[R]} = \begin{bmatrix} \{N_1'\} & \{0\} & \{0\} & \{N_2'\} & \{0\} & \{0\} & \{N_3'\} & \cdot & \cdot \\ \{0\} & \{N_1'\} & \{0\} & \{0\} & \{N_2'\} & \{0\} & \{0\} & \cdot & \cdot \\ \{0\} & \{0\} & \{N_1'\} & \{0\} & \{0\} & \{N_2'\} & \{0\} & \cdot & \cdot \end{bmatrix} \tag{3.45}$$

where

$$\{N_i'\} = \begin{Bmatrix} \dfrac{\partial N_i}{\partial \xi} \\ \dfrac{\partial N_i}{\partial \eta} \\ \dfrac{\partial N_i}{\partial \varsigma} \end{Bmatrix} \tag{3.45a}$$

Step 4

Finally, the $[B^e]$ matrix for an element is expressed as

$$\underset{6\times60}{[B^e]} = \underset{6\times9}{[P]}\ \underset{9\times9}{[Q]}\ \underset{9\times60}{[R]} \tag{3.46}$$

Once the $[B^e]$ is known, the element stiffness matrix $[K^e]$ can be evaluated as

$$[K^e] = \int_V [B(x,y,z)]^T [D][B(x,y,z)]\, dx\, dy\, dz \tag{3.47}$$

In case of isoparametric finite elements the shape functions, $[N]$, are defined for the variation of the local coordinate system from $+1$ to -1. Thus $\xi = \pm 1$, $\eta = \pm 1$ and $\varsigma = \pm 1$. For the isoparametric formulation the element stiffness matrix can be expressed as

$$\underset{60\times60}{[K^e]} = \int_{-1}^{+1}\int_{-1}^{+1}\int_{-1}^{+1} \underset{60\times6}{[B(\xi,\eta,\varsigma)]^T} [D][B(\xi,\eta,\varsigma)][J(\xi,\eta,\varsigma)]\, d\xi\, d\eta\, d\varsigma \tag{3.48}$$

In the above, the material property matrix $[D]$ for a three dimensional isotropic material can be expressed as

$$[D] = \frac{E(1-\nu)}{(1+\nu)(1-2\nu)} = \begin{bmatrix} 1 & \frac{\nu}{(1-\nu)} & \frac{\nu}{(1-\nu)} & 0 & 0 & 0 \\ & 1 & \frac{\nu}{(1-\nu)} & 0 & 0 & 0 \\ & & 1 & 0 & 0 & 0 \\ & & & \frac{1-2\nu}{2(1-\nu)} & 0 & 0 \\ & & \text{sym} & & \frac{1-2\nu}{2(1-\nu)} & 0 \\ & & & & & \frac{1-2\nu}{2(1-\nu)} \end{bmatrix}$$

Similarly, the expression for the consistent mass matrix is evaluated as

$$[M^e] = \int_V [N(x,y,z)]^T \rho [N(x,y,z)]\, dx\, dy\, dz \tag{3.49}$$

For isoparametric formulation, the above can be written as

$$\underset{60\times60}{[M^e]} = \int_{-1}^{+1}\int_{-1}^{+1}\int_{-1}^{+1} \underset{60\times6}{[N(\xi,\eta,\varsigma)]^T} \rho [N(\xi,\eta,\varsigma)][J(\xi,\eta,\varsigma)]\, d\xi\, d\eta\, d\varsigma \tag{3.50}$$

The shape function matrix used in the evaluation of the elemental consistent mass matrix above can be expressed as

$$\underset{3\times 60}{[N]} = \begin{bmatrix} N_1 & 0 & 0 & N_2 & 0 & 0 & N_3 & 0 & 0 & \cdot & \cdot \\ 0 & N_1 & 0 & 0 & N_2 & 0 & 0 & N_3 & 0 & \cdot & \cdot \\ 0 & 0 & N_1 & 0 & 0 & N_2 & 0 & 0 & N_3 & \cdot & \cdot \end{bmatrix} \tag{3.51}$$

The Jacobian matrix here relates the derivatives of the displacement field $\{u, v, w\}$ with respect to the local coordinate system $\{\xi, \eta, \varsigma\}$ to the derivatives of $\{u, v, w\}$ with respect to the global coordinate system $\{x, y, z\}$. The Jacobian matrix is evaluated by making use of the chain rule of calculus, e.g.,

$$u_{,\xi} = u_{,x}x_{,\xi} + u_{,y}y_{,\xi} + u_{,z}z_{,\xi} \tag{3.52}$$

where u is a function of x, y and z. Writing the full set of equations for u in the above in matrix form, we get

$$\begin{Bmatrix} u_{,\xi} \\ u_{,\eta} \\ u_{,\varsigma} \end{Bmatrix} = \begin{bmatrix} x_{,\xi} & y_{,\xi} & z_{,\xi} \\ x_{,\eta} & y_{,\eta} & z_{,\eta} \\ x_{,\varsigma} & y_{,\varsigma} & z_{,\varsigma} \end{bmatrix} \begin{Bmatrix} u_{,x} \\ u_{,y} \\ u_{,z} \end{Bmatrix} \tag{3.53}$$

where the Jacobian matrix is

$$[J] = \begin{bmatrix} x_{,\xi} & y_{,\xi} & z_{,\xi} \\ x_{,\eta} & y_{,\eta} & z_{,\eta} \\ x_{,\varsigma} & y_{,\varsigma} & z_{,\varsigma} \end{bmatrix} \tag{3.53a}$$

The element J_{11} of $[J]$ can be written as

$$J_{11} = x_{,\xi} = \frac{\partial}{\partial \xi} \left(\underset{1\times 20}{[N_i]^e} \right) \underset{20\times 1}{\{x\}^e} \tag{3.54}$$

Here the vector $\{x\}^e$ contains cartesian coordinates of 20-noded element along the x-direction and $e = 1, 2, \ldots, 20$. Similarly, other elements of $[J]$ can be evaluated.

The evaluation of the element stiffness matrix, equation (3.48),and element mass matrix, equation (3.50), can be achieved by numerically evaluating the volume integral in these equations. One of the most widely used techniques in integrating the elemental matrices in case of isoparametric formulation is by making use of Gaussian Quadrature [65]. Since a quadratic polynomial is used to describe each side of the 20-noded element, there would be 2 Gauss points on each side of the element. In this way a total of 8 Gauss points within the element are sufficient to integrate each element of the stiffness and mass matrices.

The final expressions for the elemental stiffness and mass matrices can be written using Gaussian quadrature as

$$[K^e] = \sum_{i=1}^{NG} \sum_{j=1}^{NG} \sum_{k=1}^{NG} w_i w_j w_k [B(\xi, \eta, \varsigma)]^T [D] [B(\xi, \eta, \varsigma)] [J(\xi, \eta, \varsigma)] \tag{3.55}$$

and

$$[M^e] = \sum_{i=1}^{NG}\sum_{j=1}^{NG}\sum_{k=1}^{NG} w_i w_j w_k [N(\xi,\eta,\varsigma)]^T \rho [N(\xi,\eta,\varsigma)][J(\xi,\eta,\varsigma)] \tag{3.56}$$

where $NG = 2$ for the case of a 20-noded element and w are weighting functions.

3.4.2 Free Vibration Analysis

After the elemental matrices are evaluated, they are assembled into global matrices and the dynamic equation of motion is written as

$$[M^G]\{\ddot{U}\} + [K^G]\{U\} = 0 \tag{3.57}$$

The natural frequencies of the system can be obtained by solving the eigenvalue problem. Since the global matrices for this three dimensional problem becomes very large, it is desirable that a coordinate reduction scheme be used to reduce the size of these matrices, without any significant loss of accuracy, in terms of the lower modes. One of these schemes is the Guyan's reduction technique [66].

In this dynamic matrix reduction scheme, one makes use of the ratios of the diagonal terms of the stiffness and mass matrices by rearranging these in terms of the ratios K_u/M_u. In this technique certain degrees of freedom are discarded and these are called the slave degrees of freedom. On the other hand the degrees of freedom which are retained in the reduced matrices are called the masters. The degree of freedom which has the largest K_u/M_u ratio is selected as the first slave. In this way one can rearrange the inertia and the stiffness matrices depending upon the number of masters and write the dynamic equation of motion in the form

$$\begin{bmatrix} [M_{mm}] & [M_{ms}] \\ [M_{ms}]^T & [M_{ss}] \end{bmatrix} \begin{Bmatrix} \{\ddot{U}_m(t)\} \\ \{\ddot{U}_s(t)\} \end{Bmatrix} + \begin{bmatrix} [K_{mm}] & [K_{ms}] \\ [K_{ms}]^T & [K_{ss}] \end{bmatrix} \begin{Bmatrix} \{U_m(t)\} \\ \{U_s(t)\} \end{Bmatrix} = 0 \tag{3.58}$$

The reduced system matrices are obtained using the following equations

$$[\overline{M}^G]_{m\times m} = [\psi]^T_{m\times n}[M^G][\psi]_{n\times m}$$

$$[\overline{K}^G]_{m\times m} = [\psi]^T_{m\times n}[K^G][\psi]_{n\times m}$$

where the transformation matrix $[\psi]$ is given by

$$[\psi] = \left[\frac{[I]}{-[K_{ss}]^{-1}[K_{ms}]^T}\right] \tag{3.60}$$

The influence of temparature on natural frequencies of turbine blades can be determined by updating the material property matrix $[D]$ in equation (3.48) at every time instant and thus calculating the elemental stiffness matrix $[K^e]$, which can then be used to solve the eigen value problem.

3.4.3 Examples

Simple Cantilever

To test the program developed on the formulation above, Bahree [67] considered a simple cantilever with the following data.

Length of the beam $= 10.0$ cm

Second moment of area $= 675.0 \times 10^{-6}$ cm^4

Modulus of elasticity $= 2.11 \times 10^{11}$ Pa

Area of cross-section $= 0.09$ cm^2

Width $= 0.30$ cm

Density $= 7860$ kg/m^3

The blade is discretized into 36 curved 20-noded isoparametric finite elements. There were in all 317 nodes, therefore 951 degrees of freedom. Using the formulation discussed above, the results obtained for the first five natural frequencies are shown in the Table 3.2 . It can be seen from this table that the natural frequencies obtained using isoparametric finite element formulation agree very well with the exact values, see [69]. This table also shows the results obtained by Reissner energy formulation [68] for this particular problem.

TABLE 3.2

Comparison of Natural Frequencies (Hz)

Mode Number	Exact Value	Vyas [68] Reissner	Bahree FEM [67]
I	251.09	248.68	252.02
II	1573.55	1558.40	1577.24
III	4405.98	4365.00	4407.58
IV	8633.95	8563.30	8648.41
V	14273.65	14468.00	14389.50

Turbine Blade Example

The discretization of the turbine blade used by Rao and Vyas [70], into finite elements is shown in Fig. 3.9. The geometric details of the blade are given later in Table 8.4. The blade material is MAR-M200 which is a superalloy of nickel. Fig. 3.9 is only a representative diagram of the turbine blade and does not take into account pre-twist, asymmetry and taper of the blade. This figure only gives a general idea of how the various elements were fitted into the turbine blade. A total of 35 elements were used to describe the blade with 7 elements across the cross-section and 5 layers

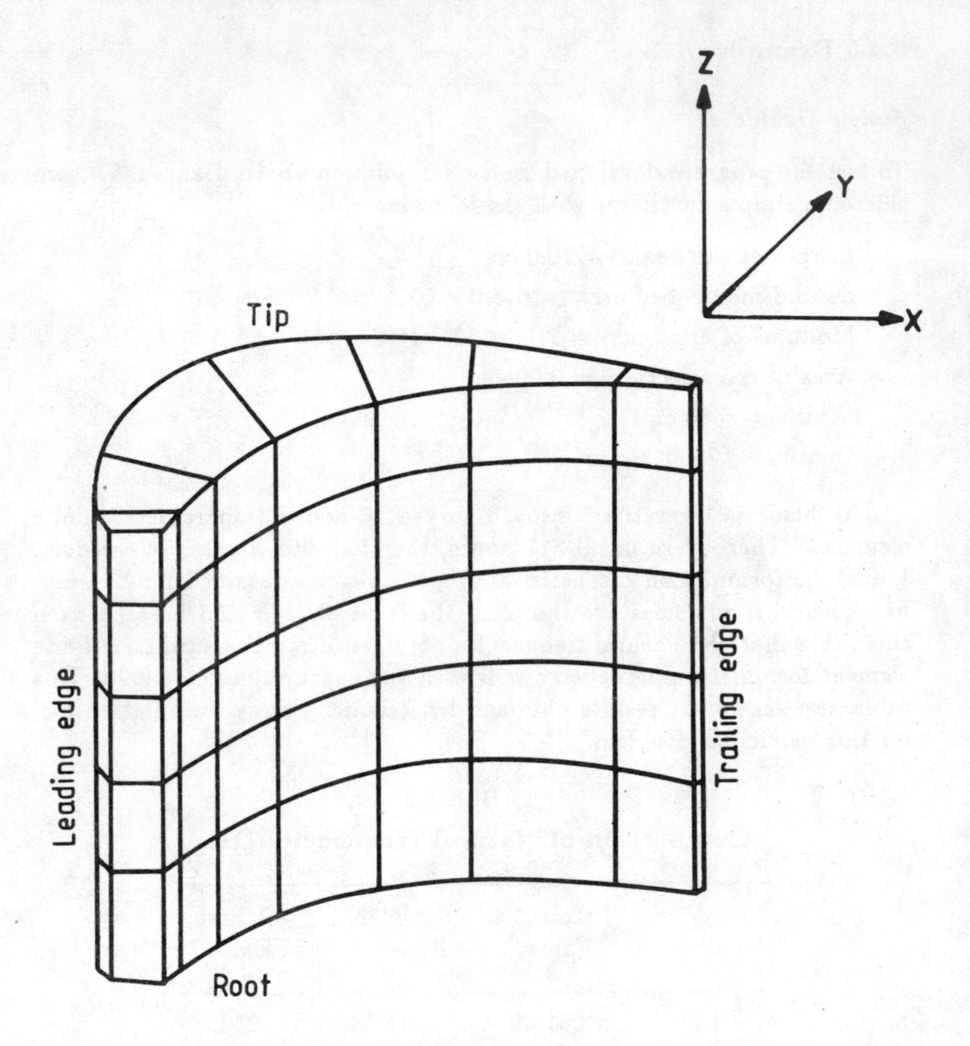

Fig. 3.9 The turbine blade finite element model

along the height. There were 308 nodes, therefore 924 degrees of freedom to represent the dynamics of the system.

Bahree [67] conducted a two dimensional heat transfer analysis by finite element method to determine temperature distribution through the elements in each layer of the blade cross-section (see Appendix 1). A typical plot of the temperature variation obtained at three different nodes on the mean chord viz., leading edge node A, middle node B and the trailing edge node C is given in Fig. 3.10, for a turbine heating rate of 56 deg C per sec, along with the gas temperature.

The variation of the Young's modulus, see [70], of the blade material with temperature, is given in Table 3.3. The natural frequencies of the blade thus vary rapidly during the transient period and in this short duration may cause resonance and low cycle fatigue failures.

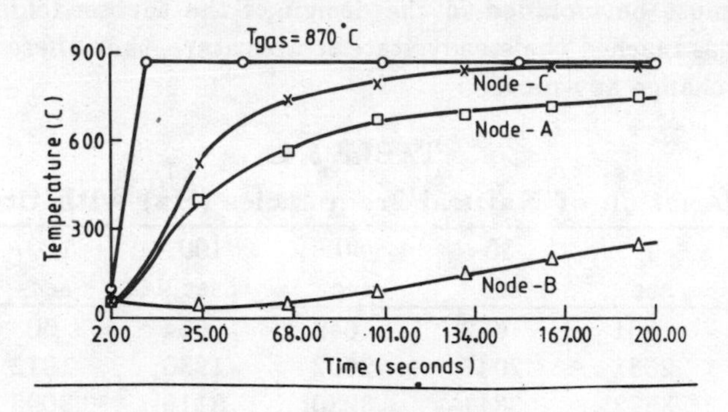

Fig. 3.10 The variation of the temperature at three nodes as a function of time corresponding to the turbine heating rate of 56° C/sec and final T_g = 870° C

TABLE 3.3
Young's Modulus as a Function of Temperature

Temperature Deg C	Young's Modulus G Pa
21	220
93	215
205	215
315	195
425	190
540	185
650	175
760	170
870	160
980	145

The natural frequencies were obtained by the FEM program taking into account the variation of temperature. As the temperature of the turbine blade changes with time, Fig. 3.10, the stiffness matrix changes resulting in the variation of the natural frequencies. The results obtained during the heating process at various instants of time are shown in the Table 3.4. In this table, the first eleven natural frequencies are shown. At t equal to 1 second, there is no significant change in the material temperature, therefore, the frequencies in the first column would be at ordinary temperatures. But as the material gets heated, each of these frequencies start decreasing due to the decrease in the value of the modulus of elasticity. Clearly there is a very significant change in the natural frequencies in the transient phase. This fact must be included in the design of the turbomachinery. Once the blade has reached the steady state temperature, then these frequencies would not change any more.

TABLE 3.4

Variation of Natural Frequencies (Hz) with time

Mode Number	1 sec	30 sec	50 sec	100 sec	120 sec	Steady State
I	661	652	649	614	603	565
II	2081	2047	2002	1930	1912	1785
III	3372	3311	3230	3116	3095	2888
IV	4230	4172	4077	3928	3900	3624
V	7159	7040	6866	6629	6588	6135
VI	8892	8763	8593	8262	8205	7619
VII	10957	10731	10487	10119	10054	9386
VIII	13302	13119	12830	12360	12280	11395
IX	15015	14778	14488	13942	13849	12862
X	14233	14012	13695	13198	13118	12192
XI	21803	21464	21048	20259	20122	18680

REFERENCES

1. Holzer, H., *Die Berechnung der Drehschwingungen*, Berlin, 1921.
2. Timoshenko, S.P., *Vibration Problems in Engineering*, Van Nostrand Co. (1955).
3. Scanlan, R.H. and Rosenbaum, R., *Introduction to the Study of Aircraft Vibration and Flutter*, MacMillan Co. (1951).
4. Rao, J.S., "Correction Factors for the Effect of Taper on the Torsional Oscillation of Cantilever Beams", *Proc. 11th Cong. ISTAM*, p. 139 (1966).
5. Rao, J.S., "Vibration of Cantilever Beams in Torsion", *JSER*, (2), p. 351 (1964).
6. Myklestad, N.O., "New Method of Calculating Natural Modes of Uncoupled Bending Vibrations", *J. Aero. Sci.*, **48**, p. 153 (1944).
7. Prohl, M.A., "A General Method of Calculating Critical Speeds of Flexible Rotors", *J. Appl. Mech., Trans. ASME*, **12**, p. A-142 (1945).
8. Rao, J.S., "A Tabular Procedure for the Determination of Uncoupled Bending Frequencies of a Cantilever Beam", *JSER*, **10** (2), p. 187 (1966).
9. Dawson, B. and Davies, M., "Automatic Root Searching Myklestad Procedure for Vibration Analysis", *Mech. Mach. Theory*, **12** (4), p. 363 (1977).
10. Mahalingam, S., "An Improvement of Myklestad Method for Flexural Vibration Problems", *J. Aero. Sci.*, **63**, p. 46 (1959).
11. Mahalingam, S., "The Iterative Solution of Flexural Vibration Problems Based on the Myklestad Method", *Intl. J. Mech. Sci.*, **4**, p. 241 (1962).
12. Housner, G.W. and Keightly, W.O., "Vibration of Linearly Tapered Cantilever Beams," *Proc. ASCE*, **88**, (EM 2), p. 95 (1962).

13. Mabie, H.H. and Rogers, C.B., "Transverse Vibrations of Tapered Cantilever Beams with End Loads", *J. Acoust. Soc. Amer.*, **36**, p. 463 (1964).
14. Targoff, W.P., "The Associated Matrices of Bending and Coupled Bending Torsion Vibration", *J. Aero. Sci.*, **14**, p. 579 (1947).
15. Rosard, D.D., "Natural Frequencies of Twisted Cantilevers", *J. Appl. Mech.*, ASME, **20**, p. 241 (1953).
16. Isakson, G. and Eisley, G.J., *Natural Frequencies in Bending of Twisted Rotating Blades*, NASA TN D-371 (1960).
17. Isakson, G. and Eisley, G.J., *Natural Frequencies in Coupled Bending and Torsion of Twisted Rotating and Nonrotating Blades*, NASA No. NSG-27-59 (1964).
18. Carnegie, W., "Solution of Equations of Motion for the Flexural Vibration of Cantilever Blades Under Rotation by the Extended Holzer Method", *BMEE*, **6**, p. 225 (1967).
19. Rao, J.S. and Carnegie, W., "Numerical Procedure for the Determination of the Frequencies and Mode Shapes of Lateral Vibration of Blades Allowing for the Effects of Pre-Twist and Rotation", *Int. Mech. Engr. Educ.*, **1**, p. 37 (1973).
20. Rao, J.S., "Coupled Bending Torsion Vibrations of Cantilever Beams", *2nd Intl. Cong. of Theory of Mechanisms and Machines*, **3**, p. 199 (1969).
21. Rao, J.S., "Flexural Vibration of Rotating Cantilever Beams", *J. Aero. Soc. Ind.*, **22**, p. 257 (1970).
22. Rao, J.S. and Banerjee, S., "Coupled Bending-Torsion Vibrations of Rotating Cantilever Blades — Method of Polynomial Frequency Equation", *Mech. Mach. Theory*, **12** (4), p. 271 (1977).
23. Krupka, R.M. and Baumanis, A.M., "Bending Bending Mode of a Rotating Tapered Twist Turbomachine Blade Including Rotary Inertia and Shear Deflection", *J. Engr. for Indus.*, ASME, **91**, p. 1017 (1969).
24. Nagaraja, J.V., Rawtani, S., and Singh, V.K., *Vibration Analysis of Turbomachine Blades*, Project Rept., Aeronautics Res. Devel. Bd., Ministry of Defence, India (1973).
25. Jacobsen, L.S. and Ayre, R.S., *Engineering Vibrations*, McGraw-Hill Book Co. (1958).
26. Duncan, W.J. and Collar A.R., "Solution of Oscillation Problems by Matrices", *Phil. Mag.*, **17**, p. 866 (1934).
27. Duncan, W.J., "Torsion and Torsional Oscillations of Blades", *Trans. NEC Inst. Engr. and Shipbuilders*, **54**, p. 301 (1938).
28. Thomson, W.T., "Matrix Solution of Vibration of Nonuniform Beams", Paper No. 49 A-11, presented to ASME (1949).
29. Thomson, W.T., "A Note on Tabular Methods for Flexural Vibrations", *J. Aero. Sci.*, **20**, p. 62 (1953).
30. Leckie, F.A. and Lindberg, G.M., "The Effect of Lumped Parameters on Beam Frequencies", *Aero. Q.*, **14**, p. 224 (1963).
31. Lindberg, G.M., "Vibration of Nonuniform Beams", *Aero. Q.*, **14**, p. 387 (1963).
32. Morris, J. and Head, J.W., "Coupled Flexural Vibrations of Blades on a Propeller and Torsional Vibrations of an Engine Crank Shaft System", *R&M*, 2011 (1942).
33. Mendelson, A. and Gendler, S., *Analytical Determination of Coupled Bending Torsion Vibrations of Cantilever Beams by Means of Station Functions*, NACA TN-2185 (1949).

34. Mendelson, A. and Gendler, S., *Analytical and Experimental Investigations of Effect of Twist on Vibration of Cantilevers*, NACA TN-2300 (1951).
35. Rao, J.S., *Rotor Dynamics*, Wiley Publication (1983).
36. Downs, B., "Transverse Vibrations of Cantilever Beams Having Unequal Breadth and Depth Tapers", *J. Appl. Mech. Trans. ASME*, **44** (4), p. 737 (1977).
37. Wilkerson, J.B., *NATFREQ — A Computer Program for Calculating the Natural Frequency of Rotating Cantilevered Beams*, Rept. No. DTNSRDC/ASED-370 (Jan. 1977).
38. Murthy, V.R. and Barna, P.S., *Vibration Analysis of Rotor Blades with an Attached Concentrated Mass*, Rept. No. NASA-CR-154987 (Apr. 1977).
39. Downs, B., "Vibration Analysis of Turbomachinery Blades Using Dedicated Discretization and Twisted Beam Theory", ASME, 79-DET-85.
40. Thomas, J., *Vibrational Characteristics of Tapered Cantilever Beams*, Ph.D. Thesis, Univ. London, England (1968).
41. Carnegie, W. and Thomas, J., "Natural Frequencies of Long Tapered Cantilevers", *Aero. Q.*, **18**, p. 309 (1967).
42. Thomas, J. and Carnegie, W., "Torsional Vibration of Tapered Beams," *Proc. 15th Cong. ISTAM*, p. 29 (1970).
43. Carnegie, W. and Thomas, J., "The Coupled Bending-Bending Vibration of Pre-Twisted Tapered Blading", *J. Engr. Indus. Trans. ASME*, **94**, p. 255 (1972).
44. Carnegie, W. and Thomas, J., "The Effect of Shear Deformation and Rotary Inertia on the Lateral Frequencies of Cantilever Beams in Bending", *J. Engr. Indus. Trans. ASME*, **94**, p. 267 (1972).
45. Carnegie, W., Stirling, C. and Fleming, J., "Vibration Characteristics of Turbine Blading under Rotation — Results of an Initial Investigation and Details of a High-Speed Test Installation", Paper No. 32, Appl. Mech. Convention held at Cambridge, England (1966).
46. Carnegie, W., Dawson, B. and Thomas, J., "Vibration Characteristics of Cantilever Blading," *Proc. Inst. Mech. Engr.*, **180** (31), p. 71 (1965-66).
47. Carnegie, W. and Dawson, B., "Vibration Characteristics of Straight Blades of Asymmetrical Airfoil Cross Section", *Aero. Q.*, **20**, p. 178 (1969).
48. Zienkiewicz, O.C., *The Finite Element Method in Engineering Science*, McGraw-Hill, London (1971).
49. Dokumaci, E., *Development and Application of the Finite Element Method to the Vibration of Beams*, Ph.D. Thesis, Univ. Surrey, England (1968).
50. Carnegie, W., Thomas, J. and Dokumaci, E., "An Improved Method of Matrix Displacement Analysis in Vibration Problems", *Aero. Q.*, **20**, p. 321 (1969).
51. Dokumaci, E., "Pre-Twisted Beam Elements Based on Approximation of Displacements in Fixed Directions", *J. Sound Vib.*, **52** (2), p. 277 (1977).
52. Nagaraj, V.T., "Coupled Bending-Bending Torsion Vibrations of Twisted Tapered Rotating Blades of Asymmetric Cross Section," *Symp. Rotor Dyn., 20th Cong. ISTAM*, Paper No. 12 (1975).
53. Singh, V.K., *The Effect of Root Flexibility on the Vibrational Characteristics of Turbomachinery Blades*, Ph.D. Thesis, Bhopal University, 1979.
54. Singh, V.K. and Rawtani, S., "Effect of Root Flexibility on Vibration Characteristics of Tapered Blades", *Symp. Dyn. Rotors 20th Cong. ISTAM*, Paper No. 8 (1975).

55. Singh, V.K. and Rawtani, S., "Natural Frequencies and Mode Shapes of Steam Turbine Blades", *Proc. BHEL Symp. Blade Des. Devel.*, Hyderabad (Sept. 1975).
56. Murthy, A.V.K. and Murthy, S.S., "Finite Element Analysis of Rotors", *Mech. Mach. Theory,* **12** (4), p. 311 (1977).
57. Gupta, R.S. and Rao, S.S., "Finite Element Eigen Value Analysis of Tapered and Twisted Timoshenko Beams," *J. Sound Vib.*, **56** (2), p. 187 (1978).
58. Abbas, B.A.H., "Simple Finite Elements for Dynamic Analysis of Thick Pre-Twisted Blades", *Aeronaut. J.*, **83** (827) p. 450 (1979).
59. Putter, S. and Manor, H., "Natural Frequencies of Radial Rotating Beams", *J. Sound Vib.*, **56** (2), p. 175 (1978).
60. Singh, V.K. and Rawtani, S., "The Effect of Root Flexibility on the Torsional Vibration of Uniform Section Blades", *Intl. J. Mech. Sci.*, **21** (3), p. 141 (1979).
61. Nagaraj, V.T., and Sahu, N., "Torsional Vibrations of Non-Uniform Rotating Blades with Attachment Flexibility", *J. Sound Vib.*, **80** (3) p. 401 (1982).
62. Banerji, S. and Rao, J.S., "Coupled Bending Torsion Vibrations of Rotating Blades", ASME, 76-GT-43.
63. Bahree, R., Sharan, A.M. and Rao, J.S., "The Design of Rotor Blades Due to the Combined Effects of Vibratory and Thermal Loads", *J. of Engg. Power and Gas Turbines, Trans. ASME,* **111**, p. 610 (1989).
64. Cook, R.D., *Concepts and Applications of Finite Element Analysis*, John Wiley and Sons (1981).
65. Huebner, K.H. and Thornton, E.A., *The Finite Element Method for Engineers,* John Wiley and Sons (1982).
66. Guyan, R.J., "Reduction of Stiffness and Mass Matrices", *AIAA J.*, **3**, No. 2, p. 380 (1965).
67. Bahree, R., *Analysis and Design of Rotor Blades due to The Transient Thermal and Vibratory Loads*, M.E. Thesis, Memorial University of Newfoundland, St. John's (1987).
68. Vyas, N.S., *Vibratory Stress Analysis and Fatigue Life Estimation of Turbine Blade*, Ph.D. Thesis, I.I.T., New Delhi (1986).
69. Rao, J.S. and Gupta, K., *Theory and Practice of Mechanical Vibrations*, John Wiley and Sons (1984).
70. Cubberly, W.H., *ASM Metals Reference Book*, American Society for Metals, Metals Park, Ohio, **3** (1980).

Chapter 4

Small Aspect Ratio Blades

4.1 INTRODUCTION

When the blade length is small and sufficiently wide along its chord, it behaves more like a plate or shell rather than a beam. The classical beam theory is not valid anymore as it cannot predict plate modes. Thus, for small aspect ratio blades, it is important to determine the plate mode frequencies. There are two methods available for this, the finite element method and the energy method. Since there are standard finite element programs like ANSYS or NASTRAN with a large library of elements, they become handy for the application to a dynamic analysis of a turbine blade. The energy methods, as in Chapter 2, require a lot of analytical background work, however they consume less computer time and allow a parametric study.

The finite element method has been used by Rawtani and Dokainish [1] to determine static deflections of a pre-twisted cantilever plate. Pre-twists were up to 90 degrees and flat plate triangular elements up to 10×10 mesh were used. This analysis was later extended to determine the natural frequencies and mode shapes of pre-twisted cantilever plates [2]. Barten, et al [3] studied stationary thin blades by the finite element method. A similar analysis has been done for thick blades by Ahmad, et al [4] and extended to rotating blades by Bossak and Zienkiwicz [5]. Dokainish and Rawtani [6] obtained modified Southwell coefficients for rotating thin plates and considered the effect of pseudo-static deformation on the natural frequencies of rotating small aspect ratio blades [7]. Hofmeister and Evensen [8] used isoparametric elements in calculating natural frequencies of non-rotating blades. Thin rotating cantilever plates have been analyzed with the finite element method by Dokainish and Gossain [9]. Thomas and Mota Soarez [10] calculated frequencies and mode shapes of thick rotating blades. Rawtani [11] studied the effects of camber on natural frequencies of small aspect ratio turbine blades. Frequencies and mode shapes of thin rotating blades have been determined and a design procedure suggested by Henry and Lalanne [12]. Lalanne and Trompette [13] used isoparametric elements in a dynamic analysis of actual turbine blade; they included the effects of root influence and temperature. A method for obtaining a mathematical model of rotating blades with Lagrange's equations and the finite element method, has been used to calculate the natural frequencies of a gas turbine blade and an air compressor blade [14]. Olson and Lindberg [15] developed a 28 degree of freedom element for cylindrical shell vibrations. MacBain [16] used four node quadrilateral plate-bending element CQUAD 2 of NASTRAN to evaluate the natural frequencies of a 2.33 ratio blade for different pre-twists and a mesh size of 11×24. The results agreed fairly well with the experimental results; however the variation of first and third mode bending results were contradictory to other results with pre-twist in the flexible direction [2]. Jumaily and Faulkner [17] considered a thin shell theory for long hollow blades and compared the results with beam theory and laboratory tests. MacBain [18] described a technique for computing the bending strain resulting from resonant modal deformation of a plate and compared the results with those from a NASTRAN finite element program. Isoparametric thick shell elements that allow arbitrary changes in shape thickness and curvature have been used in vibration analysis of impellers of rotating hydromachines [19]; good agreement with experimental results by holography was shown. Walker [20] used conforming finite shell elements for curved twisted fan blades; Mindlin's theory was used to derive the finite elements. Hoa [21] considered curved cylindrical shell finite elements for a blade with weighted edge. Triangular shell elements were used in [22] for rotating pre-twisted cantilever plates. Henry [23] used substructure techniques and wave propagation in periodic systems in conjunction with finite element method for impeller type of blades and considered several practical applications of turbine blading [24]. Fu [25] determined vibration amplitudes and stresses in gas turbine compressor blades by finite element method and holography interference technique.

Petricone and Sisto [26] used Wan's [27] strain displacement relations for a right helicoid derived from the general analysis of Reissner [28], to study the effect of pre-twist on the natural frequencies of plates by Ritz energy method. Gupta [29] used Gol' den Veizer's [30] strain displacement relations and derived general expressions for kinetic and potential energies of rotating pre-twisted shells. Levy type of analysis was conducted by Gupta and Rao to determine the torsional vibration modes [31] and bending vibration modes [32] of pre-twisted cantilever plates. Because fundamental mode frequency variation with pre-twist is controversial, they used inextensional analysis for bending vibrations [33] and showed that, for small aspect ratio blades, the natural frequency decreases with pre-twist. The analysis was extended to rotating pre-twisted shells mounted on a disc at stagger angle [29]. Leissa, et al [34] determined natural frequencies of shallow cylindrical shells with rectangular plan form with the help of classical Ritz method. The effect of rotation was included [35] along with disc radius and stagger angle [36]. A finite difference energy method working on a simple rectangular grid mapped on to the blade surface by parameter functions is given by Jansen [37].

In this chapter, the energy expressions for rotating pre-twisted shells are derived and Ritz procedure to obtain the natural frequencies is outlined. The results from different sources for bending modes, torsion modes and plate modes of rotating pre-twisted plates are presented. A finite element method using superparametric elements is also presented for dynamic analysis of cantilever plate type turbine blades.

4.2 ENERGY EXPRESSIONS AND METHOD OF SOLUTION

The differential geometry of a stationary blade is first defined and the strain displacement relations obtained. Then the potential and kinetic energies are derived. For the rotating blade, the gain in potential energy is determined next. Using these energy expressions, Ritz method is outlined to obtain the solution for natural frequencies.

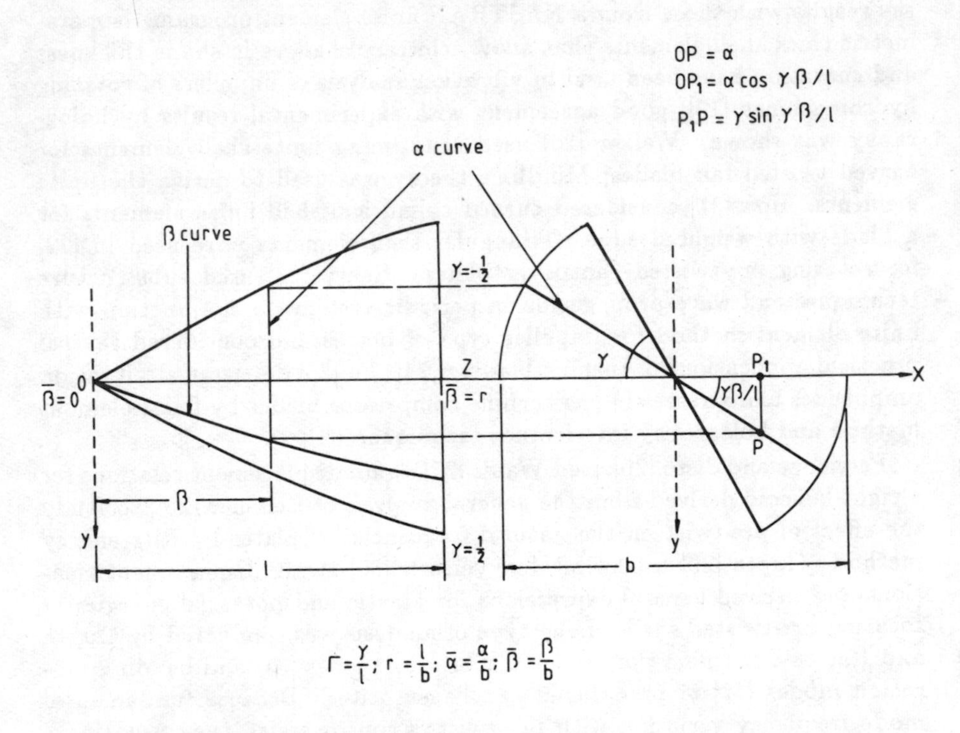

Fig. 4.1 Middle surface of a uniformly pre-twisted plate

Differential Geometry

Consider a uniformly pre-twisted plate of rectangular cross-section shown in figure 4.1. $x\,y\,z$ are the rectangular coordinates and $\alpha\,\beta$ curvilinear coordinates. The length of the plate is l and width of the plate is b. γ is pre-twist angle. The aspect ratio of the plate is $r = l/b$. The position vector of a point P on an α curve, at distance $OP = \alpha$, is

$$\overline{M}(\alpha, \beta) = (\alpha \cos \Gamma\beta)\hat{i} + (\alpha \sin \Gamma\beta)\hat{j} + (\beta)\hat{k} \tag{4.1}$$

where $\hat{i}$, $\hat{j}$ and $\hat{k}$ are unit vectors and $\Gamma = \gamma/l$ is unit pre-twist angle.

The coefficients of the first quadratic form of the undeformed middle surface are obtained by using the following relations [30].

$$\begin{aligned} A &= |\overline{M}, \alpha| = 1 \\ B &= |\overline{M}, \beta| = [1 + \Gamma^2 \alpha^2]^{\frac{1}{2}} \\ X &= \cos^{-1} \frac{\overline{M}_{,\alpha}}{A} \cdot \frac{\overline{M}_{,\beta}}{B} = \frac{\pi}{2} \end{aligned} \tag{4.2}$$

The three basic vectors of the middle surface are unit vectors $\overline{M}_{,\alpha}/A$, $\overline{M}_{,\beta}/A$ and $\hat{n}$, where

$$\hat{n} = \frac{\overline{M}_{,\alpha}}{A} \times \frac{\overline{M}_{,\beta}}{B}.$$

The basic vectors are represented in figure 4.2. Then the coefficients of the second quadratic form are

$$\begin{aligned} L &= \hat{n} \cdot \overline{M}_{,\alpha\alpha} = 0 \\ M &= \hat{n} \cdot \overline{M}_{,\alpha\beta} = -\frac{\Gamma}{B} \\ N &= \hat{n} \cdot \overline{M}_{,\beta\beta} = 0 \end{aligned} \tag{4.3}$$

The radii of curvature R_1 and R_2 in α and β directions, and the radius of torsion R_{12} are

$$\begin{aligned} \frac{1}{R_1} &= \frac{L}{A^2} = 0 \\ \frac{1}{R_2} &= \frac{N}{B^2} = 0 \\ \frac{1}{R_{12}} &= \frac{M}{AB} = \frac{\Gamma}{B^2} \end{aligned} \tag{4.4}$$

Since the curvatures along α and β directions are zero, the two curvilinear coordinates α and β coincide with real directions of asymptotes and the shell is one of negative Gaussian curvature. The Gaussian curvature for the shell surface is

$$K = \frac{LN - M^2}{A^2 B^2 \sin^2 x} = -\frac{\Gamma^2}{\left[1 + \Gamma^2 \alpha^2\right]^2} \tag{4.5}$$

The three basic vectors of figure 4.2, can be obtained from (4.1) and (4.2) as

$$\begin{aligned} \frac{\overline{M}_{,\alpha}}{A} &= (\cos \Gamma\beta)\hat{i} + (\sin \Gamma\beta)\hat{j} \\ \frac{\overline{M}_{,\beta}}{B} &= -\left(\frac{\Gamma\alpha}{B} \sin \Gamma\beta\right)\hat{i} + \left(\frac{\Gamma\alpha}{B} \cos \Gamma\beta\right)\hat{j} + \left(\frac{1}{B}\right)\hat{k} \\ \hat{n} &= \left(\frac{\sin \Gamma\beta}{B}\right)\hat{i} - \left(\frac{\cos \Gamma\beta}{B}\right)\hat{j} + \frac{\Gamma\alpha}{B}\hat{k} \end{aligned} \tag{4.6}$$

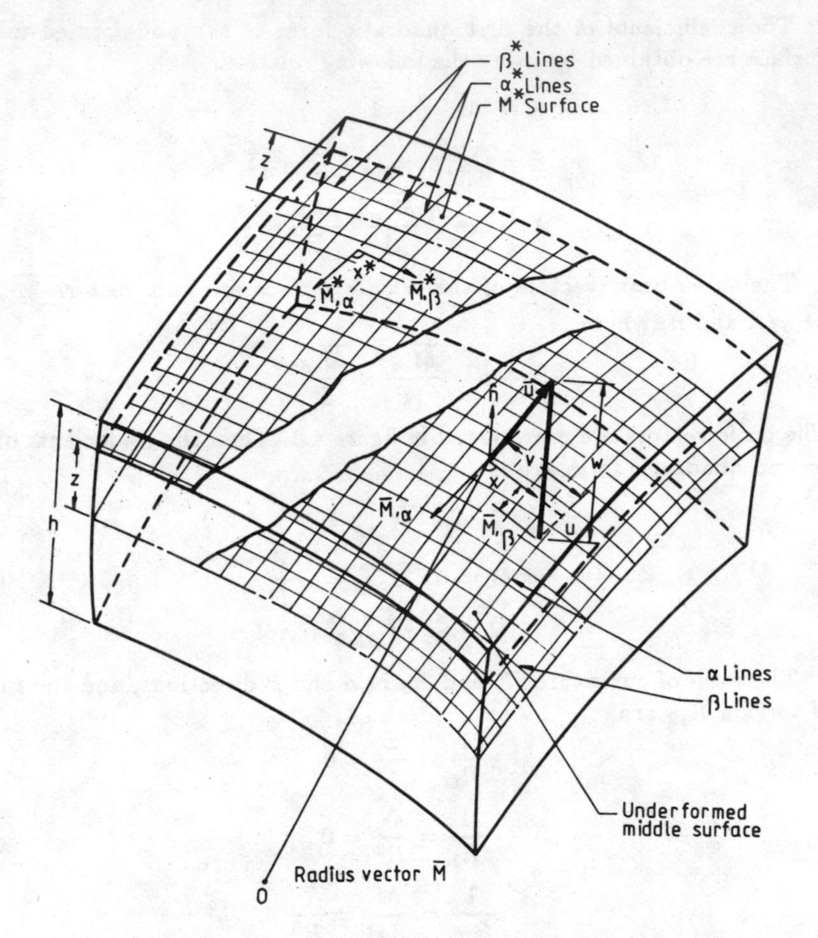

Fig. 4.2 Undeformed shell element showing M^* surface, curvilinear coordinates, basic vectors, displacement vector of middle surface

M^* is a surface, which is equidistant z from the middle surface, in the direction of unit normal vector $\hat{n}$ (Figure 4.2).

$$\overline{M}^*(\alpha, \beta, z) = \overline{M}(\alpha, \beta) + z\,\hat{n} \tag{4.7}$$

The coefficients of first quadratic form for M^* surface are

$$\begin{aligned} A^* &= \left(1 + \frac{z^2}{R_{12}^2}\right)^{\frac{1}{2}} \\ B^* &= \left(1 + \frac{z^2}{R_{12}^2}\right)^{\frac{1}{2}} \\ X^* &= \cos^{-1}\left(-\frac{2z}{R_{12}}\right) \end{aligned} \tag{4.8}$$

Strain Components

The undeformed and deformed surfaces of the shell are shown in figure 4.3. It should be noted that the α^*, β^* coordinates of M^* surface are nonorthogonal as can be seen from the expression for X^* in (4.8). The longitudinal strains ϵ^*_α, ϵ^*_β in α^* and β^* directions and shear strain $\omega^*_{\alpha\beta}$ of M^* surface are

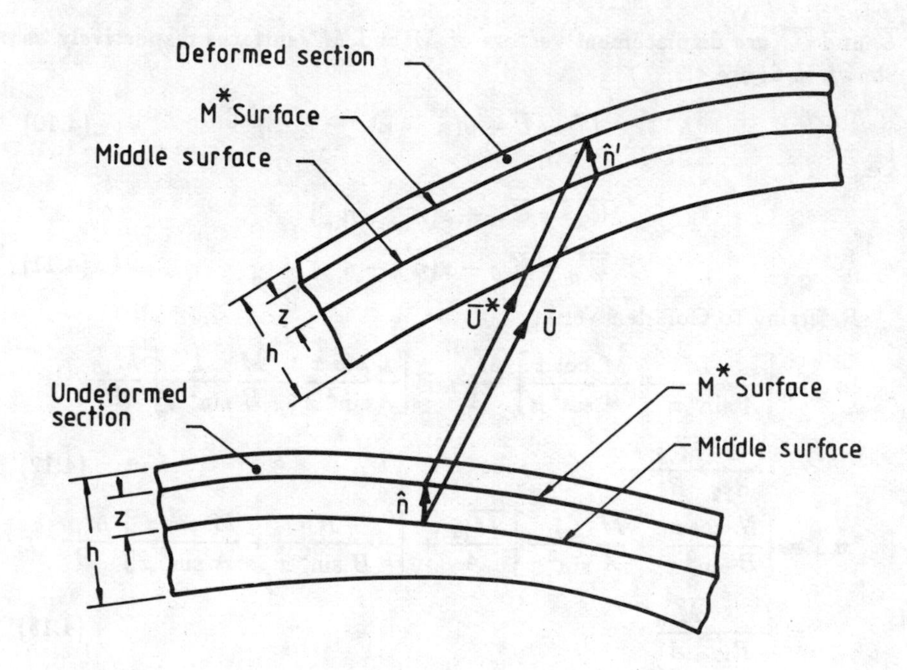

Fig. 4.3 Undeformed and deformed section

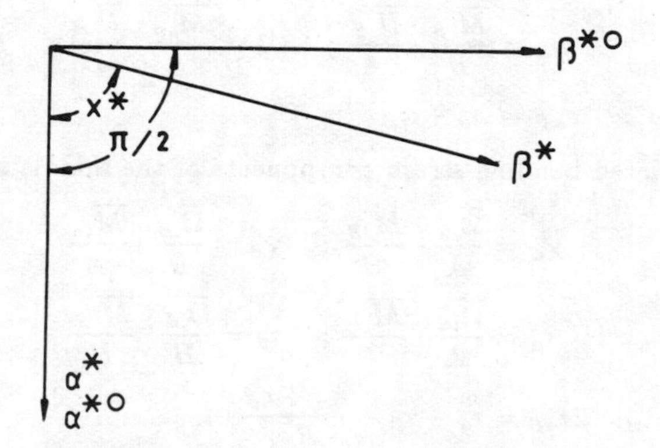

Fig. 4.4 Orthogonal coordinates

$$\epsilon_\alpha^* = \frac{\overline{M}^*_{,\alpha}}{A^*} \cdot \frac{\overline{U}^*_{,\alpha}}{A^*}$$

$$\epsilon_\beta^* = \frac{\overline{M}^*_{,\beta}}{B^*} \cdot \frac{\overline{U}^*_{,\beta}}{B^*} \tag{4.9}$$

$$\omega_{\alpha\beta}^* = \frac{\overline{M}^*_{,\alpha}}{A^*} \cdot \frac{\overline{U}^*_{,\beta}}{B^*} + \frac{\overline{M}^*_{,\beta}}{B^*} \cdot \frac{\overline{U}^*_{,\alpha}}{A^*} + \frac{2z}{R_{12}}(\epsilon_\alpha^* + \epsilon_\beta^*)$$

$\overline{U}$ and $\overline{U}^*$ are displacement vectors of M and M^* surfaces respectively as shown in figure 4.3,

$$\overline{U}^* = \overline{U} + z(\hat{n}' - \hat{n}) \tag{4.10}$$

i.e.,

$$\overline{U}^*_{,\alpha} = \overline{U}_{,\alpha} + z(\hat{n}'_{,\alpha} - \hat{n}_{,\alpha})$$

$$\overline{U}^*_{,\beta} = \overline{U}_{,\beta} + z(\hat{n}'_{,\beta} - \hat{n}_{,\beta}) \tag{4.11}$$

Referring to Gol' den Veizer,

$$\hat{n}_{,\alpha} = \left[\frac{-L}{A\sin^2 x} + \frac{M}{B}\frac{\cos x}{\sin^2 x}\right]\frac{\overline{M}_{,\alpha}}{A} + \left[\frac{L}{A}\frac{\cos x}{\sin^2 x} - \frac{M}{B}\frac{1}{\sin^2 x}\right]\frac{\overline{M}_{,\beta}}{B}$$

$$= -\frac{1}{R_{12}}\frac{\overline{M}_{,\beta}}{B} \tag{4.12}$$

$$\hat{n}_{,\beta} = \left[\frac{N}{B}\frac{\cos x}{\sin^2 x} - \frac{M}{A}\frac{1}{\sin^2 x}\right]\frac{\overline{M}_{,\alpha}}{A} + \left[-\frac{N}{B\sin^2 x} + \frac{M}{A}\frac{\cos x}{\sin^2 x}\right]\frac{\overline{M}_{,\beta}}{B}$$

$$= -\frac{1}{R_{12}}\frac{\overline{M}_{,\alpha}}{A} \tag{4.13}$$

The components of longitudinal strains of the middle surface are

$$\epsilon_\alpha = \frac{\overline{M}_{,\alpha}}{A} \cdot \frac{\overline{U}_{,\alpha}}{A}; \quad \epsilon_\beta = \frac{\overline{M}_{,\beta}}{B} \cdot \frac{\overline{U}_{,\beta}}{B}$$

$$\omega_\alpha = \frac{\overline{M}_{,\beta}}{B} \cdot \frac{\overline{U}_{,\beta}}{A}; \quad \omega_\beta = \frac{\overline{M}_{,\alpha}}{A} \cdot \frac{\overline{U}_{,\beta}}{B} \tag{4.14}$$

$$\omega_{\alpha\beta} = \omega_\alpha + \omega_\beta$$

The associated bending strain components of the middle surface are

$$\chi_\alpha = \frac{\overline{\Omega}_{,\alpha}}{A} \cdot \frac{\overline{M}_{,\beta}}{B}; \quad \chi_\beta = \frac{\overline{\Omega}_{,\beta}}{B} \cdot \frac{\overline{M}_{,\alpha}}{A} \tag{4.15}$$

$$\tau_\alpha = \frac{\overline{\Omega}_{,\alpha}}{A} \cdot \frac{\overline{M}_{,\alpha}}{A}; \quad \tau_\beta = \frac{\overline{\Omega}_{,\beta}}{B} \cdot \frac{\overline{M}_{,\beta}}{B}$$

$$2\tau_{\alpha\beta} = \tau_\alpha - \tau_\beta - \frac{\epsilon_\alpha + \epsilon_\beta}{R_{12}} \tag{4.16}$$

where $\overline{\Omega}$ is vector of elastic rotation (see Gol' den Veizer). The relations in (4.12) and (4.13) can be directly used to determine $\hat{n}'_{,\alpha}$ and $\hat{n}'_{,\beta}$, by replacing A, B, X, L, M, N with A', B', X', L', M', N' and derivatives of $\overline{M}$ by those of $\overline{M}' = \overline{M} + \overline{U}$. The coefficients of first and second quadratic forms of the deformed middle surface are related to those of undeformed middle surface through components of strain as follows:

$$
\begin{aligned}
A' &= A(1+\epsilon_\alpha)\,; & B' &= B(1+\epsilon_\beta) \\
X' &= X - \omega_{\alpha\beta}\,; & L' &= \frac{\omega_{\alpha\beta} M}{2B} - \chi_\alpha \\
M' &= M(1+\epsilon_\alpha) + B\tau_\beta\,; & N' &= \frac{MB\omega_{\alpha\beta}}{2} - B^2\chi_\beta
\end{aligned}
\tag{4.17}
$$

Therefore,

$$
\hat{n}'_{,\alpha} = \left[\frac{\omega_{\alpha\beta}}{2R_{12}} + \chi_\alpha\right]\left[\frac{\overline{M}_{,\alpha}}{A} + \frac{\overline{U}_{,\alpha}}{A}\right] + \left[\tau_\alpha - \frac{1-\epsilon_\beta}{R_{12}}\right]\left[\frac{\overline{M}_{,\beta}}{B} + \frac{\overline{U}_{,\beta}}{B}\right]
\tag{4.18}
$$

$$
\hat{n}'_{,\beta} = B\left\{\left[\frac{\omega_{\alpha\beta}}{2R_{12}} + \chi_\beta\right]\left[\frac{\overline{M}_{,\beta}}{B} + \frac{\overline{U}_{,\beta}}{B}\right] - \left[\tau_\beta + \frac{1-\epsilon_\alpha}{R_{12}}\right]\left[\frac{\overline{M}_{,\alpha}}{A} + \frac{\overline{U}_{,\alpha}}{A}\right]\right\}
\tag{4.19}
$$

Substituting (4.7) and (4.11) in (4.9), with the help of equations (4.12) – (4.16), (4.18) and (4.19), neglecting higher order terms, the following strain components are obtained:

$$
\begin{aligned}
\epsilon^*_\alpha &= \epsilon_\alpha + z\chi_\alpha - \frac{z}{R_{12}}\frac{\omega_{\alpha\beta}}{2} - \frac{z^2}{R_{12}}\tau_\alpha \\
\epsilon^*_\beta &= \epsilon_\beta + z\chi_\beta - \frac{z}{R_{12}}\frac{\omega_{\alpha\beta}}{2} - \frac{z^2}{R_{12}}\tau_\beta \\
\omega^*_{\alpha\beta} &= \omega_{\alpha\beta} + 2z\tau_{\alpha\beta} + \frac{z^2}{R_{12}}(\chi_\alpha + \chi_\beta) + \frac{2z}{R_{12}}(\epsilon_\alpha + \epsilon_\beta)
\end{aligned}
\tag{4.20}
$$

If z/R_{12} is neglected, (4.20) become simplified and one can transform them to a set of strains ϵ^{*o}_α, ϵ^{*o}_β and $\omega^{*o}_{\alpha\beta}$ referred to an orthogonal set of coordinates α^{*o}, β^{*o} shown in figure 4.4, by using appropriate transformation formulae, see Morley [38]. Neglecting terms proportional to z^2/R^2_{12}, they would become

$$
\begin{aligned}
\epsilon^{*o}_\alpha &= \epsilon^*_\alpha \\
\epsilon^{*o}_\beta &= \epsilon^*_\beta + \frac{2z}{R_{12}}\omega^*_{\alpha\beta} \\
\omega^{*o}_{\alpha\beta} &= \omega^*_{\alpha\beta} + \frac{4z}{R_{12}}\epsilon^*_\alpha
\end{aligned}
\tag{4.21}
$$

Strain Displacement Relations

The displacement vector of the middle surface is given in figure 4.2, which has components, u, v and $-w$ in $\overline{M}_{,\alpha}$, $\overline{M}_{,\beta}$ and $\hat{n}$ directions, respectively. The stretching strains ϵ_α, ϵ_β, ω_α and ω_β, bending strains χ_α, χ_β and $\tau_{\alpha\beta}$, quantities τ_α and τ_β and rotations ν_α, ν_β and ν_n about α, β and normal axis are obtained for the shell surface under consideration, with the help of general formulae of Gol' den Veizer. The physical meaning of ω_α and ω_β and rotations ν_α, ν_β and ν_n is explained in figure 4.5.

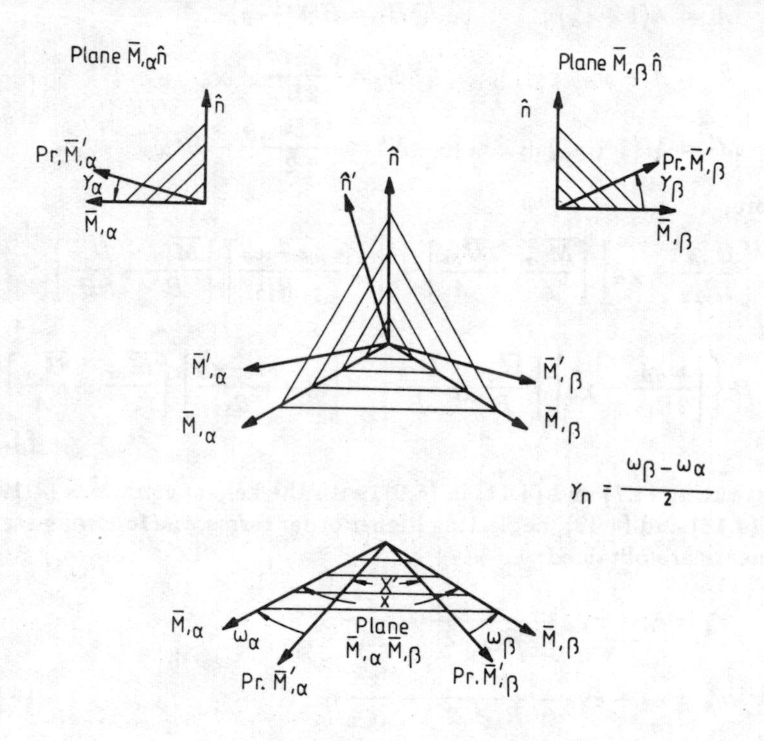

Fig. 4.5 Shear strains and rotations of element about α, β and normal axis

$$\epsilon_\alpha = u_{,\alpha} \quad ; \quad \epsilon_\beta = \frac{v_{,\beta}}{B} + \frac{\Gamma^2 \alpha u}{B^2}$$

$$\omega_\alpha = v_{,\alpha} - \frac{\Gamma w}{B^2} \quad ; \quad \omega_\beta = \frac{u_{,\beta}}{B} - \frac{\Gamma^2 \alpha v}{B^2} - \frac{\Gamma w}{B^2}$$

$$\omega_{\alpha\beta} = \omega_\alpha + \omega_\beta$$

$$\chi_\alpha = -\frac{\Gamma u_{,\beta}}{2B^3} + \frac{3}{2}\frac{\Gamma v_{,\alpha}}{B^2} - \frac{3}{2}\frac{\Gamma^3 \alpha v}{B^4} + w_{,\alpha\alpha}$$

$$\chi_\beta = \frac{\Gamma u_{,\beta}}{2B^3} - \frac{\Gamma v_{,\alpha}}{2B^2} + \frac{\Gamma^3 \alpha v}{2B^4} + \frac{\Gamma^2 \alpha w_{,\alpha}}{B^2} + \frac{w_{,\beta\beta}}{B^2}$$

$$\tau_{\alpha\beta} = -\frac{\Gamma^3 \alpha u}{B^4} + \frac{\Gamma u_{,\alpha}}{B^2} + \frac{\Gamma v_{,\beta}}{B^3} - \frac{\Gamma^2 \alpha w_{,\beta}}{B^3} + \frac{w_{,\alpha\beta}}{B}$$

$$\tau_\alpha = -\frac{2\Gamma^3 \alpha u}{B^4} + \frac{\Gamma u_{,\alpha}}{B^2} - \frac{\Gamma^2 \alpha w_{,\beta}}{B^3} + \frac{w_{,\alpha\beta}}{B}$$

$$
\begin{aligned}
\tau_\beta &= \frac{\Gamma^3 \alpha u}{B^4} - \frac{\Gamma v_{,\beta}}{B^3} + \frac{\Gamma^2 \alpha w_{,\beta}}{B^3} - \frac{w_{,\alpha\beta}}{B} \\
\nu_\alpha &= -\frac{\Gamma u}{B^2} - \frac{w_{,\beta}}{B} \\
\nu_\beta &= -\frac{\Gamma v}{B^2} - \frac{w_{,\alpha}}{B} \\
\nu_n &= \frac{1}{2}\left[u_{,\beta} - \frac{\Gamma^2 \alpha v}{B^2} - v_{,\alpha}\right]
\end{aligned} \tag{4.22}
$$

Potential Energy

The potential energy expression for the shell is

$$
V = \frac{E}{2(1-\nu^2)} \int_v \left[\epsilon_\alpha^{*\circ} + \epsilon_\beta^{*\circ} + 2\nu \epsilon_\alpha^{*\circ} \epsilon_\beta^{*\circ} + \frac{1-\nu}{2} \omega_{\alpha\beta}^{*\circ}\right] A^* B^* \, d\alpha \, d\beta \, dz \tag{4.23}
$$

Noting from (4.8) that $A^* = A$, $B^* = B$, if $(z^2/R_{12}^2) << 1$, substituting (4.20) and (4.21) in the above, integrating over the thickness of the shell, neglecting powers of z greater than 2 and that of z/R_{12} greater than or equal to 2, (i.e., using a thin shell theory), we get

$$
\begin{aligned}
V = \frac{E}{2(1-\nu^2)} \int\int \Bigg[& h\left\{(\epsilon_\alpha + \epsilon_\beta)^2 - 2(1-\nu)\left(\epsilon_\alpha \epsilon_\beta - \frac{\omega_{\alpha\beta}^2}{2}\right)\right\} \\
& + \frac{h^3}{12}\{(\chi_\alpha + \chi_\beta)^2 - 2(1-\nu)(\chi_\alpha \chi_\beta - \tau_{\alpha\beta})^2\} \\
& + \frac{h^3}{6R_{12}}\{-(2-\nu)\omega_{\alpha\beta}(\chi_\alpha + \chi_\beta) + (\epsilon_\alpha \tau_\alpha - \epsilon_\beta \tau_\beta) \\
& + \nu(\epsilon_\beta \tau_\alpha - \epsilon_\alpha \tau_\beta) - (6-2\nu)\tau_{\alpha\beta}(\epsilon_\alpha + \epsilon_\beta)\}\Bigg] AB \, d\alpha \, d\beta
\end{aligned} \tag{4.24}
$$

The first set of terms proportional to h, denote the strain energy due to stretching of the middle surface. The second and third set of terms, proportional to h^3 denote strain energy due to bending. The first and second set of terms are the conventional strain energy terms encountered in shell analysis, see Love [39]. The third set of terms is due to the retention of z/R_{12} compared to unity in the analysis. The relative order of the third set of terms in (4.24) as compared to first and second sets is h/R_{12}, see Novozhilov [40]. In (4.8) by making $A^* = A$, $B^* = B$, the inherent error in the analysis is of the order of h^2/R_{12}^2. Therefore, the third set of terms in (4.24) may be retained.

Kinetic Energy

The displacement field of any arbitrary point on M^* surface is:

$$
\begin{aligned}
u^*(\alpha, \beta, z) &= u(\alpha, \beta) - z\nu_\beta \\
v^*(\alpha, \beta, z) &= v(\alpha, \beta) - z\nu_\alpha \\
w^*(\alpha, \beta, z) &= w(\alpha, \beta)
\end{aligned} \tag{4.25}
$$

where ν_α and ν_β are rotations of the shell element about α and β axes given by

$$\nu_\alpha = \frac{\overline{U}_{,\beta}}{B} \cdot \hat{n} \quad ; \quad \nu_\beta = \frac{\overline{U}_{,\alpha}}{A} \cdot \hat{n} \tag{4.26}$$

[see (4.22)].

The kinetic energy of the vibrating shell is

$$T = \int_v \frac{1}{2} dm(\dot{u}^{*2} + \dot{v}^{*2} + \dot{w}^{*2}) \tag{4.27}$$

Noting that $dm = \rho AB\, dz\, d\alpha\, d\beta$, we have

$$T = \int\int \frac{\rho h}{2} \rho^2 \left[(u^2 + v^2 + w^2) + \frac{h^2}{12} (\nu_\alpha^2 + \nu_\beta^2) \right] AB\, d\alpha\, d\beta \tag{4.28}$$

The first set of terms is due to translatory inertia and the second set of terms proportional to h^3 is due to rotary inertia.

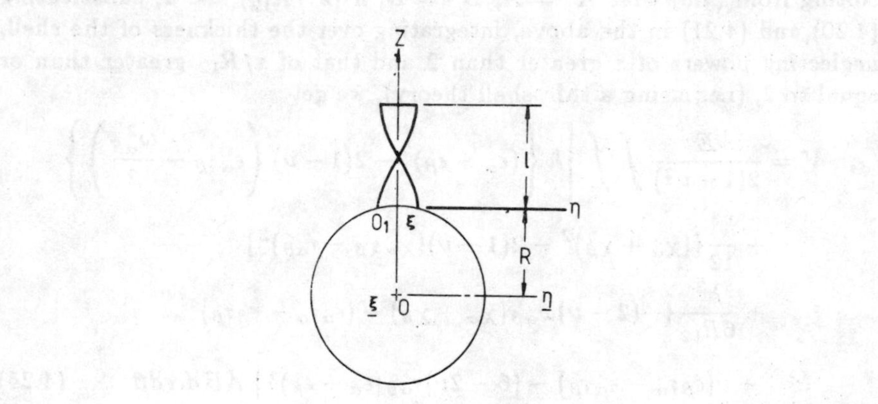

Fig. 4.6a Pretwisted blade mounted on a rotating disc

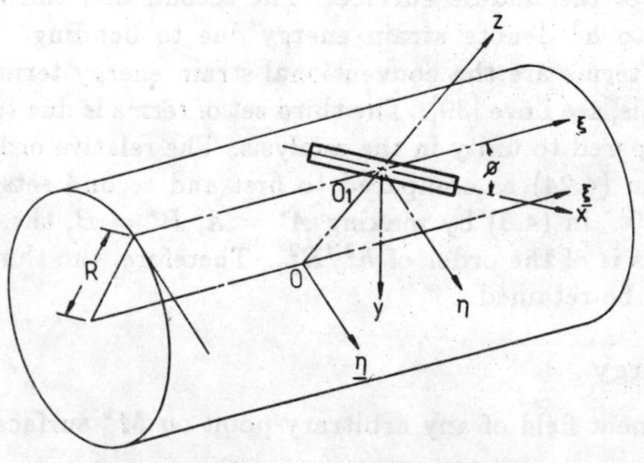

Fig. 4.6b System of rotating coordinates on the disc

Gain in Potential Energy due to Rotation

Figure 4.6a shows the twisted blade on the disc of radius R at a stagger angle ϕ. The parametric equation of the middle surface of the blade with reference to the set of coordinates XYZ (now rotating) is already given in equation (4.1). $\eta\,\xi\,z$ is another set of coordinates with η axis in the plane of rotation and ξ out of plane of the disc (i.e., turbine axis). The relations between the two sets of unit vectors of figure 4.6b are

$$\begin{aligned} \hat{\imath} &= \hat{I}\sin\phi + \hat{J}\cos\phi \\ \hat{\jmath} &= \hat{I}\cos\phi - \hat{J}\sin\phi \\ \hat{k} &= \hat{K} \end{aligned} \tag{4.29}$$

The vector equation of the surface in $\eta\,\xi\,z$ coordinates, can now be written as

$$\overline{M} = [\alpha\sin(\phi+\Gamma\beta)]\hat{I} + [\alpha\cos(\phi+\Gamma\beta)]\hat{J} + \beta\hat{K} \tag{4.30}$$

The three basic vectors in this coordinate system are

$$\begin{aligned} \frac{\overline{M}_{,\alpha}}{A} &= [\sin(\phi+\Gamma\beta)]\hat{I} + [\cos(\phi+\Gamma\beta)]\hat{J} \\ \frac{\overline{M}_{,\beta}}{B} &= -\left[\frac{\Gamma\alpha}{B}\cos(\phi+\Gamma\beta)\right]\hat{I} - \left[\frac{\Gamma\alpha}{B}\sin(\phi+\Gamma\beta)\right]\hat{J} + \frac{1}{B}\hat{K} \\ \hat{n} &= -\left[\frac{\cos(\phi+\Gamma\beta)}{B}\right]\hat{I} + \left[\frac{\sin(\phi+\Gamma\beta)}{B}\right]\hat{J} + \frac{\Gamma\alpha}{B}\hat{K} \end{aligned} \tag{4.31}$$

Let the displacement vector $\overline{U}$ be also written in the new coordinate system, then

$$\overline{U} = \eta_D\hat{I} + \xi_D\hat{J} + z_D\hat{K} \tag{4.32}$$

Since u, v and $-w$ are the components of displacement in the directions of $\overline{M}_{,\alpha}/A$, $\overline{M}_{,\beta}/B$ and $\hat{n}$ of equation (4.31), the displacement components η_D, ξ_D, z_D are

$$\begin{aligned} \eta_D &= u\sin(\phi+\Gamma\beta) + \frac{w - v\Gamma\alpha}{B}\cos(\phi+\Gamma\beta) \\ \xi_D &= u\cos(\phi+\Gamma\beta) - \frac{w + v\Gamma\alpha}{B}\sin(\phi+\Gamma\beta) \\ z_D &= \frac{v - w\Gamma\alpha}{B} \end{aligned} \tag{4.33}$$

With reference to coordinates at the centre of the disc, figure 4.6, equation (4.30) is

$$\overline{M}_0 = [\alpha\sin(\phi+\Gamma\beta)]\hat{I} + [\alpha\cos(\phi+\Gamma\beta)]\hat{J} + (R+\beta)\hat{K} \tag{4.34}$$

Using (4.32) and (4.34) the vector equation of the deformed middle surface is

$$\overline{M}' = [\alpha\sin(\phi+\Gamma\beta) + \eta_D]\hat{I} + [\alpha\cos(\phi+\Gamma\beta) + \xi_D]\hat{J} + [R+\beta+z_D]\hat{K} \tag{4.35}$$

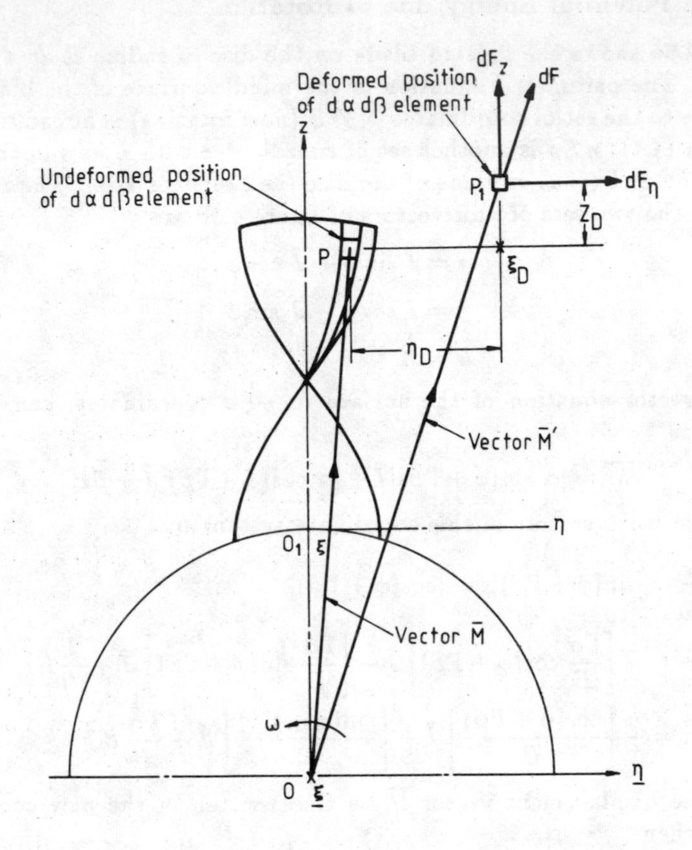

Fig. 4.7 Radius vectors for deformed and undeformed positions of $d\alpha\, d\beta$ element

The projection of $\overline{M}'$ in the plane ηz (plane of disc) is therefore, (OP_1 in figure 4.7)

$$\overline{OP}_1 = \{[\alpha \sin(\phi + \Gamma\beta) + \eta_D]^2 + [R + \beta + z_D]^2\}^{\frac{1}{2}} \tag{4.36}$$

The centrifugal force of the element $d\alpha\, d\beta$ at P, is

$$dF = AB\rho h\omega^2\, d\alpha\, d\beta \cdot \overline{OP}_1 \tag{4.37}$$

The components of dF in η and z directions are therefore

$$\begin{aligned} dF_\eta &= AB\, d\alpha\, d\beta\, \rho h\omega^2 \{\alpha \sin(\phi + \Gamma\beta) + \eta_D\} \\ dF_z &= AB\, d\alpha\, d\beta\, \rho h\omega^2 \{R + \beta + z_D\} \end{aligned} \tag{4.38}$$

The gain in strain energy of the element due to deflection η_D is

$$\begin{aligned} dU_\eta &= -\int_0^{\eta_D} dF_\eta\, d\eta_D \\ &= -AB\, d\alpha\, d\beta\, \rho h\omega^2 \left\{\eta_D \alpha \sin(\phi + \Gamma\beta) + \frac{\eta_D^2}{2}\right\} \end{aligned} \tag{4.39}$$

Similarly,

$$dU_z = -AB\,d\alpha\,d\beta\,\rho h\omega^2\left\{(R+\beta)z_D + \frac{z_D^2}{2}\right\} \tag{4.40}$$

Since the centrifugal force has no component in ξ direction, ξ_D will not result in any gain in strain energy. Hence, the net gain in the potential energy of the plate is

$$\begin{aligned} U_F &= \int\int dU_\eta + dU_z \\ &= -\omega^2 \int_{-\frac{1}{2}b}^{\frac{1}{2}b}\int_0^l \left[\eta_D \alpha \sin(\phi + \Gamma\beta) + \frac{\eta_D^2}{2}\right. \\ &\qquad \left. + (R+\beta)z_D + \frac{1}{2}z_D^2\right]\rho h AB\,d\alpha\,d\beta \end{aligned} \tag{4.41}$$

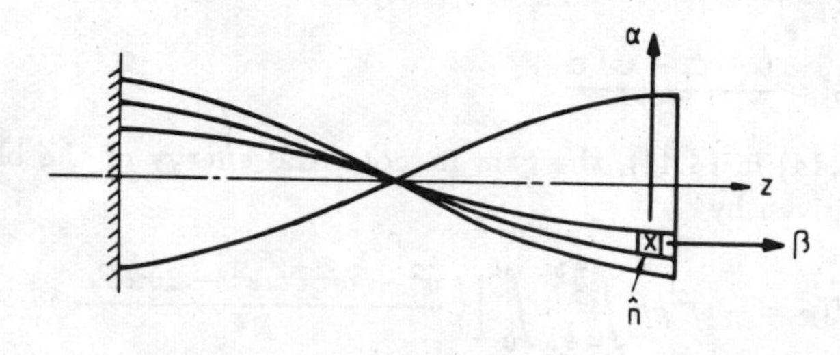

Fig. 4.8a Infinitesimal element bounded by α, β curves

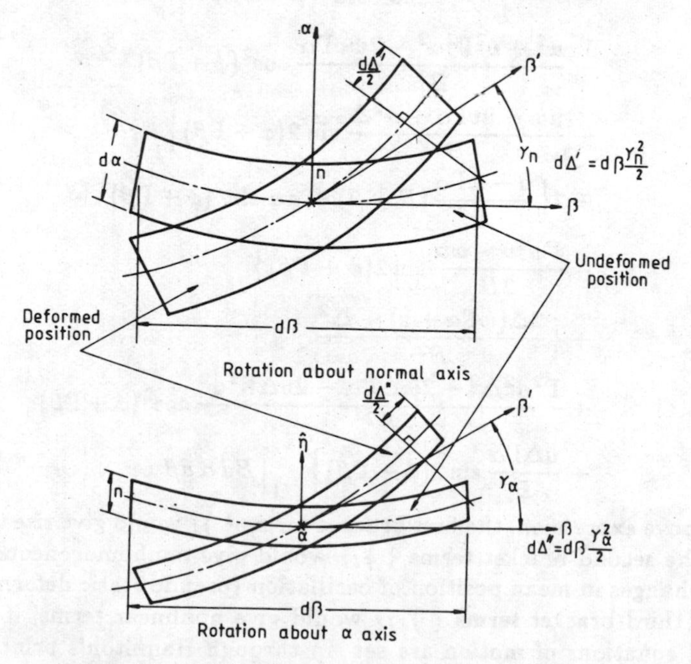

Fig. 4.8b Rotations about normal and α axes

It must be noticed that the rotations of the element $d\alpha\, d\beta$ about α and normal axis will cause some inward displacement $d\Delta$ in β direction. (See figure 4.8).

$$d\Delta = \frac{d\beta}{2}(\nu_\alpha^2 + \nu_\eta^2) \tag{4.42}$$

ν_α and ν_η are defined in (4.22). The total inward displacement of the element is therefore

$$\Delta = \frac{1}{2}\int_0^\beta (\nu_\alpha^2 + \nu_\eta^2)d\beta \tag{4.43}$$

The inward displacement is in the axial direction of plate β, and therefore v should be replaced by $(v - \Delta)$ in (4.33).

$$\begin{aligned} \eta_D &= u \sin(\phi + \Gamma\beta) + \frac{w - (v - \Delta)\Gamma\alpha}{B} \cos(\phi + \Gamma\beta) \\ \xi_D &= u \cos(\phi + \Gamma\beta) - \frac{w + (v - \Delta)\Gamma\alpha}{B} \sin(\phi + \Gamma\beta) \\ z_D &= \frac{v - \Delta - w\Gamma\alpha}{B} \end{aligned} \tag{4.44}$$

Using (4.44) in (4.41), the gain in potential energy of the blade due to rotation is given by

$$\begin{aligned} U_F = -\,\omega^2 \rho h \int_{-\frac{1}{2}b}^{\frac{1}{2}b} \int_0^l \Bigg[\Bigg\{ & \frac{v^2 + w^2\Gamma^2\alpha^2 - 2vw\Gamma\alpha}{B^2} \\ & - \frac{2\Delta(R + \beta)}{B} + u^2 \sin^2(\phi + \Gamma\beta) \\ & + \frac{w^2 + v^2\Gamma^2\alpha^2 - 2wv\Gamma\alpha}{B^2} \cos^2(\phi + \Gamma\beta) \\ & + \frac{uw - uv\Gamma\alpha + \alpha^2\Delta}{B} \sin 2(\phi + \Gamma\beta) \Bigg\}_I \\ & + 2\Bigg\{ \frac{v - w\Gamma\alpha}{B}(R + \beta) + \alpha u \sin^2(\phi + \Gamma\beta) \\ & - \frac{\Gamma\alpha^2 v - w\alpha}{2B} \sin 2(\phi + \Gamma\beta) \Bigg\}_{II} \\ & + \Bigg\{ \frac{2\Delta(w\Gamma\alpha + v) + \Delta^2}{B^2} \\ & + \frac{\Gamma^2\alpha^2\Delta^2 + 2w\Delta\Gamma\alpha - 2v\Delta\Gamma^2\alpha^2}{B^2} \cos^2(\phi + \Gamma\beta) \\ & + \frac{u\Delta\Gamma\alpha}{B} \sin 2(\phi + \Gamma\beta) \Bigg\}_{III} \Bigg] B d\alpha\, d\beta \end{aligned} \tag{4.45}$$

In the above expression, the first bracket terms $\{\ \}_I$ would give rise to linear terms, the second bracket terms $\{\ \}_{II}$ would give nonhomogeneous terms, for the changes in mean position of oscillation (pseudo static deformations) and the third bracket terms $\{\ \}_{III}$ would give nonlinear terms, if the differential equations of motion are set up through Hamilton's principle. In the analysis that follows, only the first bracket terms are considered.

Method of Solution

The Lagrangian expression $L = V + U_F - T$ can be set up by using (4.24), (4.45) and (4.28). In (4.24), only the first two sets of terms are retained, see Kalnins [41]. In (4.45) only the linear terms are retained. Thus

$$
\begin{aligned}
L = \int_{-\frac{b}{2}}^{\frac{b}{2}} \int_0^l \Bigg\{ & \frac{EB}{2(1-\nu^2)} \Bigg[h\Big\{ (\epsilon_\alpha + \epsilon_\beta)^2 - 2(1-\nu)\Big(\epsilon_\alpha \epsilon_\beta - \frac{\omega_{\alpha\beta}^2}{2} \Big) \Big\} \\
& + \frac{h^3}{12} \{ (\chi_\alpha + \chi_\beta)^2 - 2(1-\nu)(\chi_\alpha \chi_\beta - \tau_{\alpha\beta})^2 \} \Bigg] \\
& - \frac{\rho h}{2} \Bigg[\frac{\omega^2}{B} \Big\{ v^2 + \omega^2 \Gamma^2 \alpha^2 - 2v\omega\Gamma\alpha - 2\Delta(R+\beta)B \\
& + u^2 B^2 \sin^2(\phi + \Gamma\beta) + (w^2 + v^2\Gamma^2\alpha^2 - 2vw\Gamma\alpha)\cos^2(\phi + \Gamma\beta) \\
& + (uw - uv\Gamma\alpha + \alpha^2 \Gamma\Delta) B \sin 2(\phi + \Gamma\beta) \Big\} \\
& + p^2 B \Big\{ (u^2 + v^2 + w^2) + \frac{h^2}{12}(\nu_\alpha^2 + \nu_\beta^2) \Big\} \Bigg] \Bigg\} d\alpha \, d\beta
\end{aligned}
\tag{4.46}
$$

In the above equation, the strains can be replaced by the displacement field using (4.22) and B by equation (4.2), which is a function of Γ and α. The displacement field is assumed as follows:

$$
\begin{aligned}
u(\alpha, \beta) &= \sum \sum a_{ij} R_i(\alpha) \theta_j(\beta) \\
v(\alpha, \beta) &= \sum \sum b_{ij} R_i(\alpha) \theta_j(\beta) \\
w(\alpha, \beta) &= \sum \sum c_{ij} R_i(\alpha) \pi_j(\beta)
\end{aligned}
\tag{4.47}
$$

The polynomials R_i, θ_j and π_j are assumed as follows:

$$
\begin{aligned}
R_1 &= 1 \\
R_2 &= \frac{2\alpha}{b} \\
R_3 &= \Big[3\Big\{\frac{2\alpha}{b}\Big\}^2 - 1 \Big] \Big/ 2 \\
R_4 &= \Big[5\Big\{\frac{2\alpha}{b}\Big\}^3 - 3\Big\{\frac{2\alpha}{b}\Big\} \Big] \Big/ 2 \\
R_5 &= \Big[35\Big\{\frac{2\alpha}{b}\Big\}^4 - 30\Big\{\frac{2\alpha}{b}\Big\}^2 + 3 \Big] \Big/ 2 \\
R_6 &= \Big[63\Big\{\frac{2\alpha}{b}\Big\}^5 - 70\Big\{\frac{2\alpha}{b}\Big\}^3 + 15\Big\{\frac{2\alpha}{b}\Big\} \Big] \Big/ 8 \\
\theta_1 &= 1.73 \frac{\beta}{l} \\
\theta_2 &= 8.94 \Big[\Big\{\frac{\beta}{l}\Big\}^2 - \frac{3}{4}\Big\{\frac{\beta}{l}\Big\} \Big]
\end{aligned}
\tag{4.48}
$$

$$\theta_3 = 39.68\left[\left\{\frac{\beta}{l}\right\}^3 - \frac{3}{4}\left\{\frac{\beta}{l}\right\}^2 + \frac{2}{5}\left\{\frac{\beta}{l}\right\}\right]$$

$$\theta_4 = 168\left[\left\{\frac{\beta}{l}\right\}^4 - 1.815\left\{\frac{\beta}{l}\right\}^3 + 1.0714\left\{\frac{\beta}{l}\right\}^2 - 0.1785\left\{\frac{\beta}{l}\right\}\right] \tag{4.49}$$

and,

$$\pi_1 = 2.236\left\{\frac{\beta}{l}\right\}^2$$

$$\pi_2 = 15.87\left[\left\{\frac{\beta}{l}\right\}^3 - \frac{5}{6}\left\{\frac{\beta}{l}\right\}^2\right]$$

$$\pi_3 = 84\left[\left\{\frac{\beta}{l}\right\}^4 - \frac{3}{2}\left\{\frac{\beta}{l}\right\}^3 + 0.535\left\{\frac{\beta}{l}\right\}^2\right]$$

$$\pi_4 = 398\left[\left\{\frac{\beta}{l}\right\}^5 - 2.1\left\{\frac{\beta}{l}\right\}^4 + 1.4\left\{\frac{\beta}{l}\right\}^3 - 0.291\left\{\frac{\beta}{l}\right\}^2\right] \tag{4.50}$$

The above equations satisfy the geometric boundary conditions on the fixed edge $u = v = w = \nu_\alpha = 0$.

Using u, v and w and their derivatives from (4.47) in the Lagrangian expression (4.46) and minimizing it with respect to the arbitrary parameters a, b and c, we obtain the conventional eigenvalue problem.

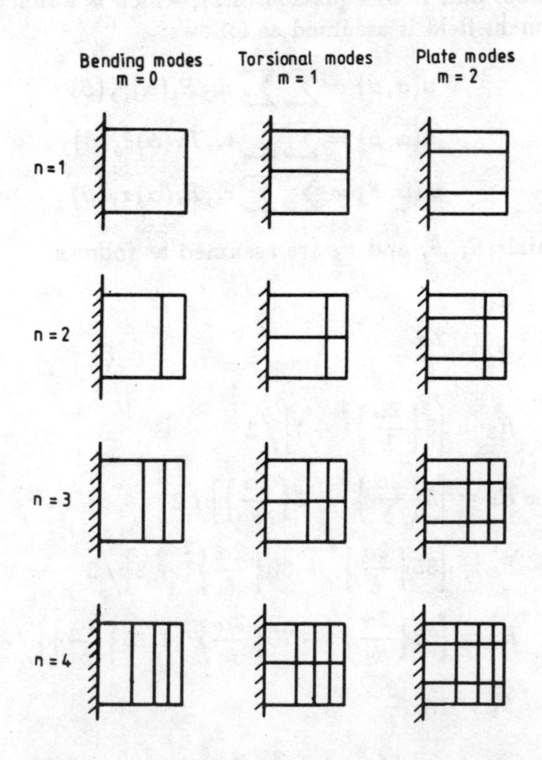

Fig. 4.9 Nodal patterns for different types of modes of vibration

Figure (4.9) gives the type of nodal patterns that are observed in cantilever plates, see Grinsted [42]. The bending modes correspond to $m = 0$, i.e., no nodal line in the direction of β. Thus $0/n$ represents bending modes and they are symmetrical about the line of axis $\alpha = 0$. Torsional modes have one nodal line along $\alpha = 0$ and they are represented by $1/n$. These are also antisymmetric modes. $m > 1$, represents plate modes with $2/n$, $4/n$ etc., as symmetric and $3/n$, $5/n$ etc., antisymmetric modes.

The symmetry and antisymmetry is governed by the function of α in the displacement field, $R_i(\alpha)$ given in equation (4.48). R_1, R_3 and R_5 are symmetric about $\alpha = 0$ and R_2, R_4 and R_6 are antisymmetric. A convergence study made by Rao and Gupta [43] indicated that a 36 term solution gives converged values upto 3/3 modes. The 36 terms correspond to 4 terms each of θ and π in (4.49) and (4.50) and 3 terms of R in (4.48), depending on symmetric or antisymmetric mode studies. This gives 12 terms each, $i = 1$, 3, 5 and $j = 1$ to 4 for symmetric and $i = 2$, 4, 6 and $j = 1$ to 4 for antisymmetric modes, for u, v and w in (4.47). Using the 36 term solution, the results obtained are discussed in the next section along with the results available in literature.

4.3 BENDING MODES

4.3.1 Stationary Blades

The following solutions are used in presenting the results.

1. Rawtani-Dokainish [1] — △ (figure 4.10)
   Flat plate triangular elements
2. Petricone and Sisto [26] — X
   Rayleigh-Ritz solution
3. MacBain [16] — □
   Quadrilateral plate element NASTRAN
4. Rao and Gupta [32] — ○
   Levy solution
5. Rao and Gupta [33] — ▽
   Inextensional solution
6. Ritz solution (section 4.2) — •

Fundamental Bending Mode (0/1)

For a beam type of blade, it is well known that the first bending frequency increases with pre-twist [44]. Rawtani [1] reported that for a plate with small aspect ratio, the frequency decreases with pre-twist, see figure 4.10. For higher aspect ratios $r = 3$, he predicted very slight increase in frequency with pre-twist like a beam. Petricone and Sisto [26] with a Rayleigh-Ritz solution gave a contradictory result, that frequency increased with pre-twist, even for small aspect ratio $r = 1$. MacBain [16] gave a finite element solution using NASTRAN and predicted similar behaviour like a beam type blade for $r = 2.33$. However, his experimental results for twisted blades were lower than the untwisted blades, though not following a regular pattern (figure 4.10).

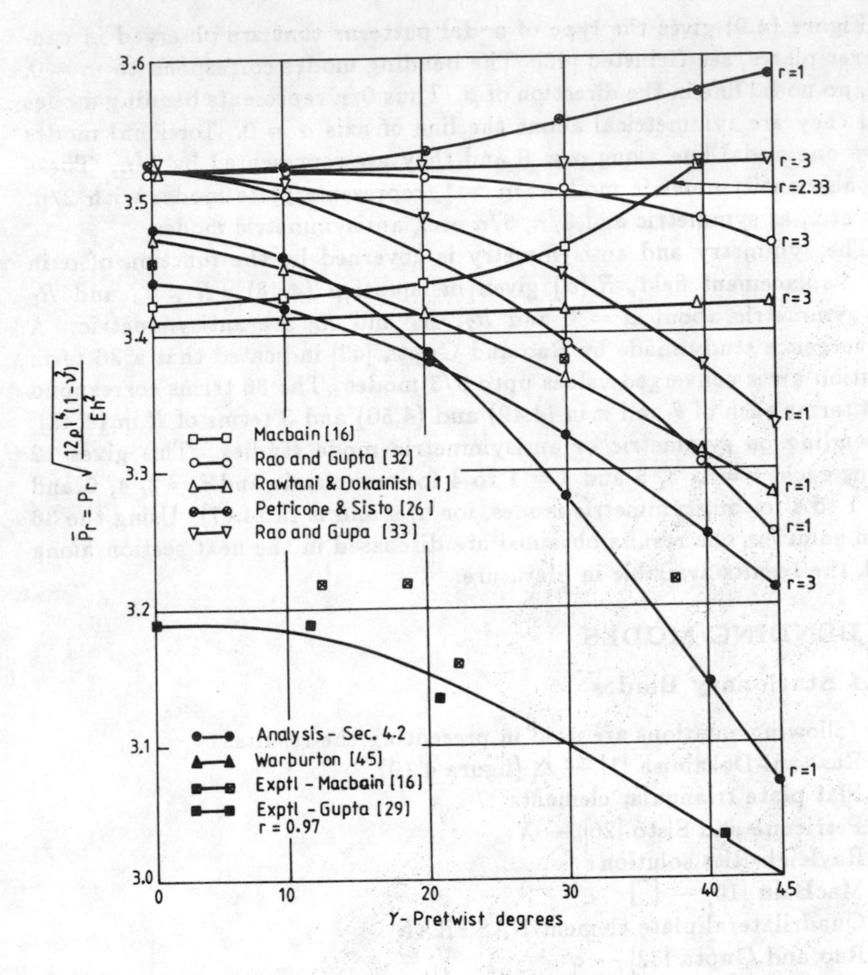

Fig. 4.10 Fundamental bending frequency of a pre-twisted plate

Rao and Gupta [32] used a Levy type of solution and predicted a decrease in frequency with pre-twist like Rawtani's results, though for $r = 3$, the frequency was decreasing still, with pre-twist. In the Levy analysis, when the terms proportional to h^3/R_{12} in (4.23) were also included, the frequency increased with pre-twist. This result is not included in figure 4.10, as all other solutions did not consider the inclusion of such terms. Further, Levy type of solution is not sufficiently accurate, as the mode shape is fixed with respect to α at the beginning itself.

An inextensional analysis [33] made by Rao and Gupta, predicted a decreasing trend for frequency with pre-twist for $r = 1$ and almost the same frequency for $r = 3$ as shown in figure 4.10. The present solution shows decrease in frequency with pre-twist for $r = 1$ and $r = 3$, and increase in frequency for $r = 6$ (not shown in figure 4.10) like in a beam. Holographic experimental results of Gupta [29] showed a consistent decrease in frequency with pre-twist as shown in figure 4.10.

The present analysis and that of Petricone and Sisto are essentially the same, though the strain displacement relations used were from different sources. Wan's [27] relations are slightly inaccurate in estimating bending strains χ_α, χ_β and $\tau_{\alpha\beta}$, thus the discrepancy between the two results following the energy method.

The inextensional theory [33] and Levy solution [32] show that the flat plate frequency is independent of aspect ratio r. The value of $\overline{p}$ for the flat plate is the same as that obtained by Warburton [45]. Martin [46] has also observed the same result. The series approach adopted by Classen and Thorne [47] has indicated a reduction in frequency for a flat plate as aspect ratio increases, see also Leissa [48]. The series approach adopted in the analysis of section 4.2 is similar to the work of Classen and Thorne, thus showing a reduction of frequency, as aspect ratio increases for a flat plate.

Judging from experimental results, it can be concluded that, for plates of small aspect ratio, the fundamental bending frequency decreases with pre-twist by about 6% in a pre-twist of 45°.

Second Bending Mode (0/2)

Figure 4.11 shows the variation of second bending frequency of a plate with pre-twist. The following observations are made:

- The frequency decreases with pre-twist and there is a predominant effect of pre-twist on the second mode, unlike the first mode.
- For low aspect ratio, pre-twist has more influence in decreasing the frequency.
- Levy's solution [32] seems to be very inaccurate.
- MacBain's finite element solution [16] for $r = 2.33$ lies in between the solutions for $r = 1$ and 3.
- Experimental values are lower than theoretical values, but predict similar trends.
- Rawtani's [1] solution predicts much lower values than MacBain's or Rao and Gupta solution.
- Petricone and Sisto's [26] solution predicted much higher values compared to the present results.
- Warburton's value [45] agrees with Levy solution [32] for flat plate.

Since Levy's solution is not accurate enough, Rawtani's solution has a lot less number of elements compared to MacBain and in view of the errors pointed out in Petricone and Sisto's solution, MacBain's [16] finite element results and the present Ritz results may be taken as reliable.

Third Bending Mode (0/3)

Figure 4.12 shows the available results. The following observations are made:

- MacBain's results for $r = 2.33$ show a decreasing trend of natural frequency with pre-twist and the experimental values are in fairly good agreement with this.

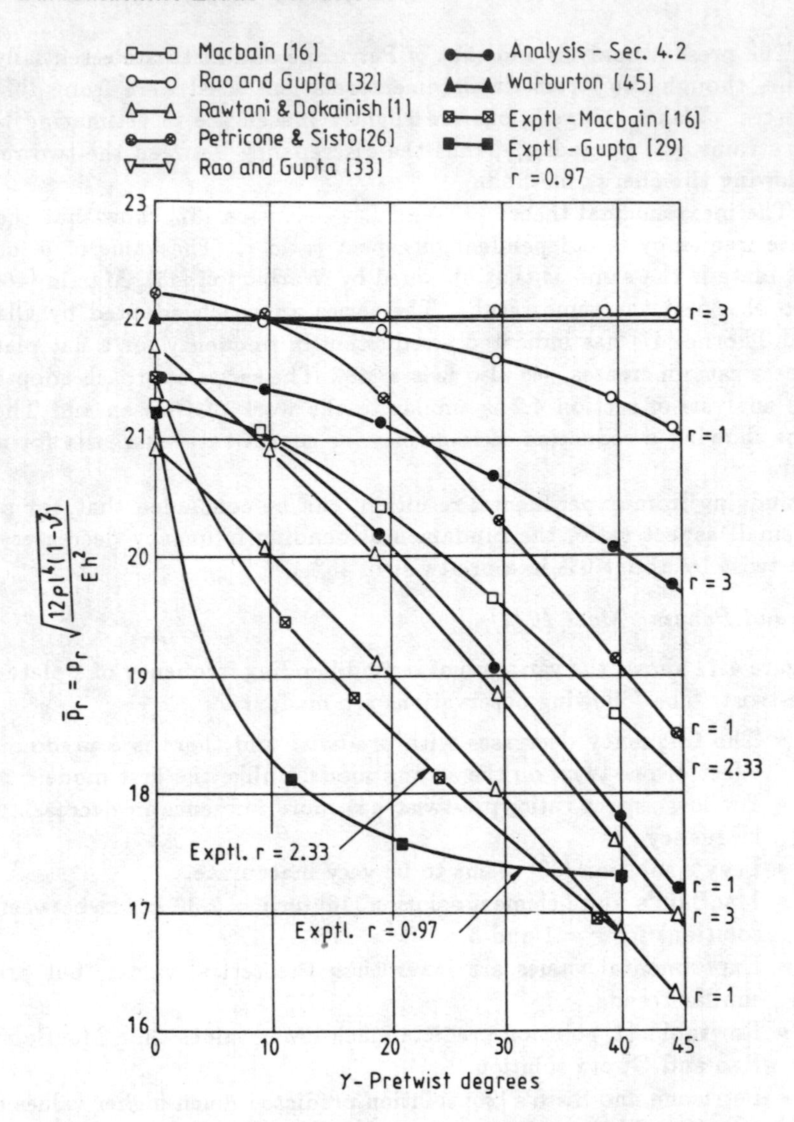

Fig. 4.11 Second bending frequency of a pre-twisted plate

- **Levy's [32] solution predicted decreasing trend of the natural frequency with pre-twist for $r = 1$ and 3.**
- **Rawtani's [1] results show a beam type trend for $r = 1$.**
- **Ritz solution shows a decrease in frequency for $r = 1$ and an increase in frequency like a beam for $r = 1.86$ and $r = 3$.**

It would have been interesting to have NASTRAN results for different values of r, to see when it predicts a reversal in the trend of plate behaviour to beam behaviour.

From figures 4.10 to 4.12, based on Ritz solution, it can be observed that transition of plate to beam type of behaviour occurs with r between 3 and 6 for 0/1 and 0/2 modes and between 1 and 1.86 for 0/3 modes.

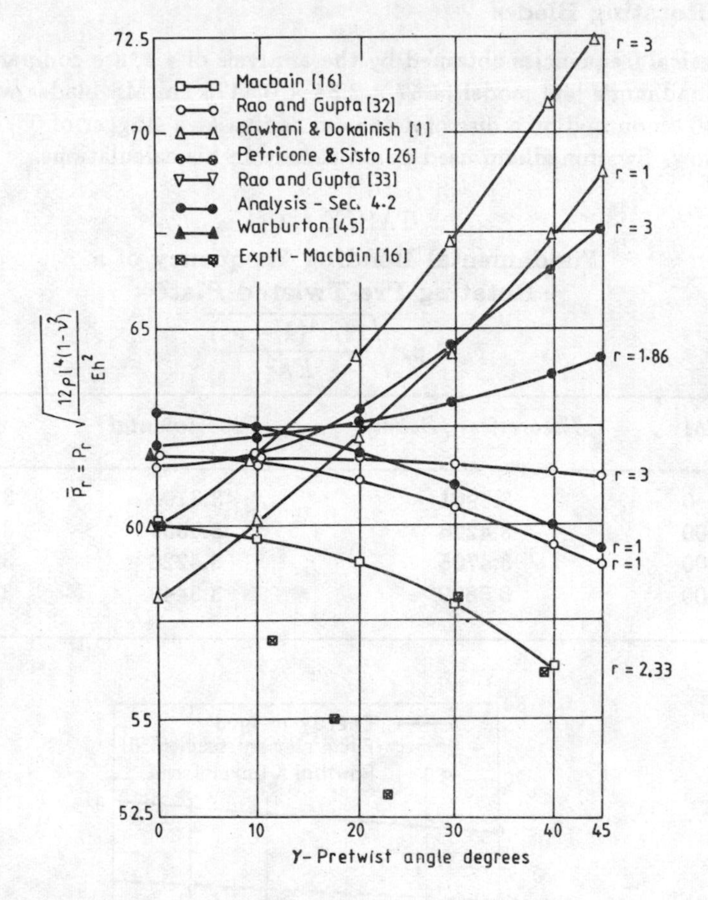

Fig. 4.12 Third bending frequency of a pre-twisted plate

Effect of Rotary Inertia

The effect of rotary inertia on the bending frequencies of a pretwisted cantilever plate is given in Table 4.1, along with experimental results of Gupta [29].

TABLE 4.1

Effect of Rotary Inertia on the Bending Modes of a Pre-Twisted Cantilever Plate [29]

$r = 0.97 \quad \gamma = 41°$

Mode	$\overline{p}$	$\overline{p}$ *Corrected*	$\overline{p}$ *Experimental*
0/1	3.1401	3.1373	3.033
0/2	17.7073	17.6494	17.263
0/3	60.4044	59.8862	—

4.3.2 Rotating Blades

Theoretical frequencies obtained by the analysis of 4.1, are compared with Swaminadham's [49] model (4.57 × 2.54 × 0.3175 cm MS blades with pre-twist 30° mounted on a disc of 4.445 cm radius at a stagger of 0°) in table 4.2 below. Swaminadham used beam theory in his calculations.

TABLE 4.2

Fundamental Bending Frequency of a Rotating Pre-Twisted Plate

$$\overline{p}_r = p_r\sqrt{\frac{12\rho l^4(1-\nu^2)}{Eh^2}}$$

RPM	*Theoretical (Beam)*	*Experimental*	*Ritz*
0	3.3961	3.3190	3.3178
2000	3.4226	3.3535	3.3262
6000	3.4705	3.4226	3.3931
10000	3.5820	3.5449	3.5225

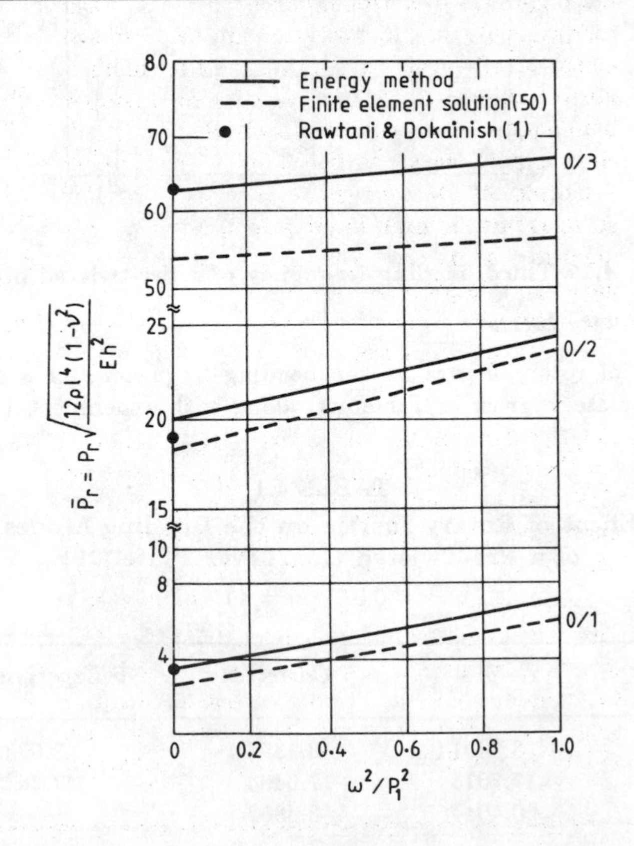

Table 4.2 shows that beam theory predicts higher natural frequencies compared to plate theory and the experimental values are closer to plate theory. Thomas and Mota Soares [50] used a super parametric element in their analysis of rotating pre-twisted shells. The data for the blade is as follows:

$$\frac{l}{b} = 2, \quad \frac{h}{b} = 0.0625, \quad \gamma = 30°, \quad \phi = 30°, \quad \frac{R}{b} = 4.$$

The results obtained by Ritz solution are given in figure 4.13, along with the results of Mota Soares [50]. The results predicted by finite element method are much lower than energy method solution. The frequency values obtained by Rawtani and Dokainish [1] for stationary blade are also shown in figure 4.13. In view of experimental results of Swaminadham [49] being more than the Ritz energy solution in Table 4.2, it may be concluded that the finite element method gave lower bound values. The trend of the relation of frequency with speed, is the same as predicted by finite element method or energy solution.

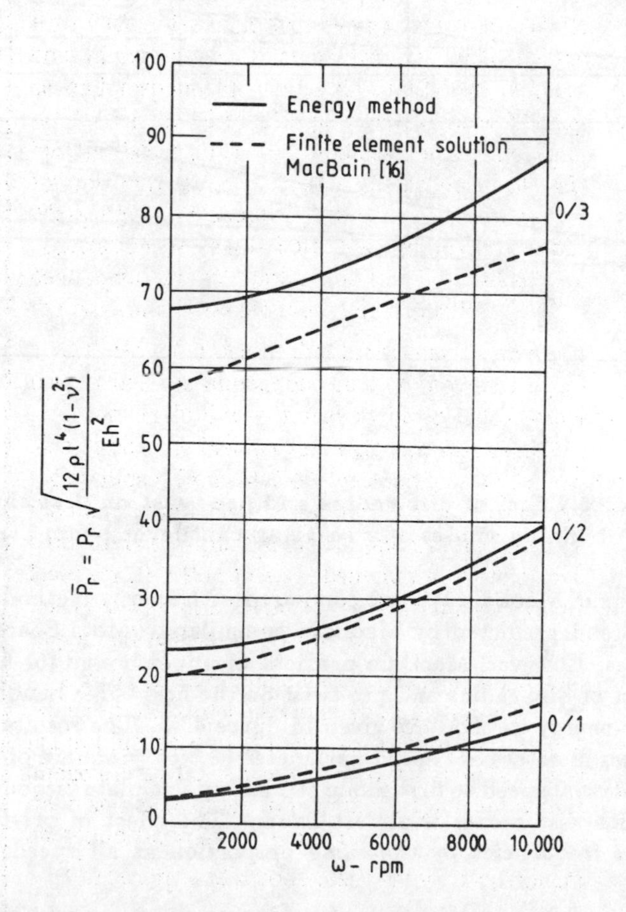

Fig. 4.14 Bending mode frequencies of a rotating plate

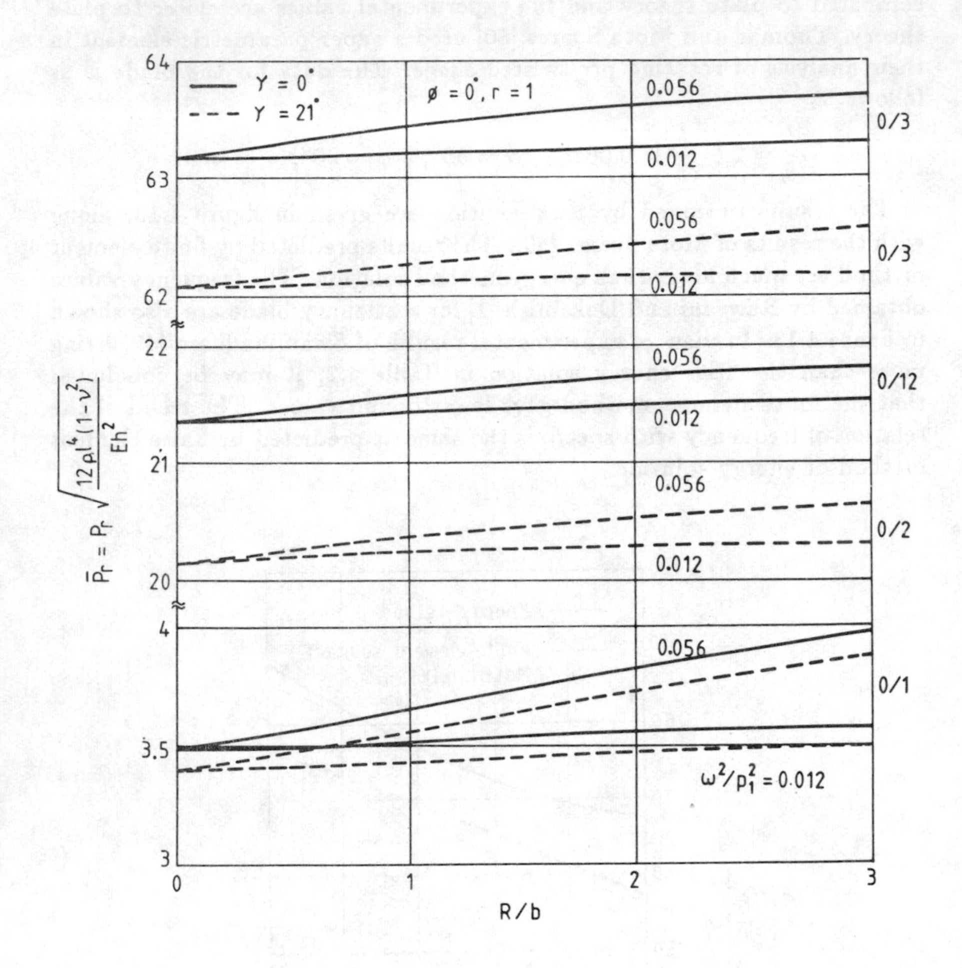

Fig. 4.15 Effect of disc radius and pre-twist on first three bending modes of a rotating cantilever beam

The results of MacBain [16] are compared with energy methods in figure 4.14. The trend predicted by MacBain is similar to Mota Soares [50] for higher modes. However, MacBain predicts an upper bound for first mode.

The effect of disc radius and pre-twist on the first three bending modes obtained by energy method are given in figure 4.15. The results are given for two values of rotational speed parameter (ω^2/p_1^2 = square of frequency ratio of rotational speed to first mode stationary flat plate frequency). The variation with disc radius is almost linear. The effect of pre-twist is to decrease the frequencies by the same proportion at all speeds and disc radii.

Figure 4.16 illustrates the effect of stagger angle on the first three bending modes for a pre-twisted plate. The percentage variation in the natural frequency of a pre-twisted rotating blade from 0–15,000 RPM, is given in Table 4.3. This shows that stagger angle has predominant influence in the first mode, for larger pre-twist angles and a larger disc radius and negligible influence on higher modes.

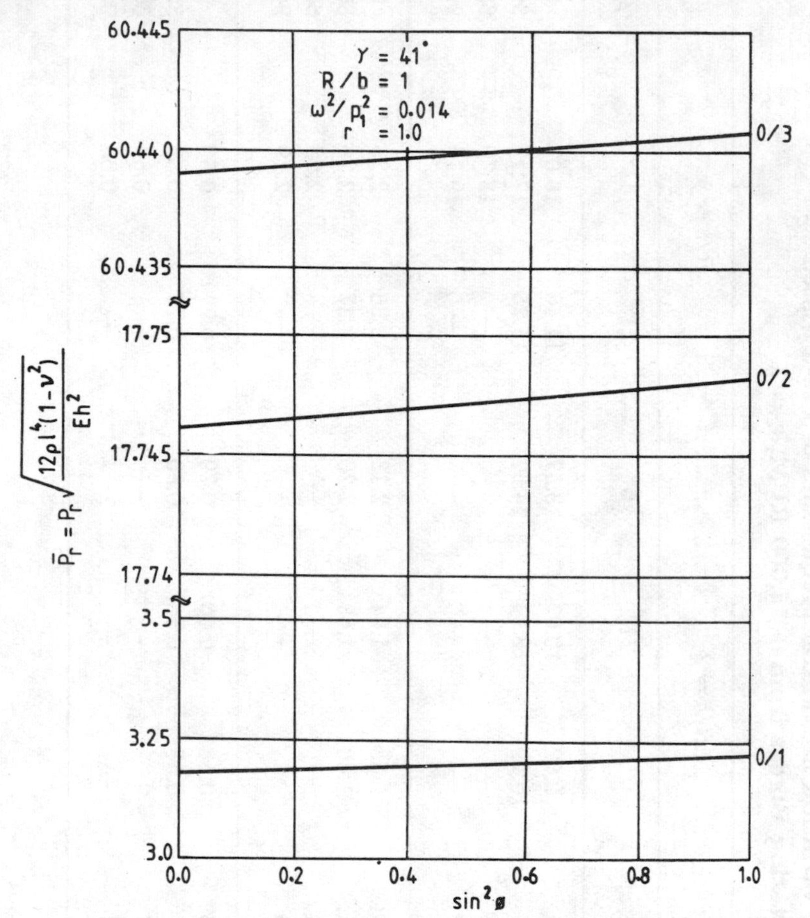

Fig. 4.16 Effect of stagger angle on the first three bending modes of a rotating pre-twisted plate

4.4 TORSION MODES

4.4.1 Stationary Blades

Effect of Aspect Ratio on Flat Plate Frequencies

Figure 4.17 gives flat plate natural frequencies as obtained by Rao and Gupta [31], compiled by Leissa [48] using Classen and Thorne's [47] results and by Warburton [45]. The flat plate results of Rao and Gupta are in good agreement with Leissa. Warburton assumed a displacement field to consist of beam function in both directions of plate and hence upper bound values. The plot in figure 4.17 shows that aspect ratios more than 4 have no effect on the nondimensional frequency and hence the behaviour will be like a beam for $r > 4$ in torsion modes.

TABLE 4.3

Percentage variation in Natural Frequencies of Pre-Twisted Rotating Blades from 0–15,000 RPM, $r = 1$

	$R/b = 1$			$R/b = 2$		$R/b = 3$			
Mode	0°	21°	41°	0°	21°	0°	21°	41°	$\leftarrow \gamma \quad \phi \downarrow$
	5.20	5.39	5.95	9.62	10.02	13.87	14.46	16.07	0°
0/1	5.89	6.25	7.15	10.29	10.84	14.52	15.25	17.17	30°
	—	—	8.84	—	—	—	—	18.70	60°
	—	—	9.32	—	—	—	—	19.16	90°
	0.98	1.01	1.01	1.58	1.64	2.18	2.26	2.25	0°
0/2	1.00	1.03	1.02	1.60	1.65	2.20	2.27	2.26	30°
	—	—	1.04	—	—	—	—	2.28	60°
	—	—	1.06	—	—	—	—	2.29	90°
	0.36	0.35	0.26	0.58	0.56	0.79	0.77	0.59	0°
0/3	0.36	0.35	0.27	0.58	0.56	0.80	0.77	0.59	30°
	—	—	0.27	—	—	—	—	0.60	60°
	—	—	0.27	—	—	—	—	0.60	90°

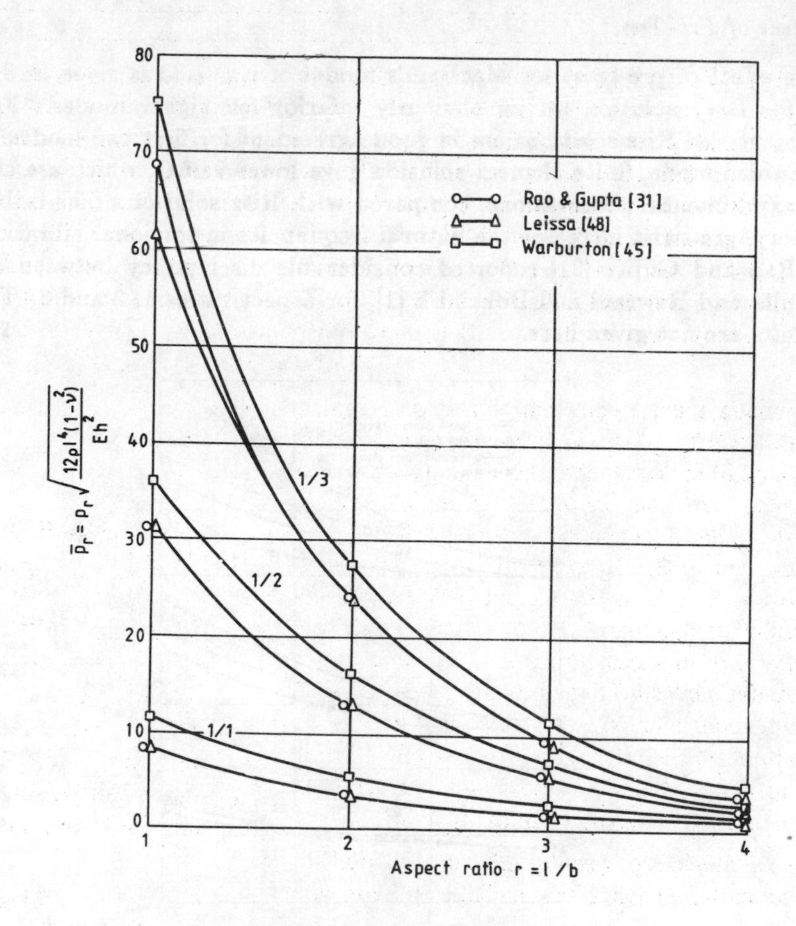

Fig. 4.17 Torsional frequencies of a flat plate for different aspect ratios

For aspect ratio 6, the plate analysis is compared with beam analysis of Rao [51], in Table 4.4 below.

TABLE 4.4
Ratio of Higher Torsional Mode Frequency with Fundamental Mode

Pre-Twist Degrees	*2nd Mode*		*3rd Mode*	
	Gupta [29]	*Rao [51]*	*Gupta [29]*	*Rao [51]*
0	3.04	3.05	5.21	5.12
30	3.04	3.07	5.21	5.20
60	3.03	3.07	5.16	5.25
90	3.02	3.07	5.12	5.16

Effect of Pre-Twist

The effect of pre-twist for MacBain's model of $r = 2.33$ is given in figure 4.18. Levy solution [31] is obviously inferior for higher modes. Finite element and Ritz solutions are in good agreement for first two modes. For the third mode, finite element solution gave lower values, which are closer to experimental observations, compared with Ritz solution. Like in beam theory, pre-twist increases the natural frequencies in torsional vibration.

Rao and Gupta [31] reported considerable discrepancy between their results and Rawtani and Dokainish [1], for aspect ratios 1, 2 and 3. These results are not given here.

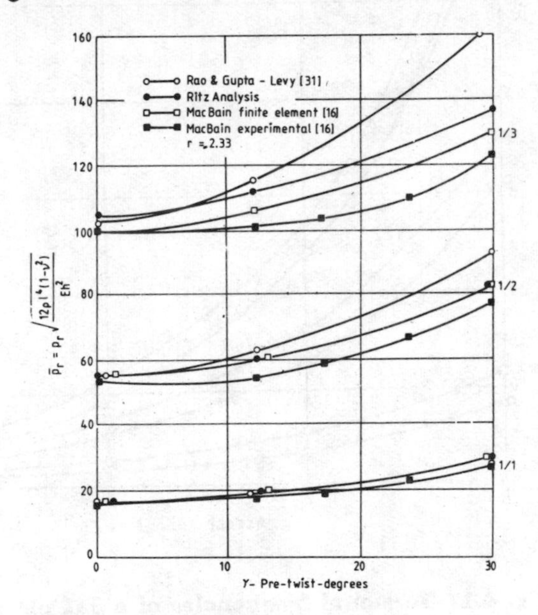

Fig. 4.18 Effect of pre-twist on torsional frequencies — MacBain's model

Effect of Rotary Inertia

Rotary inertia affects the torsional natural frequencies of plates, the results of Gupta [29] are given below in Table 4.5, for $r = 0.97$ and $\gamma = 41°$.

TABLE 4.5

Effect of Rotary Inertia on Torsional Modes of Pre-Twisted Plates

$r = 0.97 \quad \gamma = 41°$

Mode	$\bar{p}$	$\bar{p}$ *Corrected*	$\bar{p}$ *Experimental*
1/1	15.2972	15.2652	15.055
1/2	34.7125	34.4975	—
1/3	70.5457	69.6379	—

4.4.2 Rotating Blades

In beam theory, torsional modes are not affected by centrifugal forces. In plate theory, however, centrifugal forces have an influence on torsional modes, though not to the same magnitude as in bending modes.

Figure 4.19 shows the theoretical frequencies obtained for a blade of $r = 2.33$, $R/b = 1$, and $\phi = 0°$, with pre-twist $\gamma = 0°$ and $12°$. All the frequencies are influenced by rotational speed and increase as the speed is increased. Figure 4.20 shows the effect of disc radius for a pre-twisted blade. The effect of stagger angle is also given in figure 4.20, which is predominant only for high rotational speeds. The relative increase in natural frequencies of torsional modes is somewhat less compared to increase in bending modes.

4.5 PLATE MODES

4.5.1 Stationary Blades

Symmetric Modes

Figure 4.21 gives the variation of $2/n$ symmetric modes with pre-twist for MacBain's model [16]. Results of Ritz solution and Levy solution [29] are also given. Levy solution is too poor, as the number of terms used are very inadequate and the mode shape assumed is free-free beam function in α

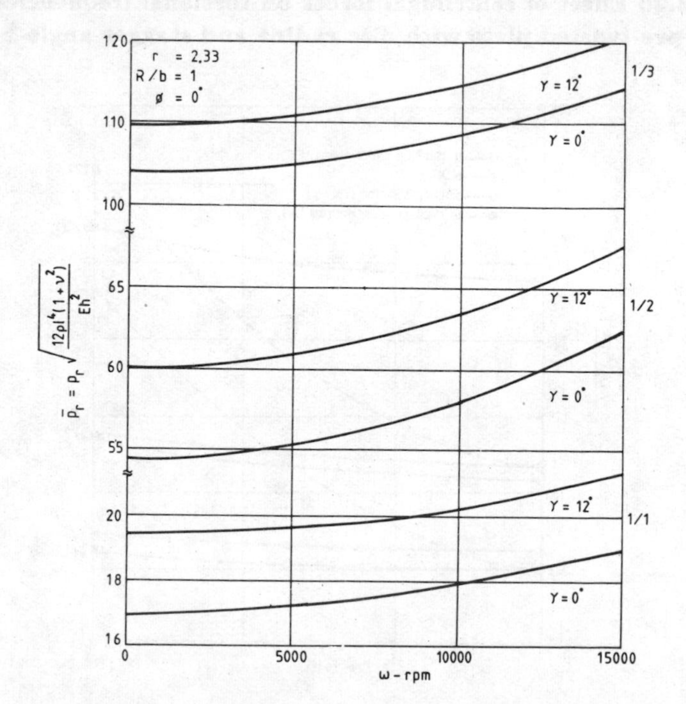

Fig. 4.19 Effect of centrifugal forces on torsional frequencies of pre-twisted cantilever plates

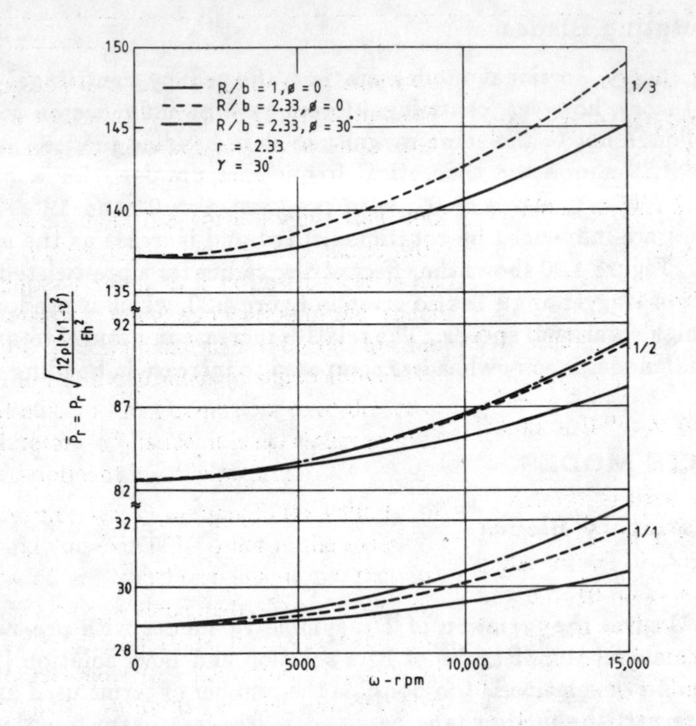

Fig. 4.20 Effect of centrifugal forces on torsional frequencies of pre-twisted plate with disc radius and stagger angle

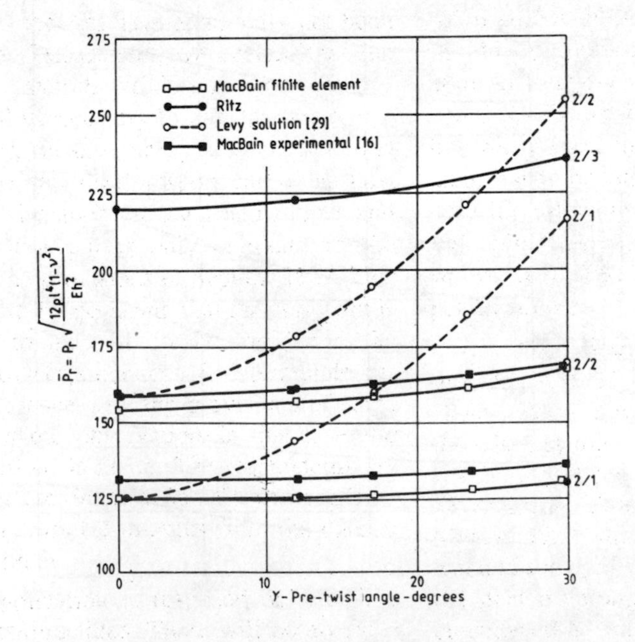

Fig. 4.21 Symmetric plate modes $2/n$ for a pre-twisted cantilever plate

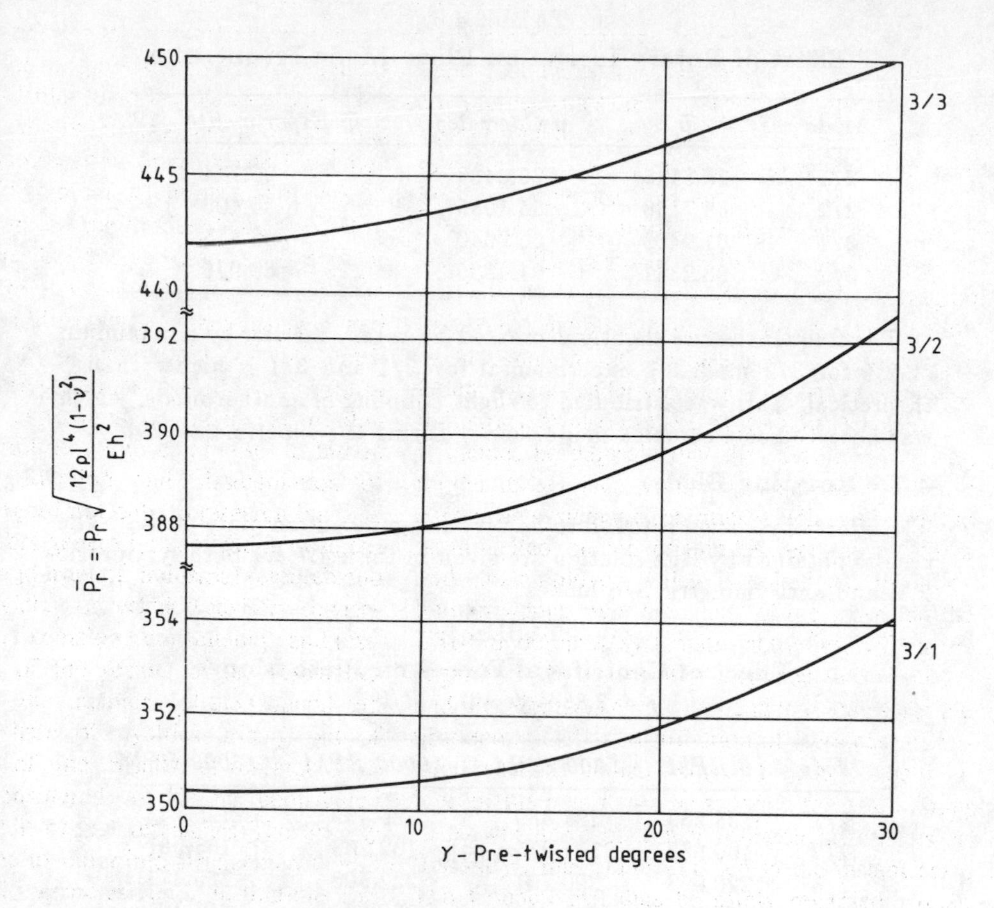

Fig. 4.22 Anti-symmetric plate modes $3/n$ for a pre-twisted cantilever plate

direction. Ritz solution and finite element solution are in good agreement for 2/1 mode. Finite element solution predicts slightly low values for 2/2 mode. Experimental results are higher than theoretical values for 2/1 mode and match well with Ritz solution for 2/2 mode. 2/3 mode obtianed by Ritz solution is also given. The increase in frequency value is between 5–10% for the three modes shown.

Asymmetric Modes

Figure 4.22 gives the variation of $3/n$ antisymmetric modes with pre-twist for MacBain's model, obtained by Ritz solution. The influence of pre-twist on these modes is negligible as can be observed from the figure.

Effect of Rotary Inertia

The effect of rotary inertia for $r = 1$, on the plate modes is given below in Table 4.6.

TABLE 4.6

Effect of Rotary Inertia on Plate Mode Frequencies

Mode	$\overline{p}$	$\overline{p}$ *Corrected*	$\overline{p}$ *Experimental [29]*
2/1	25.8145	25.6428	26.744
2/2	55.7158	55.1055	54.406
3/1	61.9409	60.9640	63.815
3/2	93.2351	91.4839	89.975

The frequencies are slightly decreased by rotary inertia, by a maximum of 2% for 3/2 mode. $\overline{p}$ experimental for 2/1 and 3/1 is higher than $\overline{p}$ theoretical. This was attributed to slight coupling of another mode, which was observed as a complex mode shape, during the experiments [29].

4.5.2 Rotating Blades

The effect of rotation is found to be very insignificant on plate modes. The results obtained by Ritz solution are given in Table 4.7, for both symmetric $2/n$ and antisymmetric $3/n$ modes.

TABLE 4.7

Effect of Centrifugal Forces on Plate Modes

$r = 2.33, \quad \phi = 0, \quad R/b = 1$

Mode	*0 RPM*	*5000 RPM*	*10000 RPM*	*15000 RPM*
2/1	125.289	125.389	125.489	125.742
2/2	160.858	161.211	162.263	163.987
2/3	220.098	220.701	222.500	225.471
3/1	350.346	350.426	350.672	351.101
3/2	387.619	387.779	388.256	389.056
3/3	442.057	442.359	443.264	444.771

The effect of disc radius is also found to be insignificant, the values of which are given in Table 4.8. Table 4.9 gives the effect of change in stagger angle, which is also found to have insignificant influence on plate modes.

TABLE 4.8

Effect of Disc Radius on Plate Mode Frequencies

$r = 2.33, \quad \gamma = 30°, \quad \phi = 0°, \quad \omega = 10,000$ RPM

Mode	*R/b = 1*	*R/b = 2.33*
2/1	130.293	130.459
2/2	169.748	170.463
2/3	237.510	238.643
3/1	354.847	355.183
3/2	393.472	393.837
3/3	451.572	452.181

TABLE 4.9
Effect of Stagger Angle on Plate Mode Frequencies
$r = 2.33, \quad \xi = 30°, \quad R/b = 2.33, \quad \omega = 10,000$ RPM

Mode	$\phi = 0$	$\phi = 30^o$
2/1	130.459	130.518
2/2	170.463	170.508
2/3	238.643	238.673
3/1	355.183	355.206
3/2	393.837	393.859
3/3	452.181	452.198

Thus, the effects of rotation can be neglected on plate mode frequencies.

4.6 FINITE ELEMENT METHOD FOR SHELLS

The analysis of shells with an arbitrary defined shape presents intractable analytical problems. An enormous amount of effort has been devoted to the development of finite elements for shell analysis. In recent years, a number of finite elements have been developed and their application spans over a wide range of geometric forms, thicknesses and severities of shell deformation.

Shell finite elements are often derived from governing equations based on classical shell theory; these are referred to as classical shell elements. Alternatively, one can obtain shell elements by modifying a continuum element to comply with shell assumptions without resorting to shell theory. These are known as degenerate shell elements.

In idealization of the shell by finite elements, a geometrical simplification of replacing the curved shell by an assembly of flat elements is most frequently used. Though many doubly curved shell elements have now been formulated, these are still less widely used than flat elements. Utku's [52] doubly curved triangular shell element, Mohr's [53] doubly curved isoparametric triangular shell element, Iron's [54] doubly curved quadrilateral semiloof element are a few examples of recently developed shell elements. However, these elements in general suffer from shortcomings inherent to many shell elements to which the element application is limited, usually within the scope of an underlying shell theory. An alternative means of obtaining shell elements is to use large curvilinear elements for three dimensional analysis based on isoparametric formulation. Elements of the type shown in figure 4.25 have been used with success for three dimensional analysis purposes as shown by Ergatoudis et al [55].

With a straightforward use of three dimensional concept, however, certain difficulties are encountered. In the first place, the retention of three degrees of freedom at each node leads to large stiffness coefficients for relative displacements along an edge corresponding to the shell thickness. This presents numerical problems and inevitably leads to ill-conditioned equations when shell thickness becomes small compared with the other dimensions of the elements.

The second factor is that of economy. The use of several nodes across shell thickness, ignores the well known fact that even for thick shells the normals to the middle surface remain practically straight after deformation. Thus an unnecessarily high number of degrees of freedom has to be carried, involving penalties of computer time.

Ahmad et al [56] presented a specialized formulation for shell problems using degenerate versions of general three dimensional elements of figure 4.23. In the degeneration process, two assumptions are made. Firstly, the original normals to the middle surface are taken as inextensible and straight and the strain energy corresponding to the stresses perpendicular to the middle surface are ignored. This means that elastic modulus in the normal direction is taken as zero.

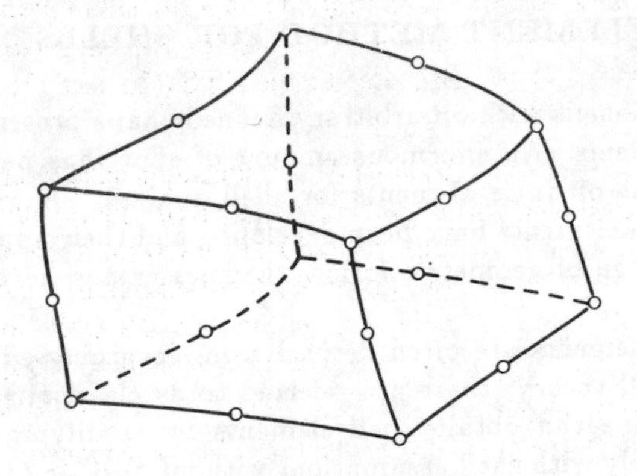

(a) Parabolic element

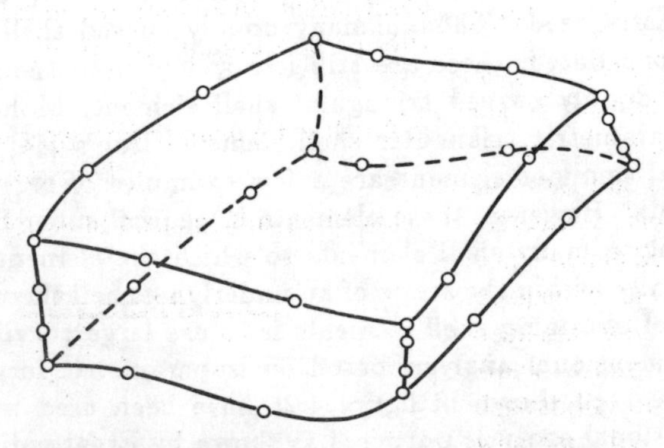

(b) Cubic element

Fig. 4.23 Three dimensional hexahedral elements parabolic and cubic type

For thick shell situations, shear deformation is significant and this is achieved by omitting the assumption that after deformation the normals remain normal to the deformed middle surface. The above assumptions are compatible with thin shell theory too. This new element is not strictly isoparametric but termed as superparametric element, because the definition of the coordinates is more general than that of the displacements.

Gupta, Ramakrishnan and Rao [57] developed a finite element program to determine the natural frequencies of a hydraulic runner blade considering the fluid structure interaction effects. The element formulated below follows closely their formulation.

4.6.1 Formulation of the Element

The basic shape of the element chosen is typical thick shell element as shown in Fig. 4.24(a). The external faces of the elements are curved, while

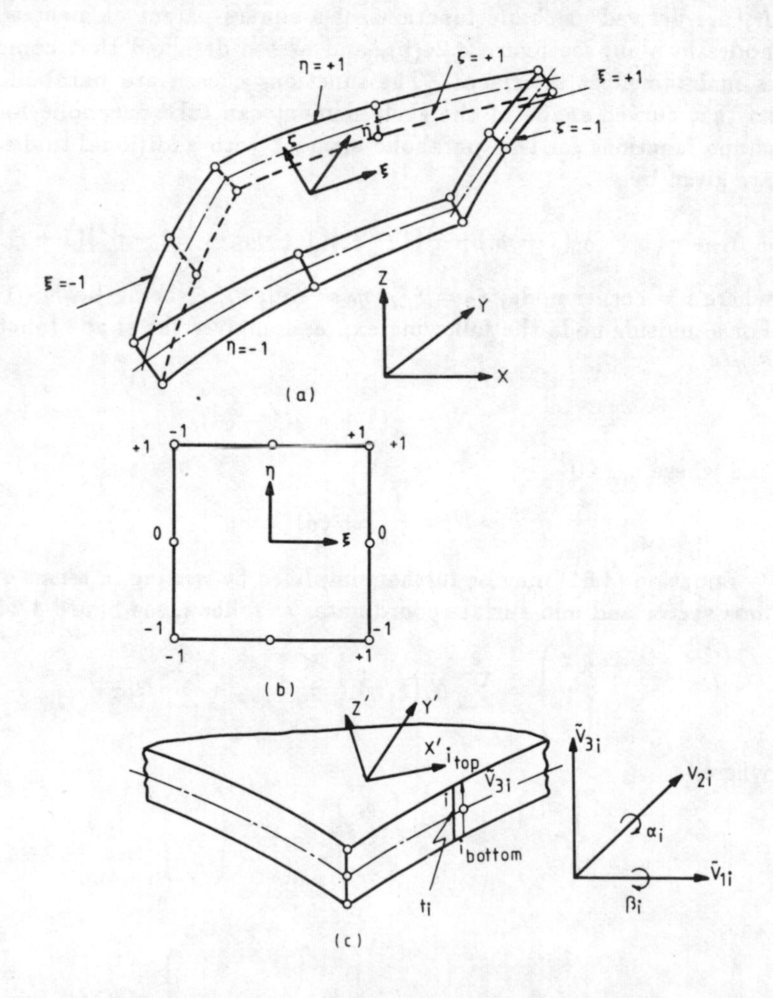

Fig. 4.24 Ahmad's 8-noded superparametric element

the sections across the thickness are generated by straight lines. The geometry of the element is described by a pair of points, i_{top}, i_{bottom} each with given cartesian coordinates. Consider ξ, η be two curvilinear coordinates in the middle plane of the shell and ς, a linear coordinate in the thickness direction. If, further, it is assumed that ξ, η, ς vary between -1 and $+1$ on the respective faces of the element, we can write a relationship between the cartesian coordinates of any point of the shell and the curvilinear coordinates in the form

$$\begin{Bmatrix} x \\ y \\ z \end{Bmatrix} = \sum_{i=1}^{8} N_i(\xi,\eta)\frac{1+\varsigma}{2}\begin{Bmatrix} x_i \\ y_i \\ z_i \end{Bmatrix}_{top} + \sum_{i=1}^{8} N_i(\xi,\eta)\frac{1-\varsigma}{2}\begin{Bmatrix} x_i \\ y_i \\ z_i \end{Bmatrix}_{bottom} \tag{4.51}$$

Here $N_i(\xi,\eta)$ is a function taking a value of unity at node i amd zero at all other nodes. For these curved space elements to fit into each other, N_i are derived as shape functions of a square parent element with eight nodes in plan, see figure 4.24(b), and are so designed that compatibility is maintained at interfaces. The functions chosen are parabolic in ξ, η so that curved shape of the shell element can take parabolic form. The shape functions for this parabolic element with additional midside nodes are given by

$$N_i = \frac{1}{4}(1+\xi_0)(1+\eta_0) - \frac{1}{4}(1-\xi^2)(1+\eta_0) - \frac{1}{4}(1-\eta^2)(1+\xi_0) \tag{4.52}$$

where i = corner node, $\xi_0 = \xi\xi_i$; $\eta_0 = \eta\eta_i$; η_i, ξ_i being both $+1$ and -1. For a midside node the following expression gives the shape function with $\xi_i = 0$.

$$N_i = \frac{1}{2}(1+\eta_0)(1-\xi^2) \tag{4.53}$$

and where $\eta_i = 0$

$$N_i = \frac{1}{2}(1+\xi_0)(1-\eta^2) \tag{4.54}$$

Equation (4.51) may be further simplified by writing in terms of a thickness vector and mid-surface coordinates as follows, see figure 4.24(c),

$$\begin{Bmatrix} x \\ y \\ z \end{Bmatrix} = \sum_{i=1}^{8} N_i(\xi,\eta)\begin{Bmatrix} x_i \\ y_i \\ z_i \end{Bmatrix}_{mid} + \sum_{i=1}^{8} N_i\frac{\varsigma}{2}\tilde{V}_{3i} \tag{4.55}$$

where,

$$\begin{Bmatrix} x_i \\ y_i \\ z_i \end{Bmatrix}_{mid} = \frac{1}{2}\begin{Bmatrix} x_i \\ y_i \\ z_i \end{Bmatrix}_{top} + \frac{1}{2}\begin{Bmatrix} x_i \\ y_i \\ z_i \end{Bmatrix}_{bottom}$$

and,

$$\tilde{V}_{3i} = \begin{Bmatrix} x_i \\ y_i \\ z_i \end{Bmatrix}_{top} - \begin{Bmatrix} x_i \\ y_i \\ z_i \end{Bmatrix}_{bottom} = \begin{Bmatrix} \Delta x_i \\ \Delta y_i \\ \Delta z_i \end{Bmatrix}$$

4.6.2 Displacement Field

As the strains in the direction normal to the mid-surface are assumed to be negligible, the displacement throughout the element will be taken to be uniquely defined by the three Cartesian components of the mid-surface node displacement i and two rotations of the nodal vector $\widetilde{V}_{3i}$ about orthogonal directions normal to it.

If two such unit orthogonal vectors are given by $\widetilde{V}_{1i}$, $\widetilde{V}_{2i}$, then we can write for the displacements in global axes x, y, z as

$$\begin{Bmatrix} u \\ v \\ w \end{Bmatrix} = \sum_{i=1}^{8} N_i \begin{Bmatrix} u_i \\ v_i \\ w_i \end{Bmatrix} + \sum_{i=1}^{8} N_i \frac{\varsigma t_i}{2} [\widetilde{V}_{1i} - \widetilde{V}_{2i}] \begin{Bmatrix} \alpha_i \\ \beta_i \end{Bmatrix} \tag{4.56}$$

where,

$$\widetilde{V}_{1i} = \begin{Bmatrix} V_{1i,x} \\ V_{1i,y} \\ V_{1i,z} \end{Bmatrix} \quad ; \quad \widetilde{V}_{2i} = \begin{Bmatrix} V_{2i,x} \\ V_{2i,y} \\ V_{2i,z} \end{Bmatrix}$$

u_i, v_i, w_i are the displacements at the mid-surface nodes and α, β are the two rotations about directions $\widetilde{V}_{2i}$, $\widetilde{V}_{1i}$ respectively, thus giving five degrees of freedom per node.

In matrix notation, Eqn. (4.56) can be written as

$$\{U\} = [N]\{a\}$$

In the above $\{a\}$ is a column vector with 40 elements and $[N]$ is a matrix 3×40 in size. Elements of $[N]$ can be easily obtained by expanding Eqn. (4.56).

An infinite number of vector directions normal to a given direction can be generated. Here a consistent definition giving unique axes is adopted. If i, for instance, is the unit vector along x-axis,

$$\widetilde{V}_{1i} = \hat{\imath} \times \widetilde{V}_{3i}$$

makes the vector $\widetilde{V}_{1i}$ perpendicular to the plane defined by direction $\widetilde{V}_{3i}$ and the x-axis.

As $\widetilde{V}_{2i}$ has to be orthogonal to both $\widetilde{V}_{3i}$ and $\widetilde{V}_{1i}$, we have

$$\widetilde{V}_{2i} = \widetilde{V}_{3i} \times \widetilde{V}_{1i}$$

To obtain unit vectors in these directions, $\widetilde{V}_{1i}$, $\widetilde{V}_{2i}$, $\widetilde{V}_{3i}$ are simply divided by their scalar lengths giving unit vectors $\widehat{V}_{1i}$, $\widehat{V}_{2i}$, $\widehat{V}_{3i}$.

The element coordinate definition is given by the relation Eqn. (4.51), which has more degrees of freedom than the definition of the displacements. The element is therefore the superparametric kind.

4.6.3 Strains and Stresses

Once the displacement field for the element is known, to derive the element properties, the essential strains and stresses are to be defined. To account for the shell assumptions, the components in directions of orthogonal axes related to the surface ς = constant are essential. Thus, if at any point on this surface we erect a normal z' with two other orthogonal axes x' and y' tangent to it, see Fig. 4.26(c), the strain components are given by

$$[\epsilon'] = \begin{Bmatrix} \epsilon_{x'} \\ \epsilon_{y'} \\ \gamma_{x'y'} \\ \gamma_{x'z'} \\ \gamma_{y'z'} \end{Bmatrix} = \begin{Bmatrix} u'_{,x'} \\ v'_{,y'} \\ u'_{,y'} + v'_{,x'} \\ w'_{,x'} + u'_{,z'} \\ w'_{,y'} + v'_{,z'} \end{Bmatrix} \tag{4.57}$$

or,

$$\{\epsilon'\} = [L]\{U'\}$$

where $[L]$ is linear operator matrix given by

$$[L] = \begin{bmatrix} ,x' & 0 & 0 \\ 0 & ,y' & 0 \\ ,y' & ,x' & 0 \\ ,z' & 0 & ,x' \\ 0 & ,z' & ,y' \end{bmatrix} \quad ; \quad \{U'\} = \begin{Bmatrix} u' \\ v' \\ w' \end{Bmatrix} \tag{4.57a}$$

From Eqn. (4.56), we can write

$$\{U'\} = [N]\{a'\} \tag{4.57b}$$

and hence,

$$\{\epsilon'\} = [B]\{a'\}$$

where,

$$[B] = [L][N] \tag{4.57c}$$

In the above, the strain in direction z' is neglected, to be consistent with the usual shell assumption. However, it is to be noted that orthogonal axes x', y', z' do not coincide with curvilinear coordinates ξ, η, ς although x', y' are in the ξ, η plane.

The stresses corresponding to these strains are related by the matrix relation

$$\{\sigma'\} = [D]\{\epsilon'\} \tag{4.58}$$

where $[D]$ is the elasticity matrix and for an isotropic material given as follows:

$$[D] = \frac{E}{1-\nu^2} \begin{bmatrix} 1 & 0 & 0 & 0 & 0 \\ & 1 & 0 & 0 & 0 \\ & & \frac{1-\nu}{2} & 0 & 0 \\ & \text{sym} & & \frac{1-\nu}{2k} & 0 \\ & & & & \frac{1-\nu}{2k} \end{bmatrix} \tag{4.59}$$

in which E and ν are Young's modulus and Poisson's ratio respectively. The factor k included in the last two shear terms is taken as 1.2 and its purpose is to improve the shear displacement approximation.

4.6.4 Element Properties

The element property matrices involve integrals over the volume of the element and are generally of the form

$$\int [S]\, dx\, dy\, dz \tag{4.60}$$

where matrix $[S]$ is a function of the global coordinates x, y, z. For the stiffness matrix

$$[S] = [B]^T[D][B] \tag{4.61}$$

with the strain given the usual definition

$$\{\epsilon\} = [B]\{a\} \quad ; \quad \{a\} = \begin{Bmatrix} a_1^e \\ a_2^e \\ \vdots \\ a_8^e \end{Bmatrix} \tag{4.62}$$

We have $[B]$ defined in terms of the displacement derivatives with respect to the local Cartesian coordinates x', y', z' as given in Eqn. (4.57) and $\{a\}^e$ is the displacement field. Two sets of transformations are therefore necessary before the element can be integrated with respect to the curvilinear coordinates ξ, η, ς.

Firstly, the transformation from the local Cartesian coordinates x', y', z' to the global coordinates x, y, z is required and the second transformation from global coordinates x, y, z to the curvilinear coordinates ξ, η, ς is done so that element properties can be defined in terms of ξ, η, ς.

Eqn. (4.56) relates the global displacements u, v, w to the curvilinear coordinates, the derivatives of these displacements with respect to the curvilinear coordinates ξ, η, ς is given as follows by the Jacobian matrix.

$$\begin{bmatrix} u_{,\xi} & v_{,\xi} & w_{,\xi} \\ u_{,\eta} & v_{,\eta} & w_{,\eta} \\ u_{,\varsigma} & v_{,\varsigma} & w_{,\varsigma} \end{bmatrix} = \begin{bmatrix} x_{,\xi} & y_{,\xi} & z_{,\xi} \\ x_{,\eta} & y_{,\eta} & z_{,\eta} \\ x_{,\varsigma} & y_{,\varsigma} & z_{,\varsigma} \end{bmatrix} \begin{bmatrix} u_{,x} & v_{,x} & w_{,x} \\ u_{,y} & v_{,y} & w_{,y} \\ u_{,z} & v_{,z} & w_{,z} \end{bmatrix} \tag{4.63}$$

or,

$$[U_d]_{\text{curvilinear}} = [J][U_d]_{\text{global}}$$

Hence, the derivatives of the displacements in global coordinates can be obtained as

$$[U_d]_{\text{global}} = [J]^{-1}[U_d]_{\text{curvilinear}} \tag{4.64}$$

The displacement field is related to the curvilinear coordinates by Eqn. (4.56) and therefore we can obtain the terms in the matrix for the derivatives of the displacements in curvilinear coordinates, i.e.,

$$u_{,\xi} = \sum N_{i,\xi} u_i + \sum N_{i,\xi} \frac{\varsigma t_i}{2} V_{1i,x} \alpha_i - \sum N_{i,\xi} \frac{\varsigma t_i}{2} V_{2i,x} \beta_i \quad \ldots \text{etc.} \tag{4.65}$$

Also, the components of the Jacobian matrix in curvilinear coordinates can be written, using Eqn. (4.51), i.e.,

$$x_{,\xi} = \sum N_{i,\xi} x_i + \sum N_{i,\xi} \frac{\varsigma}{2} V_{3i,x}$$
$$x_{,\eta} = \sum N_{i,\eta} x_i + \sum N_{i,\eta} \frac{\varsigma}{2} V_{3i,x} \tag{4.66}$$
$$x_{,\varsigma} = \frac{1}{2} \sum N_i V_{3i} \quad \ldots \text{etc.,}$$

Using relations of Eqn. (4.65) and (4.66) in Eqn. (4.64), the global derivatives of displacements for every set of curvilinear coordinates can be obtained numerically.

A further transformation from global coordinates x, y, z to local Cartesian coordinates x', y', z' will allow the strains to be evaluated. To establish the directions of local orthogonal axes x', y', z' we first obtain a vector normal to the surface $\varsigma =$ constant, by taking vector product of any two vectors tangent to the surface. Thus

$$\widetilde{V}_{z'} = \begin{Bmatrix} x_{,\xi} \\ y_{,\xi} \\ z_{,\xi} \end{Bmatrix} \times \begin{Bmatrix} x_{,\eta} \\ y_{,\eta} \\ z_{,\eta} \end{Bmatrix} \tag{4.67}$$

The other two vectors $\widetilde{V}_{x'}$, $\widetilde{V}_{y'}$, along directions x' and y' can also be obtained by the process given above and reducing these to unit magnitudes, we construct a matrix of unit vectors in x', y', z' directions

$$[\theta] = [\hat{V}_{x'}\ \hat{V}_{y'}\ \hat{V}_{z'}] \tag{4.68}$$

The global derivatives of the displacements u, v and w are now transformed to the local derivatives of the local orthogonal displacements by a standard operation.

$$\begin{bmatrix} u'_{,x'} & v'_{,x'} & w'_{,x'} \\ u'_{,y'} & v'_{,y'} & w'_{,y'} \\ u'_{,z'} & v'_{,z'} & w'_{,z'} \end{bmatrix} = [\theta]^T \begin{bmatrix} u_{,x} & v_{,x} & w_{,x} \\ u_{,y} & v_{,y} & w_{,y} \\ u_{,z} & v_{,z} & w_{,z} \end{bmatrix} \tag{4.69}$$

In Eqn. (4.69) $[\theta]$, $[\theta]^T$ are known at any point on the surface of the element and also the displacement derivatives in the global coordinates x, y, z are known from Eqn. (4.64). Hence the displacement derivatives can be obtained explicitly at any ξ, η, ς in the element. From these, the components of the strains can be found explicitly using relation Eqn. (4.57) and in matrix notation may be written as

$$\{\epsilon\} = [B] \begin{Bmatrix} a_1 \\ a_2 \\ \vdots \\ a_i \\ \vdots \\ a_8 \end{Bmatrix}^e \quad ; \quad \{a_i\} = \begin{Bmatrix} u \\ v \\ w \\ \alpha \\ \beta \end{Bmatrix}_i \tag{4.70}$$

The infinitesimal volume is given in terms of curvilinear coordinates as

$$dx\, dy\, dz = \det[J]\, d\xi\, d\eta\, d\varsigma \tag{4.71}$$

and from these two relations, the stiffness matrix is obtained as

$$[K]^e = \int_{-1}^{1}\int_{-1}^{1}\int_{-1}^{1} [B]^T[D][B]\det[J]\,d\xi\,d\eta\,d\varsigma \tag{4.72}$$

The mass matrix of the element is given as

$$[M]^e = \int_{-1}^{1}\int_{-1}^{1}\int_{-1}^{1} [N]^T\rho[N]det[J]\,d\xi\,d\eta\,d\varsigma \tag{4.73}$$

Individual mass matrices for all the finite elements are evaluated numerically using Gaussian quadrative weighted coefficients. Similarly other matrices for the element properties can be evaluated using the procedure outlined above and obtain matrices for the entire set of finite elements discretizing the continuum.

4.6.5 Equations of Motion and Solution

The principle of virtual displacements is used to obtain appropriate equations that express the equilibrium of the body as assemblage of m finite elements. According to this principle, the equilibrium of the body requires that, for any compatible, small virtual displacements imposed onto the body, the total internal virtual work is equal to the total external virtual work; i.e., we have

$$\sum_m \int \{\bar{\epsilon}\}^T[\sigma]\,dV = \sum_m \int \{\overline{U}\}^T f_B\,dV - \sum_m \int \{\overline{U}\}^T\{\dot{U}\}f_F\,dV$$
$$-\sum_m \int \{\overline{U}\}^T\rho\{\ddot{U}\}\,dV \tag{4.74}$$

The internal virtual work is given on the left side of Eqn. (4.74), and is equal to the stresses $[\sigma]$ going through the virtual strains, $\{\bar{\epsilon}\}$, that correspond to the imposed virtual displacements, $\{\overline{U}\}$. The external work is given on the right side and is equal to the forces f_B (body force); $\{\dot{U}\}f_F$ frictional force, and $\rho\{\ddot{U}\}$ (element inertia force) going through the virtual displacements, where

$$\{\overline{U}\} = \begin{Bmatrix} \bar{u} \\ \bar{v} \\ \bar{w} \end{Bmatrix} \tag{4.75}$$

$\bar{u}$, $\bar{v}$, $\bar{w}$ are virtual displacements in the x, y, z directions in the global coordinates. Assuming that virtual displacements are variations in the actual total displacements; then correspondingly, the virtual strains are variations in the actual strains. In this case we have

$$[\sigma] = [D]\{\epsilon\} \tag{4.76}$$
$$\{\overline{U}\} = \delta\{U\} \tag{4.77}$$
$$\{\bar{\epsilon}\} = \delta\{\epsilon\} \tag{4.78}$$

Substituting the above in Eqn. (4.74), we have for equilibrium of the continuum

$$\sum_m \int \delta\{\epsilon\}^T [D]\{\epsilon\}\, dV = \sum_m \int \delta\{U\}^T f_B\, dV - \sum_m \int \delta\{U\}^T \{\dot{U}\} f_F\, dV$$
$$- \sum_m \int \delta\{U\}^T \rho\{\ddot{U}\}\, dV \qquad (4.79)$$

which, because $[D]$ is a symmetric matrix, may also be written as

$$\sum_m \delta\left(\int \frac{1}{2}\{\epsilon\}^T[D]\{\epsilon\}\, dV\right) = \sum_m \int \delta\{U\}^T f_B\, dV - \sum_m \int \delta\{U\}^T \{\dot{U}\} f_F\, dV$$
$$- \sum_m \int \delta\{U\}^T \rho\{\ddot{U}\}\, dV \qquad (4.80)$$

However, the left-hand side of the above represents the variation in the strain energy, V, of the elastic continuum, and the right hand side is the variation of potential energy of the applied loads, W. Hence it is same as Hamilton's equation $\delta\pi = 0$ where

$$\pi = V - W \qquad (4.81)$$

Hence,

$$\pi = \sum_m \int \frac{1}{2}\{\epsilon\}^T[D]\{\epsilon\}\, dV + \sum_m \int \{U\}^T \rho\{\ddot{U}\}\, dV$$
$$+ \sum_m \int \{U\}^T\{\dot{U}\} f_F\, dV - \sum_m \int \{U\}^T f_B\, dV \qquad (4.82)$$

Substituting for strains and displacements from Eqns. (4.62) and (4.56), we get

$$\delta\pi = \sum_m \delta\left(\frac{1}{2}\int \{a\}^T[B]^T[D][B]\{a\}\, dV + \int \{a\}^T[N]^T\rho[N]\{\ddot{a}\}\, dV\right.$$
$$\left. + \int \{a\}^T[N]^T[N]\{\dot{a}\} f_F\, dV - \int \{a\}^T[N]^T f_B\, dV\right) = 0 \qquad (4.83)$$

Simplifying, we get

$$[M]\{\ddot{a}\} + [C]\{\dot{a}\} + [K]\{a\} = \{F_B\} \qquad (4.84)$$

where,

$$[M] = \sum_m \int [N]^T \rho [N]\, dV$$

$$[C] = \sum_m \int [N]^T [N] f_F\, dV$$

$$[K] = \sum_m \int [B]^T [D][B]\, dV$$

$$\{F\} = \sum_m \int [N]^T f_B\, dV$$

The individual element matrices given by Eqns. (4.72) and (4.73) are evaluated using numerical integration wherein

$$[K]^e = \sum_{i=1}^{n}\sum_{j=1}^{n}\sum_{k=1}^{n} H_i H_j H_k [B]^T [D][B] \det[J] \tag{4.85}$$

$$[M]^e = \sum_{i=1}^{n}\sum_{j=1}^{n}\sum_{k=1}^{n} H_i H_j H_k [N]^T \rho [N] \det[J] \tag{4.86}$$

in which H are the weight coefficients of the Gaussian quadrature, n is the order of integration, i.e., the number of points along three directions of integration.

These element matrices are evaluated using Reduced Integration Technique, in which order of integration in the thickness direction is less than the order of integration in plane ξ, η. The reason for this is that same order of integration in all the three directions result in overstiff element as shown by Zienkiewicz, Taylor and Too [58].

The total stiffness and mass matrices are

$$\begin{aligned} [K] &= \sum_m [K]^e \\ [M] &= \sum_m [M]^e \end{aligned} \tag{4.87}$$

They can be assembled using Frontal Housekeeping Algorithm, Irons [59], and the eigenvalue problem is then solved by determinant search technique. A computer program, SHELL, based on the above analysis is developed by Gupta [60] to determine the natural frequencies of small aspect ratio blades.

As a case study, consider a cantilever plate of rectangular cross-section with the following data.

Length	= 50 mm
Width	= 50 mm
Thickness	= 10 mm
Pre-twist	= 20 deg
Density	= 0.0023 kg/cu cm
E	= 7.06×10^{11} Pa

The results obtained by the above procedure are compared below with the energy method solution in the previous section.

TABLE 4.10

Non-Dimensional Frequencies of a Cantilever Plate

$$\overline{p} = p\sqrt{\frac{12\rho l^4(1-\nu^2)}{Eh^2}}$$

Mode	*FEM*	*Energy method*
1	3.2	3.5
2	21.5	21.8

The results in the above table compare well with the energy method. Gupta also analyzed a gas turbine blade using the above program SHELL. The data is:

Modulus of elasticity	$= 1.177 \times 10^{11}$ Pa
Mass density	= 0.00445 kg/cu cm
Blade length	= 200 mm
Chord length	= 96 mm
Pre-twist angle	= 57 deg

The results obtained are given in Table 4.11.

TABLE 4.11

Natural Frequencies of a Gas Turbine Blade

Mode	*Natural Frequency Hz*
1	208.16
2	1008.42
3	2185.34
4	2675.85
5	5189.23

Figures 4.25 through 4.28 give the corresponding mode shapes of the blade for modes 2 through 5.

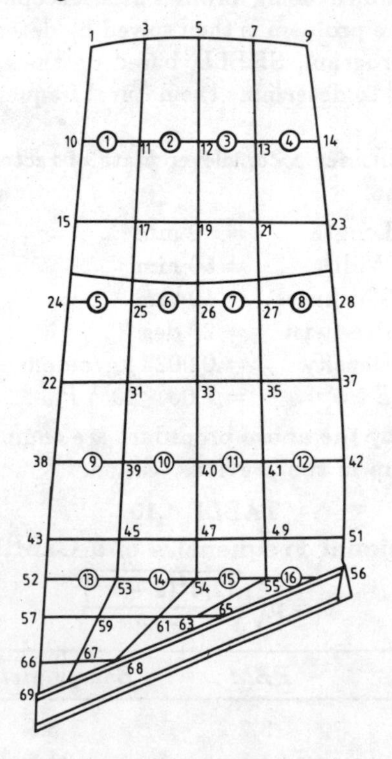

Fig. 4.25 Mode 2: second bending frequency 1008 Hz

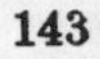

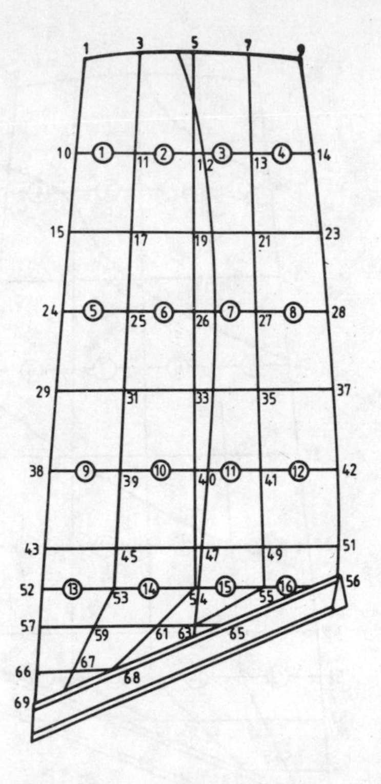

Fig. 4.26 Mode 3: first torsion frequency 2185 Hz

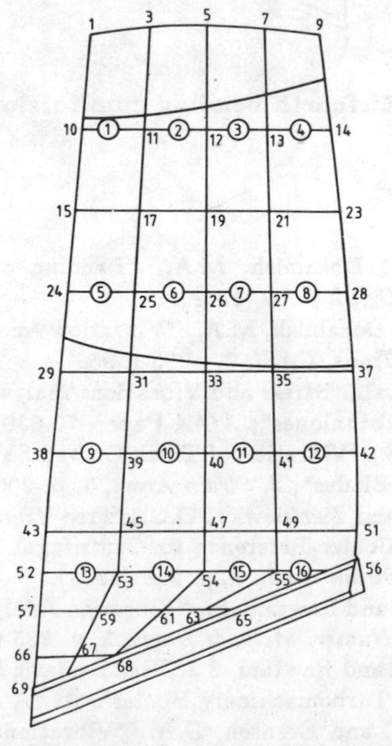

Fig. 4.27 Mode 4: third bending frequency 2676 Hz

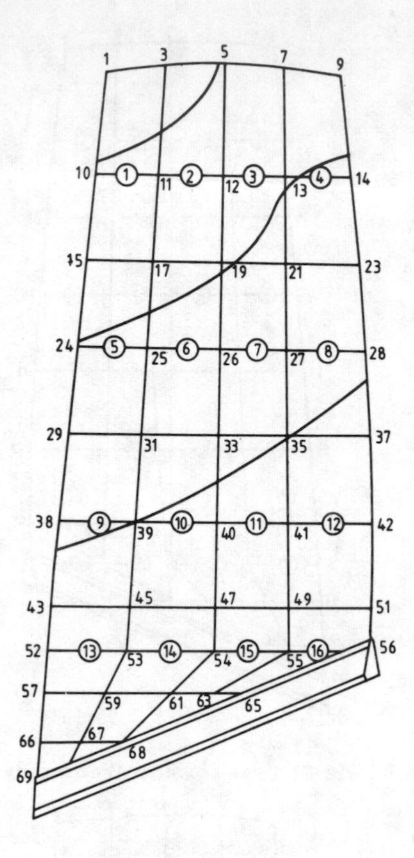

Fig. 4.28 Mode 5: fourth bending cum torsion frequency 5189 Hz

REFERENCES

1. Rawtani, S. and Dokainish, M.A., "Bending of Pre-Twisted Cantilever Plates", *Trans. CASI*, p. 89 (1969).
2. Rawtani, S. and Dokainish, M.A., "Vibration Analysis of Pre-Twisted Cantilever Plates", *Trans. CASI*, **2**, p. 95 (1969).
3. Barten, H.G., et al., "Stress and Vibration Analysis of Inducer Blades Using Finite Element Techniques", *AIAA* Paper, 70-630.
4. Ahmad, R., et al., "Vibration of Thick Curved Shells with Particular Reference to Turbine Blades", *J. Strain Anal.*, **5**, p. 200 (1970).
5. Bossak, M.A.J. and Zienkiewicz, O.C., "Free Vibrations of Initially Stressed Solids with Particular Reference to Centrifugal Force Effects in Rotating Machinery", *J. Strain Anal.*, **8**, p. 245 (1973).
6. Dokainish, M.A. and Rawtani, S., "Vibration Analysis of Rotating Cantilever Plates", *Intl. J. Numer. Methods Engr.*, **3**, p. 233 (1971).
7. Dokainish, M.A. and Rawtani, S., "Pseudo-static Deformation and Frequencies of Rotating Turbomachinery Blades", *AIAA J.*, **10**, p. 1397 (1972).
8. Hofmeister, L.D. and Evensen, D.A., "Vibration Problems Using Isoparametric Shell Elements", *Intl. J. Numer. Methods Engr.*, **5**, p. 142 (1972).

9. Dokainish, M.A. and Gossain, D.M., "Deflection and Vibration Analysis of a Flexible Fan Turbine", *J. Mech. Engr. Sci.*, **15**, p. 387 (1973).
10. Thomas, J. and Mota Soares, C.A., "Finite Element Analysis of Rotating Shells", *ASME* Paper No. 73-DET-94.
11. Rawtani, S., "The Effect of Camber on the Natural Frequencies of Low Aspect Ratio Turbomachinery Blades", *J. Aeronaut. Soc. India*, **25**, p. 119 (1973).
12. Henry, R. and Lalanne, M., "Vibration Analysis of Rotating Compressor Blades", *J. Engr. Indus., Trans. ASME*, p. 1028 (1974).
13. Lalanne, M. and Trompette, P., "Vibration Analysis of Rotating Turbine Blades", *ASME* Paper No. 74-WA/DE-23.
14. Lalanne, M., Henry, R. and Trompette, P., "Rotating Blade Analysis by Finite Element Method", *Proc. IUTAM Symp. Dyn. Rotors*, Lyngby, Denmark, p. 299 (1974).
15. Olson, M.D. and Lindberg, G.M., *A Finite Cylindrical Shell Element and the Vibrations of a Curved Fan Blade*, NRC Canada, Aeronautical Rept. LR-497 (1968).
16. MacBain, J.C., "Vibratory Behaviour of Twisted Cantilever Plates", *J. Aircraft*, **12**, p. 343 (1975).
17. Al Jumaily, A.M. and Faulkner, L.L., "Vibration Characteristics of Hollow Symmetric Blades Based on a Thin Shell Theory", *J. Mech. Des., Trans. ASME*, **100** (1), p. 183 (1978).
18. MacBain, J.C., *Quantitative Displacement and Strain Distribution of Vibrating Plate-like Structures Based on Time Average Holographic Interferometry*, Rept. No. AFAPL-TR-7744 (July 1977).
19. Nagamatsu, A., Michimura, S. and Ishihara, A., "Vibration of Impellers; Part 1: Theoretical Analysis and Experiment of Vibration of Blades", *Bull. JSME*, **20** (142), p. 411 (1977).
20. Walker, K.P., "Vibrations of Cambered Helicoidal Fan Blades", *J. Sound Vib.*, **59** (1), p. 35 (1978).
21. Hoa, S.V., "Vibration Frequency of a Curved Blade with Weighted Edge", *J. Sound Vib.*, **79** (1), p. 107 (1981).
22. Sreenivasamurty, S. and Ramamurty, V., "A Parametric Study of Vibration of Rotating Pre-twisted and Tapered Low Aspect Ratio Cantilever Plates", *J. Sound Vib.*, **76** (3), p. 311 (1981).
23. Henry, R., "Calcul des fréquences et modes des structures répétitives circulaires", *J. de Mécanique Appliquée*, **4**, p. 61 (1980).
24. Henry, R., "Contribution à l'étude dynamique des machines tournantes", D.Sc. Thesis, INSA, Lyon (1981).
25. Fu, Z.F., "Determination of Vibration Amplitudes and Stresses Using the Holography Interference Techniques and Finite Element Method", *ASME* 81-DET-132.
26. Petricone, R. and Sisto, F., "Vibration Characteristics of Low Aspect Ratio Compressor Blades", *ASME* 70-GT-94.
27. Wan, F.Y.M., "A Class of Unsymmetric Stress Distributions in Helicoidal Shells", *Qly. of Appl. Math.*, **24**, p. 374 (1967).
28. Reissner, E., "Variational Considerations for Elastic Beams and Shells", *J. of Engr. Mech., ASCE*, **88**, p. 23 (1962).
29. Gupta, K., *Vibration of Small Aspect Ratio Blades*, Ph.D. Thesis, Ind. Inst. Tech., New Delhi (1979).
30. Gol' den Veizer, A.L., *Theory of Elastic Thin Shells*, Pergamon Press (1961).

31. Gupta, K. and Rao, J.S., "Torsional Vibration of Pre-twisted Cantilever Plates", *J. Mech. Des., Trans. ASME* **100** (3), p. 528 (1978).
32. Gupta, K. and Rao, J.S., "Flexural Vibrations of Pre-twisted Cantilever Plates", *J. of Aero. Soc.*, India, **30**, p. 131 (1978).
33. Gupta, K. and Rao, J.S., "Inextensional Vibration of Pre-twisted Shells", *Proc. 7th CAN-CAM Conf.*, p. 445 (1979).
34. Leissa, A.W., Lee, J.K. and Wang, A.J., "Vibrations of Cantilevered Shallow Cylindrical Shells of Rectangular Plan Form", *J. Sound Vib.*, **78** (3), p. 311 (1981).
35. Leissa, A.W., Lee, J.K. and Wang, A.J., "Vibrations of Twisted Rotating Blades", *ASME* 81-DET-127.
36. Leissa, A.W., Lee, J.K. and Wang, A.J., "Rotating Blade Vibration Analysis Using Shells", *J. Engr. Power, ASME*, **104**, p. 296 (1982).
37. Jensen, K.R., "A Shell Analysis of Turbine Blade Vibrations", *Int. J. Mech. Sci.*, **24** (10), p. 581 (1982).
38. Morley, L.S.D., *Skew Plates and Structures*, Pergamon Press (1963).
39. Love, A.E.H., *A Treatise on Mathematical Theory of Elasticity*, Dover Publication (1944).
40. Novozhilov, V.V., *The Theory of Thin Shells*, P. Nordhoff Ltd., Netherlands (1959).
41. Kalnins, A., "Dynamic Problems of Elastic Shell", *Appl. Mech. Rev.*, **18**, p. 867 (1965).
42. Grinsted, B., "Nodal Pattern Analysis", *Proc. I. Mech. E.*, **166**, p. 309 (1952).
43. Rao, J.S. and Gupta, K., "Vibrations of Rotating Small Aspect Ratio Blades", Paper presented at the Int. Cong. of Theo. and Appl. Mech., Toronto (August 1980).
44. Rao, J.S., "Coupled Vibration of Turbine Blades", *The Shock and Vibration Bulletin*, **47** (2), p. 107 (1977).
45. Warburton, G.B., "The Vibration of Rectangular Plates", *Proc. I. Mech. E.*, **168**, p. 371 (1954).
46. Martin, A.I., "On the Vibration of a Cantilever Plate", *Qly. J. Mech. Appl. Maths.*, **9**, p. 94 (1956).
47. Classen, R.W. and Thorne, C.J., "Vibrations of a Rectangular Cantilever Plate", *J. Aerospace Sci.*, **29**, p. 1300 (1962).
48. Leissa, A.W., *Vibration of Plates*, NASA SP-160 (1969).
49. Swaminadham, M., *Vibration of Rotating Pre-Twisted and Tapered Blades*, Ph.D. Thesis, I.I.T., Kharagpur (1975).
50. Thomas, J. and Mota Soares, C.A., "Dynamic Analysis of Rotating Turbine and Compressor Blading", *Proc. Vibrations in Rotating Machinery Conf. I. Mech. E.*, p. 231 (1976-77).
51. Rao, J.S., "Torsional Vibrations of Pre-Twisted Cantilever Beams", *J. Inst. of Engrs. (India) CE Division*, **52**, p. 211 (1972).
52. Utku, S., "Stiffness Matrices for Triangular Elements of Non-zero Gaussian Curvature" *J. AIAA*, **10**, p. 1659 (1967).
53. Mohr, G., "A Doubly Curved Isoparametric Triangular Shell Element", *Computers and Structures*, **14**, p. 9 (1981).
54. Irons, B.M., *The Semiloof Shell Element, Finite Element for Thin Shells*, John Wiley (1976).

55. Ergatoudis, I., Irons, B.M. and Zienkiewicz , O.C., "Curved, Isoparametric Quadrilateral Elements for Finite Element Analysis", *Int. J. Solids and Structures*, **4**, p. 31 (1968).
56. Ahmad, S., Irons, B.M. and Zienkiewicz, O.C., "Analysis of Thick and Thin Shell Structures by Curved Finite Elements", *Int. J. Num. Methods in Engrg.*, **2**, p. 419 (1970).
57. Gupta, D.K., Ramakrishnan, C.V. and Rao, J.S., "Fluid Structure Interaction Problems in Turbine Blade Vibration", *Advances in Fluid-Structure Interaction — 1984, ASME PVP*-Vol. **78**, AMD-Vol. **64**, p. 89 (1984).
58. Zienkiewicz, O.C., Taylor, R.L. and Too, J.M., "Reduced Integration Technique in General Analysis of Plates and Shells", *Int. J. Num. Methods in Engrg.*, **3**, p. 275 (1971).
59. Irons, B.M., "A Frontal Solution Program for Finite Element Analysis", *Int. J. Num. Methods in Engrg.*, **2**, p. 5 (1970).
60. Gupta, D.K., *Vibration Analysis of Fluid Submerged Blades*, Ph.D. Thesis, I.I.T. Delhi, (1984).

Chapter 5

Blade Group Frequencies and Mode Shapes

5.1 INTRODUCTION

A blade group exhibits more complex dynamic behaviour than a free-standing blade. The cantilever mode frequencies are influenced by the neighbouring blades, the shroud mass and elasticity. In addition to the cantilever modes, fixed-pinned modes appear in groups between the cantilever modes.

Kroon [1] applied difference calculus, which was developed to study groups of similar objects in dealing with civil engineering problems, to the case of lashed blades to determine the blade stresses. No vibratory calculations were made and the deflections in the tangential directions only were considered. However, this paper is the first attempt to bring out the fact that the blade group as a whole should be considered rather than an individual blade. Myklestad's adaptation of Holzer method was used by Jarret and Warner [2] to determine the natural frequencies of rotating tapered-twisted beams; the influence of lashing wires and shroud were included as constraints. The batch modes, typical to blade groups, was not considered. Smith [3] made a two dimensional free vibrational analysis in the tangential direction using the dynamic stiffness matrix method on a six and a twenty-blade group. His contribution is most significant in that, the group frequencies and mode shapes were determined for the first time. The blade and covers were separated at their joints and equilibrium equations were written in terms of the blade tip deflections and slopes. The results of the calculations were shown on graphs in terms of three dimensionless parameters — frequency ratio, mass ratio and rigidity ratio. He also discussed the use of lacing wires in turbine blade groups. Experience and mode shapes show that lacing wires have a significant effect on supressing the second group of tangential vibrations if the wires are inserted at the proper height. Ellington and McCallion [4] simplified Smith's analysis by using finite difference calculus to the special case of a blade group with a tie wire which joins the blade tips together.

The first well-known blade group design paper was by Prohl [5]. Prohl presented a method of calculating natural frequencies, mode shapes and bending stresses for three dimensional free and forced vibrations in the tangential and axial directions. The analysis of this paper followed the approach of Smith and is extended to consider axial and torsional vibrations. Prohl used a series of concentrated masses and concentrated inertias to represent the blade group. The blade was broken into n stations and the mass of one cover section was added to the tip blade section. A modified Holzer technique was used to calculate the natural frequencies and mode shapes. This method of analysis gives all of the tangential and axial natural frequencies and mode shapes. The blades were considered to be inextensional, but the covers were extensional, and the flexibility of support hinge is accounted. In a companion paper by Weaver and Prohl [6], Prohl's method was used to calculate the natural frequencies, mode shapes, and stress levels for a simple blade group. Singh and Nandeeswariah [7] used Smith's analysis to study the tangential vibrations of a bladed group. Prohl's analysis was programmed by Rao [8] to determine the tangential and coupled axial torsion modes of blade groups.

A method of analysis for a laced group of rotating exhaust blades was developed by Deak and Baird [9]. This analysis included three dimensional coupled tangential and axial free vibrations with root stiffness. Both flexural and torsional motions were considered. This method gives all of the natural frequencies and mode shapes. An important point made in Deak and Baird's paper was that the disc effect and centrifugal stiffening effects are very important in long exhaust blades. Blade to blade coupling does occur though through the lacing wires. The frequencies calculated by this method were compared with experimental results with good correlation. Some judgement was required to define the effective point of fixity in the disc rim and the effective lacing wire constraints. The complicated root of the blade was replaced by an equivalent beam encasement.

Rieger and McCallion [10] considered two dimensional free, undamped tangential vibrations of frame structures. This work is on single-story multi-bay frameworks which represents a simplified geometry of a turbine blade group. The method separated the portal frame at the intersection of the horizontal and vertical members. Equilibrium force and moment equations are written for the intersections, to develop a dynamic stiffness matrix. Stuwing [11] used transfer matrix method for packeted blades executing coupled bending-bending-torsion modes.

Tuncel, Bueckner and Koplik [12] applied diakoptics to determine blade group frequencies in tangential vibrations. Better results were achieved on long slender blades with weak coupling than short rigid blades. Diakoptics deals with subsystems independently in the first step and includes the coupling in the second step. Diakoptics along with a method of perturbations was used to find the natural frequencies of the blade group. Provenzale and Skok [13] presented a computer program capable of calculating three dimensional coupled tangential and axial forced vibrations and stresses in blade groups. The program analyzes the complete blade including the airfoil platform, root and span constraints, and is capable of simulating elastic, damped constraints at the root, tip and mid-span locations. However, the analysis is done on a single blade with the effects of adjacent blades being introduced by restraints on the single blade and a blade to blade single phase angle.

Rao [14] made a two dimensional analysis of free vibrations in the tangential direction. The first step is to develop the potential and kinetic energies for the tangential motion of the blades and shrouds. Second, Hamilton's principle is applied to derive the differential equation of motion and the boundary conditions. Then these equations are solved to determine the natural frequencies. Rao compared his calculated frequencies for lower order vibrations with frequencies calculated by Prohl's method. The agreement was excellent.

Bajaj [15] used the finite element method to determine the natural frequencies of packeted blades in coupled bending-bending-torsion modes; 186 degrees of freedom were used. Rieger and Nowak [16] have developed a three-dimensional finite element model of the blade root and wheel root that incorporates gap elements at the root interface. This so-called super element can be used to generate automatically the mesh for the remaining blades of a group so that natural frequencies and mode shapes can be determined. A finite element model of a group of blades with rectangular cross section has been presented in the tangential mode [17]. Salama and Petyt [18] used finite element method and periodic structural analysis for the tangential vibration of packeted blades; both the positions of the lacing wire and rotation were taken into account.

Eight node isoparametric solid elements were used by Sagendorph [19] to model a fan blade with a tie-wire. Calculations were made for free and for locked shroud configurations. Comparison of the free shroud condition with experimental data gave agreement to within 10.0 per cent for the first ten modes. Holography was used to verify mode shapes, which compared favourably with the finite element method results. The locked shroud condition is not a true blade group configuration, but only a boundary condition input on a single blade. Similar work by Hall and Armstrong [20] for interlocking shrouds used the finite difference method. Srinivasan, Lionberger and Brown [21] used the component mode technique to obtain blade group natural frequencies, where the finite element method was used to obtain the motions of a single blade clamped at the root, but influenced by adjoining blades at the interlocking shroud location. Lu and Warner [22] made a statistical assessment of the effect of variable root flexibility on the vibration response of shrouded blades. The method uses an initial computation employing the full dynamic system as a base, then an approximate structural modification scheme to permit rapid and inexpensive calculations where the statistical properties of the response are to be determined.

In this chapter the continuous system analysis for packeted blades is first illustrated and then extended to discrete model for packeted and laced blades.

5.2 TWO DIMENSIONAL CONTINUOUS SYSTEM ANALYSIS

Consider a blade packet with N blades, figure 5.1, executing tangential vibrations. All the blades are assumed uniform of length L with mass per unit length $= m_b$. $E_b I_b$ is the flexural rigidity of each blade. The shroud of pitch P of mass per pitch length m_s is assumed to be replaced by a concentrated mass at each blade tip, however retaining its flexural rigidity $E_s I_s$. From figure 5.2, the shear force V_n is given by

$$V_n = \frac{M_{nr} - M_{nl}}{P} \tag{5.1}$$

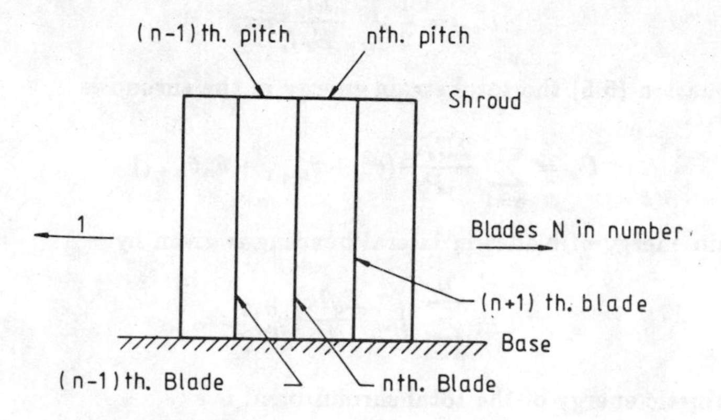

Fig. 5.1 A packet with N blades

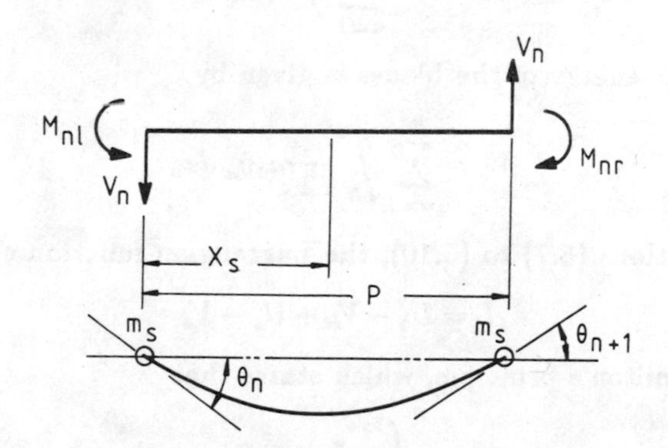

Fig. 5.2 nth pitch of the shroud

The relationship between the bending moments and the slopes are

$$M_{nl}\frac{P}{E_s I_s} + 4\theta_n + 2\theta_{n+1} = 0 \tag{5.2}$$

$$M_{nr}\frac{P}{E_s I_s} - 2\theta_n - 4\theta_{n+1} = 0 \tag{5.3}$$

Hence,

$$V_n = \frac{6E_sI_s}{P^2}(\theta_n + \theta_{n+1}) \tag{5.4}$$

Using equation (5.4), the bending moment at any distance x_s of the nth shroud is

$$\begin{aligned} M_{xs} &= M_{nl} + V_n x_s \\ &= -\frac{E_sI_s}{P}(4\theta_n + 2\theta_{n+1}) + \frac{6E_sI_s}{P^2}(\theta_n + \theta_{n+1})x_s \end{aligned} \tag{5.5}$$

The strain energy in bending of nth shroud is given by

$$U_{sn} = \frac{1}{2}\int_0^P \frac{M_{xs}^2}{E_sI_s} dx_s \tag{5.6}$$

Using equation (5.5) the total strain energy in the shroud is

$$U_s = \sum_{n=1}^{N-1} \frac{2E_sI_s}{P}(\theta_n^2 + \theta_{n+1}^2 + \theta_n\theta_{n+1}) \tag{5.7}$$

The strain energy of blades in lateral bending is given by

$$U_b = \sum_{n=1}^{N} \int_0^L \frac{E_bI_b}{2}(y_n'')^2\, dx_b \tag{5.8}$$

The kinetic energy of the total shroud band is

$$V_s = \sum_{n=1}^{N} \frac{1}{2} m_s \dot{y}_{nl}^2 \tag{5.9}$$

and the kinetic energy of the blades is given by

$$V_b = \sum_{n=1}^{N} \int_0^L \frac{1}{2} m_b \dot{y}_n^2\, dx_b \tag{5.10}$$

Using equations (5.7) to (5.10), the Lagrangian function can be set up

$$L = U_b - V_b + U_s - V_s \tag{5.11}$$

Applying Hamilton's principle, which states that

$$\delta \int_{t_1}^{t_2} L\, dt = 0 \tag{5.12}$$

and noting that $\theta_n = y_{nl}'$ etc., the following differential equations and corresponding restraint conditions are obtained:

$$(E_bI_by_n'')'' + m_b\ddot{y}_n = 0 \quad n = 1, 2, \ldots, N \tag{5.13}$$

At $x_b = 0$

$$E_bI_b\left[y_n''\,\delta y_n' - y_n'''\,\delta y_n\right] = 0 \quad n = 1, 2, \ldots, N \tag{5.14}$$

At $x_b = L$

$$\delta y_1'\left[E_bI_by_1'' + \frac{2E_sI_s}{P}(2y_1' + y_2')\right] = 0$$
$$\delta y_1\left[-E_bI_by_1''' + m_s\ddot{y}_1\right] = 0 \tag{5.15}$$
$$\delta y_n'\left[E_bI_by_n'' + \frac{2E_sI_s}{P}(y_{n-1}' + 4y_n' + y_{n+1}')\right] = 0$$
$$\delta y_n\left[-E_bI_by_n''' + m_s\ddot{y}_n\right] = 0 \quad n = 2, 3, \ldots, N-1 \tag{5.16}$$
$$\delta y_N'\left[E_bI_by_N'' + \frac{2E_sI_s}{P}(2y_N' + y_{N-1}')\right] = 0$$
$$\delta y_N\left[-E_bI_by_N''' + m_s\ddot{y}_N\right] = 0 \tag{5.17}$$

For packeted blades fixed at the base, equation (5.14) gives

$$\begin{aligned} y_n &= 0 \\ y_n' &= 0 \qquad n = 1, 2, \ldots, N \end{aligned} \tag{5.18}$$

Depending upon the boundary conditions at $x_b = L$, the packet will execute cantilever modes, fixed-supported modes or fixed-fixed modes. Fixed-fixed modes do not occur in practice, as the shroud band is not heavy enough, and the first two kinds of modes are analyzed below.

Cantilever modes

For cantilever modes, the boundary conditions from equations (5.15) to (5.17) at $x_b = L$ become

$$E_bI_by_1'' + \frac{2E_sI_s}{P}(2y_1' + y_2') = 0$$
$$E_bI_by_1''' + m_s\omega^2 y_1 = 0 \tag{5.19}$$
$$E_bI_by_n'' + \frac{2E_sI_s}{P}(y_{n-1}' + 4y_n' + y_{n+1}') = 0$$
$$E_bI_by_n''' + m_s\omega^2 y_n = 0 \quad n = 2, 3, \ldots, N-1 \tag{5.20}$$
$$E_bI_by_N'' + \frac{2E_sI_s}{P}(y_{N-1}' + 2y_N') = 0$$
$$E_bI_by_N''' + m_s\omega^2 y_N = 0 \tag{5.21}$$

For uniform blading, solution of equations (5.13) satisfying the boundary conditions given in equation (5.14) for $x_b = 0$, is

$$y_n = A_n(\cos mx_b - \cosh mx_b) + B_n(\sin mx_b - \sinh mx_b) \tag{5.22}$$

where,

$$m^4 = \frac{m_b\omega^2}{E_bI_b} \tag{5.23}$$

From equation (5.22), the derivatives of y_n at $x_b = L$ can be obtained

$$\begin{aligned} y_{nl}' &= [-A_n(s + sh) + B_n(c - ch)]m \\ y_{nl}'' &= m^2[-A_n(c + ch) - B_n(s + sh)] \\ y_{nl}''' &= m^3[A_n(s - sh) - B_n(c + ch)] \end{aligned} \tag{5.24}$$

where $s = \sin mL$, $sh = \sinh mL$, etc.

The second boundary conditions in equations (5.19) to (5.21) gives

$$B_n = RA_n \tag{5.25}$$

where,

$$R = \frac{(s - sh) + \beta(c - ch)}{(c + ch) - \beta(s - sh)}, \quad \beta = \frac{m_s \omega^2}{E_b I_b m^3} \tag{5.26}$$

With the help of equation (5.25) and the first boundary conditions in equations (5.19) to (5.21), the frequency determinant can be obtained:

$$\Delta = \begin{vmatrix} t_0 & t_1 & 0 & 0 & 0 & \cdots & 0 & 0 & 0 \\ t_1 & t_2 & t_1 & 0 & 0 & \cdots & 0 & 0 & 0 \\ 0 & t_1 & t_2 & t_1 & 0 & \cdots & 0 & 0 & 0 \\ \cdots & \cdots & \cdots & \cdots & \cdots & \cdots & \cdots & \cdots & \cdots \\ 0 & 0 & 0 & 0 & 0 & \cdots & t_1 & t_2 & t_1 \\ 0 & 0 & 0 & 0 & 0 & \cdots & 0 & t_1 & t_0 \end{vmatrix} = 0 \tag{5.27}$$

where,

$$\begin{aligned} t_1 &= \frac{2r}{mL}[-(s + sh) + R(c - ch)] \\ t_2 &= 4t_1 - [(c + ch) + R(s + sh)] \\ t_0 &= 3t_1 - [(c + ch) + R(s + sh)] \\ r &= \frac{E_s I_s L}{E_b I_b P} \end{aligned} \tag{5.28}$$

Fixed-supported modes

For these modes, the second boundary condition in equations (5.19) to (5.21) will be simplified as

$$Y_{nL} = 0 \tag{5.29}$$

The above equation gives

$$R = \frac{(c - ch)}{(s - sh)} \tag{5.30}$$

The frequency determinant will still be given by equation (5.27), with the above value of R.

Consider the following data of a uniform bladed packet:

$$\begin{aligned} E_b = E_s &= 1.96 \times 10^{11} \text{ Pa} \\ I_b = I_s &= 0.66 \times 10^{-3} \text{ cm}^4 \\ P &= 1.4 \text{ cm} \\ L &= 20 \text{ cm} \\ m_b &= 2.462 \times 10^{-3} \text{ kg/cm} \\ m_s &= 3.840 \times 10^{-3} \text{ kg} \\ N &= 5 \end{aligned}$$

For the five bladed packet, equation (5.27) can be expanded to give

$$\Delta = (t_1^2 - t_0 t_2)[2t_0 t_1^2 + t_2(t_1^2 - t_0 t_2)]$$

A root searching technique can be used to determine the natural frequency ω, which satisfies $\Delta = 0$. The results obtained are given in Table 5.1, along with those obtained by Chattopadhyay [23] with Prohl's method.

TABLE 5.1

No.	*Natural Frequencies — rad/sec.*		*Remarks*
	Present Analysis	*Prohl's Method*	
1	290	289.644	I Cantilever mode
2	1230	1233.000	I Fixed-supported modes
3	1232	1233.436	-do-
4	1257	1256.796	-do-
5	1265	1263.849	-do-
6	1601	1620.564	II Cantilever mode

Smith [3] gave results for 6 and 20 blade packets in a general form. The ratio of packet frequency to the fundamental free standing blade frequency is plotted as a function of ratio of flexural rigidities of one pitch of the shroud to the blade for different ratios of total mass of shroud to total mass of the blades as shown in figure 5.3.

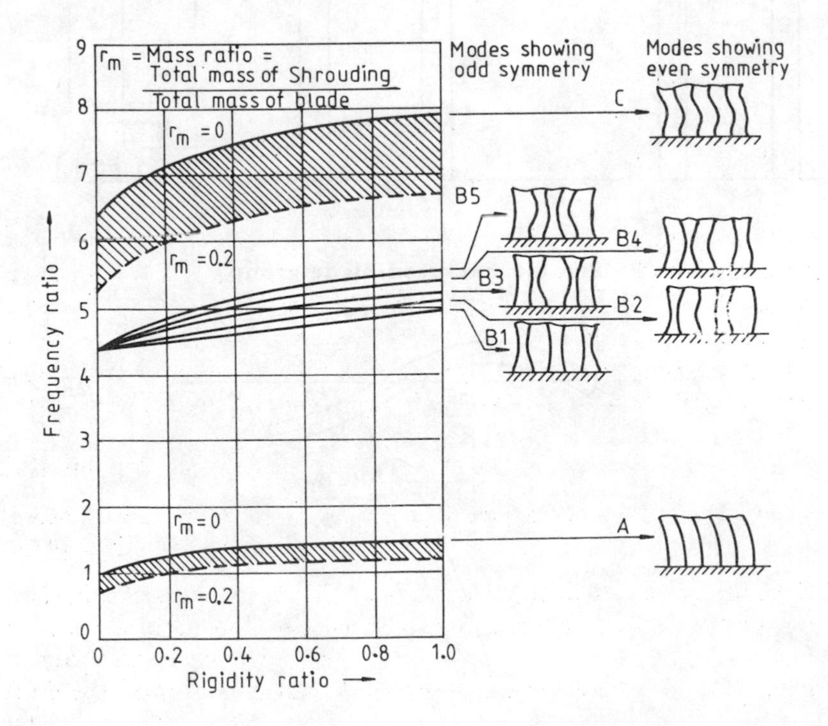

Fig. 5.3 Vibrations of packet of 6 blades

The following may be observed:

- The frequencies increase with the flexural rigidity of the shroud.
- The frequencies decrease with the shroud mass.
- There are 5 fixed-pinned batch modes, three of them showing odd symmetry and two even symmetry. (There are $N-1$ batch modes for an N bladed packet).

For a 20 bladed packet, an almost identical figure as in figure 5.3 is obtained by Smith, excepting that the batch modes are 19 in number, closely packed in the same batch mode domain B.

5.3 PROHL'S ANALYSIS

The blade packet consists of a number of blades which are identical, evenly spaced and joined together by a band at the free end as shown in figure 5.4.

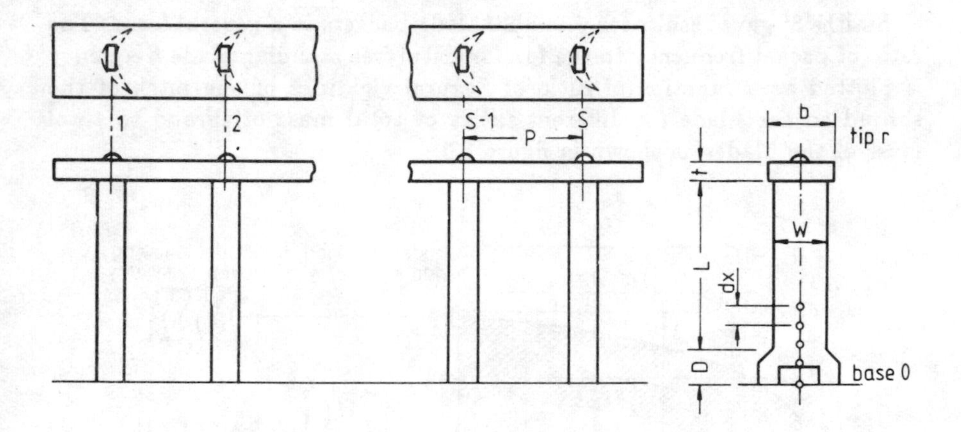

Fig. 5.4 Packeted blade group

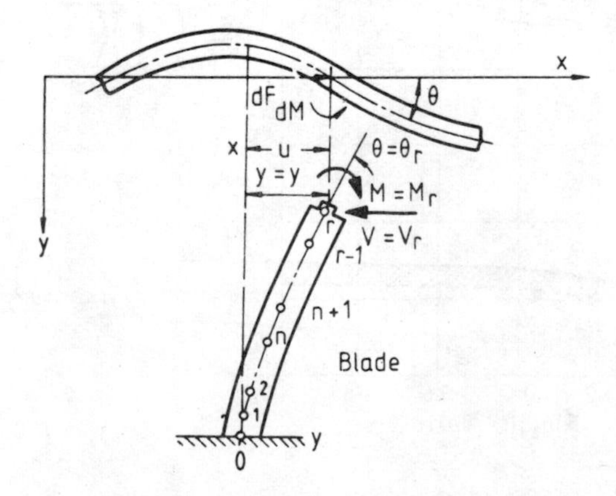

Fig. 5.5 Relations at band attachment

The connection between the band and the blade is assumed to be tight; however, the connection of the blade to the base can provide for a hinge. The blades are assumed stationary, i.e., the rotating effects are neglected. Conventional bending and torsional theories are assumed and higher order effects are disregarded. The blade is assumed inextensional whereas it is accounted for in the band. The principal axes of all bucket cross-sections are assumed to be in the tangential and axial directions respectively. Coupling between bending and torsion motions of the blade is disregarded; however, this coupling through the band is accounted in the analysis. The blade is represented by a series of concentrated masses connected by massless elastic elements as shown in figure 5.5. The notations are given in these figures.

Blade Flexure Relations

For a frequency of vibration p, of the blade packet, the blade relations are

$$\begin{aligned} M_{n+1} &= M_n + dx_n V_n \\ \theta_{n+1} &= \theta_n + a_n M_n + b_n V_n \\ Y_{n+1} &= Y_n + dx_n \theta_n + c_n M_n + d_n M_n \\ V_{n+1} &= V_n + m_{n+1} p^2 Y_{n+1} \end{aligned} \tag{5.31}$$

where,

$$\begin{aligned} a_n &= \int_{x_n}^{x_{n+1}} \frac{dx}{EI(x)} \\ b_n &= \int_{x_n}^{x_{n+1}} \frac{x\,dx}{EI(x)} \\ c_n &= a_n\,dx_n - b_n \quad ; \quad d_n = b_n\,dx_n - g_n \\ g_n &= \int_{x_n}^{x_{n+1}} \frac{x^2\,dx}{EI(x)} \end{aligned} \tag{5.32}$$

Blade Torsion Relations

The blade is represented by a series of mass moments of inertia connected by massless torsional springs. Then,

$$\begin{aligned} \phi_{n+1} &= \phi_n + e_n T_n \\ T_{n+1} &= T_n - J_{n+1} p^2 \phi_{n+1} \end{aligned} \tag{5.33}$$

where,

$$e_n = \int_{x_n}^{x_{n+1}} \frac{dx}{GC} \tag{5.34}$$

Calculations for Tangential Vibration

The complete flexural calculations of the blade are performed by a step by step application of equations (5.31), starting with station O at the base and terminating with station r at the tip. These calculations are made in two parts for the initial values at station O, with $V_0' = 1$, $M_0' = \theta_0' = Y_0' = 0$ and $M_0'' = 1$, $\theta_0'' = Y_0'' = V_0'' = 0$. At the base we can use $\theta_0'' = A_t M_0''$ in the second part to account for flexibility of the hinge in the tangential vibration. Then at the blade tip,

$$\begin{aligned} M_r &= M_r' V_0 + M_r'' M_0 \\ \theta_r &= \theta_r' V_0 + \theta_r'' M_0 \\ Y_r &= Y_r' V_0 + Y_r'' M_0 \\ V_r &= V_r' V_0 + V_r'' M_0 \end{aligned} \tag{5.35}$$

where V_0 and M_0 are the unknown shear force and bending moment at the base. Equations (5.35) apply to any blade, but V_0 and M_0 may differ from blade to blade in the packet.

Considering the flexure and extension of band elements, the following relations from strength of materials can be derived:

$$\begin{aligned} &a_t\, dM_1 + 4\theta_1 + 2\theta_2 = 0 \\ &a_t\, dM_2 + 2\theta_1 + 8\theta_2 + 2\theta_3 = 0 \\ &a_t\, dM_3 + 2\theta_2 + 8\theta_3 + 2\theta_4 = 0 \quad \text{etc.} \end{aligned} \tag{5.36}$$

where,

$$a_t = \frac{P}{E_b I_t} \tag{5.37}$$

and

$$\begin{aligned} &h\, dF_1 - U_1 + U_2 = 0 \\ &h\, dF_2 + U_1 - 2U_2 + U_3 = 0 \\ &h\, dF_3 + U_2 - 2U_3 + U_4 = 0 \quad \text{etc.} \end{aligned} \tag{5.38}$$

where,

$$h = \frac{P}{E_b A} \tag{5.39}$$

With the help of equation (5.35) and figure 5.5, we have

$$\begin{aligned} dM_N &= M_{rn} = M_r' V_{ON} + M_r'' M_{ON} \\ \theta_N &= \theta_{rN} = \theta_r' V_{ON} + \theta_r'' M_{ON} \\ dF_n &= V_{rn} = V_r' V_{ON} + V_r'' M_{ON} \\ U_N &= Y_{rn} = Y_r' V_{ON} + Y_r'' M_{ON} \end{aligned} \tag{5.40}$$

Substituting equations (5.40) in equations (5.36) and (5.38), two sets of equations can be obtained, see equations (5.41), which are similar for all interior blades and particular in nature for the first and last blades in the packet. This set of equations will lead to the frequency determinant, as will be demonstrated shortly. The size of this determinant can be considerably reduced by considering even or odd symmetry with respect to the middle of the bucket. These two separate cases are discussed as follows, for a packet with S number of buckets.

Band Station	V_{01}	M_{01}	V_{02}	M_{02}	V_{03}	M_{03}	V_{04}	M_{04}	
1	$a_t M_r' + 4\theta_r'$	$a_t M_r'' + 4\theta_r''$	$2\theta_r'$	$2\theta_r''$	—	—	—	—	$= 0$
	$hV_r' - Y_r'$	$hV_r'' - Y_r''$	Y_r'	Y_r''	—	—	—	—	$= 0$
2	$2\theta_r'$	$2\theta_r''$	$a_t M_r' + 8\theta_r'$	$a_t M_r'' + 8\theta_r''$	$2\theta_r'$	$2\theta_r''$	—	—	$= 0$
	Y_r'	Y_r''	$hV_r' - 2Y_r'$	$hV_r'' - 2Y_r''$	Y_r'	Y_r''	—	—	$= 0$
3	—	—	$2\theta_r'$	$2\theta_r''$	$a_t M_r' + 8\theta_r'$	$a_t M_r'' + 8\theta_r''$	$2\theta_r'$	$2\theta_r''$	$= 0$
	—	—	Y_r'	Y_r''	$hV_r' - 2Y_r'$	$hV_r'' - 2Y_r''$	Y_r'	Y_r''	$= 0$
etc.									

(5.41)

A. Even Number of Blades

Let the blade adjacent to the middle location be $N^* = S/2$.

(i) *Even symmetry*

$$V_{ON^*+1} = V_{ON^*} \quad ; \quad M_{ON^*+1} = M_{ON^*} \tag{5.42}$$

Hence the equations for N^* blade will be

$$\begin{array}{cccc} V_{ON^*-1} & M_{ON^*-1} & V_{ON^*} & M_{ON^*} \\ 2\theta'_r & 2\theta''_r & a_t M'_r + 10\theta'_r & a_t M''_r + 10\theta''_r = 0 \\ Y'_r & Y''_r & hV'_r - Y'_r & hV''_r - Y''_r = 0 \end{array} \tag{5.43}$$

There will be a total of $S = 2N^*$ equations in the determinantal set.

(ii) *Odd symmetry*

For this case

$$V_{ON^*-1} = -V_{ON^*} \quad ; \quad M_{ON^*-1} = -M_{ON^*} \tag{5.44}$$

and equations for N^* blade will become

$$\begin{array}{cccc} V_{ON^*-1} & M_{ON^*-1} & V_{ON^*} & M_{ON^*} \\ 2\theta'_r & 2\theta''_r & a_t M'_r + 6\theta'_r & a_t M''_r + 6\theta''_r = 0 \\ Y'_r & Y''_r & hV'_r - 3Y'_r & hV''_r - Y''_r = 0 \end{array} \tag{5.45}$$

B. Odd Number of Blades

Let the middle blade be $N^* = (S+1)/2$.

(i) *Even symmetry*

For this case

$$V_{ON^*+1} = V_{ON^*-1} \quad ; \quad M_{ON^*+1} = M_{ON^*-1} \tag{5.46}$$

Equations for N^* blade are

$$\begin{array}{cccc} V_{ON^*-1} & M_{ON^*-1} & V_{ON^*} & M_{ON^*} \\ 4\theta'_r & 4\theta''_r & a_t M'_r + 8\theta'_r & a_t M''_r - 8\theta''_r = 0 \\ 2Y'_r & 2Y''_r & hV'_r - 2Y'_r & hV''_r - 2Y''_r = 0 \end{array} \tag{5.47}$$

There will be a total of $S + 1$ equations in this case.

(ii) *Odd symmetry*

$$\begin{gathered} V_{ON^*+1} = -V_{ON^*-1} \quad ; \quad M_{ON^*+1} = -M_{ON^*-1} \\ V_{ON^*} = M_{ON^*} = 0 \end{gathered} \tag{5.48}$$

In this case, equations for N^* blade drop out, and for $N^* - 1$ blade we have

$$\begin{array}{cccc} V_{ON^*-2} & M_{ON^*-2} & V_{ON^*-1} & M_{ON^*-1} \\ 2\theta'_r & 2\theta''_r & a_t M'_r + 8\theta'_r & a_t M''_r + 8\theta''_r = 0 \\ Y'_r & Y''_r & hV'_r - 2Y'_r & hV''_r - 2Y''_r = 0 \end{array} \tag{5.49}$$

There will be a total of $2(N^* - 1) = S - 1$ equations in the set.

The condition for a trial frequency to be a natural frequency of the blade packet is that the determinant of the set of equations (5.41), with appropriate boundary conditions given in equations (5.43), (5.45), (5.47) and (5.49) for different cases discussed, be equal to zero. Successive interpolation technique can be used to determine the required natural frequencies. Once the natural frequency is determined, the relative values of M_{01}, V_{02}, M_{02}, ... in terms of V_{01} from equations (5.41) can be determined by back substitution and hence the mode shapes from equation (5.31).

Calculations for Coupled Axial and Torsional Motion

As mentioned before, the calculations for flexure and torsion are carried out for the blade independently, since there is no coupling assumed between them in the blade. The calculation for axial flexure is just the same as that for tangential flexure, except that for determining the flexibility factors, the second moment of area of cross section I_A for axial bending is used in equations (5.32), and that the flexibility factor A_a is used for the support hinge in axial flexure. Following similar procedure for the blade in tangential vibration, the equations for the tip condition, identical to those given in equations (5.35), are obtained.

$$\begin{aligned} M_r &= M_r' V_0 + M_r'' M_0 \\ \theta_r &= \theta_r' V_0 + \theta_r'' M_0 \\ Y_r &= Y_r' V_0 + Y_r'' M_0 \\ V_r &= V_r' V_0 + V_r'' M_0 \end{aligned} \tag{5.50}$$

In torsional calculations, choose the same stations as for bending and lump the mass moments of inertia J_n and use the initial conditions

$$\begin{aligned} T_0' &= 1 \\ \phi_0' &= A_r T_0 \end{aligned} \tag{5.51}$$

The calculations proceed from station to station, with the help of equations (5.33) until the tip is reached, whence,

$$\begin{aligned} \phi_r &= \phi_r' T_0 \\ T_r &= T_r' T_0 \end{aligned} \tag{5.52}$$

As before, the elastic restraint offered by the band is considered, since the mass effect is included in the blade. The details of relationships of deflections, slopes, forces and moments are given in figure 5.6. The axial and torsional motions Y and ϕ are coupled through the band, this coupling is more predominant than the coupling due to centre flexure and centre of gravity of the cross-section of the blade not being the same. The moment dM can be shown to be related to the slopes θ and displacements Y of the band as

$$\begin{aligned} &C_a\, dM_1 + 2P\theta_1 + P\theta_2 + 3Y_1 - 3Y_2 = 0 \\ &C_a\, dM_2 + P\theta_1 + 4P\theta_2 + P\theta_3 + 3Y_1 - 3Y_3 = 0 \\ &C_a\, dM_3 + P\theta_2 + 4P\theta_3 + P\theta_4 + 3Y_2 - 3Y_4 = 0 \\ &\text{etc.} \end{aligned} \tag{5.53}$$

where,

$$C_a = \frac{P^2}{2E_b I_a} \tag{5.54}$$

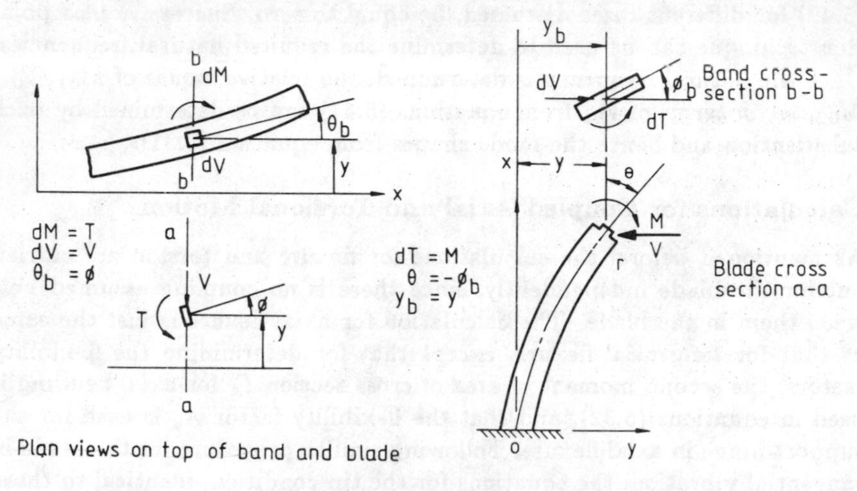

Fig. 5.6 Relationships at the point of attachment for coupled motion

The transverse force dV is also related to θ and Y as below

$$\begin{aligned} d_a\, dV_1 - P\theta_1 - P\theta_2 - 2Y_1 + 2Y_2 &= 0 \\ d_a\, dV_2 + P\theta_1 - P\theta_3 + 2Y_1 - 4Y_2 + 2Y_3 &= 0 \\ d_a\, dV_3 + P\theta_2 - P\theta_4 + 2Y_2 - 4Y_3 + 2Y_4 &= 0 \end{aligned} \tag{5.55}$$

etc.

where,

$$d_a = \frac{P^3}{6E_b I_a} \tag{5.56}$$

Also,

$$\begin{aligned} e\, dT_1 - \phi_{b1} + \phi_{b2} &= 0 \\ e\, dT_2 + \phi_{b1} - \phi_{b2} + \phi_{b3} &= 0 \\ e\, dT_3 + \phi_{b2} - \phi_{b3} + \phi_{b4} &= 0 \end{aligned} \tag{5.57}$$

etc.

where,

$$e = \frac{P}{G_b C} \tag{5.58}$$

With the help of equations (5.50) and (5.52) and figure 5.6, we have for blade N,

$$dM_N = T_r' T_{ON}\,; \quad dV_N = V_r' V_{ON} + V_r'' M_{ON}\,; \quad \theta_N = \phi_r' T_{ON}$$
$$Y_N = Y_r' V_{ON} + Y_r'' M_{ON}\,; \quad dT_N = M_r' V_{ON} + M_r'' M_{ON}$$
$$\phi_N = -\theta_r' V_{ON} - \theta_r'' M_{ON} \tag{5.59}$$

Hence, the equations (5.60) result.

	T_{01}	V_{01}	M_{01}	T_{02}	V_{02}	M_{02}	T_{03}	V_{03}	M_{03}	
	$C_aT'_r + 2P\phi'_r$	$3Y'_r$	$3Y''_r$	$P\phi'_r$	$-3Y'_r$	$-3Y''_r$	—	—	—	$= 0$
1	$-P\phi'_r$	$d_AV'_r - 2Y'_r$	$d_AV''_r - 2Y''_r$	$-P\phi'_r$	$2Y'_r$	$2Y''_r$	—	—	—	$= 0$
	—	$eM'_r + \theta'_r$	$eM''_r + \theta''_r$	—	$-\theta'_r$	$-\theta''_r$	—	—	—	$= 0$
	$P\phi'_r$	$3Y'_r$	$3Y''_r$	$C_aT'_r + 4P\phi'_r$	—	—	$P\phi'_r$	$-3Y'_r$	$-3Y''_r$	$= 0$
2	$P\phi'_r$	$2Y'_r$	$2Y''_r$	—	$d_AV'_r - 4Y'_r$	$d_AV''_r - 4Y''_r$	$-P\phi'_r$	$2Y'_r$	$2Y''_r$	$= 0$
	—	$-\theta'_r$	$-\theta''_r$	—	$eM'_r + 2\theta'_r$	$eM''_r + 2\theta''_r$	—	$-\theta'_r$	$-\theta''_r$	$= 0$
etc.										

(5.60)

The following cases for N^* blade are considered.

A. Even number of blades

$$N^* = \frac{S}{2}$$

(i) *Even symmetry*

Equations for N^* blade are

$$\begin{array}{cccccccc}
T_{ON^*-1} & V_{ON^*-1} & M_{ON^*-1} & T_{ON^*} & V_{ON^*} & M_{ON^*} & & \\
P\phi'_r & 3Y'_r & 3Y''_r & C_aT'_r + 3P\phi'_r & -3Y'_r & -3Y''_r & = 0 & \\
P\phi'_r & 2Y'_r & 2Y''_r & P\phi'_r & d_aV'_r - 2Y'_r & d_aV''_r - 2Y''_r & = 0 & \\
- & -\theta'_r & -\theta''_r & - & eM'_r + \theta'_r & eM''_r + \theta''_r & = 0 &
\end{array} \qquad (5.61)$$

(ii) *Odd symmetry*

Equations for N^* blade are

$$\begin{array}{cccccccc}
T_{ON^*-1} & V_{ON^*-1} & M_{ON^*-1} & T_{ON^*} & V_{ON^*} & M_{ON^*} & & \\
P\phi'_r & 3Y'_r & 3Y''_r & C_aT'_r + 5P\phi'_r & 3Y'_r & 3Y''_r & = 0 & \\
P\phi'_r & 2Y'_r & 2Y''_r & -P\phi'_r & d_aV'_r - 6Y'_r & d_aV''_r - 6Y''_r & = 0 & \\
- & -\theta'_r & -\theta''_r & - & eM'_r + 3\theta'_r & eM''_r + 3\theta''_r & = 0 &
\end{array} \qquad (5.62)$$

There will be a total of $3S/2$ equations in both the above cases, in the determinantal set.

B. Odd number of blades

$$N^* = \frac{1}{2}(S + 1)$$

(i) *Even symmetry*

$$T_{ON^*} = 0; \quad \text{Number of equations} = 3N^* - 1 = \tfrac{1}{2}(3S + 1).$$

[See equation (5.63) on page 165].

(ii) *Odd symmetry*

$$V_{ON^*} = M_{ON^*} = 0$$

Number of equations $= 3N^* - 2 = \frac{1}{2}(3S - 1)$. Equations for $N^* - 1$ and N^* are given in equation (5.64) [page 165].

	T_{ON^*-2}	V_{ON^*-2}	M_{ON^*-2}	T_{ON^*-1}	V_{ON^*-1}	M_{ON^*-1}	V_{ON^*}	M_{ON^*}	
	$P\phi'_r$	$3Y'_r$	$3Y''_r$	$C_aT'_r + 4P\phi'_r$	—	—	$-3Y'_r$	$-3Y''_r$	$= 0$
$N^* - 1$	$P\phi'_r$	$2Y'_r$	$2Y''_r$	—	$d_aV'_r - 4Y'_r$	$d_aV''_r - 4Y''_r$	$2Y'_r$	$2Y''_r$	$= 0$
	—	$-\theta'_r$	$-\theta''_r$	—	$eM'_r + 2\theta'_r$	$eM''_r + 2\theta''_r$	$-\theta'_r$	$-\theta''_r$	$= 0$
	—	—	—	$2P\phi'_r$	$4Y'_r$	$4Y''_r$	$d_aV'_r - 4Y'_r$	$d_aV''_r - 4Y''_r$	$= 0$
N^*									
	—	—	—	—	$-2\theta'_r$	$-2\theta''_r$	$eM'_r + 2\theta'_r$	$eM''_r + 2\theta''_r$	$= 0$

(5.63)

T_{ON^*-2}	V_{ON^*-2}	M_{ON^*-2}	T_{ON^*-1}	V_{ON^*-1}	M_{ON^*-1}	T_{ON^*}	
$P\phi'_r$	$3Y'_r$	$2Y''_r$	$C_aT'_r + 4P\phi'_r$	—	—	$P\phi'_r$	$= 0$
$P\phi'_r$	$2Y'_r$	$2Y''_r$	—	$d_aV'_r - 4Y'_r$	$d_aV''_r - 4Y''_r$	$-P\phi'_r$	$= 0$
—	$-\theta'_r$	$-\theta''_r$	—	$eM'_r + 2\theta'_r$	$eM''_r + 2\theta''_r$	—	$= 0$
—	—	—	$2P\phi'_r$	$6Y'_r$	$6Y''_r$	$C_aT'_r + 4P\phi'_r$	$= 0$

(5.64)

The natural frequencies and corresponding mode shapes are determined in a similar manner to that of the case of tangential vibrations considered earlier.

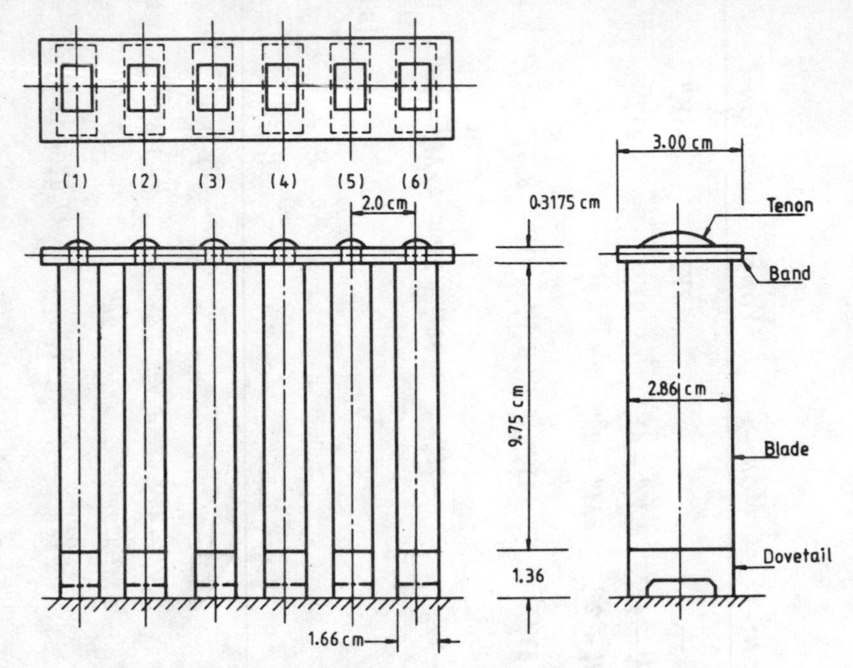

Fig. 5.7 Schematic representation of a six blade group

As an example consider the data of Weaver and Prohl [6]. The blade packet is schematically given in figure 5.7; it has the typical cross-section of an impulse blade with the areas of minimum and maximum moment of inertia very nearly perpendicular and parallel to the plane of the wheel respectively. Additional data for the blade is as follows.

Tangential moment of inertia	$= 0.3213$ cm^4
Axial moment of inertia	$= 0.7384$ cm^4
Cross-section area	$= 1.9226$ cm^4
Torsional stiffness constant	$= 0.3858$ cm^4

The frequencies and mode shapes obtained by Weaver are given in figure 5.8 for tangential mode and in figure 5.9 for coupled axial torsional mode.

The mode shapes are labeled according to the following scheme

T_{ij} = tangential

A_{ij} = Axial

R_{ij} = Torsional

i = number of nodes in the blade

j = number of nodes in the blade cover, 0 to $N-1$

The resonant response factors in figures 5.8 and 5.9 are discussed later in Section 8.3.

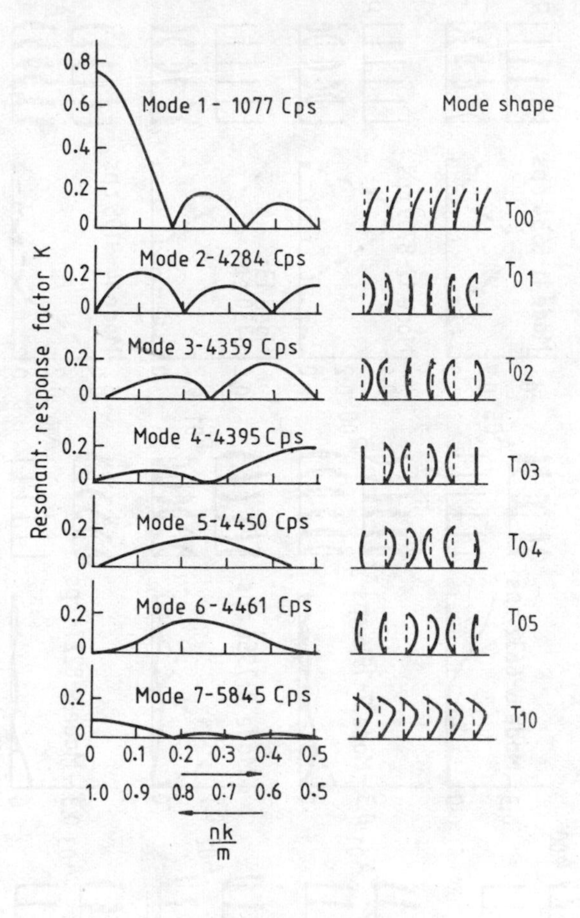

Fig. 5.8 Tangential vibration with resonant response factors

5.4 COMPLEX GROUP WITH LACING WIRES — DEAK AND BAIRD ANALYSIS

A complex group of blades with lacing wires used for the analysis is shown in figure 5.10. The blades are all set at zero stagger angle, on a rotating disc and the figure illustrates the coordinate axes and displacements. As in Prohl's method, the coupling between bending and torsion motions is through the lacing wires.

Let the number of blades in the packet be M. The notations for kth blade in the packet are given in the figure. The number of spans in each blade are n. The distance from the centre of the disc to each blade span beginning is denoted by x_j and the total length of blade from disc centre is L. For harmonic motion, with frequency ω

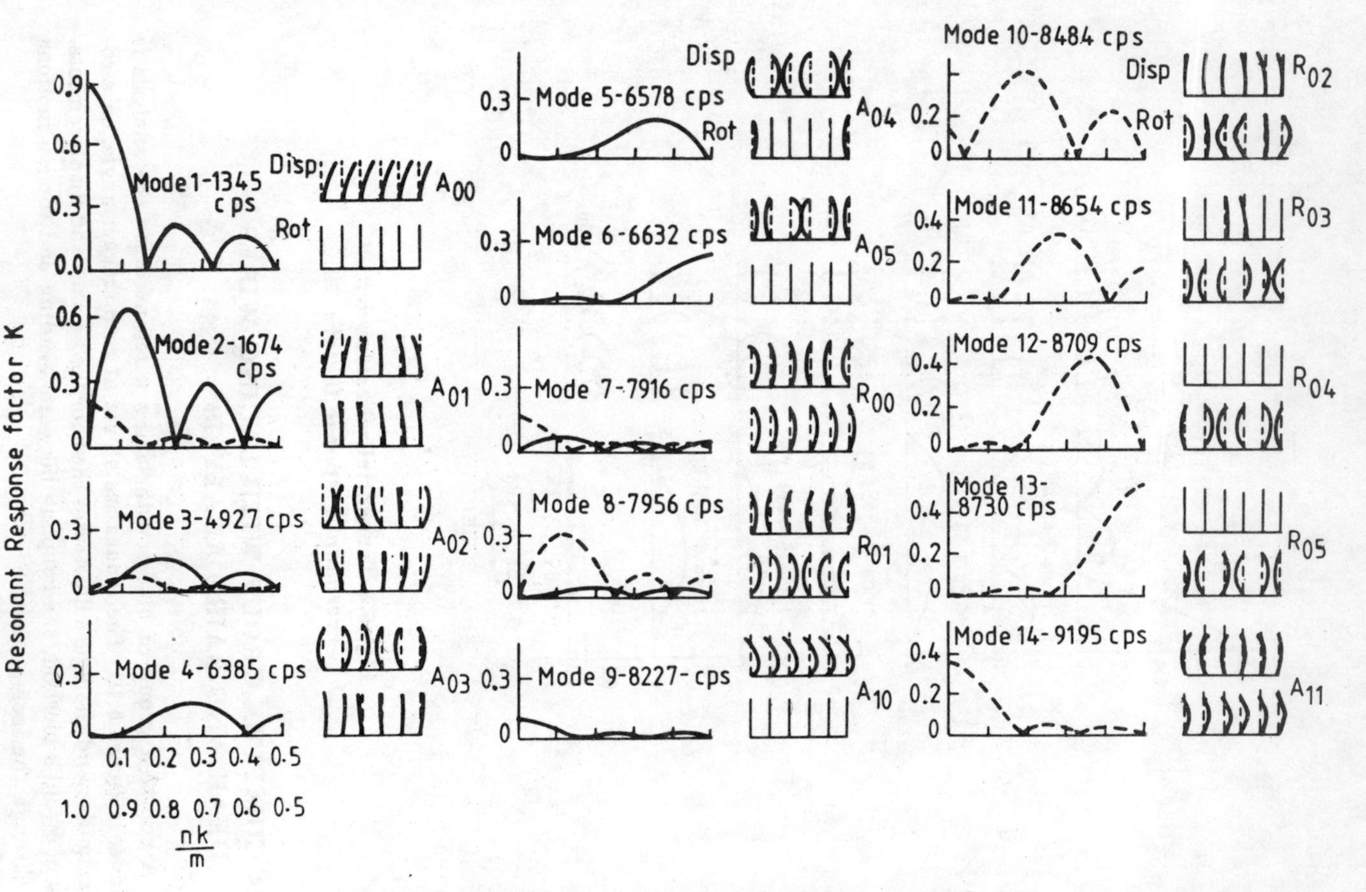

Fig. 5.9 Axial vibrations with resonant response factors

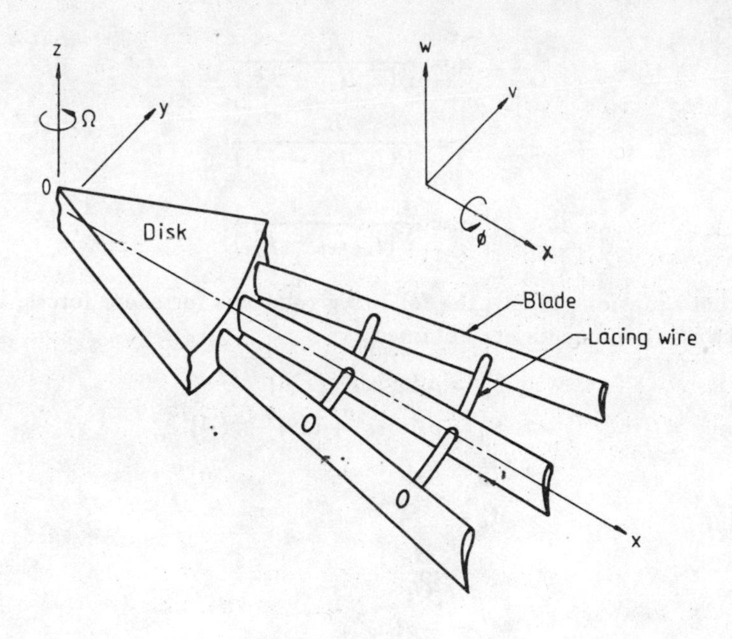

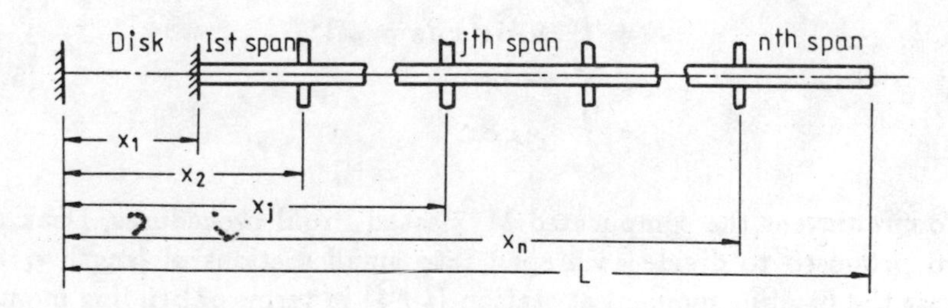

Fig. 5.10 Laced blade packet; *k*th blade with *n* spans

$$-[EI_{yz}v'' + EI_{yy}w'']'' + \rho A\omega^2 w + \left[w' \int_x^L \rho A\Omega^2 x\,dx\right]' = 0$$

$$-[EI_{zz}v'' + EI_{yz}w'']'' + \rho Av(\omega^2 + \Omega^2) + \left[v' \int_x^L \rho A\Omega^2 x\,dx\right]' = 0$$

$$\rho I_p\omega^2\phi + [C_t\phi']' = 0 \tag{5.65}$$

Let

$$v' = \theta\,; \quad w' = \mu \tag{5.66}$$

then the relations for a pre-twisted blade are

$$\begin{aligned} v'' &= \theta' = \alpha M_z + \beta M_y \\ w'' &= \mu' = -\gamma M_y - \beta M_z \end{aligned} \tag{5.67}$$

where,

$$\alpha = \frac{I_{yy}}{E(I_{yy}I_{zz} - I_{yz}^2)}$$
$$\beta = \frac{I_{yz}}{E(I_{yy}I_{zz} - I_{yz}^2)} \tag{5.68}$$
$$\gamma = \frac{I_{zz}}{E(I_{yy}I_{zz} - I_{yz}^2)}$$

From equations (5.65), the following relations for shear forces, bending and twisting moments are obtained

$$\begin{aligned}
S_z' &= \rho A\omega^2 w - (F\mu)' \\
S_y' &= \rho A v(\omega^2 + \Omega^2) - (F\theta)' \\
M_y' &= S_z \\
M_z' &= -S_y \\
\phi' &= \frac{M_x}{C_t} \\
M_x' &= -\rho I_p \omega^2 \phi
\end{aligned} \tag{5.69}$$

where,

$$\begin{aligned}
F &= \int_x^L \rho A\Omega^2 x\,dx = \rho\Omega^2 f \\
f &= \int_x^L Ax\,dx
\end{aligned} \tag{5.70}$$

To circumvent the complicated Myklestad-Prohl procedures, Deak and Baird proposed to divide each span into small sections of length ϵ_i and express the bending moment at station $(i+1)$ in terms of bending moment at station i, as

$$M_{i+1} = M_i + M_i'\epsilon_i + \frac{1}{2}M_i''\epsilon_i^2 \tag{5.71}$$

Other quantities are transferred from station to station by simple integration. The following transfer relations are thus obtained.

$$\begin{aligned}
M_{y,i+1} &= M_{y,i} + S_{z,i}\epsilon_i + \frac{1}{2}[-\rho A_i\omega^2 w_i + \rho\Omega^2 f_i(\gamma_i M_{yi} + \beta_i M_{zi}) \\
&\quad - \rho\Omega^2 f_i'\mu_i]\epsilon_i^2 \\
M_{z,i+1} &= M_{z,i} - S_{y,i}\epsilon_i + \frac{1}{2}[\rho A_i v_i(\omega^2 + \Omega^2) + \rho\Omega^2 f_i(\alpha_i M_{zi} + \beta_i M_{yi}) \\
&\quad + \rho\Omega^2 f_i'\theta_i]\epsilon_i^2 \\
\theta_{i+1} &= \theta_i + \frac{1}{2}[\alpha_{i+1}M_{z,i+1} + \alpha_i M_{zi}]\epsilon_i + \frac{1}{2}[\beta_{i+1}M_{y,i+1} + \beta_i M_{yi}]\epsilon_i \\
\mu_{i+1} &= \mu_i - \frac{1}{2}[\gamma_{i+1}M_{y,i+1} + \gamma_i M_{yi}]\epsilon_i - \frac{1}{2}[\beta_{i+1}M_{z,i+1} + \beta_i M_{zi}]\epsilon_i
\end{aligned}$$

$$v_{i+1} = v_i + \frac{1}{2}(\theta_{i+1} + \theta_i)\epsilon_i$$
$$w_{i+1} = w_i + \frac{1}{2}(\mu_{i+1} + \mu_i)\epsilon_i$$
$$S_{z,i+1} = S_{z,i} - \frac{1}{2}\rho\omega^2(A_{i+1}w_{i+1} + A_i w_i)\epsilon_i - \rho\Omega^2(f_{i+1}\mu_{i+1} - f_i\mu_i)$$
$$S_{y,i+1} = S_{y,i} - \frac{1}{2}\rho(\omega^2 + \Omega^2)(A_{i+1}v_{i+1} + A_i v_i)\epsilon_i - \rho\Omega^2(f_{i+1}\theta_{i+1} - f_i\theta_i)$$
$$M_{x,i+1} = M_{x,i} - \rho\omega^2 I_{pi}\phi_i\epsilon_i$$
$$\phi_{i+1} = \phi_i + \frac{1}{2}\left(\frac{M_{xi+1}}{C_{t,i+1}} + \frac{M_{x,i}}{C_{ti}}\right)\epsilon_i \tag{5.72}$$

A Prohl type of formulation is used with 10 fundamental solutions. The initial conditions for these 10 solutions are I : $M_y = 1$; II : $M_z = 1$; III : $S_y = 1$; IV : $S_z = 1$; V : $\theta = 1$; VI : $\mu = 1$; VII : $v = 1$; VIII : $w = 1$; IX : $\phi = 1$; X : $M_x = 1$. In all these fundamental solutions, denoted by I, II, etc., X, the remaining state vector quantities other than those mentioned are set zero. Hence for jth span of kth blade, the state quantities can be transferred from the first station to the last station and for each fundamental solution, e.g., the end quantities for I fundamental solution are $M_y^I, M_z^I, S_y^I, \ldots, w^I$ (eight for bending). Since there is no coupling between bending and torsion motions in the blade, ninth and tenth fundamental solutions give torsion motion separately. By using superposition principle, the end quantities of jth span in kth blade are

$$M_{yj}^K(x) = C_{11}M_y^I + C_{21}M_y^{II} + C_{31}M_y^{III} + \cdots + C_{81}M_y^{VIII}$$
$$M_{zj}^K(x) = C_{11}M_z^I + C_{21}M_z^{II} + C_{31}M_z^{III} + \cdots + C_{81}M_z^{VIII}$$
$$S_{yj}^K(x) = C_{11}S_y^I + C_{21}S_y^{II} + C_{31}S_y^{III} + \cdots + C_{81}S_y^{VIII}$$
$$S_{zj}^K(x) = C_{11}S_z^I + C_{21}S_z^{II} + C_{31}S_z^{III} + \cdots + C_{81}S_z^{VIII}$$
$$\theta_j^K(x) = C_{11}\theta^I + C_{21}\theta^{II} + C_{31}\theta^{III} + \cdots + C_{81}\theta^{VIII}$$
$$\mu_j^K(x) = C_{11}\mu^I + C_{21}\mu^{II} + C_{31}\mu^{III} + \cdots + C_{81}\mu^{VIII}$$
$$v_j^K(x) = C_{11}v^I + C_{21}v^{II} + C_{31}v^{III} + \cdots + C_{81}v^{VIII}$$
$$w_j^K(x) = C_{11}w^I + C_{21}w^{II} + C_{31}w^{III} + \cdots + C_{81}w^{VIII}$$
$$\phi_j^K(x) = C_{91}\phi^{IX} + C_{10,1}\phi^X$$
$$M_{xj}^K(x) = C_{91}M_x^{IX} + C_{10,1}M_x^X \tag{5.73}$$

For the first span, i.e., $j = 1$, only five fundamental solutions are needed, since $\theta = \mu = v = w = \phi = 0$ at the root. Equation (5.73) can be represented as

$$\{F_{r1}(x)\}_j^K = [f_{rs}(x)]_j^K \{C_{s1}\}_j^K \quad r = 1, 2, \ldots, 10 \quad s = 1, 2, \ldots, 10 \tag{5.74}$$

$$f_{r1}(x_j + 0) = \begin{cases} 1 & r = s \\ 0 & r \neq s \end{cases}$$

The lacing wires are lumped at corresponding blade sections and are treated as massless beam elements with extensional, flexural and torsional flexibility. Internal boundary conditions x_{j+1} for 1st, Kth and Mth blade are

$$\{F_{r1}(x_{j+1}+0)\}_{j+1}^{1} =$$

$$\begin{bmatrix} F_{11}-gF_{61} \\ F_{21}-2a_zF_{51} \\ F_{31}+(b-m\omega^2)F_{71} \\ F_{41}+\dfrac{3a_x}{l}F_{91}+\left(\dfrac{6a_x}{l^2}-m\omega^2\right)F_{81} \\ F_{51} \\ F_{61} \\ F_{71} \\ F_{81} \\ F_{91} \\ F_{10,1}+\dfrac{3a_x}{l}F_{81}+(2a_x-h\omega^2)F_{91} \end{bmatrix}_{j}^{1,\;x_{j+1}-0} + \begin{bmatrix} gF_{61} \\ a_zF_{51} \\ -bF_{71} \\ \dfrac{3a_x}{l}F_{91}-\dfrac{6a_x}{l^2}F_{81} \\ 0 \\ 0 \\ 0 \\ 0 \\ 0 \\ a_xF_{91}-\dfrac{3a_x}{l}F_{81} \end{bmatrix}_{j}^{2} \tag{5.75}$$

$$\{F_{r1}(x_{j+1}+0)\}_{j+1}^{K} = \begin{bmatrix} gF_{61} \\ a_zF_{51} \\ -bF_{71} \\ \dfrac{-3a_x}{l}F_{91}-\dfrac{6a_x}{l^2}F_{81} \\ 0 \\ 0 \\ 0 \\ 0 \\ 0 \\ a_xF_{91}+\dfrac{3a_x}{l}F_{81} \end{bmatrix}_{j}^{K+1,\;x_{j+1}-0}$$

$$+\begin{bmatrix} F_{11}-2gF_{61} \\ F_{21}+4a_zF_{51} \\ F_{31}+(2b-m\omega^2)F_{71} \\ F_{41}+\left(\dfrac{12a_x}{l^2}-m\omega^2\right)F_{81} \\ F_{51} \\ F_{61} \\ F_{71} \\ F_{81} \\ F_{91} \\ F_{10,1}+(4a_x-h\omega^2)F_{91} \end{bmatrix}_{j}^{K,\;x_{j+1}-0} + \begin{bmatrix} gF_{61} \\ a_zF_{51} \\ -bF_{71} \\ \dfrac{3a_x}{l}F_{91}-\dfrac{6a_x}{l^2}F_{81} \\ 0 \\ 0 \\ 0 \\ 0 \\ 0 \\ a_xF_{91}-\dfrac{3a_x}{l}F_{81} \end{bmatrix}_{j}^{K+1,\;x_{j+1}-0} \tag{5.76}$$

$$\{F_{r1}(x_{j+1}+0)\}_{j+1}^{M} =$$

$$\begin{bmatrix} gF_{61} \\ a_z F_{51} \\ -bF_{71} \\ \dfrac{-3a_x}{l}F_{91} - \dfrac{6a_x}{l^2}F_{81} \\ 0 \\ 0 \\ 0 \\ 0 \\ 0 \\ a_x F_{91} + \dfrac{3a_x}{l}F_{81} \end{bmatrix}_j^{M-1} \begin{matrix} \\ x_{j+1}-0 \end{matrix} + \begin{bmatrix} F_{11} - gF_{61} \\ F_{21} + 2a_z F_{51} \\ F_{31} + (b - m\omega^2)F_{71} \\ F_{41} - \dfrac{3a_x}{l}F_{91} + \left(\dfrac{6a_x}{l^2} - m\omega^2\right)F_{81} \\ F_{51} \\ F_{61} \\ F_{71} \\ F_{81} \\ F_{91} \\ F_{10,1} - \dfrac{3a_x}{l}F_{81} + (2a_x - h\omega^2)F_{91} \end{bmatrix}_j^{M} \quad (5.77)$$

(both column matrices evaluated at $x_{j+1}-0$)

where,

$$\begin{aligned} a_x &= 2E_l\frac{I_{xxl}}{l} \\ a_z &= 2E_l\frac{I_{zzl}}{l} \\ b &= E_l\frac{A_l}{l} \\ g &= \frac{I_{xxl} + I_{zzl}G_l}{l} \\ h &= m\lambda^2 \end{aligned} \quad (5.78)$$

m is lumped lacing wire mass, λ is the radius of gyration of lumped lacing wire mass, l is lacing wire length and suffix l denotes the lacing wire quantities. Equations (5.75) to (5.77) can be rewritten as

$$\begin{aligned} \{F_{r1}(x_{j+1}+0)\}_{j+1}^{K} &= [f_{rs}(x_{j+1}+0)]_{j+1}\{C_{s1}\}_{j+1}^{K} \\ &= [a_{sp}]_j^{K-1}\{C_{p1}\}_j^{K-1} + [b_{sp}]_j^{K}\{C_{p1}\}_j^{K} \\ &\quad + [d_{sp}]_j^{K+1}\{C_{p1}\}_j^{K+1} \end{aligned} \quad (5.79)$$

Since $[f_{rs}(x_j+0)]$ is a unit matrix, we have the following recursive relationship relating the C's of $(j+1)$ span of Kth blade to the C's of jth span of the $(K-1)$, K and $(K+1)$ blades.

$$\begin{aligned} \{C_{s1}\}_{j+1}^{K} &= [a_{sr}]_j^{K-1}\{C_{r1}\}_j^{K-1} + [b_{3r}]_j^{K}\{C_{r1}\}_j^{K} \\ &\quad + [d_{sr}]_j^{K+1}\{C_{r1}\}_j^{K+1} \end{aligned} \quad (5.80)$$

In the above, the components of $\left[a_{sr}\right]_j^K$ matrix are

$$
\begin{aligned}
a_{1r} &= g f_{6r}(x_{j+1}) \\
a_{2r} &= a_z f_{5r}(x_{j+1}) \\
a_{3r} &= -b f_{7r}(x_{j+1}) \\
a_{4r} &= -\frac{6a_x}{l^2} f_{8r}(x_{j+1}) \\
a_{10r} &= \frac{3a_x}{l} f_{8r}(x_{j+1}) \quad r = 1, 2, \ldots, 8 \\
a_{4r} &= -\frac{3a_x}{l} f_{9r}(x_{j+1}) \\
a_{10r} &= a_x f_{9r}(x_{j+1}) \quad r = 9, 10 \\
a_{rs} &= 0 \quad \text{for all other } r \text{ and } s
\end{aligned}
\tag{5.81}
$$

The components of $\left[b_{sr}\right]_j^1$ matrix are

$$
\begin{aligned}
b_{1r} &= f_{1r}(x_{j+1}) - g f_{6r}(x_{j+1}) \\
b_{2r} &= f_{2r}(x_{j+1}) + 2a_z f_{5r}(x_{j+1}) \\
b_{3r} &= f_{3r}(x_{j+1}) + (b - m\omega^2) f_{7r}(x_{j+1}) \\
b_{4r} &= f_{4r}(x_{j+1}) + \left(\frac{6a_x}{l^2} - m\omega^2\right) f_{8r}(x_{j+1}) \\
b_{10r} &= \frac{3a_x}{l} f_{8r}(x_{j+1}) \quad r = 1, 2, \ldots, 8 \\
b_{4r} &= \frac{3a_x}{l} f_{9r}(x_{j+1}) \\
b_{9r} &= f_{9r}(x_{j+1}) \\
b_{10,r} &= f_{10,r}(x_{j+1}) + (2a_x - h\omega^2) f_{9r}(x_{j+1}) \quad r = 9, 10 \\
b_{sr} &= f_{sr}(x_{j+1}) \quad s = 5, 6, 7, 8;\ r = 1, 2, \ldots, 8 \\
b_{sr} &= 0 \quad \text{for all other } r \text{ and } s
\end{aligned}
\tag{5.82}
$$

The components of $\left[b_{sr}\right]_j^K$ matrix are

$$
\begin{aligned}
b_{1r} &= f_{1r}(x_{j+1}) - 2g f_{6r}(x_{j+1}) \\
b_{2r} &= f_{2r}(x_{j+1}) + 4a_z f_{5r}(x_{j+1}) \\
b_{3r} &= f_{3r}(x_{j+1}) + (2b - m\omega^2) f_{7r}(x_{j+1}) \\
b_{4r} &= f_{4r}(x_{j+1}) + \left(\frac{12a_x}{l^2} - m\omega^2\right) f_{8r}(x_{j+1}) \quad r = 1, 2, \ldots, 8 \\
b_{9r} &= f_{9r}(x_{j+1}) \\
b_{10r} &= f_{10r}(x_{j+1}) + (4a_x - h\omega^2) f_{9r}(x_{j+1}) \quad r = 9, 10 \\
b_{sr} &= f_{sr}(x_{j+1}) \quad s = 5, 6, 7, 8;\ r = 1, 2, \ldots, 8 \\
b_{sr} &= 0 \quad \text{for all other } r \text{ and } s
\end{aligned}
\tag{5.83}
$$

The components of $\left[b_{sr}\right]_j^M$ matrix are

$$
\begin{aligned}
b_{1r} &= f_{1r}(x_{j+1}) - g f_{6r}(x_{j+1}) \\
b_{2r} &= f_{2r}(x_{j+1}) + 2a_z f_{5r}(x_{j+1}) \\
b_{3r} &= f_{3r}(x_{j+1}) + (b - m\omega^2) f_{7r}(x_{j+1}) \\
b_{4r} &= f_{4r}(x_{j+1}) + \left(\frac{6a_x}{l^2} - m\omega^2\right) f_{8r}(x_{j+1}) \\
b_{10r} &= -\frac{3a_x}{l} f_{8r}(x_{j+1}) \quad r = 1, 2, \ldots, 8 \\
b_{4r} &= -\frac{3a_x}{l} f_{9r}(x_{j+1}) \\
b_{9r} &= f_{9r}(x_{j+1}) \\
b_{10,r} &= f_{10,r}(x_{j+1}) + (2a_x - h\omega^2) f_{9r}(x_{j+1}) \quad r = 9, 10 \\
b_{sr} &= f_{sr}(x_{j+1}) \quad s = 5, 6, 7, 8;\ r = 1, 2, \ldots, 8 \\
b_{sr} &= 0 \quad \text{for all other } r \text{ and } s
\end{aligned}
\tag{5.84}
$$

$\left[d_{sr}\right]_j^K$ is

$$
\begin{aligned}
d_{1r} &= g f_{61}(x_{j+1}) \\
d_{2r} &= a_z f_{5r}(x_{j+1}) \\
d_{3r} &= -b f_{7r}(x_{j+1}) \\
d_{4r} &= -\frac{6a_x}{l^2} f_{8r}(x_{j+1}) \\
d_{10,r} &= -\frac{3a_x}{l} f_{8r}(x_{j+1}) \quad r = 1, 2, \ldots, 8 \\
d_{4r} &= \frac{3a_x}{l} f_{9r}(x_{j+1}) \\
d_{10,r} &= a_x f_{9r}(x_{j+1}) \quad r = 9, 10 \\
d_{sr} &= 0 \quad \text{for rest of } r \text{ and } s
\end{aligned}
\tag{5.85}
$$

For the first span of the Kth blade, from equation (5.74)

$$
\left\{F_{r1}(x)\right\}_1^K = \left[f_{rs}(x)\right]_1 \left\{C_s\right\}_1^K \tag{5.86}
$$

where $\left[f_{rs}(x)\right]_1$ is given by equation (5.74) for the first part of any blade at the end of $p1$ stations.

$\left\{F_{r1}(x)\right\}_1^K$ represents the values of state vector at $x = x_2 - 0$. From equation (5.79) we get for the beginning of second part

$$
\left\{F_{r1}(x_2 + 0)\right\}_2^K = \left[f_{rs}(x_2 + 0)\right]_2 \left\{C_{s1}\right\}_2^K \tag{5.87}
$$

In the above $\left[f_{rs}(x_2 + 0)\right]_2$ is a unit matrix and the values of state vector at $(x_2 + 0)$ are unknowns given by $\left\{C_{s1}\right\}_2^K$. This vector is to be determined in terms of $\left\{C_{s1}\right\}_1^{K-1}$, $\left\{C_{s1}\right\}_1^K$ and $\left\{C_{s1}\right\}_1^{K+1}$. This is given by equation (5.80).

$$\{C_{s1}\}_2^K = [a_{sr}]_1^K \{C_{r1}\}_1^{K-1} + [b_{sr}]_1^K \{C_{r1}\}_1^K + [d_{sr}]_1^{K+1} \{C_{r1}\}_1^{K+1}$$

$$= \underset{(10 \times 30)}{\left[[a]^{K-1}\,[b]^K\,[d]^{K+1}\right]_1} \underset{(30 \times 1)}{\begin{bmatrix} \{C\}^{K-1} \\ \{C\}^K \\ \{C\}^{K+1} \end{bmatrix}_1} \tag{5.88}$$

For the second span,

$$\{F_{r1}(x_2)\}_2^K = [f_{rs}(x_2)]\,\{C_{s1}\}_2^K \tag{5.89}$$

where $[f_{rs}(x_2)]_2$ is given by (5.74) for second span of any blade with $p2$ stations and $\{C_{s1}\}_2^K$ is given by equation (5.88). Hence,

$$\{F_{r1}(x_2)\}_2^K = [f_{rs}(x_2)]_2 \left[[a]\,[b]\,[d]\right]_1 \begin{bmatrix} \{C\}^{K-1} \\ \{C\}^K \\ \{C\}^{K+1} \end{bmatrix} \tag{5.90}$$

If there is only one lacing wire, the following equations for free ends of all the M blades are obtained

$$\begin{aligned}
\{F_{r1}(x)\}_2^1 &= [f_{rs}(x)]_2 \quad \left[0\,[b]^1\,[d]^2\right]_1 \begin{bmatrix} \{0\} \\ \{C\}^1 \\ \{C\}^2 \end{bmatrix}_1 \\
\{F_{r1}(x)\}_2^K &= [f_{rs}(x)]_2 \quad \left[[a]^{K-1}\,[b]^K\,[d]^{K+1}\right]_1 \begin{bmatrix} \{C\}^{K-1} \\ \{C\}^K \\ \{C\}^{K+1} \end{bmatrix} \\
\{F_{r1}(x)\}_2^M &= [f_{rs}(x)]_2 \quad \left[[a]^{M-1}\,[b]^M\,0\right]_1 \begin{bmatrix} \{C\}^{M-1} \\ \{C\}^M \\ \{0\} \end{bmatrix} \\
(10 \times 1) \quad & \quad (10 \times 10) \quad\quad (10 \times 30) \quad\quad (30 \times 1)
\end{aligned} \tag{5.91}$$

In general, equation (5.91) can be written as

$$\begin{bmatrix} M_y \\ M_z \\ S_y \\ S_z \\ \theta \\ \mu \\ v \\ w \\ \phi \\ M_x \end{bmatrix}_2^K = \begin{bmatrix} S_{1,1} & S_{1,2} & - & S_{1,30} \\ S_{2,1} & S_{2,2} & - & S_{2,30} \\ S_{3,1} & & - & S_{3,30} \\ S_{4,1} & & - & S_{4,30} \\ S_{5,1} & & - & S_{5,30} \\ S_{6,1} & & - & S_{6,30} \\ S_{7,1} & & - & S_{7,30} \\ S_{8,1} & & - & S_{8,30} \\ S_{9,1} & & - & S_{9,30} \\ S_{10,1} & & - & S_{10,30} \end{bmatrix}_1^K \begin{bmatrix} \begin{bmatrix} C_{11} \\ \vdots \\ C_{10,1} \end{bmatrix}^{K-1} \\ \begin{bmatrix} C_{11} \\ \vdots \\ C_{10,1} \end{bmatrix}^{K} \\ \begin{bmatrix} C_{11} \\ C_{21} \\ \vdots \\ C_{10,1} \end{bmatrix}^{K+1} \end{bmatrix}_1 \tag{5.92}$$

Noting that at free end, $M_y = M_z = S_y = S_z = M_x = 0$ and at fixed end $C_{51} = C_{61} = C_{71} = C_{81} = C_{91} = 0$,

$$\begin{aligned} & S_{r1}C_{11}^{K-1} + S_{r2}C_{21}^{K-1} + S_{r3}C_{31}^{K-1} + S_{r4}C_{41}^{K-1} + S_{r,10}C_{10,1}^{K-1} \\ & + S_{r,11}C_{11}^{K} + S_{r,12}C_{21}^{K} + S_{r,13}C_{31}^{K} + S_{r,14}C_{41}^{K} + S_{r,20}C_{10,1}^{K} \\ & + S_{r,21}C_{11}^{K+1} + S_{r,22}C_{21}^{K+1} + S_{r,23}C_{31}^{K+1} + S_{r,24}C_{41}^{K+1} + S_{r,30}C_{10,1}^{K+1} \\ & = 0 \qquad r = 1, 2, 3, 4, 10 \end{aligned} \tag{5.93}$$

If there are M blades, we will have $5M$ such equations. It should be remembered that S is the same for all blades excepting for the first and last blades. This system determinant must be zero for a natural frequency of the blade packet. Symmetry or antisymmetry properties can be used to reduce the size of the determinant.

If there are two lacing wires, for third span, equation (5.80) gives

$$\{C_{31}\}_3^K = \left[[a_{sr}]^{K-1} \, [b_{sr}]^{K} \, [d_{sr}]^{K+1} \right]_2 \begin{bmatrix} \{C\}^{K-1} \\ \{C\}^{K} \\ \{C\}^{K+1} \end{bmatrix}_2 \tag{5.94}$$

Substituting for $\{C_{s1}\}_2$ from equation (5.88)

$$\underset{(10\times 1)}{\{C\}_3^K} = \underset{(10\times 30)}{\left[[a]\,[b]\,[d] \right]_2} \underset{(30\times 50)}{\begin{bmatrix} [a] & [b] & [d] & [0] & [0] \\ [0] & [a] & [b] & [d] & [0] \\ [0] & [0] & [a] & [b] & [d] \end{bmatrix}_1} \underset{(50\times 1)}{\begin{bmatrix} \{C\}^{K-2} \\ \{C\}^{K-1} \\ \{C\}^{K} \\ \{C\}^{K+1} \\ \{C\}^{K+2} \end{bmatrix}_1} \tag{5.95}$$

Hence,

$$\underset{(10\times 1)}{\{C\}_3^K} = \underset{(10\times 50)}{[T_c]_2} \underset{(50\times 1)}{\begin{bmatrix}\{C\}^{K-2}\\ \{C\}^{K-1}\\ \{C\}^{K}\\ \{C\}^{K+1}\\ \{C\}^{K+2}\end{bmatrix}_1} \tag{5.96}$$

From the state vector for the third span

$$\underset{(10\times 1)}{\{F_{r1}(x_3)\}_3^K} = \underset{(10\times 50)}{[T]_2} \underset{(50\times 1)}{\begin{bmatrix}\{C\}^{K-2}\\ \{C\}^{K-1}\\ \{C\}^{K}\\ \{C\}^{K+1}\\ \{C\}^{K+2}\end{bmatrix}_1} \tag{5.97}$$

where

$$[T]_2 = [f_{rs}(x_3)]_3\,[T_c]_2$$

Using boundary conditions,

$$\begin{aligned}
&T_{r1}C_{11}^{K-2} + T_{r2}C_{21}^{K-2} + T_{r3}C_{31}^{K-2} + T_{r4}C_{41}^{K-2} + T_{r,10}C_{10,1}^{K-2}\\
&+ T_{r,11}C_{11}^{K-1} + T_{r,12}C_{21}^{K-1} + T_{r,13}C_{31}^{K-1} + T_{r,14}C_{41}^{K-1} + T_{r,20}C_{10,1}^{K-1}\\
&+ T_{r,21}C_{11}^{K} + T_{r,22}C_{21}^{K} + T_{r,23}C_{31}^{K} + T_{r,24}C_{41}^{K} + T_{r,30}C_{10,1}^{K}\\
&+ T_{r,31}C_{11}^{K+1} + T_{r,32}C_{21}^{K+1} + T_{r,33}C_{31}^{K+1} + T_{r,34}C_{41}^{K+1} + T_{r,40}C_{10,1}^{K+1}\\
&+ T_{r,41}C_{11}^{K+2} + T_{r,42}C_{21}^{K+2} + T_{r,43}C_{31}^{K+2} + T_{r,44}C_{41}^{K+2} + T_{r,50}C_{10,1}^{K+2}\\
&= 0 \qquad r = 1,2,3,4,10
\end{aligned} \tag{5.98}$$

There are M blades. Hence there are, as before $5M$ equations. Again, elements of T are the same for all blades excepting for first and Mth blades. The size of the frequency determinant can be reduced as before, by considering the symmetry and antisymmetry of the packet. To find the mode shape, assign any quantity, say $C_{11}^1 = 1.0$ and determine the rest of the fundamental solutions at the roots of all blades by solution of simultaneous equations of the system. Once the root values of state vector are known, the deflections at all other stations can be determined. Deak and Baird [9] used this method to calculate the natural frequencies of a four bladed packet and found good agreement with test results (see Fig. 5.11).

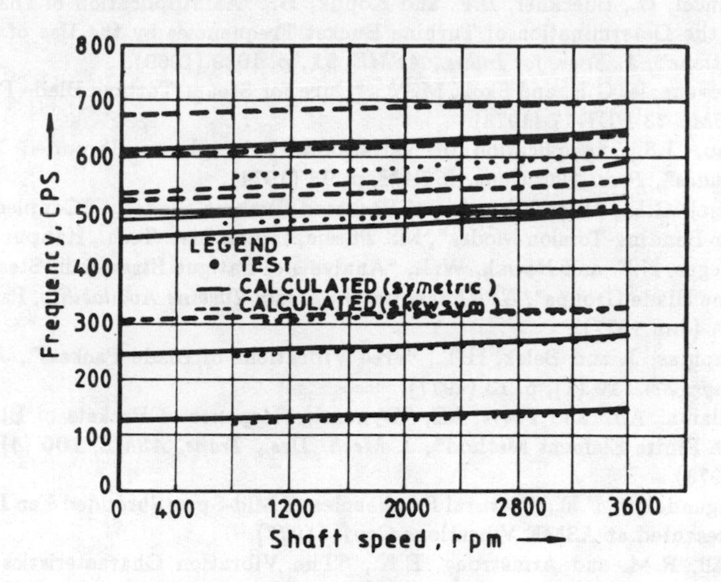

Fig. 5.11 Campbell diagram of a four-blade doubly laced packet

REFERENCES

1. Kroon, R.P., "Influence of Lashing and Centrifugal Force on Turbine-Blade Stresses", *Trans. ASME*, **56**, p. 109 (1934).
2. Jarret, G.W. and Warner, P.C., "Vibration of Rotating Tapered Twisted Beams", *J. Appl. Mech., ASME*, **20**, p. 381 (1953).
3. Smith, D.M., "Vibration of Turbine Blades in Packets", *Proc. 7th Conf. for Appl. Mech.*, **3**, p. 178 (1948).
4. Ellington, J.P. and McCallion, H., "The Vibration of Laced Turbine Blades", *J. Roy. Aero. Soc.*, **61**, p. 563 (1957).
5. Prohl, M.A., "A Method for Calculating Vibration Frequency and Stress of a Banded Group of Turbine Buckets", *Trans. ASME*, **80**, p. 169 (1958).
6. Weaver, F.L. and Prohl, M.A., "High Frequency Vibration of Steam Turbine Buckets", *Trans. ASME*, **80**, p. 181 (1958).
7. Singh, B.R. and Nandeeswariah, N.S., "Vibration Analysis of Shrouded Turbine Blades", *J. Inst. Engr. Ind.*, **40**, p. 16 (1959).
8. Rao, J.S., *Blade Group Forced Vibration — Computer Program*, Tech. Memo. 76 WRL M23, Rochester Institute of Technology, Rochester, New York (1976).
9. Deak, A.L. and Baird, R.D., "A Procedure for Calculating the Frequencies of Steam Turbine Exhaust Blades", *Trans. ASME*, **85**, p. 324 (1963).
10. Rieger, N.F. and McCallion, H., "The Natural Frequencies of Portable Frames II", *Int. J. Mech. Sci.*, **7**, p. 263 (1965).
11. Stuwing, D., "Zur Bereschnung der Eigen Frequenzen von Schaufelpaketen Axialer Turbomaschinen", *Maschinenbautechnik*, **17**, p. 457 (1968).

12. Tuncel, O., Bueckner, H.F. and Koplik, B., "An Application of Diakoptics in the Determination of Turbine Bucket Frequencies by the Use of Perturbations", *J. Engr. for Indus., ASME*, **91**, p. 1029 (1969).
13. Provenzale, G.E. and Skok, M.W., "Cure for Steam-Turbine Blade Failure", *ASME* 73-PET-17 (1973).
14. Rao, J.S., "Application of Variational Principle to Shrouded Turbine Blades", *Proc. 19th Cong. ISTAM*, p. 93 (1974).
15. Bajaj, G.R., "Free Vibration of Packeted Turbine Blades — Coupled Bending-Bending-Torsion Modes", MS Thesis, Indian Inst. Tech., Kanpur (1974).
16. Rieger, N.F. and Nowak, W.J., "Analysis of Fatigue Stresses in Steam Turbine Blade Groups", *EPRI Seminar on Steam Turbine Availability*, Palo Alto. CA (Jan 1977).
17. Thomas, J. and Belek, H.T., "Free Vibrations of Blade Packets", *J. Mech. Engr. Sci.*, **19** (1), p. 13 (1977).
18. Salama, A.L. and Petyt, M., "Dynamic Response of Packets of Blades by the Finite Element Method", *J. Mech. Des., Trans. ASME*, **100** (4), p. 660 (1978).
19. Sagendorph, F.E., "Natural Frequencies of Mid-Span Shrouded Fan Blades", Presented at ASME Vibrations Conf. (1977).
20. Hall, R.M. and Armstrong, E.K., "The Vibration Characteristics of and Assembly of Interlock Shrouded Turbine Blades", Presented at ASME Vibrations Conf. (1977).
21. Srinivasan, A.V., Lionberger, S.R. and Brown, K.W., "Dynamic Analysis of an Assembly of Shrouded Blades Using Component Modes", *J. Mech. Des., ASME*, **100** (3), p. 520 (1978).
22. Lu, L.K.H. and Warner, P.C., "A Statistical Assessment of the Effect of Variable Root Flexibility on the Vibration Response of Shrouded Blades", *Vibrations of Bladed Disc Assemblies*, ASME Publication (1983).
23. Chattopadhyay, A.K., "Vibration of Shrouded Turbine Blade Packets", M. Tech. Thesis, Ind. Inst. of Tech., Kharagpur (1974).

Chapter 6

Excitation

6.1 INTRODUCTION

The non-steady forces acting in a turbine stage are primarily due to flow interaction between the stationary diaphragms and rotating blades. Such an interaction produces excitation at Nozzle Passing Frequency (NPF) and its harmonics. An appropriate approach to the problem of design of a turbomachine blade is to study the nature of these excitation forces and determine the dynamic stresses arising out of them. The important interactions which cause these excitation forces are the potential flow and wake interactions. This chapter is concerned with the determination of non-steady forces due to NPF excitation in a turbomachine stage.

Basic Isolated Airfoil Theories

Karman and Sears [1] represented an airfoil by a vortex sheet and pointed out the mechanism by which a wake of vorticity is produced by an airfoil in non-uniform motion. The flow was assumed to be potential and two dimensional and the wake behind was considered to be rectilinear. They derived methods to determine both lift and moment acting on the airfoil and the general results obtained were applied to the case of an oscillating airfoil and then to the problem of a plane airfoil entering a sharp edged gust. Silverstein, Katzoff and Bullivant [2] studied the turbulent wake of an airfoil experimentally and suggested the governing equations for wake profile in terms of the distance behind airfoil trailing edge and profile drag coefficient. They assumed that the wake consists of symmetrical shear perturbations of the undisturbed stream. A more systematic investigation of the mean velocity characterisitics of the wake behind an isolated airfoil was carried out by Preston and Sweeting [3]. They observed similarity in velocity profiles and provided a general expression of the decay of velocity defect which is independent of the shape of the body or the loading on the body. Wake studies are also made by Hah and Lakshminarayana [4] who have conducted experimental investigations of the near wake of a thin airfoil at various incidence angles. The effect of free stream turbulence on the development of a three dimensional wake of a compressor rotor blade was also studied analytically and experimentally by them [5].

Sears [6] extended the theory of thin airfoils proposed by him and Karman to obtain the lift and moment for a rigid airfoil passing through a vertical gust pattern having a sinusoidal variation. The expressions derived by him essentially apply to a convecting gust in which the velocity perturbation moves past the airfoil with a velocity equal to flow velocity. Later Kemp [7] generalized the results of Sears [6] for an airfoil flying through a non-convecting gust in which the velocity of flow perturbation is not equal to the flow velocity. Geissing et al [8] reviewed the Sears function for the lift on a two dimensional airfoil induced by a harmonic gust field in an incompressible flow and defined a simpler function by transforming the gust reference point from the airfoil mid chord to its leading edge.

Until late 1960's, the analytical approach to the basic thin airfoil studies in non-uniform motion was mainly restricted to the case of velocity perturbations perpendicular to the airfoil chord. Horlock [9] used the linearized form of the unsteady Bernoulli equation to define the lift in terms of vorticity, instead of Karman and Sears [1] whose arguments were based on the moment of a vortex pair, to determine the response of an airfoil subjected to a streamwise gust. He obtained a Fourier series expression for the quasi steady vorticity and obtained the unsteady lift through a series of integrals which could be evaluated. By combining his analysis with established theory for transverse gusts, Horlock suggested simple design rules for obtaining fluctuating lift on a rotating fan or a compressor blade moving through a flow disturbance. Holmes [10] found gust response functions for some specific cases, e.g., flat plate airfoil subjected to non-convecting streamwise flow perturbation, cambered airfoil profile subjected to convective flow perturbations in both streamwise and chordwise directions, etc. His analysis was similar to that of Horlock [9] in that Fourier series expression was obtained through a series of integrals. He also conducted wind tunnel tests to determine the lift and pressure fluctuations on an isolated airfoil subjected to transverse and streamwise gusts.

Naumann and Yeh [11] obtained closed form solutions for the unsteady lift of a cambered airfoil with angle of attack, moving through both longitudinal and transverse gusts. The unsteady lift was expressed as sum of three terms by them representing the influence of (1) camber and longitudinal gust, (2) angle of attack and longitudinal gust and (3) a transverse gust. These results were compared with Sears and Horlock functions. Rao [12,13] developed expressions for the lift and moment fluctuations of a general camber airfoil subjected to non-convecting and convecting streamwise gusts, as an extension of Holmes [10] work, who considered cambered blades in streamwise convecting gust only. Rao's analysis was based on linearized potential theory and carried on similar lines to that of Horlock [9].

Graham [14] constructed similarity rules for load distribution induced on a thin two dimensional wing at subsonic speeds by sinusoidal gusts, whose wave fronts are at angle to the leading edge of the wing. Using the method of solution of integral equation developed for the incompressible oblique gust, he computed some values of the induced lift coefficient for the compressible two dimensional gust at Mach numbers 0 to 0.8 and has shown that the zero Mach number values agreed with those of Sears [6]. Amiet and Sears [15] used the method of matched asymptotic expansions to simplify the calculations of noise produced by aerodynamic flows involving small perturbations of a stream of a negligible subsonic Mach number. Osborne [16] developed closed form expressions for the forces on a two dimensional thin airfoil in subsonic flow that is subjected to a general type of oscillating upwash whose x dependance may be expanded in a cosine series, under certain frequency and Mach number restrictions. By relating pressures in both compressible and incompressible planes, he brought upwash expressions in the compressible plane into the forms of Sears and Kemp-type upwashes and determined the lift and moments acting upon the airfoils. A significant outcome of the work is that the forces are shown to depend on all Glauert coefficients, whereas Sears has shown that they depend upon only first three coefficients. However Miles [18] expressed doubts earlier itself about the procedure adopted by Osborne.

Kemp [19] reviewed Osborne's analysis and suggested a method by means of which closed form solutions can be obtained for all types of upwash discussed in the analysis. Amiet [20] derived an analytical expression based on a previous solution by Miles, for the pressure, lift and moment on a two dimensional airfoil encountering a sinusoidal gust in compressible subsonic flow. The solution is closely related to that of Osborne [17]. Later Kemp [21] derived the lift and moment expressions for any arbitrary integer power law upwash, using algebraic methods. Amiet [22] derived an approximate solution for the response function of an infinite span airfoil in a three dimensional (the gust wave front skewed relative to the airfoil leading edge) gust convecting with the free stream incompressible flow.

Supersonic flows over isolated airfoils were investigated by several authors, one of the recent ones being that of Timothy, Lewis and Sirovitch [23]. The approximation produced sufficiently accurate results for a wide range of Mach numbers.

Flow Interference in a Turbomachine Stage

The most promising approach to the problem of predicting unsteady aerodynamic forces in the blade rows has been made by Kemp and Sears [24]. They applied the results of the theory of single thin airfoils to calculate the unsteady lifts and moments of an elementary stage. The analysis made by them considers a single stator row followed by a single rotor row and the flow was assumed incompressible and inviscid, except that the wakes are considered as regions of velocity defect. Further, it was assumed that the unsteady effects at any given blade are influenced by steady circulation only but not by non-steady effects of other blades. However for rotor blades, the effects of their passage through the non-steady vortex wakes and also steady viscous wakes, shed by the stator are also included. Rao [25] developed a computer program to calculate the unsteady lift forces for the first two harmonics acting on stationary and moving blades including the viscous effects considered by Kemp and Sears [26].

Lots and Raabe [27] used a direct integral equation approach to determine the lift forces in an axial turbomachine stage with similar assumptions of Kemp and Sears. The integral equation for the unknown unsteady vortex strength was obtained by distributing vortices along the blades in the row of interest and imposing no penetration condition at a blade in that row. While arriving at the upwash imposed at a blade, they included the effects of the unsteady bound vortices on the blades in the same row and the effects of vortex wakes emanating from those blades, whereas in Kemp and Sears analysis only the effects of the wake from the blade under consideration is taken into account. Mani [28] used the methods of linearized, subsonic, plane unsteady flow to include the compressibility effects in Kemp and Sears interference problem.

Osborne [16] extended Kemp and Sears problem to subsonic compressible flow by using Prandtl-Glauert transformation to correct the steady potential flow field. Adachi [29] made an experimental study to measure the fluctuating forces on the stationary blades in a turbomachine stage, for various axial distances between blade rows and compared his results from those from Kemp and Sears theory. Kemp and Sears analysis was also used by Rao et al [30] to present an analysis for determining the unsteady lift and moment acting on generalized camber blades of an elementary turbomachine stage taking into account both transverse and chordwise sinusoidal flow perturbations. Numerical results for three types of blades viz., flat blades with incidence and cambered blades of parabolic and skew type were obtained.

Sisto [31] determined unsteady aerodynamic reactions with small incidence, small camber, taking into account the disturbances of neighbouring blades, according to a potential theory in a cascade, with all blades vibrating in the same mode and with a constant phase angle. Chang and Chu [32] used conformal mapping to determine the aerodynamic load on cascades pitching and flapping in synchronized harmonic oscillation. The wake interference considered in the cascade resulted in a modified Theodorsen's function. Sebatiuk and Sisto [33] conducted a survey of aerodynamic excitation problems in turbomachines.

Whitehead [34] used flat plate theory with inviscid and incompressible flow, to determine the lift and moment coefficients for vibrating airfoils in a cascade. Effects due to blade camber and thickness were neglected and no stall is assumed. The amplitude of vibration was assumed to be small so that the wakes of the blades are considered rectilinear. Later Whitehead [35] considered both upwash and downwash due to vortices (replacing blades), concentrated at several points along the length of the chord, allowing for continuous vorticity shed in the blade wakes, and determined the unsteady lift on a cascade of two dimensional flat plate airfoils.

Tanabe and Horlock [36] developed a simpler potential theory for flow through the cascades of low solidity. Parker [37] made use of finite difference method for computation for the potential flow through multiple cascades of blades with relative movement of cascades and estimated the pressure fluctuations which are responsible for the generation of unsteady forces on blades with small blade row spacings. The flow was assumed to be incompressible. Parker indicates through his analysis that the potential flow interaction effects will be greater than the wake effects except when the blade row spacings are large. Parker [38] also investigated the potential flow interaction effects analytically and experimentally and established that the potential flow interaction between blade rows is a significant source of generation of fluctuating pressures and of blade vibration excitation. He has also shown that the passing of rotor blades behind guide vanes produce large pressure fluctuations on the surfaces of guide vanes. Partington [39] conducted strain guage tests for the last three stages of a low pressure steam turbine to determine the sources of low harmonic blade vibration excitation.

Horlock [40] combined his analysis for the lift fluctuation on an airfoil due to gust parallel to the undisturbed flow with an already established theory for transverse gusts to obtain the total response of an airfoil encountering both streamwise and transverse gusts. The analysis considered flat plate airfoils with non-zero angle attack. Naumann and Yeh [11] extended the analysis to cambered airfoils with angle of attack moving through both longitudinal and transverse gusts and made use of their analysis to obtain the unsteady lift in an axial flow turbomachine stage. A qualitative idea of how the angle of attack, camber and cascade geometry can influence the design is presented. Gearhart et al [41] presented a quasi steady analysis useful for the design of rotating turbomachinery.

Kaji and Okajaki [42] used acceleration potential method to study the generation of sound by rotor stator interaction. The method adopted consists of distributing the acceleration potential doublets over the airfoils for both steady and unsteady flow. By satisfying no penetration condition for the steady flow and including the boundary condition that in the stationary coordinate system, the flow for upstream does not deviate from the flow for downstream, they obtained two single, singular integral equations for the unknown rotor and stator steady acceleration doublet distributions. Numerical techniques were used for obtaining final results. Murata and Tsujimoto [43] used the acceleration potential method in combination with conformal mapping to determine the lift fluctuations on flat plate airfoils in cascade due to transverse and chordwise gusts. A new mapping function was introduced by them [44] to take into account the phase difference of the gusts between adjacent blades.

Henderson and Daneshyar [45] simulated the effect of bound vorticity of the neighbouring blades, by a vortex at their quarter chord point, to determine the fluctuating lift, generated by a moving blade row, interacting with the potential flow disturbances, of an upstream blade row. Airfoils are assumed to be thin and slightly cambered and results are obtained for circular arc cambered airfoils with incidence. Henderson and Horlock [46] suggested an approximate analysis for determining the unsteady lift on airfoils in moving cascades, subject to disturbances in the inlet axial flow. The equations of motion are averaged across the pitch and the lift is determined by integration of the instantaneous pressure differences obtained across the surface of the blade. The pitch is made small in comparison with the wave length of the disturbance as in Whitehead's analysis, but pitch to chord ratio is made small compared to a finite value used by Whitehead.

The mechanism of interaction between the blade rows in a turbomachine was discussed by Parker and Watson [47] in detail, and some considerations involved in the design of axial flow fans and compressors were listed by him. Walker and Oliver [48] conducted several experiments to measure the effects of interaction between the wakes from one row of blades and the boundary layers and wakes of next row of blades downstream.

Rao et al [49] considered a flat plate stage in subsonic compressible flow, including both upwash and downwash effects, following the Kemp and Sears procedure. The analytical results obtained are presented graphically as a function of stage gap ratio, blade spacing ratio and Mach number. A compilation of established turbine calculations and a convolution scheme is used by Bedrosyan, Maier and Waggot [50] to determine the unsteady turbine blade forces. The results of experiments of the exciting force on typical rotor blades have been given by Tanaka et al [51]. Rao and Rao [52] extended their previous analysis [49] to generalised camber blades.

Hydraulic Analogy and Its Applications

The most significant investigations as to how far the hydraulic analogy theories are valid were made by Preiswerk [53, 54]. He discussed the applications of methods of gas dynamics to water flows with no energy dissipation and flows with momentum discontinuities (flows involving shocks), on completely mathematical grounds. The analogy between free surface incompressible water flows and isentropic compressible gas flows was established for steady conditions by Loh [55] and for unsteady conditions by Bryant [56], when the specific heat ratio of the gas is equal to 2.

However, no naturally occurring gas has a specific heat ratio equal to 2 and the classical analogy remained a qualitative tool for a long time. Several investigators considered this aspect and tried to modify the analogy by appropriate correction factors. Harlemann and Ippen [57] in their transonic flow experiments, using towed models, adopted a method in which the hydraulic depth ratio is assumed to be equal to the density ratio for a gas with specific heat ratio = 2, which is in exact correspondence according to the classical analogy. They then used the appropriate gas equations with specific heat ratio of 2 to calculate the parameters such as pressure ratios and local Mach numbers. The results are then applied to airflows about the same profiles by means of transonic similarity laws of Von Karman [58].

To account for variations of the specific heat ratio in a gas system from the ideal analogous value of specific heat ratio 2, Byrd and Williams [59] proposed some correction factors for pressure, density and temperature ratios. Adams [60] presented analytical methods to relate the hydraulic analogy to axisymmetric compressible gas systems. Two axisymmetric rocket nozzles operating with heated air were simulated by him on the water table. He applied correction factor for the Froude number obtained on the water table in order to take into account the variation of specific heat ratio from the analogous value 2. The corrected Mach number thus obtained was used by Adams along with the correction factors suggested by Byrd and Williams for obtaining corrected wall pressure distributions which are compared with the corresponding values in the gas nozzles. He also reviewed the methods to minimize or account for non-analogous effects such as three dimensional water flow, wave propagation and viscosity in the system. Schorr [61] suggested a correction factor for pressure ratio purely based on isentropic relations. Rao [62] made a detailed analytical study of compressible gas flows in an axisymmetric converging nozzle and derived correction factors for the Mach number, pressure and temperature ratios and discussed their application in hydraulic analogy. He demonstrated that the hydraulic analogy can be used as a powerful tool to study the isentropic flow of gases in converging nozzles.

An alternative procedure for modelling a gas system by hydraulic analogy was suggested by Loh [63], who has examined one dimensional unsteady flow equation, that the cross-section of the channel should be varied from rectangular section. Bryant [64] examined the basic equations for one and two dimensional analogies and has shown that a non-rectangular channel can be used only to give a mathematical analogy with a one dimensional gas flow. If the flow is two dismensional, a rectangular channel must be used for the mathematical analogy to be valid. The requirements of non-rectangular channels and the difficulties involved in the development of modelling techniques for practical applications have not allowed one dimensional analogy to make real progress in any direction.

Rao [65] in his analysis, modelled compressible gas flow through converging-diverging nozzles whose walls are defined by linear and circular segments. Isentropic shock free flows are accounted and the pressure and temperature correction factors are generalized and through comparisons of gas dynamic solutions and analytical water table solutions, he demonstrated the usefulness of hydraulic analogy to model compressible isentropic gas flows in converging-diverging nozzles.

Gilmore, Plesset and Crossley [66] discussed at length the hydraulic jumps in liquids and shock waves in gases. He conducted experiments to study the effects of interactions and reflections of shocks, and concluded that the deviation from the simple theory of hydraulic jump interactions limit the use and validity of hydraulic analogy as a means for studying compression shock interactions in gases. Laitone [67] conducted experiments to study the transonic gas flows by the hydraulic analogy. He showed that an analogy with the two dimensional flow of a perfect gas exists only if the water depth is small, around 6 mm, the model is fairly large and shock waves or hydraulic jumps present are of the weak type, having a negligible increase in entropy. Later along with Neilson [68] he conducted experimental studies of the transonic flow about symmetrical slender wedge profiles, in order to check the validity of the simplifying assumptions of the small perturbation theory, for determining the location of the detached bow shock wave as the Mach number approached unity.

Bryant and Grant [69] conducted studies of shock wave detachment. They towed their models at zero incidence in a 5 mm depth of water, with the surfaces of the model treated to minimize the influence of surface energy. To determine the unsteady effects, the models were towed with constant velocity or with uniform rate of acceleration (retardation). They suggested further fundamental work on the analogy before the history of accelerated motion of a model in shallow water can be transferred qualitatively to a corresponding history for a similar affine shape in free flight. Saiva [70] studied transonic flow over two dimensional wedge profiles through the use of hydraulic analogy by adopting the technique of distorted dissimilar modelling and through the application of transonic similarity laws. Klein [71] made studies using hydraulic analogy to compressible gas flows of two dimensional flow patterns for wedge in supersonic motion subjected to weak and strong shocks. He compared his experimental hydraulic flow patterns with theoretical aerodynamic features and obtained good overall agreement.

The normal shock in a gas flow and the corresponding hydraulic jump in water flow were examined by Rao [72] in detail, to assess the conditions under which the hydraulic analogy can be used to model nozzle flows with a reasonable accuracy. A numerical example of a rocket nozzle was considered by him to study the errors in the Mach number, pressure and temperature distribution in the subsonic flow following the normal shock. It was shown through the analysis that the errors induced in the pressure distribution are very small for flows with Froude numbers as large as 3. For Froude numbers greater than 1.5, the errors in the temperature distribution were large. The flows discussed analytically by Rao [65, 72] were simulated in an equivalent 2-D model of the nozzle on the water table by Rao et al [73]. Reasonably good agreement between the results obtained through modified analogy, in which corrections are applied to water table data, and gas dynamic solution for isentropic flows was found. They also indicated that for non-isentropic flows, the deviations in the post shock flow can be kept to within reasonable results, if only weak shocks are considered and care is taken to define the location of hydraulic jump correctly on the water table.

Rao [74] also made analytical studies regarding over and under expanded nozzle exit supersonic flows. For an over expanded exit flow, he showed that the flow inside the nozzle is modelled exactly by hydraulic analogy, but that there will be second order errors in determining the flow properties subsequent to the oblique shock. The effect of straight oblique shock waves on the accuracy of the hydraulic analogy was discussed by Rao [75], while conducting analytical studies regarding modelling of supersonic compressible gas flows on the water table. He developed a computer program to determine the flow properties across the shock and derived corresponding relations for a slant hydraulic jump in two dimensional incompressible gas flow. From a comparison of results of the gas and water flow, it was concluded that the presence of weak shocks does not significantly affect the hydraulic analogy. The presence of multiple shocks followed by a normal shock was also studied by Rao and it was shown that the analogy can be used to a first approximation for low Mach numbers in the approach supersonic stream and small flow turning angles.

Bryant [56] discussed some aspects of the size of airfoil models for qualitative hydraulic analogy research. According to him, the boundary layer, as well as producing some small discrepancies near the boundaries not compatible with gas flow, gives an effective model profile which is not affine transformation of the physical model tested. He suggested first order corrections based on the boundary layer displacement thickness to overcome the difficulty. He [76] also discussed some important non-analogous effects which may occur in two dimensional flow analogy and also surface contamination effects [56]. Certain limitations to hydraulic analogy concerning problems with vertical accelerations and surface wave propagations were discussed by Laitone [77]. Gupta [78] elaborated the concept of optimum depth introduced by Laitone [67] and through analytical considerations arrived at an optimum depth for hydraulic analogy experiments.

The principles of hydraulic analogy were used by several investigators for studying flow problems in turbomachine stages, e.g., Owczarek [79] used a radial in-flow water table to study pressure wave interaction phenomena between rotor row and a stator row of a turbine. Giraud [80] using the apparatus developed by Harleman and Ippen group, made a large number of studies on supersonic compressor cascades. Rieger et al [81] developed a large rotating water table apparatus which allows the gas flow conditions in a turbomachine stage to be modelled by the flow of water. They used such a table to measure force variation which occurs on a moving blade with partial admission inlet flow and other turbine stage geometries. The forces were measured through instrumented blade rods and the signals were recorded on a strip chart record. They compared the results obtained from a practical turbine stage by reducing the stage geometry to the equivalent flat plate Kemp and Sears problem, with the results from the computer program of Rao [25]. The comparison between the theoretical and experimental results was not highly satisfactory. In all the above rotating water table studies, the effect of specific heat ratio not equal to 2, and compressibility effects were not considered in modelling a turbine stage.

Rao et al [82] built a similar table as that of Rieger et al [81], but used modified hydraulic analogy to model a flat plate turbine stage. This has significantly altered the scope of application of hydraulic analogy to study the blade interference problem of a simple turbomachine stage. Encourged by the improvement in the comparison of the results [49] for a simple stage, an Orpheus engine stage was modelled by Sandhu [83]. Theoretical results obtained by program developed by Rao [84] agreed closely with the hydraulic analogy experiment in [83].

Thick Airfoil Stages

When the blades are thick as in high pressure stage steam turbines, thin airfoil theories may be inadequate. Numerical techniques involving considerable computation timebecome necessary. One such program is developed by Mylonas [85]. In this, the blade surfaces are modelled by a vortex distribution with a net of singularities surrounding them. These singularities consist of two components, a source strength to simulate compressible effects and a vortex strength to model the unsteady viscous wakes. Thus the unsteady flow field consists of a transition flow, a vortex distribution on profile surfaces, a vortex distribution in the net points and a source distribution in the net points. The basic work of representing the blade surfaces by vortex rows due to Martensen [86] and Jacob [87] is employed by Lienhart [88], who derived expressions for induced velocity components at any point in the field due to the infinity rows of vortices or sources. Mylonas used a similar procedure to that of Pfost [89], wherein the compressible effects are simulated by means of a source distribution in a net of points surrounding the profiles and vortices are disposed in the same net points to simulate the motion of unsteady viscous wakes similar to that of Krammer [90]. These vortices are generated in the net points near the separation onset in the trailing edge region on the suction and pressure side of the profiles and the separation point on the blade surface is determined according to Truckenbrodt [91].

Using the condition that the blade is impermeable, i.e., the relative velocity outside the profile is tangential, a system of equations is set up containing the known influences of surface vortices, unknown vortices, known transition flow and the unknown source and vortex distributions in the grid points. First the steady incompressible solution is obtained with the source and vortex distributions in the net points set to zero. Then the steady pressure field is obtained by solving Euler's equation. The source distribution is computed iteratively using the isentropic equation starting with the incompressible solution for the velocity field. Unsteady wake vortices are shed in the trailing edge region of the profiles, which are carried downstream by the main flow and change their strength according to the circulation transport equation. A time marching technique is adopted at each time step. Contrarotating vortices are generated in two grid points approximately 10% of the chord length downstream of the trailing edge of each blade row. Hubensteiner and Mylonas [92] compared the theoretical values obtained from such a program with the experimental results of a stator blade.

Yet another approach to study the rotor-stator interaction is to solve the unsteady thin layer Navier-Stokes equations using a system of patched and overlaid grids, see Rai [93] and Gundy-Burlet et al [94]. They derived the necessary equations for an accurate transfer of information between the several grids and developed an iterative implicit algorithm. The computational time is however enormous. Experimental data for an axial compressor stage was obtained by Rai [93] and Stauter et al [95].

In this chapter we will study the two dimensional thin air-foil theories as applicable to turbomachinery stages and discuss the use of hydraulic analogy to simulate such flows to obtain the nonsteady forces.

6.2 BASIC ISOLATED AIRFOIL THEORIES

In this section, the theory to determine the response of isolated airfoils subjected to simple flow perturbations is described.

Effect of Transverse Gust

The airfoil can be visualized as being replaced by a plane vortex sheet as shown in Fig. 6.1, whose strength $\gamma(x,t)$ is such at every instant as to cancel the relative velocity component normal to the airfoil surface, thus satisfying the boundary condition there. The general expressions for the lift and moment acting upon an airfoil entering a sharp edged transverse gust are

$$ {}_tL(t) = {}_qL(t) + {}_1L(t) + {}_2L(t) \tag{6.1} $$

$$ {}_tM(t) = {}_qM(t) + {}_1M(t) + {}_2M(t) \tag{6.2} $$

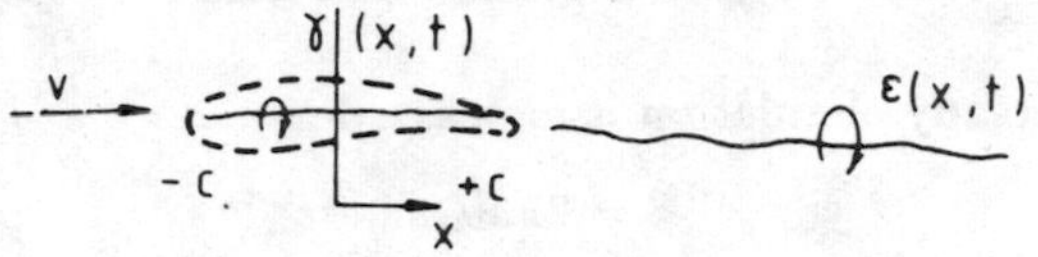

Fig. 6.1 Airfoil and wake in unsteady motion represented as vortex sheets

The first terms on right hand side in the above expressions give quasi-steady lift and moment, second terms give lift and moment due to apparent mass of the airfoil, and the third terms give lift and moment considering all wake effects. If the motion of airfoil is such that its quasi-steady circulation is given by

$$ {}_q\Gamma(t) = Fe^{i\nu t} \tag{6.3} $$

then Karman and Sears have shown that

$$ {}_qL(t) + {}_2L(t) = {}_qL(t)C(\omega) \tag{6.4} $$

where the Theodorsen's function is given by

$$ C(\omega) = \frac{K_1(i\omega)}{K_0(i\omega) + K_1(i\omega)} \tag{6.5} $$

If an airfoil operates in a sinusoidal gust pattern such that the relative upwash at the airfoil is given by, see Fig. 6.2,

$$ v(x,t) = v_0 e^{i\nu(t-x/V)} \tag{6.6} $$

then the total unsteady lift and moment for a chord of length $2c$ are

$$ {}_tL_v = 2\pi\rho cV v_0 S_L(\omega)e^{i\nu t} \tag{6.7} $$

$$ {}_tM_v = {}_tL_v \frac{1}{2}c \tag{6.8} $$

where the Sears function in (6.7) is given by

$$ S_L(\omega) = \frac{1}{i\omega\{K_0(i\omega) + K_1(i\omega)\}} \tag{6.9} $$

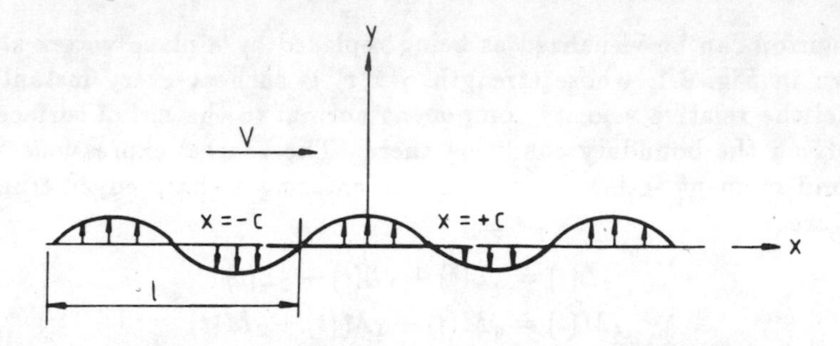

Fig. 6.2 Transverse gust

The quasi-steady circulation is expressed as

$$_q\Gamma^{(t)} = 2\pi c v_0 J(\omega) e^{i\nu t} \tag{6.10}$$

For a general sinusoidal motion in which the quasi-steady circulation is given in the form of equation (6.3), the strength of the vortex distribution in the wake of the airfoil is given as

$$\epsilon(x,t) = -\frac{F}{c} i\omega S_L(\omega) e^{i\nu(t-x/V)} \tag{6.11}$$

For an airfoil with chord length $2c$, subjected to a generalized upwash in the form

$$v(x,t) = v_0 e^{i\left(\nu t - \mu \frac{x}{V}\right)} \tag{6.12}$$

the lift and moment expressions are given by

$$_tL_v = 2\pi\rho c V v_0 S_L(\omega,\lambda) e^{i\nu t} \tag{6.13}$$

$$_tM_v = \pi\rho c^2 V v_0 S_M(\omega,\lambda) e^{i\nu t} \tag{6.14}$$

where,

$$S_L(\omega,\lambda) = J(\lambda)C(\omega) + i\left(\frac{\omega}{\lambda}\right) J_1(\lambda) \tag{6.15}$$

$$S_M(\omega,\lambda) = J(\lambda)\{C(\omega) - 1\} + \left(\frac{\omega}{\lambda}\right) J_0(\lambda) + \frac{2}{\lambda}\left(1 - \frac{\omega}{\lambda}\right) J_1(\lambda) \tag{6.16}$$

The corresponding quasi-steady state circulation for this case is

$$_q\Gamma(t) = 2\pi c v_0 J(\lambda) e^{i\nu t} \tag{6.17}$$

with the wake strength given by

$$\epsilon(x,t) = -2\pi v_0 J(\lambda) i\omega S_L(\omega) e^{i\nu\left(t - \frac{x}{V}\right)} \tag{6.18}$$

Effect of Gust in the Direction Parallel to the Mean Flow

Referring to Fig. 6.3, consider a variation in the velocity parallel to the blade of the form

$$u = u_0 e^{i\nu\left(t - \frac{x}{V}\right)} \tag{6.19}$$

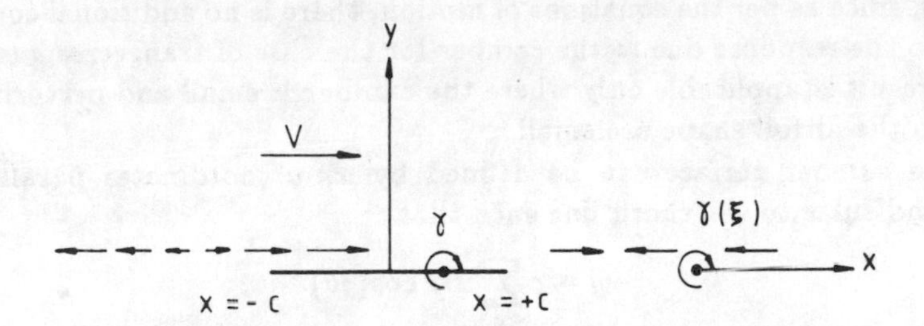

Fig. 6.3 Gust parallel to undisturbed flow

The u velocity fluctuation varies with time at any point distant x from the central point of the airfoil which has a semichord length c and has a distributed bound vorticity $\gamma(x)$ in the steady flow. The final expression obtained by Horlock for the fluctuating lift on a flat blade airfoil with an angle of incidence subjected to chordwise velocity disturbance is of the form

$$ {}_tL_u = 2\pi\rho c V u_o \, \delta T(\omega) e^{i\nu t} \tag{6.20} $$

where the Horlock's function is given by

$$ T(\omega) = X_1 + iY_1 \tag{6.21} $$

where,

$$ X_1 = (2 - a)J_0(\omega) - b^* J_1(\omega) \tag{6.22} $$

$$ Y_1 = (1 + a)J_1(\omega) - b^* J_0(\omega) \tag{6.23} $$

The constants in the above equation depend upon the Theodorsen's function, fixed by the relation

$$ a + ib^* = 1 - C(\omega) \tag{6.24} $$

For flat plate airfoil subjected to non-convecting streamwise gust, in which the velocity distribution is of the form

$$ u(x,t) = u_0 e^{i\left(\nu t - \mu \frac{x}{V}\right)} \tag{6.25} $$

the lift fluctuation is given by

$$ {}_tL_u = 2\pi\rho c V u_0 \, \delta T(\omega, \lambda) e^{i\nu t} \tag{6.26} $$

where,

$$ \begin{aligned} T(\omega,\lambda) = {} & 2J_0(\lambda) - aJ_0(\lambda) - b^* J_1(\lambda) \\ & + i\left\{ aJ_1(\lambda) - b^* J_0(\lambda) + \frac{\omega}{\lambda} J_1(\lambda) \right\} \end{aligned} \tag{6.27} $$

The above can also be expressed as

$$ T(\omega,\lambda) = S_L(\omega,\lambda) + J_0(\lambda) + iJ_1(\lambda) \tag{6.28} $$

Extension to Cambered Airfoils

The general solution for the response of an airfoil with camber, encountering a transverse gust, does not change from that related to a flat plate airfoil, since as per the equations of motion, there is no additional contribution to the response due to the camber for the case of transverse gust flow. This result is applicable only where the camber is small and perturbations due to the airfoil shape are small.

The camber surface can be defined by x, y coordinates parallel and perpendicular to the chord line such that

$$y = c \sum_j B_j \cos(j\theta) \tag{6.29}$$

$$x = -c \cos\theta \tag{6.30}$$

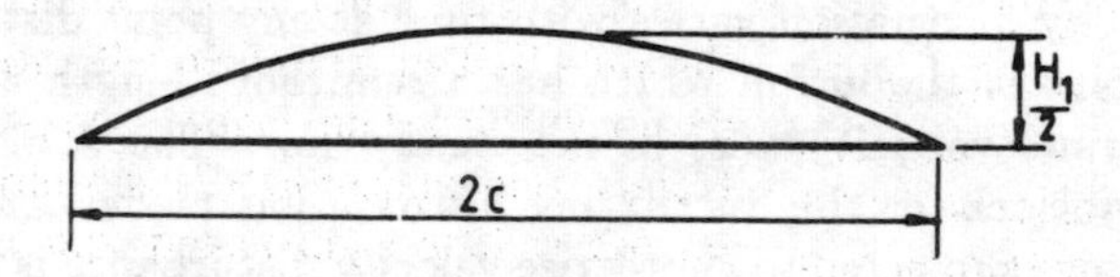

Fig. 6.4 Airfoil with parabolic mean line

Holmes' solution for the lift fluctuation on an airfoil with a parabolic or circular arc mean line of the form shown in Fig. 6.4, encountering a streamwise gust is

$$_tL_u = 2\pi\rho c V u_0 F^* C_a(\omega) e^{i\nu t} \tag{6.31}$$

where,

$$C_a(\omega) = X_1 + iY_1 \tag{6.32}$$

$$X_1 = 2[J_0(\omega) + J_2(\omega)] - [a\{J_0(\omega) - J_2(\omega)\} + 2b^* J_1(\omega)] \tag{6.33}$$

$$Y_1 = -2J_1(\omega) + [2aJ_1(\omega) - b^*\{J_0(\omega) - J_2(\omega)\}] \tag{6.34}$$

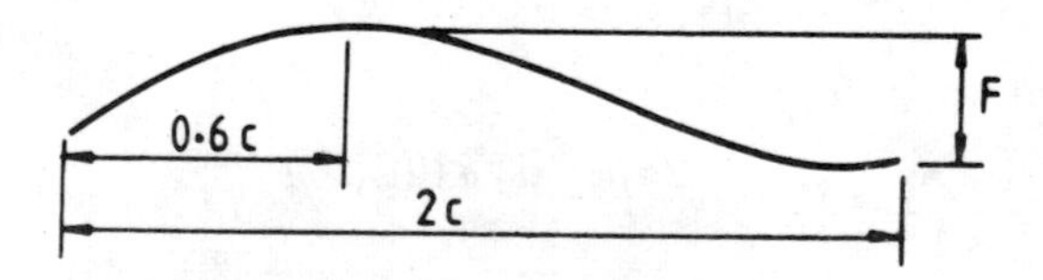

Fig. 6.5 Airfoil with skewed mean line

The solution for the lift fluctuation on an airfoil with a skewed mean line, see Fig. 6.5, in a streamwise convecting gust is of the form

$$_tL_u = \pi\rho c V u_0 2\frac{F^*}{c} C'_a(\omega) e^{i\nu t} \tag{6.35}$$

where,

$$C'_a(\omega) = X_1 + iY_1 \tag{6.36}$$

$$X_1 = 0.4\{0.8(2-a)J_0(\omega) - b^* J_1(\omega) + (7.6 - 1.6a)J_2(\omega) - 1.8b^* J_3(\omega)\} \tag{6.37}$$

$$Y_1 = 0.4\{-0.8b^* J_0(\omega) + (0.2+a)J_1(\omega) - 1.6b^* J_2(\omega) + (5.4 + 1.8a)J_3(\omega)\} \tag{6.38}$$

For a symmetrical flat plate airfoil at an angle of incidence to the gust and mean flow, the expression for the fluctuating moment given by Holmes is

$$_tM_u = \pi\rho c^2 V u_0\, \delta M_a(\omega) e^{i\nu t} \tag{6.39}$$

where,

$$M_a(\omega) = P_1 + iQ_1 \tag{6.40}$$

$$P_1 = (3-a)J_0(\omega) - (1+b^*)J_1(\omega) - J_2(\omega) \tag{6.41}$$

$$Q_1 = -b^* J_0(\omega) + (2+a)J_1(\omega) \tag{6.42}$$

Effect of Incidence and Camber

The expressions given above for the cambered blades, Eqns. (6.31) and (6.35) pertain to the case where the mean flow is aligned with the chord of the airfoil and the gust acting in the mean flow direction. In order to obtain a more realistic prediction of the resultant lift fluctuation due to camber and incidence, the governing equations can be considered to be linear in their dependence upon incidence and camber. Thus the lift on the airfoil with both the incidence and camber are simply derived by the super-position of the responses which are derived from considering the effects separately.

Naumann and Yeh [11] extended the analysis of Kemp and Sears and Horlock and obtained closed form solutions for cambered airfoils which have incidence and moving through both longitudinal and transverse gusts (see Fig. 6.6). For the case of cambered airfoil blades, the total lift is given by

$$_tL = {_tL_1} + {_tL_2} + {_tL_3} \tag{6.43}$$

The first term on the right hand side represents that part of the lift fluctuation due to the camber and longitudinal gust and is given by

$$_tL_1 = 2\pi\rho c V u_0 f \left\{ F(\omega) + \frac{4}{\omega} J_1(\omega) \right\} \tag{6.44}$$

where f is the ratio of maximum camber over half chord length and

$$F(\omega) = F'(\omega)\left\{ J_0(\omega) - \frac{J_1(\omega)}{\omega} - iJ_1(\omega) \right\} - \left\{ J_0(\omega) - \frac{J_1(\omega)}{\omega} + iJ_1(\omega) \right\} \tag{6.45}$$

$$F'(\omega) = \frac{H_0^{(2)}(\omega) + iH_1^{(2)}(\omega)}{-H_0^{(2)}(\omega) + iH_1^{(2)}(\omega)} \tag{6.46}$$

In the above H are Hankel functions of second kind of zero and first order.

The second term on the right hand side of the equation (6.43) gives that part of the lift fluctuation due to the angle of attack and longitudinal gust.

$$ {}_tL_2 = 2\pi\rho c V u_0\, \delta F_\delta(\omega) e^{i\nu t} \tag{6.47}$$

where,

$$F_\delta(\omega) = J_0(\omega) + iJ_1(\omega) \tag{6.48}$$

The third term on the right hand side of equation (6.43) is the lift fluctuation due to a transverse gust and is given by the same equation as that of equation (6.7).

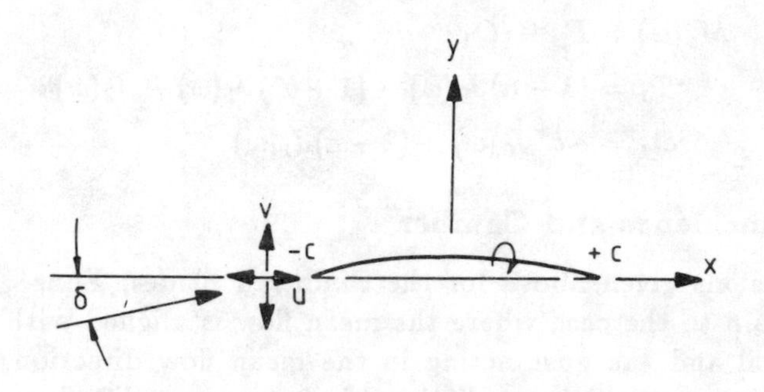

Fig. 6.6 Orientation of airfoil subjected to both chordwise and transverse gusts

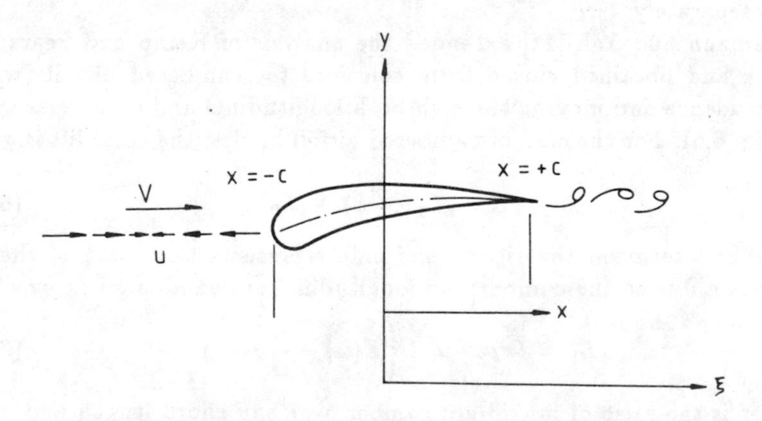

Fig. 6.7 General camber airfoil subjected to non-convecting streamwise gust

General Camber Airfoil in Non-Convecting Streamwise Gust

Rao's analysis [13] gave expressions for unsteady lift and moment of a general camber isolated airfoil subjected to non-convecting streamwise gust.

Referring to Fig. 6.7, the airfoil mean line is defined as

$$x = -c\cos\theta \tag{6.49}$$

and,

$$\frac{dy}{dx} = \sum_m H_m \cos(m\theta) \tag{6.50}$$

The lift expression is obtained as

$$_tL_u = \pi\rho c V u_0 H^*(\omega,\lambda)e^{i\nu t} \tag{6.51}$$

The gust response function is

$$H^*(\omega,\lambda) = X_1 + iY_1 \tag{6.52}$$

$$\begin{aligned}
X_1 =& 2H_0\{aJ_0(\lambda) + b^*J_1(\lambda) - 2J_0(\lambda)\} \\
&+ H_1\left[2\left\{J_0(\lambda) + \frac{\omega}{\lambda}J_2(\lambda)\right\} - \{a[J_0(\lambda) - J_2(\lambda)] + 2b^*J_1(\lambda)\}\right] \\
&+ \sum_{m=2}^{\infty} H_m\Bigg[J_m(\lambda)\left\{\sin\left(m\frac{\pi}{2}\right)(\omega - 2b^*) + 2\cos\left(m\frac{\pi}{2}\right)(a-1)\right\} \\
&+ J_{m+1}(\lambda)\left\{b^*\sin\left[(m+1)\frac{\pi}{2}\right] - a\cos\left[(m+1)\frac{\pi}{2}\right]\right\} \\
&+ J_{m-1}(\lambda)\left\{(2-a)\cos\left[(m-1)\frac{\pi}{2}\right] + b^*\sin\left[(m-1)\frac{\pi}{2}\right]\right\} \\
&- \frac{\omega}{2}J_{m+2}(\lambda)\sin\left\{(m+2)\frac{\pi}{2}\right\} - \frac{\omega}{2}J_{m-2}(\lambda)\sin\left\{(m-2)\frac{\pi}{2}\right\}\Bigg]
\end{aligned} \tag{6.53}$$

$$\begin{aligned}
Y_1 =& 2H_0\left\{-J_1(\lambda)\left(\frac{\omega}{\lambda} + a\right) + b^*J_0(\lambda)\right\} \\
&+ H_1\{-2J_1(\lambda) + 2aJ_1(\lambda) - b^*[J_0(\lambda) - J_2(\lambda)]\} \\
&+ \sum_{m=2}^{\infty} H_m\Bigg[J_m(\lambda)\left\{\cos\left(m\frac{\pi}{2}\right)(2b^* - \omega) + 2\sin\left(m\frac{\pi}{2}\right)(a-1)\right\} \\
&+ J_{m+1}(\lambda)\left\{-b^*\cos\left[(m+1)\frac{\pi}{2}\right] - a\sin\left[(m+1)\frac{\pi}{2}\right]\right\} \\
&+ J_{m-1}(\lambda)\left\{(2-a)\sin\left[(m-1)\frac{\pi}{2}\right] - b^*\cos\left[(m-1)\frac{\pi}{2}\right]\right\} \\
&+ \frac{\omega}{2}J_{m+2}(\lambda)\cos\left\{(m+2)\frac{\pi}{2}\right\} + \frac{\omega}{2}J_{m-2}(\lambda)\cos\left\{(m-2)\frac{\pi}{2}\right\}\Bigg]
\end{aligned} \tag{6.54}$$

The moment expression for a general camber airfoil subjected to a non-convecting streamwise gust is

$$_tM_u = \pi\rho c^2 V u_0 M^*(\omega,\lambda)e^{i\nu t} \tag{6.55}$$

where the moment response function is

$$M^*(\omega,\lambda) = P_1 + iQ_1 \tag{6.56}$$

$$
\begin{aligned}
P_1 =& H_0\Big[\frac{\omega}{\lambda}J_2(\lambda) - 2J_0(\lambda) + aJ_0(\lambda) + b^*J_1(\lambda) - J_0(\lambda) + J_1(\lambda)\Big] \\
&+ \frac{H_1}{2}\big[a\{J_2(\lambda) - J_0(\lambda)\} - 2b^*J_1(\lambda)\big] \\
&+ \sum_{m=2}^{\infty} H_m\Big[J_m(\lambda)\Big\{(a-1)\cos\Big(m\frac{\pi}{2}\Big) - b^*\sin\Big(m\frac{\pi}{2}\Big)\Big\} \\
&+ J_{m+1}(\lambda)\Big\{\Big(\frac{\omega}{8}+\frac{b^*}{2}\Big)\sin\Big[(m+1)\frac{\pi}{2}\Big] - \frac{a}{2}\cos\Big[(m+1)\frac{\pi}{2}\Big]\Big\} \\
&+ J_{m-1}(\lambda)\Big\{\Big(\frac{\omega}{8}+\frac{b^*}{2}\Big)\sin\Big[(m-1)\frac{\pi}{2}\Big] - \frac{a}{2}\cos\Big[(m-1)\frac{\pi}{2}\Big]\Big\} \\
&+ J_{m-2}(\lambda)\cos\Big\{(m-2)\frac{\pi}{2}\Big\} - J_{m+3}(\lambda)\frac{\omega}{8}\sin\Big\{(m+3)\frac{\pi}{2}\Big\} \\
&- J_{m-3}(\lambda)\frac{\omega}{8}\sin\Big\{(m-3)\frac{\pi}{2}\Big\}\Big]
\end{aligned}
\qquad (6.57)
$$

$$
\begin{aligned}
Q_1 =& H_0\big[-2J_1(\lambda) + b^*J_0(\lambda) - aJ_1(\lambda)\big] \\
&+ \frac{H_1}{2}\Big\{-b^*J_0(\lambda) - 2J_1(\lambda)\Big(\frac{\omega}{\lambda} - a\Big) + J_2(\lambda)\Big(6\frac{\omega}{\lambda^2} + b^*\Big)\Big\} \\
&+ \sum_{m=2}^{\infty} H_m\Big[J_m(\lambda)\Big\{(a-1)\sin\Big(m\frac{\pi}{2}\Big) + b^*\cos\Big(m\frac{\pi}{2}\Big)\Big\} \\
&- J_{m+1}(\lambda)\Big\{\Big(\frac{\omega}{8}+\frac{b^*}{2}\Big)\cos\Big[(m+1)\frac{\pi}{2}\Big] + \frac{a}{2}\sin\Big[(m+1)\frac{\pi}{2}\Big]\Big\} \\
&- J_{m-1}(\lambda)\Big\{\Big(\frac{\omega}{8}+\frac{b^*}{2}\Big)\cos\Big[(m-1)\frac{\pi}{2}\Big] + \frac{a}{2}\sin\Big[(m-1)\frac{\pi}{2}\Big]\Big\} \\
&+ J_{m-2}(\lambda)\sin\Big\{(m-2)\frac{\pi}{2}\Big\} + J_{m+3}(\lambda)\frac{\omega}{8}\cos\Big\{(m+3)\frac{\pi}{2}\Big\} \\
&+ J_{m-3}(\lambda)\frac{\omega}{8}\cos\Big\{(m-3)\frac{\pi}{2}\Big\}\Big]
\end{aligned}
\qquad (6.58)
$$

The lift expression for a general camber airfoil subjected to a downwash of the form given by equation (6.25) with an angle of incidence is

$$ {}_tL_i = 2\pi\rho V c u_0 H_L(\omega,\lambda)e^{i\nu t} \qquad (6.59) $$

where,

$$ H_L(\omega,\lambda) = \frac{2}{\pi\lambda}\Big(1 - \frac{\omega}{\lambda}\Big)\sin(\lambda) + (\delta - H_0)\{J_0(\lambda) + iJ_1(\lambda)\} \qquad (6.60) $$

The corresponding moment expression is

$$ {}_tM_i = \pi\rho c^2 V u_0 M_L(\omega,\lambda)e^{i\nu t} \qquad (6.61) $$

where,

$$
\begin{aligned}
M_L(\omega,\lambda) =& 2(\delta - H_0)\Big\{J_0(\lambda) + iJ_1(\lambda) - \frac{J_1(\lambda)}{\lambda}\Big\} \\
&- \frac{i}{\pi}\frac{4}{\lambda}\Big(1 - \frac{\omega}{\lambda}\Big)\Big(\cos\lambda - \frac{\sin\lambda}{\lambda}\Big)
\end{aligned}
\qquad (6.62)
$$

Thus the general expressions given by equations (6.51) and (6.59), for a general camber airfoil subjected to a non-convecting streamwise gust with an angle of incidence can be used to obtain the fluctuating lift of an airfoil of any shape viz., flat, parabolic or skew type, by simply changing the H parameters in the appropriate equations. Similarly the general expressions given by equations (6.55) and (6.61) can be used to obtain the fluctuating moment on the airfoil of any shape.

Isolated Airfoil in Compressible Flow

Osborne [17] presented an analytical approximation for the lift of an airfoil in a two dimensional compressible subsonic flow with an imposed upwash sinusoidal in time. With certain frequency and Mach number restrictions, he obtained closed form expressions for the forces on a two dimensional thin airfoil in subsonic flow, that is subjected to a general upwash whose x dependence may be expanded in a cosine series. The basic theory depends upon the work of Amiet and Sears [15] who extended the familiar Prandtl-Glauert transformation for steady subsonic flow to the unsteady case.

The theory consists of applying the combined Galilean-Lorentz transformation to the small perturbation equation, resulting in the wave equation. The method of Matched Asymptotic Expansions is used to treat an inner region where the semi-chord is the length scale and an outer region, where the acoustic wave length is the length scale. However, Osborne used slightly different scales. He used c/β^* as an inner length scale and $2\pi\beta^*/\nu$ as time scale. The wave equation is nondimensionalized in each region according to the relevant scale. In the inner region Laplace's equation is obtained, if the square of a certain unsteadiness parameter, the ratio of inner to the outer length scales,

$$\epsilon = \frac{\nu c}{2\pi a_0} = M\frac{\omega}{2\pi} \tag{6.63}$$

is small.

After obtaining the Laplace's equation in the equivalent incompressible plane the load distribution was obtained in the incompressible plane by relating the pressure between the compressible and the equivalent incompressible planes. Once the load distributions are known the forces are determined. Osborne imposed the condition in the equivalent incompressible flow in the inner region that

$$\frac{\epsilon}{\beta^{*2}} \ll 1 \tag{6.64}$$

The general lift and moment expressions given by Osborne for a Kemp type upwash given in equation (6.12) are

$$ {}_tL_v = \frac{2\pi}{\beta}\rho c V v_0 e^{i\nu t}\Big[J(\Lambda)\{C(\omega^i)J(M^2\omega^i) + iJ_1(M^2\omega^i)\}$$

$$-\frac{2i}{\Lambda}\left(1-\frac{\omega^i}{\Lambda}\right)\Big\{\sum_n nJ_n(\Lambda)J_n(M^2\omega^i)\Big\}$$

$$+\frac{i\omega^i}{\Lambda}\{J_0(M^2\omega^i)J_1(\Lambda)-J_1(M^2\omega^i)J_0(\Lambda)\}\Bigg] \tag{6.65}$$

$$
\begin{aligned}
{}_tM_v = \frac{\pi}{\beta}\rho c^2 V v_0 e^{i\nu t}\Big[& J(\Lambda)\{[C(\omega^i)-1][J_0(M^2\omega^i)-J_2(M^2\omega^i)] \\
& -2iC(\omega^i)J_1(M^2\omega^i)\} + J_0(M^2\omega^i)\Big\{\frac{\omega^i}{\Lambda}J_0(\Lambda)+\frac{2}{\Lambda}\Big(1-\frac{\omega^i}{\Lambda}\Big)J_1(\Lambda)\Big\} \\
& +\frac{\omega^i}{\Lambda}\{2J_1(\lambda)J_1(M^2\omega^i)-J_0(\Lambda)J_2(M^2\omega^i)\} \\
& +\frac{2}{\Lambda}\Big(1-\frac{\omega^i}{\Lambda}\Big)\sum_n\{(n+1)J_{n+1}(\Lambda)-(n-1)J_{n-1}(\Lambda)\}J_n(M^2\omega^i)\Big]
\end{aligned} \tag{6.66}
$$

where,

$$\Lambda = \lambda + \lambda^i \tag{6.66a}$$

6.3 INTERFERENCE EFFECTS IN A TURBOMACHINE STAGE

A significant extension of the basic thin isolated airfoil theory to the problem of calculating unsteady lifts and moments caused by the relative motion of the blades in a simple turbomachine stage, was made by Kemp and Sears [24]. Each airfoil in the cascade is considered to be acting in a velocity field induced by the following: its own wake, the variable bound vortices of fellow members of its own blade row, the wakes of fellow member blades, the variable bound vortices of members of other blade row and the wakes of other blade row.

As the problem is very complex, a successive approximation scheme is adopted, in which the assumption is made that the entire unsteady effect on the circulation of any blade is small compared to the steady circulation carried by the blade and thus the effect of steady circulation alone on the adjacent blade row is taken into account. This is valid as long as the cascade is of low solidity. The aerodynamic problem is ultimately reduced to the determination of unsteady upwash at the blade due to the interference of adjacent blades. The unsteady upwash expression is finally brought to the form of general velocity perturbation given by equation (6.12) and the single thin airfoil theory is applied to determine the lift and moment expressions.

Osborne [16] presented an analytical approximation for the lift and moment of an airfoil in a 2D compressible subsonic flow with the interferences considered as unsteady upwashes which are sinusoidal in time. His analysis is restricted to the flat plate blade stage taking into account only the upwash components. Rao et al [30] considered a stage with general cambered airfoils taking into account both the upwash and downwash components, however the compressibility effects were ignored. Rao et al [49] presented an extension of Osborne's analysis taking into account both the upwash and downwash components. Rao [84] presented a more general analysis by combining compressibility effects with the influence of both the upwash and downwash components on a cambered airfoil turbomachine stage. The following analysis closely follows this work.

The assumptions briefly are:

1. The airfoil is replaced by a distribution of vortices such that no penetration condition is satisfied.
2. The airfoil is oriented in such a way that the trailing vorticity lies along the positive x-axis, i.e., the wake is rectilinear.
3. Each blade row is taken as an infinite 2D cascade and only one stage, viz., a stator row followed by a rotor row is considered.
4. Flow is assumed to be 2D and inviscid.
5. Thin airfoil theory is used and the effect of free stream vorticity has been ignored.
6. Fluctuations in the blade circulation are small compared with the mean value.
7. The flow is assumed to be compressible and subsonic and the induced velocities are small compared with the flow velocities over the blade.

Interference of Rotor on Stator

Consider now a simple turbomachine stage with general cambered thin airfoils as given in Fig. 6.8. The blades are spaced at distances d_s, d_r respectively for the stator and rotor. The Mach numbers of the flows over the blades are M_s, M_r and c_s, c_r are semi-chords of the stator and rotor blades. The first step in the analysis is to obtain the upwash and downwash components induced at stator blades due to potential interaction of rotor blades and upwash and downwash components at rotor blades due to potential and vortex wakes interactions of stator blades in the compressible plane.

Figure 6.9a shows the blade configuration in the compressible plane, with the origin of the coordinate axis at the centre of the zeroth rotor blade. The Prandtl-Glauert equation for the steady compressible flow field is given by

$$(1 - M_r^2)\phi_{,x_r^0,x_r^0} + \phi_{,y_r^0,y_r^0} = 0 \tag{6.67}$$

M_r is given by

$$M_r = \frac{V_r}{a_0} \tag{6.68}$$

and the no penetration boundary condition will be

$$\phi_{,y_r^0} = V_r \frac{d}{dx_r^0}\{{}_rY(x_r^0)\} \tag{6.69}$$

on $y_x^0 \sim 0$ in the range of $-c_r \leq x_r^0 \leq +c_r$.

In equation (6.69) ${}_rY(x_r^0)$ is the shape of the rotor blade in the compressible plane.

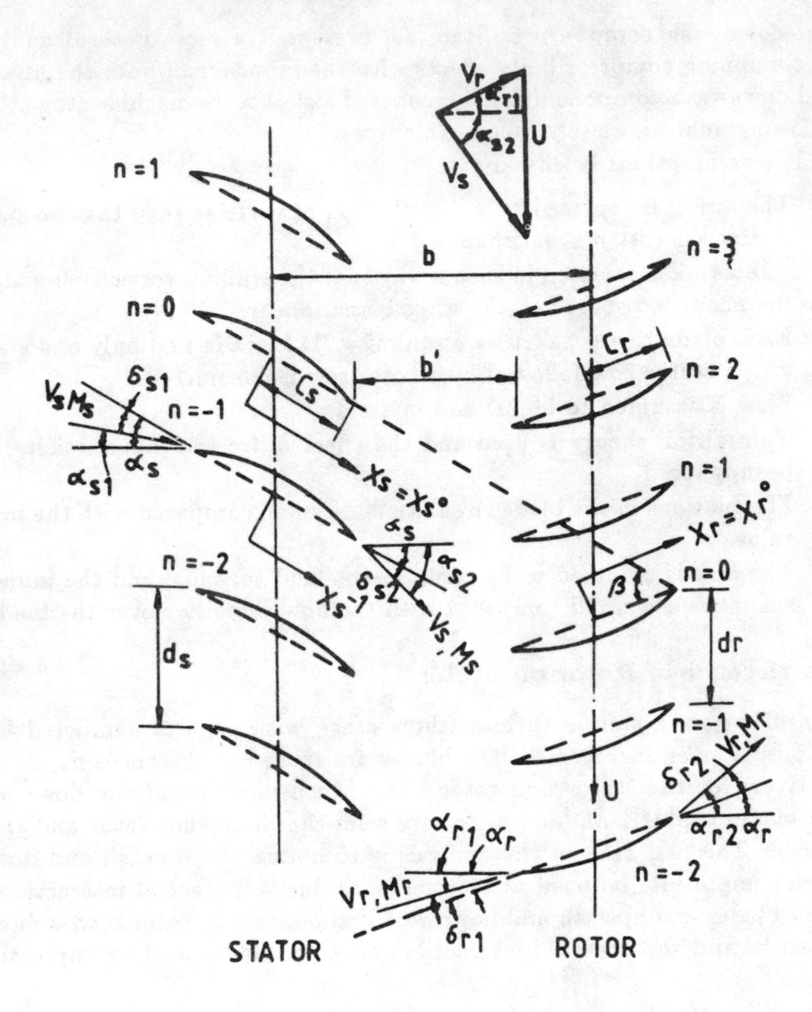

Fig. 6.8 Configuration of turbomachine stage (compressible plane)

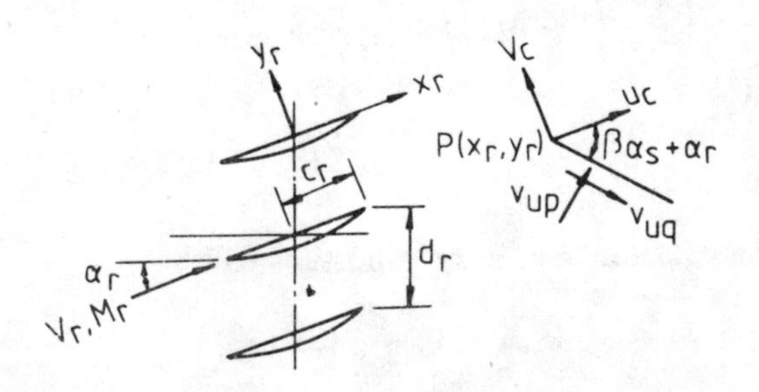

Fig. 6.9(a) Rotor blades in compressible plane

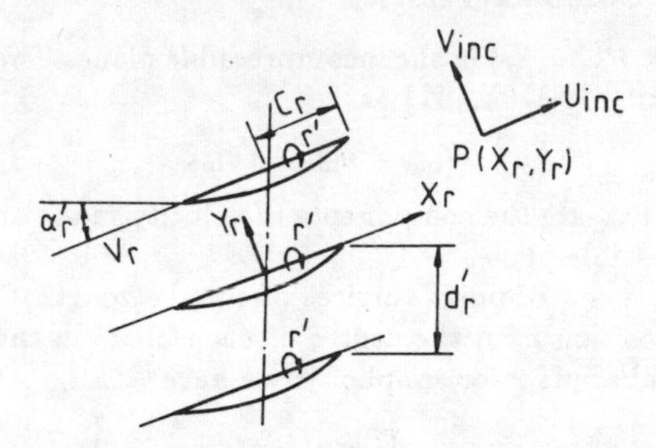

Fig. 6.9(b) Rotor blades in the equivalent incompressible plane

By applying the Prandtl-Glauert transformation related to the flow direction x_r^0, the compressible plane problem is reduced to that of incompressible plane. Then

$$x_r^0 = \beta_r^* X_r^0$$
$$y_r^0 = Y_r^0$$

and

$$\phi = a\Phi \tag{6.70}$$

Choosing the constant a to be

$$a = \frac{1}{\beta_r^*} = \frac{1}{\sqrt{1 - M_r^2}} \tag{6.71}$$

equation (6.67) reduces to the form

$$\Phi_{,x_r^0,x_r^0} + \Phi_{,y_r^0,y_r^0} = 0 \tag{6.72}$$

The boundary condition will be

$$\Phi_{,y_r^0} = V_r \frac{d}{dX_r^0}\{{}_r\overline{Y}(X_r^0)\} \tag{6.73}$$

on $Y_r \sim 0$ with $-c_r' \leq X_r^0 \leq +c_r'$.

In the above ${}_r\overline{Y}(X_r^0)$ is the new rotor blade shape with its chord c_r'

$${}_r\overline{Y}(X_r^0) = {}_rY(X_r^0 \beta_r^*) = {}_rY(x_r^0) \tag{6.74}$$

In the above transformation the blade chord, gap and stagger angle change to new values corresponding to the incompressible plane.

$$c_r' = \frac{c_r}{\beta_r^*}$$

$$d_r' = d_r \sqrt{\frac{1 - M_r^2 \cos^2 \alpha_r}{\beta_r^*}} \tag{6.75}$$

$$\tan \alpha_r' = \frac{\tan \alpha_r}{\beta_r^*}$$

The new configuration is shown in Fig. 6.9b.

Derivation of Complex Velocity

Consider a point $P(X_r, Y_r)$ in the incompressible plane. Then we can define a complex velocity, $W_{\text{inc}}(X_r, Y_r)$ as

$$W_{\text{inc}} = v_{\text{inc}} + iu_{\text{inc}} \tag{6.76}$$

where u_{inc} and v_{inc} are the components of velocity in X_r and Y_r directions in the incompressible plane.

By assuming a row of point vortices of equal strength Γ' located at the same relative position from the centre of each blade in the rotor cascade, and dropping subscipts r for simplicity, we have

$$v_{\text{inc}}, u_{\text{inc}} = \Re, \Im \left[-\frac{\Gamma'}{2\pi} \sum_n{}' \frac{1}{X + iY - \varsigma_n} \right] \tag{6.76(a)}$$

where ς_n are the positions of point vortices given by

$$\varsigma_n = \xi + ind'_r e^{-i\alpha'_r} \tag{6.76(b)}$$

Hence,

$$W_{\text{inc}}(X, Y) = \frac{\Gamma'}{2\pi} \sum_n{}' \frac{1}{X + iY - \xi - ind'_r e^{-i\alpha'_r}} \tag{6.76(c)}$$

Using the following identity

$$\coth Z = \sum_n{}' \frac{1}{Z - in\pi}$$

the above equation becomes

$$W_{\text{inc}}(X, Y) = \frac{-\Gamma' e^{i\alpha'_r}}{2d'_r} \coth \left[\frac{\pi e^{i\alpha'_r}}{d'_r} (X - \xi + iY) \right] \tag{6.76(d)}$$

Using another identity,

$$\coth Z = 1 + 2 \sum_m e^{-2mZ}, \quad \Re(Z) > 0$$

equation (6.76(d)) can be written as

$$W_{\text{inc}}(X, Y) = -\frac{\Gamma'}{2d'_r} e^{i\alpha'_r} \left[1 + 2 \sum_m \exp\left\{ -\frac{2\pi m}{d'_r} e^{i\alpha'_r} (X - \xi + iY) \right\} \right] \tag{6.76(e)}$$

Replacing Γ' by ${}_0\overline{\gamma}^r \, d\xi$ and integrating over the chord

$$W_{\text{inc}}(X, Y) = \frac{{}_0\overline{\Gamma}^r}{2d'_r} e^{i\alpha'_r} \left[1 + 2 \sum_m \exp\left\{ \frac{2\pi m}{d'_r} e^{i\alpha'_r} (X + iY) \right\} \right.$$

$$\left. \int_{-c'_r}^{+c'_r} \frac{{}_0\overline{\gamma}^r}{{}_0\overline{\Gamma}^r} \exp\left\{ -\frac{2\pi m}{d'_r} e^{i\alpha'_r} \xi \right\} d\xi \right] \tag{6.76(f)}$$

where

$${}_0\overline{\Gamma}^r = \int_{-c'_r}^{+c'_r} {}_0\overline{\gamma}^r(\xi) \, d\xi \tag{6.76(g)}$$

The integral in equation (6.76(f)) is removed by using Glauert expansion

$$_0\overline{\gamma}^r(\xi) = 2V_r\left[\frac{1-\cos\theta}{\sin\theta}\overline{A}\,^r_0 + \sum_n \overline{A}\,^r_n \sin n\theta\right] \qquad (6.76(h))$$

where,

$$\xi = c'_r \cos\theta$$

Hence the circulation given by equation (6.76(g)) can be written as

$$_0\overline{\Gamma}^r = 2\pi c'_r V_r(\overline{A}\,^r_0 + \overline{A}\,^r_1) \qquad (6.76(g'))$$

Using the identity

$$2\sum_n \overline{A}\,^r_n \sin n\theta \sin\theta = \sum_n(\overline{A}\,^r_{n+1} - \overline{A}\,^r_{n-1})\cos n\theta + \overline{A}\,^r_1 + \overline{A}\,^r_0 \cos\theta$$

the integral in equation (6.76(f)) can be written as

$$\begin{aligned}\overline{H}^r_m &= \int_{-c'_r}^{+c'_r} \frac{_0\overline{\gamma}^r(\xi)}{_0\Gamma^r}\exp\left(-\frac{2\pi m}{d'_r}e^{i\alpha'_r}\xi\right)d\xi \\ &= \frac{1}{\pi}\int_0^{\pi}\left[1+\sum_n \frac{\overline{A}\,^r_{n+1} - \overline{A}\,^r_{n-1}}{\overline{A}\,^r_0 + \overline{A}\,^r_1}\cos n\theta\right]\exp(-\pi m e^{i\alpha'_r}\sigma'_r \cos\theta)\,d\theta\end{aligned} \qquad (6.76(i))$$

where,

$$\sigma'_r = 2\frac{c'_r}{d'_r}$$

Replacing,

$$\pi\sigma'_r e^{i\alpha'_r} = i\pi e^{-i\left(\frac{\pi}{2}-\alpha'_r\right)}\sigma'_r$$

a typical integral in equation (6.76(i)) is

$$\frac{1}{\pi}\int_0^{\pi} e^{-iZ\cos\theta}\cos n\theta\, d\theta = (-i)^n J_n(Z)$$

with

$$Z = \pi m\sigma'_r e^{-i\left(\frac{\pi}{2}-\alpha'_r\right)}$$

Hence equation (6.76(i)) can be written as

$$\begin{aligned}\overline{H}^r_m &= J_0\left[\pi m\sigma'_r e^{-i\left(\frac{\pi}{2}-\alpha'_r\right)}\right] \\ &\quad + \sum_n (-i)^n \frac{\overline{A}\,^r_{n+1} - \overline{A}\,^r_{n-1}}{\overline{A}\,^r_0 + \overline{A}\,^r_1} J_n\left[\pi m\sigma'_r e^{-i\left(\frac{\pi}{2}-\alpha'_r\right)}\right]\end{aligned} \qquad (6.77)$$

Hence the complex velocity given by equation (6.76(f)) becomes

$$W_{\text{inc}}(X,Y) = \frac{_0\overline{\Gamma}^r}{2d'_r}e^{i\alpha'_r}\left[1 + 2\sum_n \overline{H}^r_m \exp\left\{\frac{2\pi m}{d'_r}e^{i\alpha'_r}(X+iY)\right\}\right] \qquad (6.78)$$

The velocity components in the compressible plane are

$$u_c = \frac{u_{\text{inc}}}{\beta_r^{*2}} \tag{6.79}$$

$$v_c = \frac{v_{\text{inc}}}{\beta_r^{*2}} \tag{6.80}$$

The complex velocity in the compressible plane is then given by

$$W_c = v_c + iu_c \tag{6.81}$$

Expressing the above in terms of the incompressible plane components,

$$W_c = \frac{1}{\beta_r^{*2}}\left[W_{\text{inc}} - (1 - \beta_r^*)\Re(W_{\text{inc}})\right] \tag{6.82}$$

Now a specific direction is chosen in the compressible plane and two velocities V_{up}, V_{uq} perpendicular and parallel to this direction are determined. The direction is so chosen that it is at an angle β to the x_r^0 axis, where

$$\beta = \alpha_s + \alpha_r \tag{6.83}$$

α_s and α_r being stagger angles of the stator and rotor cascades respectively in the compressible plane. The velocities V_{up} and V_{uq} can be expressed as

$$V_{up} = v_c \cos\beta + u_c \sin\beta = \Re\left[W_c e^{-i\beta}\right] \tag{6.84}$$

and

$$V_{uq} = -v_c \sin\beta + u_c \cos\beta = \Im\left[W_c e^{-i\beta}\right] \tag{6.85}$$

Equation (6.84) can be written as

$$V_{up} = \Re\left[\frac{e^{-i\beta}}{\beta_r^{*2}} W_{\text{inc}}\right] - \Re\left[\frac{1 - \beta_r^*}{\beta_r^{*2}} e^{-i\beta}\Re(W_{\text{inc}})\right] \tag{6.86}$$

Writing

$$e^{-i\beta}\Re(W_{\text{inc}}) = \cos\beta\, W_{\text{inc}}$$

$$V_{up} = \frac{e^{-i\beta}}{\beta_r^{*2}} W_{\text{inc}}\left[1 - (1 - \beta_r^*)e^{i\beta}\cos\beta\right] \tag{6.87}$$

in which the real part only is implied, the above can be written as

$$V_{up} = \left[1 - (1 - \beta_r^*)e^{i\beta}\cos\beta\right]V_1 \tag{6.88}$$

where,

$$V_1 = W_{\text{inc}}\frac{e^{-i\beta}}{\beta_r^{*2}} \tag{6.89}$$

Similarly, we can obtain

$$V_{uq} = \Im\left[W_{\text{inc}}\frac{e^{-i\beta}}{\beta_r^{*2}}\right] + \Im\left[\frac{i(1 - \beta_r^*)}{\beta_r^{*2}} e^{-i\beta}\Im(W_{\text{inc}})\right] \tag{6.90}$$

The above can be simplified to

$$V_{uq} = U_1\left[1 + \frac{1 - \beta_r^*}{\beta_r^*} e^{i\beta}\cos\beta\right] \tag{6.91}$$

where,

$$U_1 = W_{\text{inc}} \frac{e^{-i\beta}}{\beta_r^*} \tag{6.92}$$

Using the expressions given in equation (6.77) and the relations in (6.70), equations (6.89) and (6.92) become

$$V_1(x_r, y_r) = \frac{e^{-i\beta}}{\beta_r^{*2}} \frac{{}_0\overline{\Gamma}^r e^{i\alpha'_r}}{2d'_r} \Bigg[1 + 2\sum_m \overline{H}_m^r \exp\Big\{\frac{2\pi m}{d'_r} e^{i\alpha'_r}\Big(\frac{x_r}{\beta_r^*} + iy_r\Big)\Big\}\Bigg] \tag{6.93}$$

$$U_1(x_r, y_r) = \frac{e^{-i\beta}}{\beta_r^*} \frac{{}_0\overline{\Gamma}^r e^{i\alpha'_r}}{2d'_r} \Bigg[1 + 2\sum_m \overline{H}_m^r \exp\Big\{\frac{2\pi m}{d'_r} e^{i\alpha'_r}\Big(\frac{x_r}{\beta_r^*} + iy_r\Big)\Big\}\Bigg] \tag{6.94}$$

The above are the upwash and downwash components perpendicular and parallel to the stator blade in terms of coordinates of rotor blade in the compressible plane. In order to obtain the expressions for these components in terms of stator coordinates, the following relation can be used.

$$x_r + iy_r = (-b + ih_r^*)e^{-i\alpha_r} + x_s e^{-i\beta} \tag{6.95}$$

where the orientation parameter defined in Fig. 6.10 can be expressed as

$$h_r^* = b \tan\alpha_s + c_r \frac{\sin\beta}{\cos\alpha_s} + Ut \tag{6.96}$$

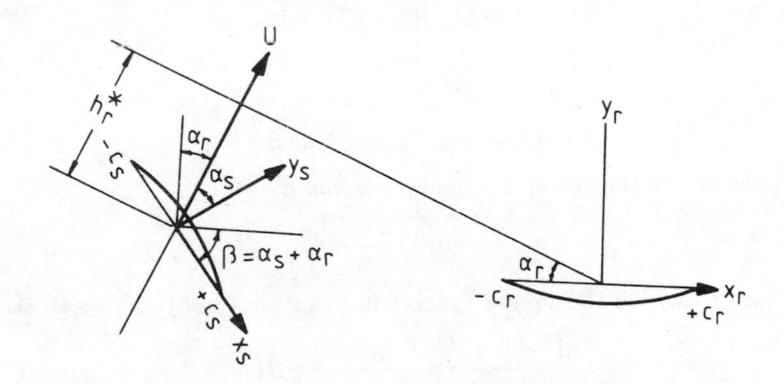

Fig. 6.10 Orientation of stator blade relative to rotor blade

From equation (6.95), we have

$$x_r = -b\cos\alpha_r + h_r^* \sin\alpha_r + x_s \cos\beta$$
$$y_r = b\sin\alpha_r + h_r^* \cos\alpha_r - x_s \sin\beta$$

Hence

$$\frac{x_r}{\beta_r^*} + iy_r = \frac{1}{\beta_r^*}[x_r + iy_r - i(1-\beta_r^*)y_r]$$

With the help of equations (6.95) and (6.96) and the above we can obtain the following.

$$\begin{aligned}\frac{x_r}{\beta_r^*} + iy_r = \frac{1}{\beta_r^*}\Big[& iUt\{e^{-i\alpha_r} - (1-\beta_r^*)\cos\alpha_r\} \\ & - b\{e^{-i\alpha_r} + i(1-\beta_r^*)\sin\alpha_r\} \\ & + bi\{e^{-i\alpha_r} - (1-\beta_r^*)\cos\alpha_r\}\tan\alpha_s \\ & + \frac{ic_r\sin\beta}{\cos\alpha_s}\{e^{-i\alpha_r} - (1-\beta_r^*)\cos\alpha_r\} \\ & + x_s\{e^{-i\beta} + i(1-\beta_r^*)\sin\beta\}\Big]\end{aligned}$$

Hence the exponential term in equation (6.93) can be written as

$$\begin{aligned}\exp\left[\frac{2\pi m}{d_r'}e^{-i\alpha_r'}\left\{\frac{x_r}{\beta_r^*} + iy_r\right\}\right] = e^{i\nu_s mt}\exp\Big[-\pi m\sigma_r\Big\{\frac{b}{c_r}(1 - i\tan\alpha_s)H(M_r) \\ - \frac{i\sin\beta}{\cos\alpha_s}\Big\}\Big]\exp\left[\frac{2\pi m}{d_r}e^{-i\alpha_s}I(M_r)x_s\right] \quad (6.97)\end{aligned}$$

where,

$$H(M_r) = E(M_r)G(M_r) \quad (6.98)$$

$$E(M_r) = \frac{1-(1-\beta_r^*)\cos\alpha_r\, e^{-i\alpha_r}}{1 - M_r^2\cos^2\alpha_r} \quad (6.99)$$

$$G(M_r) = 1 + i(1-\beta_r^*)\sin\beta\, e^{i\alpha_s} \quad (6.100)$$

and

$$I(M_r) = E(M_r)F(M_r) \quad (6.101)$$

$$F(M_r) = 1 + i(1-\beta_r^*)\sin\beta\, e^{i\beta} \quad (6.102)$$

In the above the stator frequency is given by

$$\nu_s = \frac{2\pi U}{d_r} \quad (6.103)$$

The factor outside the square bracket in equation (6.93) can be written as

$$\frac{{}_0\overline{\Gamma}^r e^{i\alpha_r'}}{2d_r'}\frac{e^{-i\beta}}{\beta_r^{*2}} = \frac{{}_0\Gamma^r e^{-i\alpha_s}}{2d_r}J(M_r) \quad (6.104)$$

where,

$$J(M_r) = \frac{E(M_r)}{\beta_r^*} \quad (6.105)$$

Similarly in equation (6.94), we have

$$\frac{{}_0\overline{\Gamma}^r e^{i\alpha_r'}}{2d_r'}\frac{e^{-i\beta}}{\beta_r^{*2}} = \frac{{}_0\Gamma^r e^{-i\alpha_s}}{2d_r}E(M_r) \quad (6.106)$$

Using equations (6.104), (6.97) and (6.93), we can write the final expression for V_1 given by equation (6.89) as

$$V_1(x_s,t) = \frac{{}_0\Gamma^r e^{-i\alpha_s}}{2d_r} J(M_r) + \frac{{}_0\Gamma^r}{2\pi c_s} \sum_m \overline{G}^s_m \exp\left[\frac{2\pi m}{d_r} e^{-i\alpha_s} I(M_r) x_s\right] e^{i\nu_s mt} \quad (6.107)$$

The velocity coefficient induced at the blade in the above is

$$\overline{G}^s_m = \pi\sigma_s e^{-i\alpha_s} \frac{d_s}{d_r} J(M_r) \overline{H}^r_m \exp\left[-\pi m\sigma_r \left\{\frac{b}{c_r}\left(1 + i\tan\alpha_r - \frac{iU}{V_s \cos\alpha_s}\right) H(M_r) - \frac{i\sin\beta}{\cos\alpha_s}\right\}\right] \quad (6.108)$$

Similarly,

$$U_1(x_s,t) = \frac{{}_0\Gamma^r e^{-i\alpha_s}}{2d_r} E(M_r) + \frac{{}_0\Gamma^r}{2\pi c_s} \sum_m \overline{Q}^s_m \exp\left[\frac{2\pi m}{d_r} e^{-i\alpha_s} I(M_r) x_s\right] e^{i\nu_s mt} \quad (6.109)$$

$$\overline{Q}^s_m = \pi\sigma_s e^{-i\alpha_s} \frac{d_s}{d_r} E(M_r) \overline{H}^r_m \exp\left[-\pi m\sigma_r \left\{\frac{b}{c_r}\left(1 + i\tan\alpha_r - \frac{iU}{V_s \cos\alpha_s}\right) H(M_r) - \frac{i\sin\beta}{\cos\alpha_s}\right\}\right] \quad (6.110)$$

$\overline{H}^r_m$ in equations (6.108) and (6.110) is given by equation (6.77). In the $\overline{H}^r_m$ equation $\overline{A}^r_n$ can be replaced by A^r_n, which are the coefficients in the Glauert's expansion of the vortex distribution of actual stator cascade. The argument in the Bessel function in equation (6.78) can be expressed, using equations (6.75) and (6.99) as

$$z_r = \pi m\sigma_r e^{-i(\frac{\pi}{2}-\alpha_r)} E(M_r)$$

Hence the simplified expression for $\overline{H}^r_m$ becomes

$$\overline{H}^r_m = J_0\left[\pi m\sigma_r e^{-i(\frac{\pi}{2}-\alpha_r)} E(M_r)\right] + \sum_n (-i)^n \frac{A^r_{n+1} - A^r_{n-1}}{A^r_0 + A^r_1} J_n\left[\pi m\sigma_r e^{-i(\frac{\pi}{2}-\alpha_r)} E(M_r)\right] \quad (6.111)$$

If the first two Glauert's constants only are considered, the above becomes

$$\overline{H}^r_m = J_0(z_r) + \frac{iA^r_0}{A^r_0 + A^r_1} J_1(z_r) + \frac{A^r_1}{A^r_0 + A^r_1}\left[\frac{2J_1(z_r)}{\pi m\sigma_r e^{-i(\frac{\pi}{2}-\alpha_r)} E(M_r)} - J_0(z_r)\right] \quad (6.112)$$

Considering only the unsteady effects, from equations (6.88), (6.107), (6.91) and (6.109), we can now obtain the final expressions for upwash and downwash induced on the stator blade due to the interference of the rotor cascade as

$$V^s_{up}(x_s,t) = \left[1-(1-\beta^*_r)\cos\beta e^{i\beta}\right] \cdot \frac{{}_0\Gamma^r}{2\pi c_s}\sum_m \overline{G}^s_m \exp\left[\frac{2\pi m}{d_r}e^{-i\alpha_s}I(M_r)x_s\right]e^{i\nu_s mt} \quad (6.113)$$

$$V^s_{uq}(x_s,t) = \left[1+\frac{1-\beta^*_r}{\beta^*_r}\cos\beta\, e^{i\beta}\right] \cdot \frac{{}_0\Gamma^r}{2\pi c_s}\sum_m \overline{Q}^s_m \exp\left[\frac{2\pi m}{d_r}e^{-i\alpha_s}I(M_r)x_s\right]e^{i\nu_s mt} \quad (6.114)$$

Interference of Stator on Rotor

The analysis for this case is similar to that of the case of interference of rotor on stator considered in the section above. The similarity between the two cases enables one to write down simply the equations analogous to those given for rotor interference by changing subscripts on geometric parameters and by interchanging $-\alpha_s$ with α_r.

As in the previous case, the steady flow field over the stator cascade is corrected for compressibility by Prandtl-Glauert transformation. The coordinate axes are so chosen that the origin will be at the centre of zeroth stator blade. The blade configuration in the compressible plane is shown in Fig. 6.11a. By applying the transformations related to the flow direction x^0_s, the compressible plane problem is reduced to that of incompressible plane as shown in Fig. 6.11b, and the analysis is carried out on similar lines to that of stator, to find ultimately the upwash and downwash induced at the rotor blade perpendicular and parallel to the rotor blade chord. The final velocity expressions thus obtained are given below.

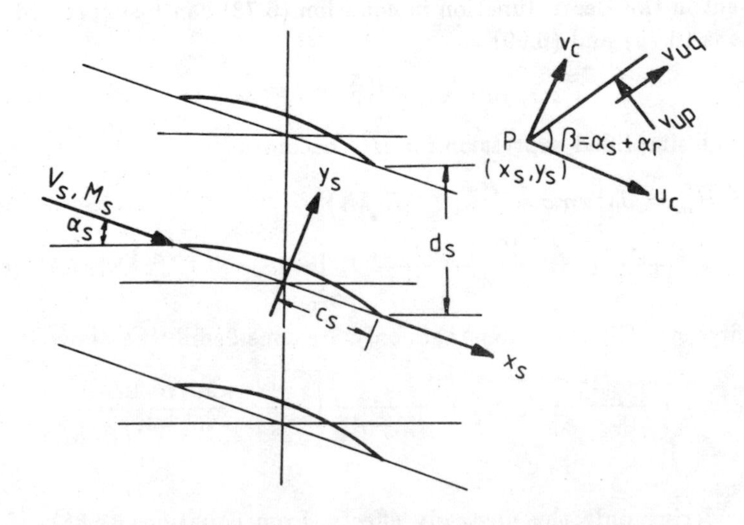

Fig. 6.11(a) Stator blades in compressible plane

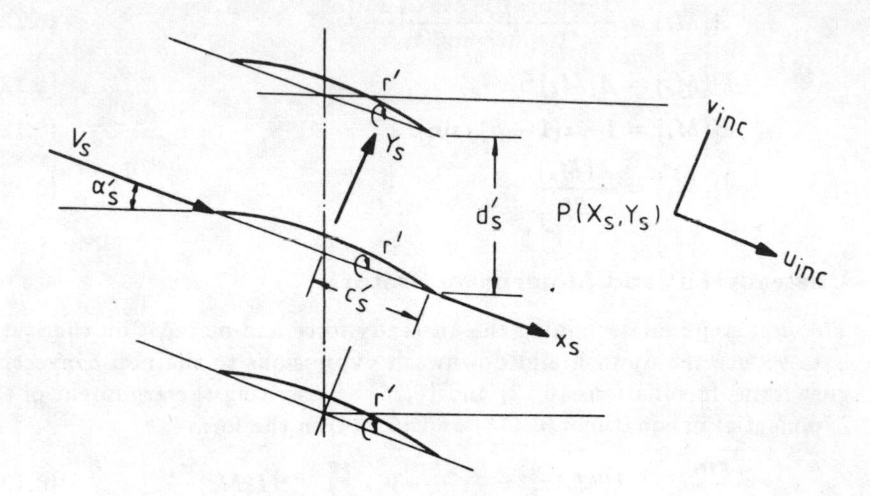

Fig. 6.11(b) Stator blades in the equivalent incompressible plane

$$V^{r1}_{up}(x_r,t) = \left[1-(1-\beta^*_s)e^{-i\beta}\cos\beta\right]$$

$$\cdot\frac{{}_0\Gamma^s}{2\pi c_r}\sum_m \overline{G}^{r1}_m \exp\left[-\frac{2\pi m}{d_s}e^{i\alpha_r}C(M_s)x_r\right]e^{i\nu_r mt} \qquad (6.115)$$

$$\overline{G}^{r1}_m = -\pi\sigma_r e^{i\alpha_r}\frac{d_r}{d_s}D(M_s)\overline{H}^s_m \exp\left[-\pi m\sigma_r\frac{d_r}{d_s}\left\{\frac{b}{c_r}\left(1\right.\right.\right.$$

$$\left.\left.\left.+i\tan\alpha_r-\frac{iU}{V_s\cos\alpha_s}\right)A(M_s)-\frac{i\sin\beta}{\cos\alpha_s}\right\}\right] \qquad (6.116)$$

and

$$V^{r1}_{uq}(x_r,t) = \left[1+\frac{1-\beta^*_s}{\beta^*_s}e^{-i\beta}\cos\beta\right]$$

$$\cdot\frac{{}_0\Gamma^s}{2\pi c_r}\sum_m \overline{Q}^{r1}_m \exp\left[-\frac{2\pi m}{d_s}e^{i\alpha_r}C(M_s)x_r\right]e^{i\nu_r mt} \qquad (6.117)$$

$$\overline{Q}^{r1}_m = -\pi\sigma_r e^{i\alpha_r}\frac{d_r}{d_s}A(M_s)\overline{H}^s_m \exp\left[-\pi m\sigma_r\frac{d_r}{d_s}\left\{\frac{b}{c_r}\left(1\right.\right.\right.$$

$$\left.\left.\left.+i\tan\alpha_r-\frac{iU}{V_s\cos\alpha_s}\right)A(M_s)-\frac{i\sin\beta}{\cos\alpha_s}\right\}\right] \qquad (6.118)$$

where,

$$\overline{H}^s_m = J_0(z_s)+\frac{iA^s_0}{A^s_0+A^s_1}J_1(z_s)$$

$$+\frac{A^s_1}{A^s_0+A^s_1}\left\{\frac{2J_1(z_s)}{\pi m\sigma_s A(M_s)}e^{-i\left(\frac{\pi}{2}-\alpha_s\right)}-J_0(z_s)\right\} \qquad (6.119)$$

$$z_s = \pi m\sigma_s A(M_s)e^{i\left(\frac{\pi}{2}-\alpha_s\right)} \qquad (6.119a)$$

$$A(M_s) = \frac{1-(1-\beta_s^*)\cos\alpha_s\, e^{i\alpha_s}}{1-M_s^2\cos^2\alpha_s} \tag{6.120}$$

$$C(M_s) = A(M_s)B(M_s) \tag{6.121}$$

$$B(M_s) = 1 - i(1-\beta_s^*)\sin\beta\, e^{-i\beta} \tag{6.122}$$

$$D(M_s) = \frac{A(M_s)}{\beta_s^*} \tag{6.123}$$

Unsteady Lift and Moment on Stator

The first step in determining the unsteady force and moment on the stator is to reduce the upwash and downwash expressions to the non convecting gust forms in equations (6.12) and (6.25). By writing the argument of the exponential in equations (6.113) and (6.114) in the form

$$\frac{2\pi m}{d_r} e^{-i\alpha_s} I(M_r) x_s = -i\frac{2\pi m}{d_r} V_s e^{i\left(\frac{\pi}{2}-\alpha_s\right)} I(M_r)\frac{x_s}{V_s} \tag{6.124}$$

we can identify the following in equations (6.12) and (6.25).

$$\mu_s = \frac{2\pi V_s}{d_r} e^{i\left(\frac{\pi}{2}-\alpha_s\right)} I(M_r) \tag{6.125}$$

$$\lambda_s = \frac{\mu_s c_s}{V_s} = \frac{2\pi c_s}{d_r} e^{i\left(\frac{\pi}{2}-\alpha_s\right)} I(M_r) \tag{6.126}$$

where λ_s is the reduced non-convecting frequency for stator blades. The reduced convecting frequency is given by

$$\omega_s = \frac{\nu_s c_s}{V_s} \tag{6.127}$$

Using equations (6.13) and (6.14), the lift and moment expressions due to upwash are given by

$$\frac{{}_tL_v^s}{{}_0L^s} = \frac{{}_0\Gamma^r}{{}_0\Gamma^s}\{1-(1-\beta_r^*)e^{i\beta}\cos\beta\} \sum_m \overline{G}_m^s S_L(\omega_s m, \lambda_s m) e^{i\nu_s mt} \tag{6.128}$$

$$\frac{{}_tM_v^s}{{}_0L^s 2c_s} = \frac{{}_0\Gamma^r}{{}_0\Gamma^s}\{1-(1-\beta_r^*)e^{i\beta}\cos\beta\} \sum_m \frac{1}{4}\overline{G}_m^s S_M(\omega_s m, \lambda_s m) e^{i\nu_s mt} \tag{6.129}$$

In the above ${}_0L^s$ is the steady stator lift in the incompressible flow. The gust response functions S_L, S_M are given in equations (6.15) and (6.16).

The lift and moment for an airfoil subjected to a downwash given by equation (6.25) are given by (6.51) and (6.55). Considering the real part, we get the following expressions for the downwash.

$$\frac{{}_tL_u^s}{{}_0L^s} = -\frac{{}_0\Gamma^r}{{}_0\Gamma^s}\left\{1+\frac{1-\beta_r^*}{\beta_r^*}e^{i\beta}\cos\beta\right\} \frac{1}{2}\sum_m \overline{Q}_m^s e^{i\frac{\pi}{2}} H^*(\omega_s m, \lambda_s m) e^{i\nu_s mt} \tag{6.130}$$

$$\frac{{}_tM_u^s}{{}_0L^s 2c_s} = -\frac{{}_0\Gamma^r}{{}_0\Gamma^s}\left\{1 + \frac{1-\beta_r^*}{\beta_r^*}e^{i\beta}\cos\beta\right\} \cdot \frac{1}{4}\sum_m \overline{Q}_m^s e^{i\frac{\pi}{2}} M^*(\omega_s m, \lambda_s m) e^{i\nu_s mt} \tag{6.131}$$

The gust response functions H^*, M^* are given by equations (6.52) and (6.56).

The components of lift and moment due to the angle of incidence are obtained from equations (6.59) and (6.61) by reducing the downwash of equation (6.114) into the form given by equation (6.25). Then

$$\frac{{}_tL_i^s}{{}_0L^s} = -\frac{{}_0\Gamma^r}{{}_0\Gamma^s}\left\{1 + \frac{1-\beta_r^*}{\beta_r^*}e^{i\beta}\cos\beta\right\} \cdot \sum_m \overline{Q}_m^s e^{i\frac{\pi}{2}} H_L(\omega_s m, \lambda_s m) e^{i\nu_s mt} \tag{6.132}$$

$$\frac{{}_tM_i^s}{{}_0L^s 2c_s} = -\frac{{}_0\Gamma^r}{{}_0\Gamma^s}\left\{1 + \frac{1-\beta_r^*}{\beta_r^*}e^{i\beta}\cos\beta\right\} \cdot \frac{1}{4}\sum_m \overline{Q}_m^s e^{i\frac{\pi}{2}} M_L(\omega_s m, \lambda_s m) e^{i\nu_s mt} \tag{6.133}$$

where the response functions H_L, M_L are given by equations (6.60) and (6.62).

Combining the upwash, downwash and incidence effects, we get

$$\frac{{}_tL^s}{{}_0L^s} = \frac{{}_tL_v^s}{{}_0L^s} + \frac{{}_tL_u^s}{{}_0L^s} + \frac{{}_tL_i^s}{{}_0L^s} \tag{6.134}$$

i.e.,

$$\frac{{}_tL^s}{{}_0L^s} = \frac{{}_0\Gamma^r}{{}_0\Gamma^s}\Bigg[\{1-(1-\beta_r^*)e^{i\beta}\cos\beta\}\sum_m \overline{G}_m^s S_L(\omega_s m, \lambda_s m) - \left\{1 + \frac{1-\beta_r^*}{\beta_r^*}e^{i\beta}\cos\beta\right\}\sum_m \overline{Q}_m^s\left\{H_L(\omega_s m, \lambda_s m) + \frac{H^*}{2}(\omega_s m, \lambda_s m)\right\}e^{i\frac{\pi}{2}}\Bigg]e^{i\nu_s mt} \tag{6.135}$$

$$\frac{{}_tM^s}{{}_0L^s 2c_s} = \frac{{}_tM_v^s}{{}_0L^s 2c_s} + \frac{{}_tM_u^s}{{}_0L^s 2c_s} + \frac{{}_tM_i^s}{{}_0L^s 2c_s} \tag{6.136}$$

i.e.,

$$\frac{{}_tM^s}{{}_0L^s 2c_s} = \frac{1}{4}\frac{{}_0\Gamma^r}{{}_0\Gamma^s}\Bigg[\{1-(1-\beta_r^*)e^{i\beta}\cos\beta\}\sum_m \overline{G}_m^s S_M(\omega_s m, \lambda_s m) - \left\{1 + \frac{1-\beta_r^*}{\beta_r^*}e^{i\beta}\cos\beta\right\}\sum_m \overline{Q}_m^s\left\{M_L(\omega_s m, \lambda_s m) + M^*(\omega_s m, \lambda_s m)\right\}e^{i\frac{\pi}{2}}\Bigg]e^{i\nu_s mt} \tag{6.137}$$

To determine the unsteady lift and moment on the rotor, we write the argument of the exponential in equations (6.115) and (6.117) as

$$-\frac{2\pi m}{d_s}e^{i\alpha_r}C(M_s)x_r = -i\frac{2\pi m}{d_s}C(M_s)e^{-i(\frac{\pi}{2}-\alpha_r)}x_r \tag{6.138}$$

Then from equations (6.12) and (6.25) we get

$$\mu_r = \frac{2\pi V_r}{d_s}e^{-i(\frac{\pi}{2}-\alpha_r)}C(M_s) \tag{6.139}$$

$$\lambda_r = \frac{\mu_r c_r}{V_r} = \frac{2\pi c_r}{d_s}C(M_s)e^{-i(\frac{\pi}{2}-\alpha_r)} \tag{6.140}$$

$$\omega_r = \frac{\nu_r c_r}{V_r} \tag{6.141}$$

Following a similar procedure for the stator components, we can obtain the following expressions for the total unsteady lift on the rotor due to upwash, downwash and incidence effects of interference of stator on rotor.

$$\begin{aligned}\frac{{}_tL^{r1}}{{}_0L^r} = \frac{{}_0\Gamma^s}{{}_0\Gamma^r}\Big[&\{1-(1-\beta_s^*)e^{-i\beta}\cos\beta\}\sum_m \overline{G}_m^{r1} S_L(\omega_r m, \lambda_r m)\\ &-\Big\{1+\frac{1-\beta_s^*}{\beta_s^*}e^{-i\beta}\cos\beta\Big\}\sum_m \overline{Q}_m^{r1}\Big\{H_L(\omega_r m, \lambda_r m)\\ &+\frac{H^*}{2}(\omega_r m, \lambda_r m)\Big\}e^{i\frac{\pi}{2}}\Big]e^{i\nu_r mt}\end{aligned} \tag{6.142}$$

$$\begin{aligned}\frac{{}_tM^{r1}}{{}_0L^r 2c_r} = \frac{{}_0\Gamma^s}{{}_0\Gamma^r}\frac{1}{4}\Big[&\{1-(1-\beta_s^*)e^{-i\beta}\cos\beta\}\sum_m \overline{G}_m^{r1} S_M(\omega_r m, \lambda_r m)\\ &-\Big\{1+\frac{1-\beta_s^*}{\beta_s^*}e^{-i\beta}\cos\beta\Big\}\sum_m \overline{Q}_m^{r1}\Big\{M^*(\omega_r m, \lambda_r m)\\ &+M_L(\omega_r m, \lambda_r m)\Big\}e^{i\frac{\pi}{2}}\Big]e^{i\nu_r mt}\end{aligned} \tag{6.143}$$

6.4 Effect of Vortex Wakes Shed by Stator

Let the velocities V_{up} and V_{uq} in equations (6.113) and (6.114) be denoted by v^c and u^c. Then we can write the following.

$$w^c = v^c + iu^c \tag{6.144}$$

$$\begin{aligned}= \Big[&\{1-(1-\beta_r^*)e^{i\beta}\cos\beta\}\frac{{}_0\Gamma^r}{2\pi c_s}\sum_m \overline{G}_m^s\\ &+i\Big\{1+\frac{1-\beta_r^*}{\beta_r^*}e^{i\beta}\cos\beta\Big\}\frac{{}_0\Gamma^r}{2\pi c_s}\sum_m \overline{Q}_m^s\Big]\\ &\exp\Big\{\frac{2\pi m}{d_r}e^{-i\alpha_s}I(M_r)x_s\Big\}e^{i\nu_s mt}\end{aligned} \tag{6.145}$$

With the help of equations (6.12) and (6.18), the wake vortex strength can be written as

$$\epsilon^s(x_s^0,t) = -\frac{{}_0\Gamma^r}{c_s} i\omega_s m S_L(\omega_s m) e^{i\nu_s m\left(t-\frac{x_s^0}{V_s}\right)} \Bigg[\{1-(1-\beta_r^*)e^{i\beta}\cos\beta\}$$

$$\sum_m F_m^s + i\left\{1+\frac{1-\beta_r^*}{\beta_r^*}e^{i\beta}\cos\beta\right\}\sum_m F_m^{\prime s}\Bigg] \tag{6.146}$$

where

$$F_m^s = \overline{G}_m^s J(\lambda_s m) \tag{6.147}$$

$$F_m^{\prime s} = \overline{Q}_m^s J(\lambda_s m) \tag{6.148}$$

The above is the strength of the vortex wake shed by the zeroth stator blade. The strength of the vortex wakes of the other stator blades can be found using certain phase relationships given by Osborne [16]. The vortex sheets are assumed to stretch from − to + infinity instead of originating from the stator trailing edge. By using Poisson summation formula to convert the resulting expressions to the convenient complex Fourier series which is treated by Sear's theory for obtaining upwash at the rotor blade and extending it to obtain the downwash also, the final results for upwash and downwash at a rotor blade due to the vortex wakes of the stator cascade are expressed as

$$V_{up}^{r2}(x_r,t) = \{1-(1-\beta_r^*)\cos\beta\, e^{i\beta}\}\frac{{}_0\Gamma^r}{2\pi c_r}\sum_m{}' \overline{G}_m^{r2} e^{i\nu_r m\left(t-\frac{x_r^0}{V_r}Z_1^s\right)} \tag{6.149}$$

$$V_{uq}^{r2}(x_r,t) = \left\{1+\frac{1-\beta_r^*}{\beta_r^*}\cos\beta\, e^{i\beta}\right\}\frac{{}_0\Gamma^r}{2\pi c_r}\sum_m{}' \overline{G}_m^{\prime r2} e^{i\nu_r m\left(t-\frac{x_r^0}{V_r}Z_1^s\right)} \tag{6.150}$$

where

$$\overline{G}_m^{r2} = -\frac{\pi\sigma_r d_r}{d_s\cos\alpha_s}\sum_k P_k^s \overline{Q}_{mk}^r e^{i\omega_r m Z_1^s} \tag{6.151}$$

$$\overline{P}_k^s = \overline{F}_k^s S_L(\omega_s k)\exp\left\{-i\pi\sigma_r k\left(\frac{bU}{V_s c_r\cos\alpha_s}+\frac{U}{V_r}\right)\right\} \tag{6.152}$$

$$\overline{F}_k^s = \overline{G}_k^s J(\lambda_s k) \tag{6.153}$$

$$\overline{Q}_{mk}^r = \frac{\cos\beta - \sin\beta\left\{\frac{V_s - U\sin\alpha_s}{U\cos\alpha_s} - \frac{\pi m\sigma_s}{k\omega_s\cos\alpha_s}\right\}}{1+\left\{\frac{V_s - U\sin\alpha_s}{U\cos\alpha_s} - \frac{\pi m\sigma_s}{k\omega_s\cos\alpha_s}\right\}^2} \tag{6.154}$$

$$Z_1^s = \frac{\sin\beta}{\cos\alpha_s}\frac{V_r}{U} \tag{6.155}$$

The expression for G'^{r2}_m is obtained from equations (6.151), (6.152) and (6.153) by replacing $\overline{G}^s_k$ with $\overline{Q}^s_k$. Also for $\overline{G}$ and $\overline{Q}$, see equations (6.108) and (6.110).

Equations (6.149) and (6.150) are the upwash and downwash components at the rotor blade induced by the stator wakes when the rotor blade passes through them. These are reduced to the convecting gust forms given in equations (6.6) and (6.19). Further, using equations (6.13) and (6.14) for the upwash given in equation (6.149), and equations (6.51), (6.55) and (6.59), (6.61) after reducing the expressions for the case of a convecting gust, for the downwash in equation (6.150), the following equations are obtained for the lift and moment.

$$\frac{{}_tL^{r2}}{{}_0L^r} = \Bigg[\left\{1-(1-\beta_r^*)\cos\beta\, e^{i\beta}\right\}\sum_m{}' \overline{G}^{r2}_m S_L(\omega_r m) - \left\{1+\frac{1-\beta_r^*}{\beta_r^*}\cos\beta\, e^{i\beta}\right\}\sum_m{}' \overline{G}'^{r2}_m \left\{\frac{H^*}{2}(\omega_r m) + H_L(\omega_r m)\right\} e^{i\frac{\pi}{2}}\Bigg] e^{i\nu_r m t} \tag{6.156}$$

$$\frac{{}_tM^{r2}}{{}_0L^r 2c_r} = \frac{1}{4}\Bigg[\left\{1-(1-\beta_r^*)\cos\beta\, e^{i\beta}\right\}\sum_m{}' \overline{G}^{r2}_m S_M(\omega_r m) - \left\{1+\frac{1-\beta_r^*}{\beta_r^*}\cos\beta\, e^{i\beta}\right\}\sum_m{}' \overline{G}'^{r2}_m \left\{M^*(\omega_r m) + M_L(\omega_r m)\right\} e^{i\frac{\pi}{2}}\Bigg] e^{i\nu_r m t} \tag{6.157}$$

6.5 EFFECT OF VISCOUS WAKE INTERACTION

Besides the potential interaction and vortex wakes interaction considered earlier, the viscous wakes shed by the stator blade will also have an influence on the unsteady lift and moment of the rotor blades. Taking into account the small perturbation type of approximation that is inherent in thin airfoil theory, the lift and moment coefficients can be added in their proper phase to obtain the total non-steady force and moments.

The first step is to determine the wake geometry and velocity profile of the model as a function of Mach number. This information is not easily available, particularly as a function of Mach number and hence the incompressible model of Kemp and Sears is used for this purpose. The model adopted is based on the experimental results of Silverstein et al [2]. Since the wakes are relatively narrow compared to the blade spacing, it can be assumed that the wake behind each blade is same as the wake behind an isolated airfoil and further that the wake consists of only an inviscid, symmetrical shear perturbation of the undisturbed stream.

Referring to Fig. 6.12, the half width of the wake behind an airfoil can be taken as

$$Y_w = 0.68\sqrt{2}\,C_D^{\frac{1}{2}}\left[c\left(\frac{x}{c} - 0.7\right)\right]^{\frac{1}{2}} \tag{6.158}$$

where C_D is the profile drag coefficient. If u denotes the perturbation velocity in the x direction, assuming that $u \ll V$, the wake velocity profile is given by

$$\frac{u_e}{V} = -\frac{2.42\sqrt{C_D}}{\dfrac{x}{c} - 0.4} \tag{6.159}$$

and

$$\frac{u}{u_e} = \cos^2\left(\frac{\pi}{2}\frac{y}{Y_w}\right) \tag{6.160}$$

$$|y| \leq Y_w$$

where u_e is the perturbation velocity at the centre of the wake.

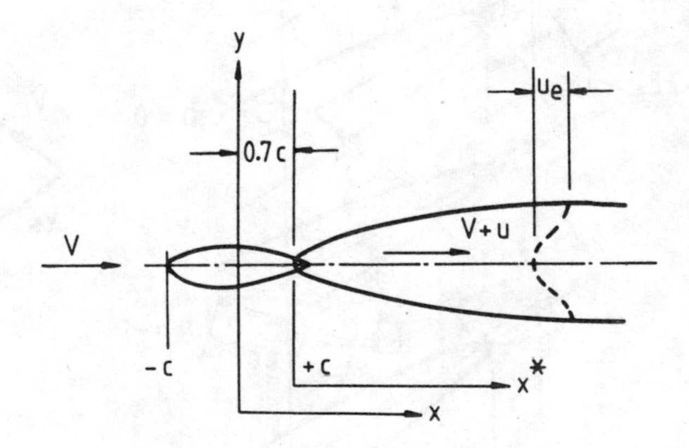

Fig. 6.12 Viscous wake of an airfoil

The cosine wake profile is replaced by a profile in the form of a Gaussian error curve that has the same momentum flow which simplifies the mathematical analysis. The momentum flow in the wake is proportional to square of the velocity, i.e., $(V + u)^2 \approx V^2 + 2uV$. The momentum flow across the section depends on $\int u\,dy$ taken across the wake and this integral is made same for the cosine profile and the assumed Gaussian profile. By this approach, one gets

$$\frac{u}{u_e} = \exp\left\{-\pi\left(\frac{y}{Y_w}\right)^2\right\} \tag{6.161}$$

which approximates equation (6.160) in the range $-Y_w \leq y \leq Y_w$.

For convenience a new coordinate $x^* = x - 0.7c$ is introduced, and in terms of this, the wake half width and the velocity at the centre are

$$Y_w = 0.68\sqrt{2c}\left(\frac{C_D x^*}{c}\right)^{\frac{1}{2}} \tag{6.162}$$

$$\frac{u_e}{V} = -\frac{2.42\sqrt{C_D}}{\dfrac{x^*}{c} + 0.3} \tag{6.163}$$

The above three equations completely describe the viscous wake model of the upstream row of stator blades.

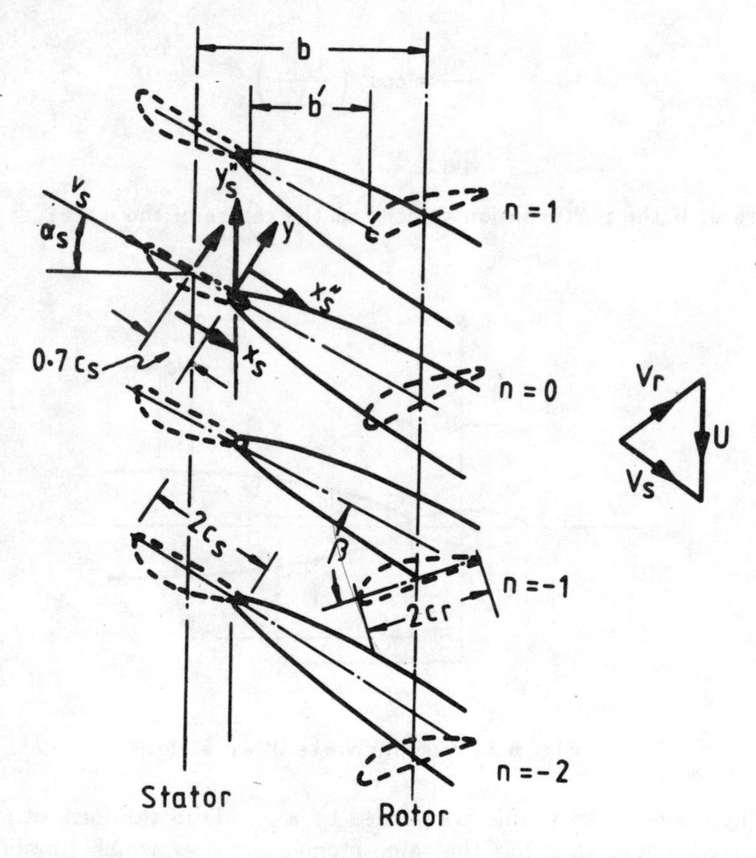

Fig. 6.13 Rotor cascade in stator viscous wakes

Since the rotor blades move along a line oblique to the x_s or x_s^* axis, oblique coordinates are introduced as shown in Fig. 6.13 where y'' is parallel to the direction of rotor motion. Hence

$$x^* \sim x'' \tag{6.164}$$

and

$$y_s \sim y_s'' \cos\alpha_s \tag{6.165}$$

Therefore

$$Y_w = 0.68\sqrt{2c_s}\left(\frac{C_D^s x''}{c_s}\right)^{\frac{1}{2}} \tag{6.166}$$

$$\frac{u_e}{V_s} = -\frac{2.42\sqrt{C_D^s}}{\dfrac{x''}{c_s} + 0.3} \tag{6.167}$$

and

$$\frac{u}{u_e} = \exp\left[-\pi\left\{\frac{y'' \cos\alpha_s}{Y_w}\right\}^2\right] \tag{6.168}$$

The above equation gives the velocity perturbation due to zeroth stator blade. Hence, the total velocity perturbation is

$$\frac{u_T}{u_e} = \sum_{n=-\infty}^{\infty} \exp\left[-\frac{\pi\cos^2\alpha_s}{Y_w^2}(y_s'' - nd_s)^2\right] \tag{6.169}$$

By applying Poisson summation formula to the above to convert it into a complete Fourier series, we get

$$\frac{u_T}{u_e} = \frac{\sqrt{\pi}}{K}\sum_m{}' \exp\left[-2\pi i m\frac{y''}{d_s} - \frac{\pi^2 m^2}{K^2}\right] \tag{6.170}$$

where,

$$K^2 = \pi\cos^2\alpha_s\left(\frac{d_s}{Y_w}\right)^2 \tag{6.171}$$

After dropping the constant term, which is the average value of u_r/u_e in the above, we get

$$\frac{u_T}{u_e} = \frac{2\sqrt{\pi}}{K}\sum_m \exp\left[-2\pi i m\frac{y''}{d_s} - \frac{\pi^2 m^2}{K^2}\right] \tag{6.172}$$

The real part is only implied from now on. The time dependence of the y'' coordinate of a fixed point x_r of the rotor blade can be obtained from Fig. 6.13. Hence

$$y'' = U\left(\frac{x_r}{V_r} - t\right) \tag{6.173}$$

The velocity u_T is in the direction x_s, which makes an angle β with the rotor blade so that component giving upwash at the blade is $-u_T \sin\beta$. Hence, the upwash induced at the rotor blade can be written as

$$V_{up}^{rv} = -u_e \sin\beta\frac{2\sqrt{\pi}}{K}\sum_m \exp\left[-\frac{\pi^2 m^2}{K^2}\right] e^{i\nu_r m\left(t - \frac{x_r}{V_r}\right)} \tag{6.174}$$

The relation between x'' and x_r is given by

$$x'' = b\sec\alpha_s - 0.7c_s + x_r\frac{V_s}{V_r} \tag{6.175}$$

This gives

$$\frac{x^*}{c_s} = \frac{c_r}{c_s}\left\{\frac{b}{c_r \cos\alpha_s} + \frac{x_r}{c_r}\frac{V_s}{V_r}\right\} - 0.7 \tag{6.176}$$

To simplify the problem, it is assumed here that x_r does not have significant influence on u_e and K. Towards this, a fixed value for x_r, say, x_{r0} is used in equation (6.176) to obtain x_0''. Using equations (6.166), (6.167) and (6.171) in equation (6.174), the final expression for the velocity induced in the direction perpendicular to the chord is

$$V_{up}^{rv}(x_r, t) = \frac{1}{2\pi}V_r \sum_m \overline{G_m^{rv}} e^{i\nu_r m\left(t - \frac{x_r}{V_r}\right)} \tag{6.177}$$

where

$$\overline{G_m^{rv}} = 4\pi\frac{V_s}{V_r}\frac{2.42\sqrt{C_D^s}\sin\beta}{\frac{x_0''}{c_s} + 0.3}\frac{0.68\sigma_s}{\sqrt{2}\cos\alpha_s}\sqrt{\frac{C_D^s x_0''}{c_s}}.$$

$$\exp\left[-\pi m^2\left\{\frac{0.68\sigma_s}{\sqrt{2}\cos\alpha_s}\right\}^2\frac{C_D^s x_0''}{c_s}\right] \tag{6.178}$$

The velocity expression given above is of the same form as equation (6.6) and hence

$$C_L^{rv} = \frac{{}_tL_v^{rv}}{0.5\rho V_r^2 2c_r} = \sum_{m=1}^{\infty}\overline{G_m^{rv}} S_L(\omega_r m) \tag{6.179}$$

$$C_M^{rv} = \frac{{}_tM_v^{rv}}{0.5\rho V_r^2(2c_r)^2} \tag{6.180}$$

The steady rotor lift is

$${}_0L^r = C_L^r 0.5\rho V_r^2 2c_r \tag{6.181}$$

Hence

$$\frac{{}_tL_v^{rv}}{{}_0L^r} = \frac{1}{C_L^r}\sum_m \overline{G_m^{rv}} S_L(\omega_r m) e^{i\nu_r m t} \tag{6.182}$$

and the moment ratio is

$$\frac{{}_tM_v^{rv}}{{}_0M^r 2c_r} = \frac{1}{4}\frac{{}_tL_v^{rv}}{{}_0L^r} \tag{6.183}$$

If the flow is considered to be at an angle of incidence there will be a component of velocity induced in the chordwise direction of the rotor blade due to the viscous wakes shed by the stator blade which is

$$V_{up}^{rv}(x_r, t) = -\frac{V_r}{2\pi}\sum_m \overline{G_m^{rv}} e^{i\nu_r m\left(t - \frac{x_r}{V_r}\right)} e^{i\frac{\pi}{2}} \tag{6.184}$$

The above is similar in form to u given by equation (6.19) for a convecting type of flow perturbation. Hence equations (6.51) and (6.59), modified to apply for the convecting gust, are used to obtain the unsteady lift due to downwash induced and the angle of incidence. The resulting expression is then combined with equation (6.182) to obtain the total unsteady lift on a rotor blade due to viscous wake effects of the stator.

$$\frac{{}_tL^{rv}}{{}_0L^r} = \frac{1}{C_L^r}\sum_m \overline{G}_m^{rv}\Big[S_L(\omega_r m) - \Big\{H_L(\omega_r m) + \frac{H^*}{2}(\omega_r m)\Big\}e^{i\frac{\pi}{2}}\Big]e^{i\nu_r m t} \quad (6.185)$$

The lift expressed by the above equation is a sum of harmonic components and can be added to the lift components given in equations (6.142) and (6.156), in order to obtain the total lift on the rotor blade due to potential, vortex and viscous wakes interactions. It should be mentioned here that these quantities are added in proper phase, that is they are superimposed as complex numbers before taking the real parts. Absolute values of such results are presented in the next section.

6.6 SOME RESULTS

A general computer program can be developed based on the analysis in the above sections to calculate the unsteady lifts and moments in a general cambered airfoil stage. The input to the program would consist of the following:

- σ_r Solidity ratio of rotor cascade
- σ_s Solidity ratio of stator cascade
- α_r Stagger angle of rotor cascade
- α_s Stagger angle of stator cascade
- V_s Velocity in the flow over stator blade
- V_r Velocity in the flow over rotor blade
- U Velocity of the rotor cascade
- b'/c_r Axial spacing/chord ratio
- d_r/d_s Pitch ratio
- H Blade parameter constants
- m Harmonic number
- M_r Mach number in the flow over rotor
- M_s Mach number in the flow over stator
- x_r^0/c_r Ratio for viscous wakes
- C_D^s Drag coefficient, stator
- C_L^r Drag coefficient, rotor
- A_i^r Glauert constants for rotor vorticity distribution $i = 0, 1, 2$
- A_i^s Glauert constants for stator vorticity distribution $i = 0, 1, 2$

In the program, it can be assumed that the flow through the stage undergoes no turning, i.e.,

$$\frac{{}_0\Gamma^r}{{}_0\Gamma^s} = -\frac{d_r}{d_s}$$

The stator and rotor lifts are to be taken as

$$\frac{{}_tL^s}{{}_0L^s} = \sum_m \frac{{}_tL^s}{{}_0L^s} e^{i\nu_s mt}$$

$$\frac{{}_tL^{r1}}{{}_0L^r} = \sum_m \frac{{}_tL^{r1}}{{}_0L^r} e^{i\nu_r mt}$$

$$\frac{{}_tL^{r2}}{{}_0L^r} = \sum_m \frac{{}_tL^{r2}_m + {}_t\overline{L}^{r2}_{-m}}{{}_0L^r} e^{i\nu_r mt}$$

where bar represents the complex conjugate taken and

$$\frac{{}_tL^{rv}}{{}_0L^r} = \sum_m \frac{{}_tL^{rv}}{{}_0L^r} e^{i\nu_r mt}$$

Typical results for three different cases are presented below from [49] and [52]. Both flat plate and parabolic blades are considered with case A: Elliptical distribution on stator and flat plate distribution on rotor and case B: Elliptical distribution on stator and rotor.

6.6.1 Flat Plate Blades with Incidence

The following data is used.

$$\sigma_r = \sigma_s = 1.0$$

$$\alpha_r = \alpha_s = 45^\circ$$

$$\delta_{s1} = \delta_{r1} = 6^\circ$$

$$V_s = 1.0, \quad V_r = 0.91, \quad U = 1.28 \qquad \text{(from Fig. 6.8)}$$

$$H_0 = -\delta, \quad H_1 = H_2 = 0.0 \qquad \text{(Equation 6.50)}$$

$$\frac{x_r^0}{c_r} = \frac{1}{2}, \quad C_L^R = 1.0, \quad C_D^s = 0.01$$

Glauert constants are evaluated from

$$\delta - A_0 = \frac{1}{\pi}\int_0^\pi \frac{dy}{dx}\, d\theta$$

$$A_n = \frac{1}{\pi}\int_0^\pi \frac{dy}{dx} \cos n\theta \, d\theta$$

$$A_n = 0, \quad n \geq 2$$

For elliptical steady lift distribution $A_0 = 0$, and for flat plate steady lift distribution $A_1 = 0$, are considered.

Case A: For elliptical distribution on stator $A_1^s = 1.0$ and for flat plate load distribution on rotor $A_0^r = 0.1051$ for $\delta = 6^\circ$.

Case B: $A_1^r = A_1^s = 1.0$ with the rest of the components made zero.

In Figs. 6.14 to 6.22, T indicates that only the effect of upwash component is considered and $T+C$ indicates that both the upwash and downwash effects are included. Fig. 6.14 shows the variation of stator lift with b'/c_r

for case A, for four different Mach numbers. Both the upwash and downwash effects are included. As rotor Mach number increases, the unsteady stator lift decreases at all b'/c_r. It can also be observed that the unsteady lift decreases as the axial stage gap increases.

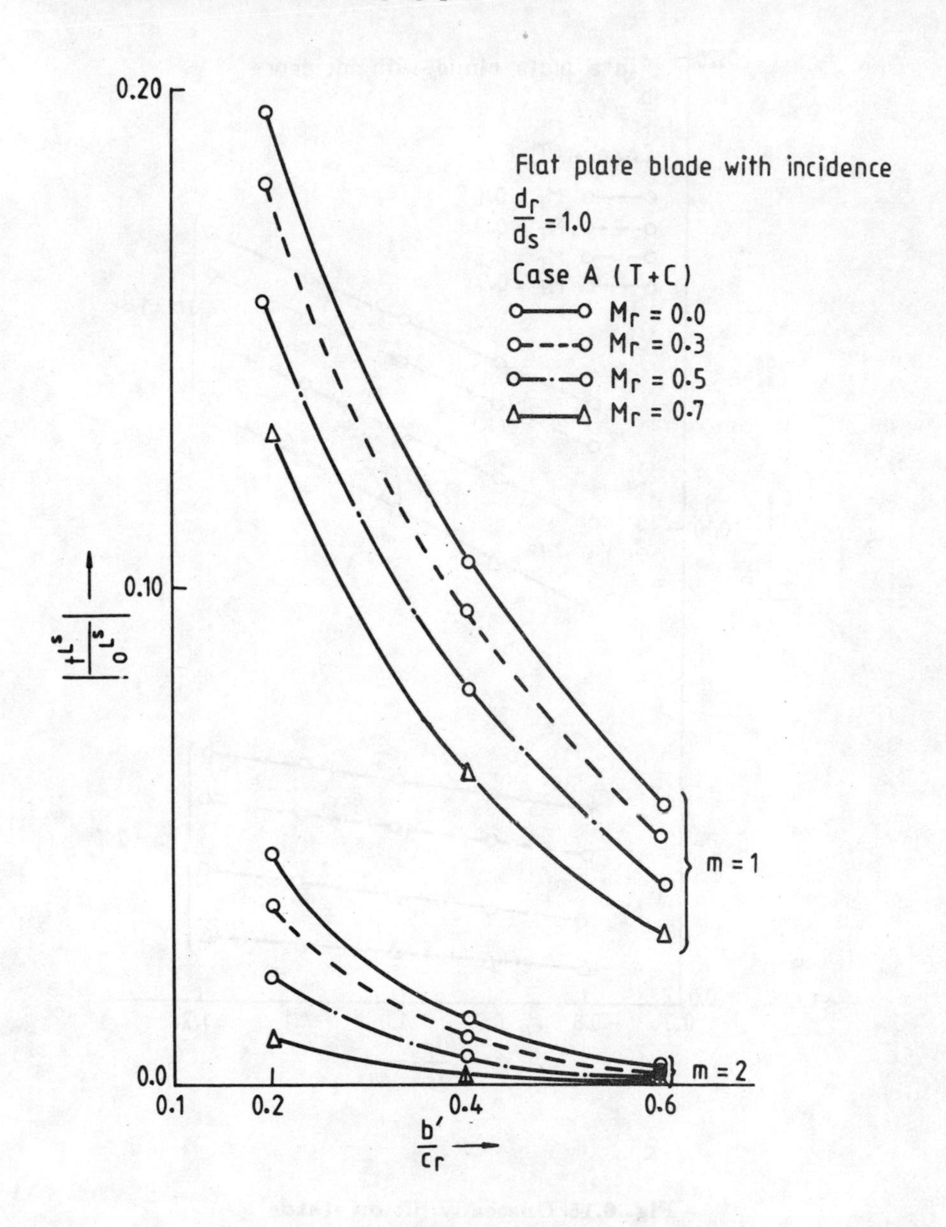

Fig. 6.14 Unsteady lift on stator (potential interaction)

For the same case A, Fig. 6.15 shows the variation of unsteady stator lift with d_r/d_s values. As before the unsteady stator lift decreases with rotor Mach number, and increases almost linearly with d_r/d_s.

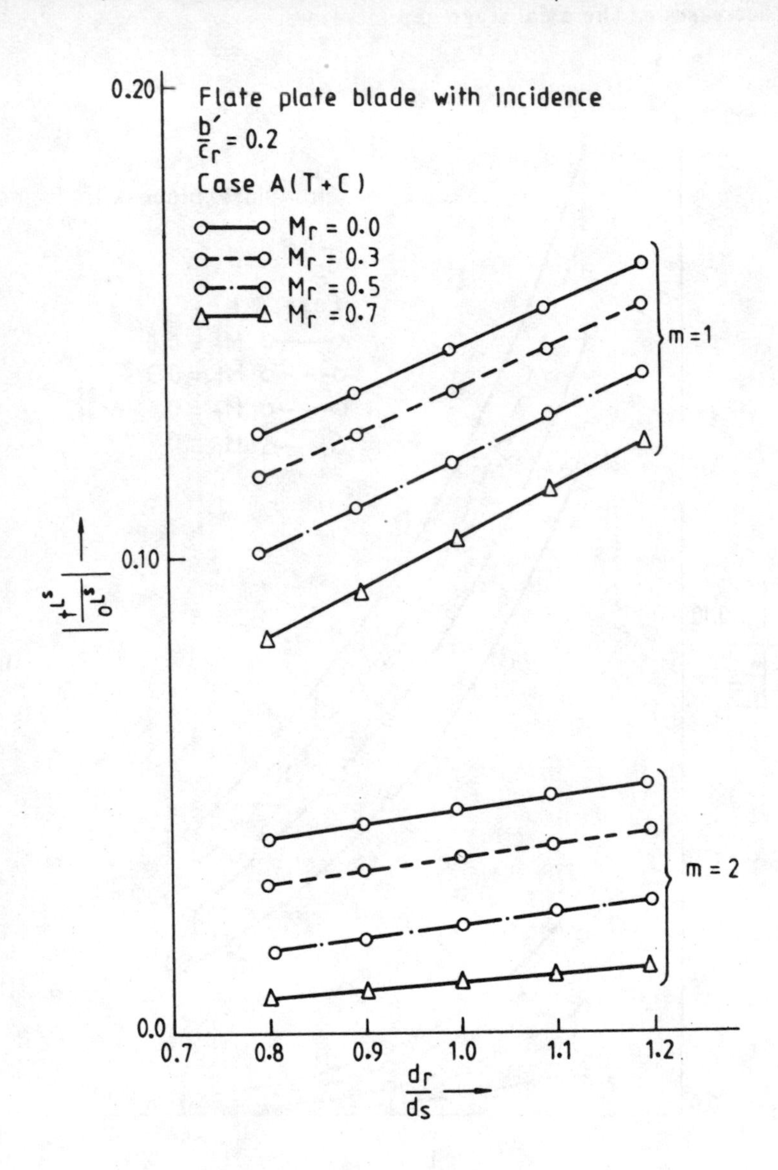

Fig. 6.15 Unsteady lift on stator (potential interaction)

Variation of rotor lift due to potential interaction alone is shown as a function of axial stage gap in Fig. 6.16. Effects of both upwash and **downwash are** considered. The unsteady lift decreases with the axial stage **gap and increases** with the stator Mach number.

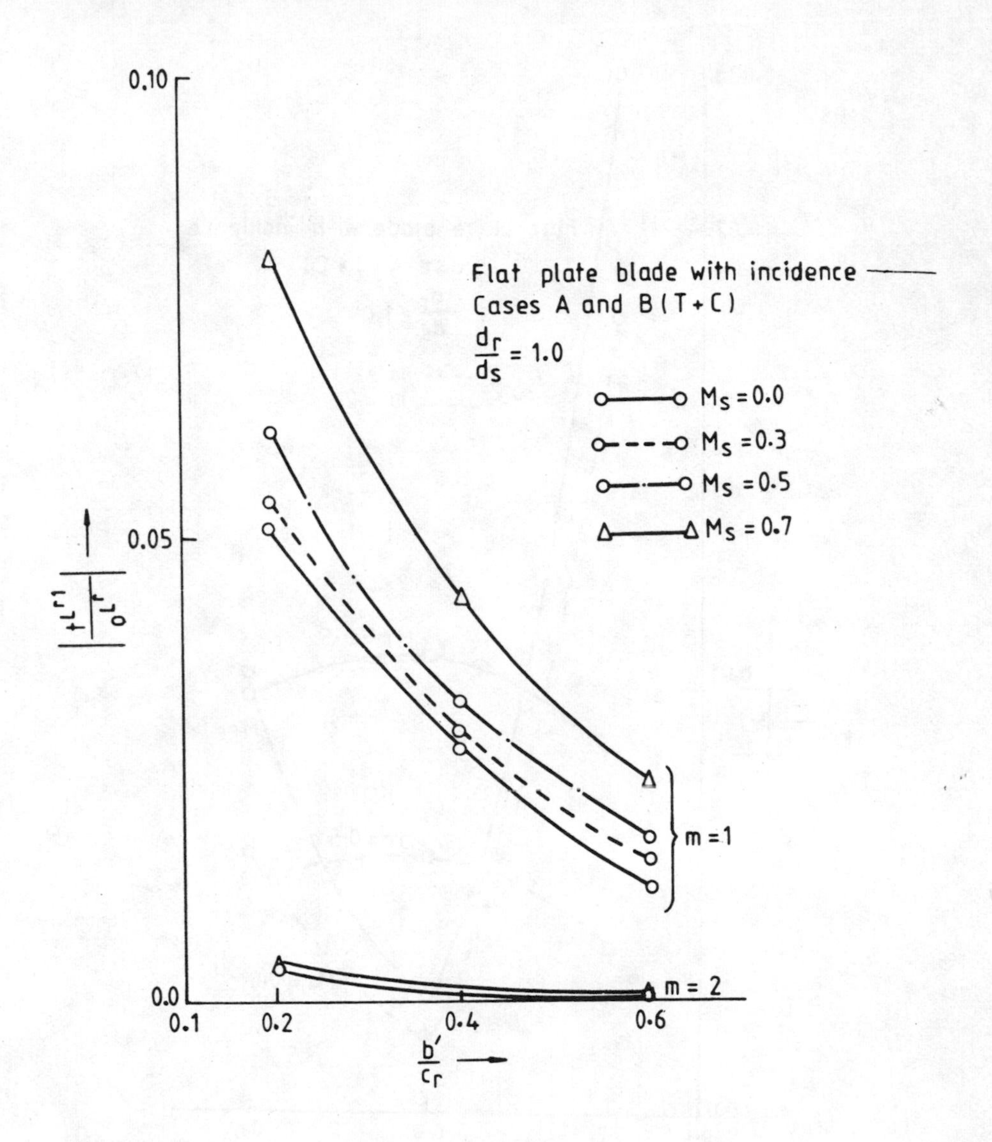

Fig. 6.16 Unsteady lift on rotor (potential interaction)

Fig. 6.17 shows the variation of lift with the axial stage gap for two different Mach numbers, due to vortex wake interaction. Fig. 6.18 shows the effects of combining the lifts due to potential, vortex and viscous wake interactions. The visous effects are significant when the stage gap is large.

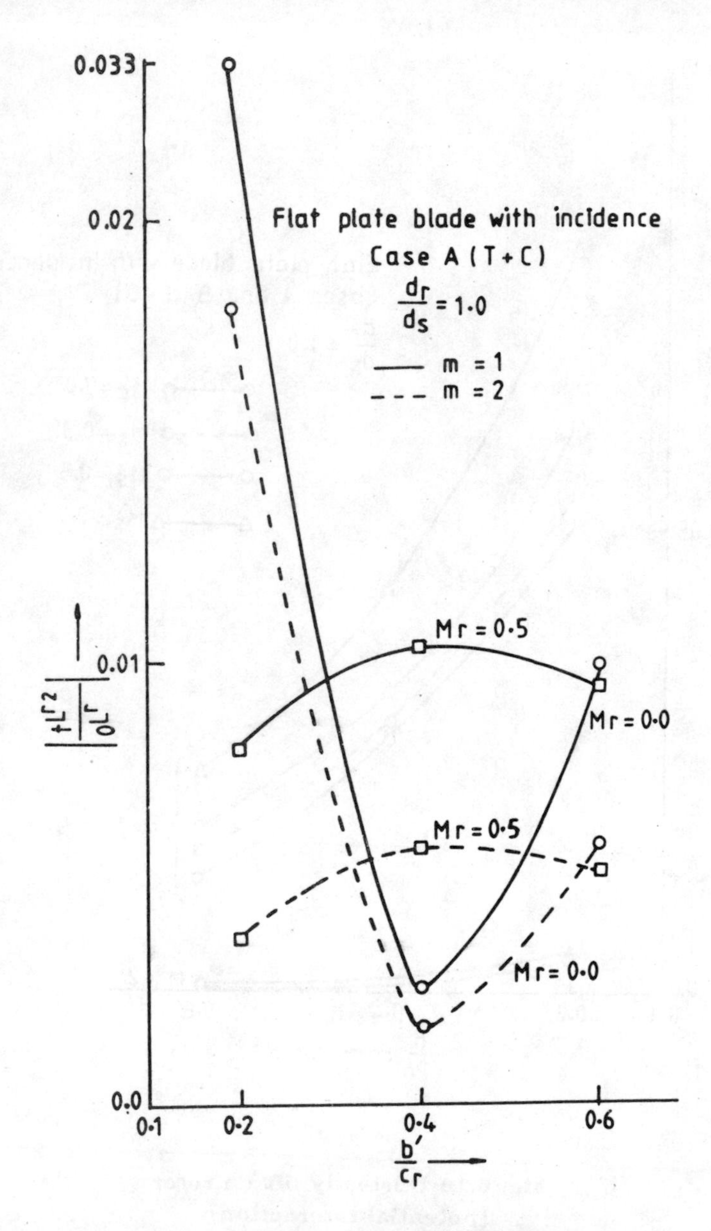

Fig. 6.17 Unsteady lift on rotor (vortex wake interaction)

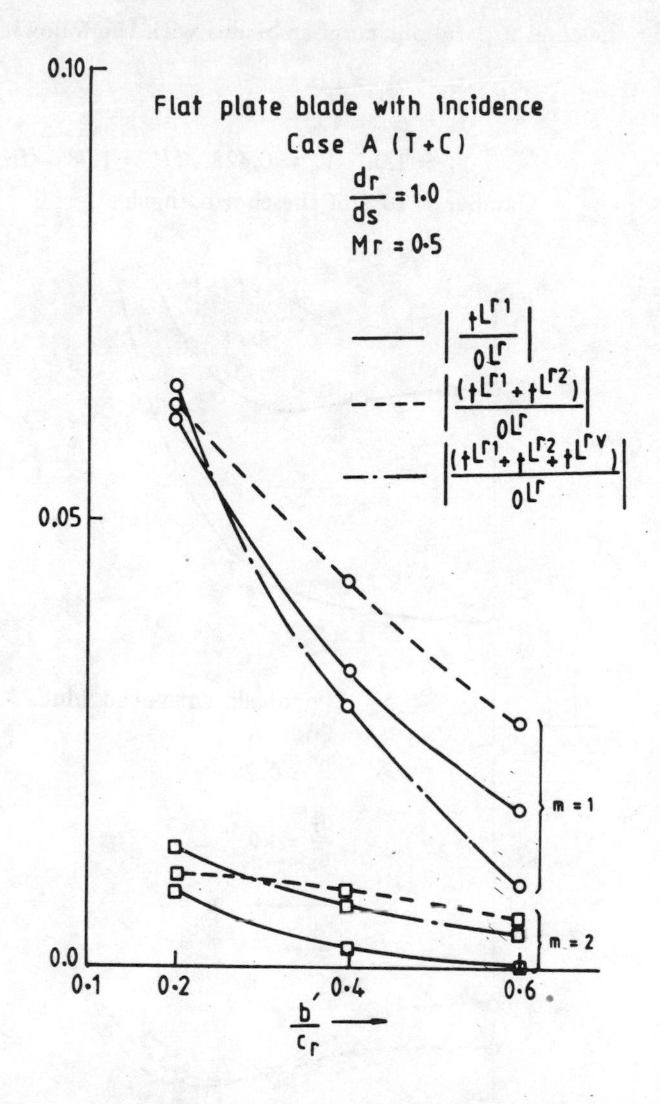

Fig. 6.18 Unsteady lift on rotor (potential, vortex and viscous wake interactions)

The results shown here only illustrate certain aspects and are not intended for a thorough study of various aspects. In fact most of such studies are of analytical importance, as the actual blades are skewed with a general camber.

6.6.2 Parabolic Camber Blades

Consider the case of parabolic camber blades with the following data.

$$\sigma_r = \sigma_s = 1.0$$
$$\alpha_r = \alpha_s = 45^\circ$$
$$V_s = 1.0, \quad V_r = 0.428, \quad U = 1.08 \quad \text{(from Fig. 6.8)}$$

Camber = 10% of the chord length

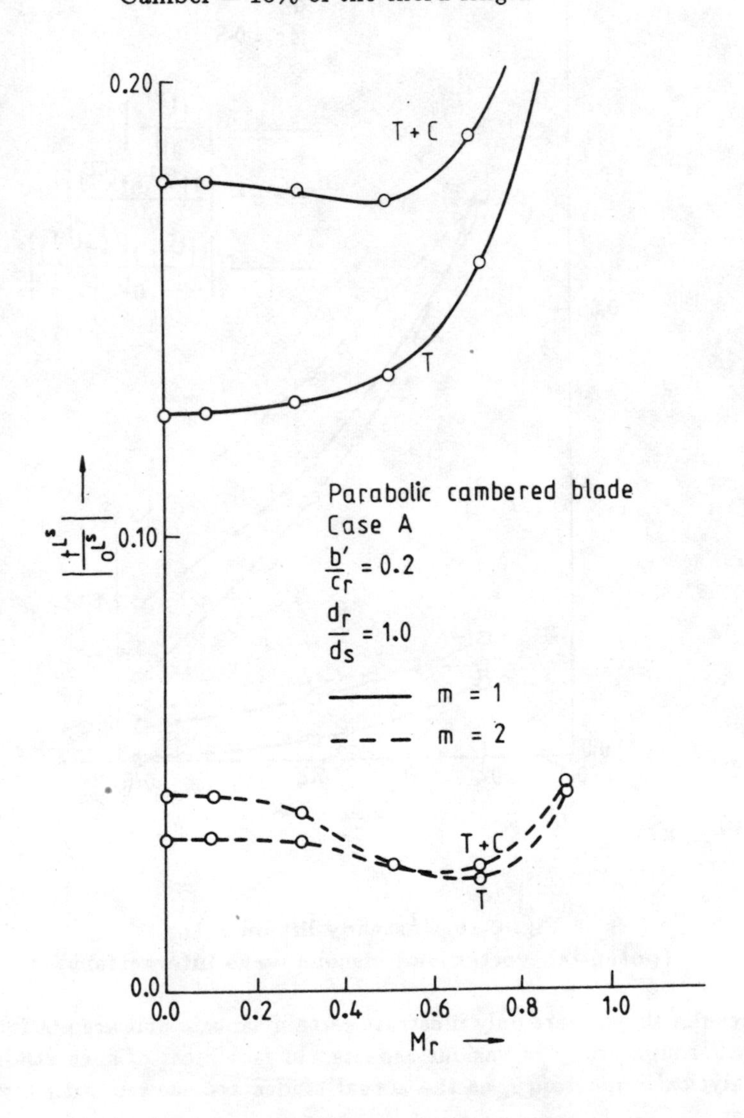

Fig. 6.19 Unsteady lift on stator (potential interaction)

The parabola is defined as

$$y = \frac{F^*}{2}(1 - \cos 2\theta)$$

$$\frac{dy}{dx} = \frac{2F^*}{c} \cos\theta$$

Hence for 10% camber, $F^* = 0.2c$, the blade profile constants are $H_1 = 0.4$, $H_0 = H_2 = \cdots = 0.0$. Hence

$$\frac{dy}{dx} = 0.4 \cos\theta$$

and therefore the angle of incidence is 22° 55′.

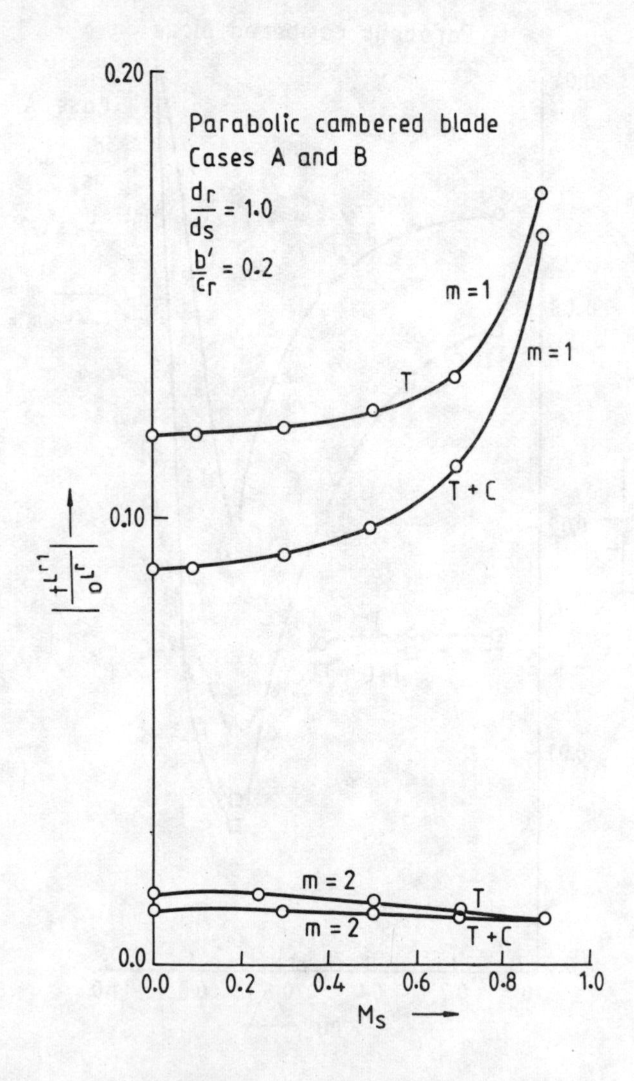

Fig. 6.20 Unsteady lift on rotor (potential interaction)

The Glauert constants are

Case A: For elliptical distribution on stator, $A_1^s = 0.2$, and for flat plate distribution on rotor $A_0^r = 0.40$.

Case B: For elliptical distribution on both stator and rotor $A_1^r = A_1^s = 0.2$, the rest of the components are zero.

The unsteady lift on stator due to potential interaction is shown in Fig. 6.19 as a function of rotor Mach number. The influence of downwash component is shown by addition to the upwash component results. As Prandtl-Glauert transformation is generally valid for Mach numbers upto 0.7, the results beyond this range of 0.7 may not be considered accurate.

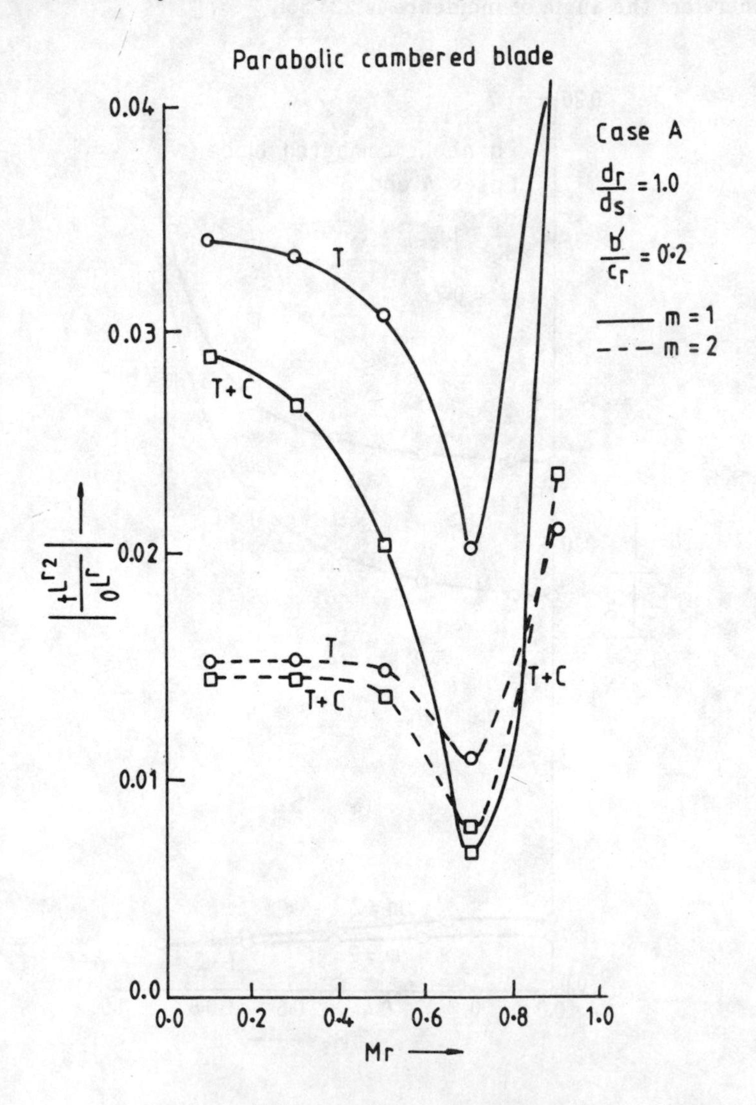

Fig. 6.21 unsteady lift on rotor (vortex wake interaction)

Fig. 6.20 shows the unsteady lift of the rotor due to potential interaction as a function of stator Mach number. The influence of downwash component in this case is to reduce the lift values for both the harmonics. The effect of rotor Mach number on the unsteady rotor lift due to vortex wake interaction is shown in Fig. 6.21. In this case the downwash component increases the unsteady lift considerably. The effect of addition of both the potential and vortex wake interactions is shown in Fig. 6.22 for the rotor lift as a function of axial gap of the stage.

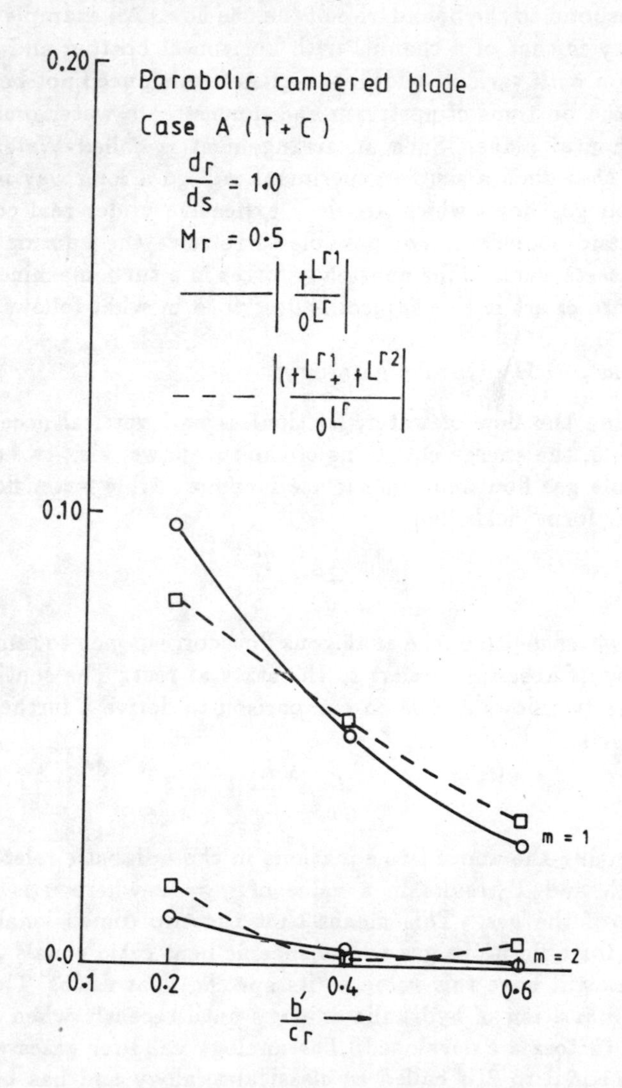

Fig. 6.22 Unsteady lift on rotor (potential and vortex wake interactions)

6.7 HYDRAULIC ANALOGY

The existence of a mathematical analogy between the frictionless flow of a liquid over a horizontal free surface and the isentropic two-dimensional compressible flow of a gas has been known since a long time. It is possible to produce an analogous flow to the two dimensional flow of a compressible gas with water flowing over a horizontal bottom under the effect of gravity with a free surface and at sides bounded by vertical walls [53]. The fixed vertical walls correspond to the boundaries of the gas flow. An example of this type of boundary is that of a channel with horizontal bottom and rectangular cross-section with variable width, the axis of which need not be rectilinear. However, the bottoms of upstream and downstream water must be in the same horizontal plane. Such an arrangement is called Water Table. It is obvious that such a simple experiment will go a long way in modelling complicated gas flows which are very expensive under real conditions of operation and sometimes not possible to retrieve the information desired from such tests, such as the nonsteady forces in a turbomachine stage. The present state of art in this subject is described in what follows.

6.7.1 Classical Hydraulic Analogy

By assuming the flow of water, frictionless and vertical accelerations to be negligible, the energy equations of the two flows, viz., two dimensional compressible gas flow and free surface incompressible water flow, brought into similar form yields [55],

$$\frac{h}{h_0} = \frac{T}{T_0} \tag{6.186}$$

Or the water depth in the analogous flow corresponds to temperature in the gas flow. Subscript 0 refers to the state at rest. The continuity equations of the two flows enable on comparison to derive a further analogous condition, viz.,

$$\frac{\rho}{\rho_0} = \frac{h}{h_0} \tag{6.187}$$

Substituting the above two equations in the adiabatic relation of a gas between T_0 and T results in a value of $\gamma = 2$ where γ is the specific heat ratio of the gas. This means that the two dimensional analogy is valid only for a fictitious gas whose specific heat ratio equals 2. However, no real gas will have this value as its specific heat ratio. This has been the main limitation of hydraulic analogy until recently when appropriate correction factors are developed. The analogy valid for gases with specific heat ratio equal to 2 is called as classical analogy and has been used to make qualitative studies of gas flows using a water table.

In the classical analogy, writing the equation of state

$$P = \rho RT \tag{6.188}$$

for an arbitrary state and state of rest and dividing one by another, the analogous quantity for pressure ratio in the gas flow is obtained as

$$\frac{P}{P_0} = \left\{ \frac{h}{h_0} \right\}^2 \tag{6.189}$$

By introducing the condition of irrotationality and comparing the differential equation for the velocity potential of the ideal free surface water flow over a horizontal bottom with the equation for the velocity potential of the two dimensional compressible gas flow, it is possible to show that the basic wave velocity in shallow water represented by $\sqrt{gh}$ is analogous to the sonic velocity a_0 in the gas flow. On the basis of this wave velocity, the water flow may be described as shooting or streaming depending upon whether wave velocity is less or greater than the water velocity.

In shooting water under certain conditions the velocity may strongly decrease over short distances and the water depth suddenly increases, a phenomena known as hydraulic jump (or shock). With such jumps in the water flow, the kinetic energy is converted to heat which has to be treated as a lost energy in the water flow whereas, in a gas flow involving shock the lost energy again enters thermodynamically into the computation. Hence the hydraulic analogy fails when flows with a shock are considered. If the shocks are weak, the analogy may be valid to a first approximation. A summary of analogous quantities for the classical analogy is given in the table below.

TABLE 6.1

Analogous Quantities	*Classical Analogy*
Two Dimensional Gas Flow	Free Surface Incompressible Water Flow
Side Boundaries Geometrically Similar	Side Boundary Vertical, Bottom Horizontal
Velocity $V/V_{\max}$	Velocity $V/V_{\max}$
Temperature Ratio T/T_0	Water Depth Ratio h/h_0
Density Ratio ρ/ρ_0	Water Depth Ratio h/h_0
Pressure Ratio P/P_0	Square of Water Depth Ratio $(h/h_0)^2$
Sound Velocity a_0	Wave Velocity $\sqrt{gh}$
Mach Number V/a_0	Froude Number $V/\sqrt{gh}$
Subsonic Flow	Streaming Water
Supersonic Flow	Shooting Water
Compressive Shock	Hydraulic Jump

6.7.2 Limitations of Hydraulic Analogy

Besides the major problem of the classical analogy on the specific heat ratio, and the compressible shock, some other second order limitations are discussed below.

Three Dimensional Flow on the Table

In deriving equations for a two dimensional flow, an assumption is made that the vertical accelerations are negligible, i.e., the horizontal motion in a shallow water flow over a plane horizontal bottom will be same at all depths. All infinitesimal disturbances move at a constant velocity, the velocity of propagation being given by $\sqrt{gh}$, and this motion is directly analogous to a two dimensional compressible gas flow where $\sqrt{gh}$ corresponds to speed of sound a_0. However, three dimensional flows occur in the water flow and there will be vertical accelerations present, specially when the depths are considerable. This depth can be minimized by choosing shallow depth on the water table. This should be kept in mind, while modelling any flow on water table.

Surface Tension and Capillary Waves

The vertical accelerations are not negligible, even when the frictional drag on the bottom is neglected, unless the wave length $\lambda_1 \gg h$ or h tends to 0 [67]. But, when h tends to 0, the surface tension effects become predominant, which may impose severe inaccuracies in the analogy flow conditions by forming capillary waves which propagate at a different speed than the analogous gravity waves, causing a change in the wave propagation velocity. Hence proper precaution should be taken in modelling that the depth at any place does not become too shallow, say, less than 5 mm.

Boundary Layer Effects on the Walls

It is to be noted that the growth of boundary layer in the free surface water flows as well as in the analogous compressible gas flows will not be identical, since the two flows are physically different. This effect will appear both on the side walls, as well as on the bottom surface. It tends to make the flow three dimensional on the water table, and hence the analogy will become poorer. Precautions may have to be taken, where necessary, to remove such boundary layers or prevent their growth.

6.7.3 Modified Analogy

To remove the major limitation due to the specific heat ratio on the classical analogy, several attempts have been made to develop correction factors for the flow quantities. For example, Harlemann and Ippen [57] derived their correction factors using transonic similarity laws of von Karman for a family of airfoils which are useful to model flow over an isolated airfoil.

To account for the variations of the specific heat ratio in a gas system from the ideal analogous value of 2, Bird and Williams [59] proposed certain correction factors for the pressure, density and temperature ratios so that classical hydraulic analogy relations apply to any real gas with specific heat ratio γ of any value at any Mach number. However, they have assumed that the Froude number obtained on the water table is equal to the Mach number in the gas flow, which is not true. Adams [60], by assuming that the area relationship A/A^* equal in both the analogous systems, suggested a correction factor for the Froude number on the water table and used the resulting Mach number with the relations proposed by Bird and Williams to obtain pressure and temperature ratios.

Rao, through a series of studies [62, 65, 72, 74, 75], established a modified analogy for simulating a 2 dimensional isentropic compressible gas flow on a water table. Salient details of this study are given below.

6.7.4 Rao's Correction Factors

(a) *Mach Number*

For a compressible gas flow through a converging diverging nozzle, the cross sectional area of the nozzle A^* at which the Mach number becomes unity is given by

$$\frac{A}{A^*} = \frac{\left\{\left(\frac{2}{\gamma+1}\right)\left(1+\frac{\gamma-1}{2}M^2\right)\right\}^{\frac{\gamma+1}{2(\gamma-1)}}}{M} \tag{6.190}$$

If $\gamma = 2$, then

$$\frac{A}{A^*} = \frac{\left\{\frac{2}{3}\left(1+\frac{\overline{M}^2}{2}\right)\right\}^{1.5}}{\overline{M}} \tag{6.191}$$

where,

$$\overline{M} = M_{\gamma=2}$$

Since A/A^* is same for a gas with specific heat ratio 2, and a gas with any specific heat ratio, equating (6.190) and (6.191)

$$M = M_c = C_M\overline{M} \tag{6.192}$$

where,

$$C_M = 1.8371\frac{\left\{\left(\frac{2}{\gamma+1}\right)\left(1+\frac{\gamma-1}{2}M^2\right)\right\}^{\frac{\gamma+1}{2(\gamma-1)}}}{\left\{1+\frac{\overline{M}^2}{2}\right\}^{1.5}} \tag{6.193}$$

M_c is the corrected value of $\overline{M}$, which is the required Mach number of actual flow of gas through the nozzle.

(b) *Pressure Ratio*

For an isentropic flow through the nozzle, the stagnation pressure P_t at any point is given by

$$P_t = P\left[1+\frac{\gamma-1}{2}M^2\right]^{\frac{\gamma}{\gamma-1}} \tag{6.194}$$

The stagnation pressure for $\gamma = 2$ is

$$P_t = P\left\{1 + \frac{\overline{M}^2}{2}\right\}^2 \tag{6.195}$$

For $\gamma \neq 2$, using M_c from equation (6.192), (6.194) becomes

$$P_t = P_c\left\{1 + \frac{\gamma - 1}{2}M_c^2\right\}^{\frac{\gamma}{\gamma-1}} \tag{6.196}$$

where P_c is the corrected pressure at any point corresponding to M_c. For the stagnation pressure with $\gamma = 2$ and $\gamma \neq 2$, to be equal we get

$$P_c = P\frac{\left\{1 + \frac{\overline{M}^2}{2}\right\}^2}{\left\{1 + \frac{\gamma - 1}{2}M_c^2\right\}^{\frac{\gamma}{\gamma-1}}} \tag{6.197}$$

or,

$$P_c' = P'\frac{\left\{1 + \frac{\overline{M}'^2}{2}\right\}^2}{\left\{1 + \frac{\gamma - 1}{2}M_c'^2\right\}^{\frac{\gamma}{\gamma-1}}} \tag{6.198}$$

where $'$ represents a chosen reference section. The corrected pressure ratio can be obtained from the above as

$$\frac{P_c}{P_c'} = C_P\frac{P}{P'} \tag{6.199}$$

where,

$$C_P = \left\{\frac{1 + \frac{\overline{M}^2}{2}}{1 + \frac{\overline{M}'^2}{2}}\right\}^2 \left\{\frac{1 + \frac{\gamma - 1}{2}M_c'^2}{1 + \frac{\gamma - 1}{2}M_c^2}\right\}^{\frac{\gamma}{\gamma-1}} \tag{6.200}$$

(c) *Temperature Ratio*

For an isentropic flow through the nozzle, the stagnation temperature at any point is given by

$$T_t = T\left(1 + \frac{\gamma - 1}{2}M^2\right) \tag{6.201}$$

For $\gamma = 2$

$$T_t = T\left(1 + \frac{\overline{M}^2}{2}\right) \tag{6.202}$$

Using M_c for corrected Mach number from equation (6.192), equation (6.201) becomes for $\gamma \neq 2$

$$T_t = T_c\left(1 + \frac{\gamma - 1}{2} M_c^2\right) \tag{6.203}$$

Equating the stagnation temperatures in both classical and modified analogies

$$T_c = T\frac{1 + \dfrac{\overline{M}^2}{2}}{1 + \dfrac{\gamma - 1}{2} M_c^2} \tag{6.204}$$

or,

$$T_c' = T'\frac{1 + \dfrac{\overline{M}'^2}{2}}{1 + \dfrac{\gamma - 1}{2} M_c'^2} \tag{6.205}$$

Hence

$$\frac{T_c}{T_c'} = C_T \frac{T}{T'} \tag{6.206}$$

where,

$$C_T = \frac{1 + \dfrac{\overline{M}^2}{2}}{1 + \dfrac{\overline{M}'^2}{2}} \; \frac{1 + \dfrac{\gamma - 1}{2} M_c'^2}{1 + \dfrac{\gamma - 1}{2} M_c^2} \tag{6.207}$$

Equations (6.193), (6.200) and (6.207) give the correction factors for obtaining Mach number, pressure and temperature ratios in the modified analogy from the corresponding values in classical analogy. Rao, considering the nozzle of Adams [60], has shown the exact validity of the modified analogy in his analytical work.

This analogy has been verified on a flat water table for isentropic subsonic, transonic and supersonic flows by Rao et al [82] and is shown to be in fairly good agreement with theoretical results for supersonic flows with weak shocks [73]. Rao [72] has shown that the errors in the post shock flow can be considered as second order for Froude numbers less than 1.7. While errors in temperature ratios are of higher order for Froude numbers greater than 1.7, the errors in pressure ratios are very low, within 6%, even for Froude numbers as high as 3. The error in temperature ratio is almost constant in the entire downstream flow. However, the error in pressure ratio decreases rapidly with downstream, from the normal shock position to the exit. The pressure ratio error is least at the exit for a given location of the normal shock.

6.7.5 Rotating Water Table

The description of the rotating water table described here closely follows references [82, 83]. The schematic diagram of the rotating water table for simulation of a turbine stage is shown in Fig. 6.23.

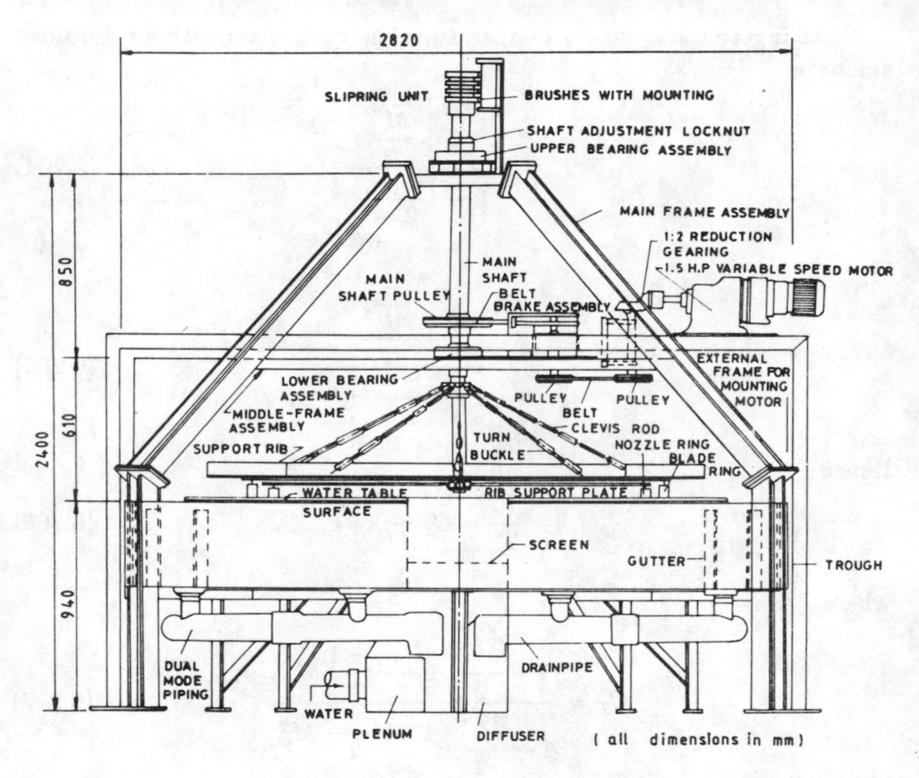

Fig. 6.23 Rotating water table for simulation of turbine stages

The rotor assembly consists of vertical main shaft supported by upper and lower ball bearings which in turn are supported by upper and middle frame assemblies which are welded steel structures. In the upper most portion of the shaft, a phosphor bronze slip ring unit is mounted. A pulley is fitted on the main shaft for receiving the drive from the drive mechanism. The shaft carries a prespex ring through eight radial ribs attached with clevis support rods to which moving blades are attached through blade rods above the water table surface. The levelling of the rotor support ring and adjustments for the blade clearances are obtained by the turnbuckles arranged in the clevis support rods. The rotor blades rotate concentric to the stationary nozzle blades at the speed assigned by the drive mechanism.

The nozzle assembly consists of a perspex ring to which the nozzle or stator blades are attached. The ring along with the nozzle blades is supported by the water table surface. The actual water table on which the turbine stage is simulated rests on a hexagonal welded structure in the lower frame assembly. The upper, middle and lower frame assemblies are also supported through the welded joints by the main frame assembly, which is a large steel structure. The water table is made of optically flat glass sheet having an outer diameter of 250 cm, with inner diameter of 60 cm. A number of levelling screws arranged beneath the table enable it to be levelled accurately. A trough system consisting of a concentric channel surrounding the edge of the table is provided to collect the water flowing on the table. The trough system is connected to the sump through dual mode piping system. The rotor assembly is driven by a variable speed motor which is mounted on an external frame isolated from the main structure. The drive is transmitted from the motor shaft through bevel gearing to the viscous drive or brake mechanism which ultimately drives the rotor assembly.

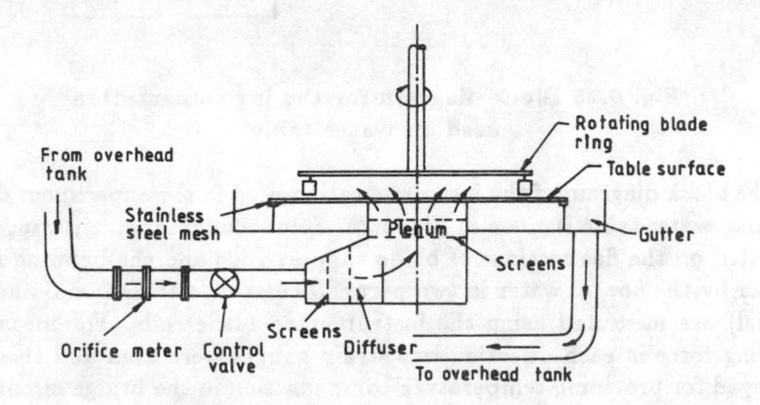

Fig. 6.24 Flow circuit for the water table

The flow circuit for the water table is shown in Fig. 6.24. Water is supplied to water table apparatus from an over head tank through an orifice meter designed and calibrated and is incorporated in the supply line to meter the amount of water passing on to the water table surface. The water, after flowing through the orifice meter, a control valve and a filter fitted in the pipe line, enters a plenum chamber (a vertical settling chamber) through a diffuser. Fine mesh screens fitted in the plenum and diffuser settle the disturbance in the incoming water. The water enters the table at the centre from the plenum chamber and flows on the levelled water table surface radially. The nozzle blades fixed to the nozzle ring direct the oncoming radial flow on to the concentric rotating blades. At the edge of the water table a fine stainless steel mesh is arranged for controlling back pressure. The water, after passing through the back pressure screen, is collected into the trough system to be transmitted to the sump via dual mode piping, from where it is pumped to the over head tank for recirculation. The depths on the water table are obtained by providing 3 mm wall taps in the table and connecting them to an inclined tube manometer. Wall taps are provided on 3 equiangularly located radial lines. The tappings are made before the nozzle and after the rotor blades. Additional tappings are also provided at the exit of the nozzles. The measured levels in the manometer before and after flowing the water on the table indicate the flow depths on the water table.

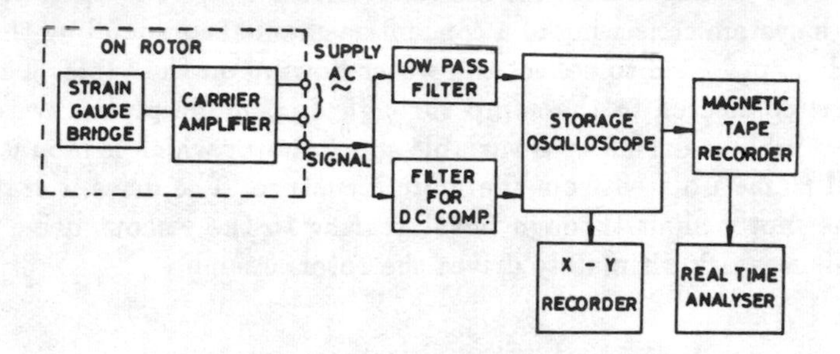

Fig. 6.25 Block diagram for the instrumentation used on water table

The block diagram of the instrumentation used in the experiment on the rotating water table is given in Fig. 6.25. Semi-conductor strain gauges are mounted on the flat sections of blade support rods and the bending forces caused by the flow of water in two perpendicular directions (axial and tangential) are measured using the instrumented blade rods. For measuring bending force in each direction two strain gauges were used and these are arranged for providing temperature compensation in the bridge circuit of a carrier amplifier unit. The carrier amplifier system is mounted on the rotor itself and consists of fully transistorized bridge amplifiers and demodulators which effect mechanical to electrical conversion of resistive transducer signals. During the conversion process, the electrical signals are amplified to a level adequate for driving, indicating or recording equipment. The resistive strain gauge elements are connected to circuit elements in the amplifier and demodulator units to form a full bridge network. The output signal for the amplifier unit is collected through the phosphor bronze slip ring unit and is passed on to a low pass active filter in order to pass only the desired low frequency signal (nozzle passing frequency). The filter is a two pole VCVS low pass filter using a precision op amp followed by a single pole integrator. The average DC level of the strain signal is obtained using a low pass active filter which produces a steady DC voltage representing the average value of the strain signal.

The output signals from the filter circuits are fed to a storage oscilloscope. The signals are analysed on the screen of the oscilloscope, stored and then transferred to an *X*-*Y* recorder for continuous recording of the signals on the charts. The signals are also passed on to a multi channel tape recorder for continuous recording and further analysis. The signals recorded on the tape recorder are finally analysed in a narrow band real time analyser. The result of such an analysis of a typical signal is shown in Fig. 6.26.

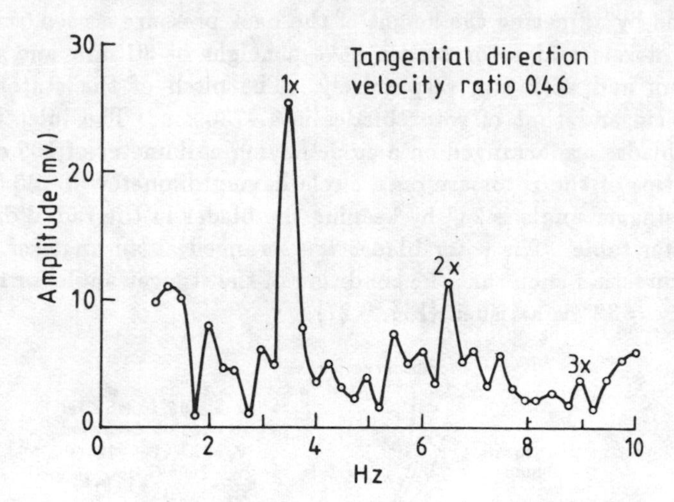

Fig. 6.26 Typical frequency spectrum of the non-steady force

The gas turbine stage of an Orpheus engine was modelled on the water table, with the following data.

σ_s	0.4094
σ_r	0.5856
c_r/c_s	0.7616
b'/c_r	1.90
V_s/U	1.5746
α_s	24°
α_r	32°
d_r/d_s	0.5324
M_s	0.798

Number of stator blades = 63
Number of rotor blades = 125
Pressure ratio of the stage = 1.9754
Nozzle inlet Mach number = 0.254
Rotor exit Mach number = 0.535

The Froude number on the water table corresponding to Mach number 0.798 at the nozzle exit in the gas flow is obtained from equation (6.192) using an iterative procedure, which is 0.781. The flow on the water table is to be adjusted with this value of Froude number at the exit of the nozzle blades.

The Froude numbers corresponding to the Mach numbers 0.254 and 0.535 at nozzle inlet and rotor exit conditions are obtained from equation (6.192) again as 0.239 and 0.512. With the help of these values and equation

(6.199), the height ratio corresponding to the pressure ratio of the stage is obtained as 1.4339. The required pressure ratio on the water table is obtained by adjusting the height of the back pressure screen/strip.

The nozzle and rotor blades have a height of 30 mm and a chord of 34.4 mm and 26.2 mm respectively. The pitch of the stator blades is 8.4025 cm and that of rotor blades is 4.4736 cm. The inlet tips of the stator blades are arranged on a circle having a diameter of 165 cm and the outlet tips of the rotor are on a circle having diameter of 180.5 cm. The stator stagger angle is 24° by keeping the blades in the radial direction on the water table. The rotor blades are arranged at an angle of 32° to the radial direction such that the condition of the stagger angle for rotor blade to be $\alpha_r = 32°$ is satisfied (Fig. 6.27).

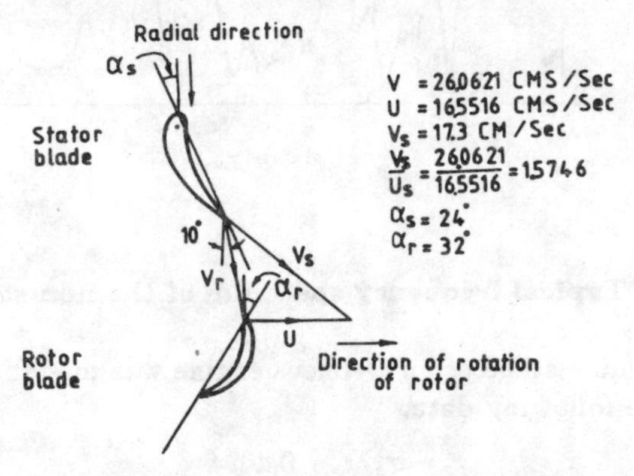

Fig. 6.27 Velocity triangle at stator exit and rotor entry

The nozzle exit Froude number of the value 0.781 corresponding to the Mach number of 0.798 in the gas flow is achieved on the water table by adjusting the flow to 15.846 kg/sec (measured through orifice meter). The depth obtained at stator exit with this flow is 1.1351 cm with a velocity of 26.0621 cm/sec, and corresponding depth at rotor outlet is obtained as 0.81 cm. The depth of water at stator inlet for this flow is 1.195 cm.

Glauert constants for the stage are determined to be

$$A_s^0 = 0, \quad A_s^1 = 0.4294, \quad A_s^2 = 0.0973,$$
$$A_r^0 = 1.053, \quad A_r^1 = 0, \quad A_r^2 = 0.09714$$

The rotor blade profile constants are

$$H(1) = 0.05376, \quad H(2) = 0.8588, \quad H(3) = 0.1943$$

The non-steady force ratio of the rotor blade in the tangential direction for the first harmonic obtained from the experiment is 0.01884. The theoretical value for the lift ratio obtained is 0.0202. The component of this force in the tangential direction is $0.0202 \cos 10° = 0.01989$ which compares well with the experimental value 0.01884.

The results obtained above show that water table can be used to determine non-steady force ratios with a good accuracy using the modified analogy. This will become very useful for stage geometries with thick blades, where the analysis has limitations because of 2 dimensional thin airfoil theory.

REFERENCES

1. Karman, T. and Sears, W.R., "Airfoil Theory for Non-Uniform Motion", *J. of Aero. Sci.*, **5**, p. 379 (1938).
2. Silverstein, A., Katzoff, S. and Bullivant, W.K., "Downwash and Wake Behind Plain and Flapped Airfoils", NACA TR 651 (1939).
3. Preston, J.H. and Sweeting, N.E., "The Experimental Determination of the Boundary Layer and Wake Characteristics of a Simple Joukouski Airfoil with Particular Reference to the Trailing Edge Region", ARC R&M, 1998 (1934).
4. Hah, C. and Lakshminarayana, B., "Measurement and Prediction of Mean Velocity and Turbulence Structure in the Near Wake of an Airfoil", *J. of Fluid Mechanics*, **115**, p. 251 (1982).
5. Hah, C. and Lakshminarayana, B., "Free Stream Turbulence Effects on the Development of a Rotor Wake", *AIAA J.*, **19**, #6 (1981).
6. Sears, W.R., "Some Aspects of Non-Steady Airfoil Theory and Its Practical Applications", *J. of Aero. Sci.*, **8**, p. 104 (1941)
7. Kemp, N.H., "On the Lift and Circulation of Airfoils in Some Unsteady Motion", *J. of Aero. Sci.*, **19**, p. 713 (1952).
8. Geissing, J.P., Rodden, W.P. and Stahl, B., "Sears Function and Lifting Surface Theory for Harmonic Gust Fields", *AIAA J. of Aircraft*, **7**, #3 (1970).
9. Horlock, J.H., "Fluctuating Lift on Airfoils Moving Through Transverse and Chordwise Gusts", *J. Basic Engrg., ASME*, **90**, p. 494 (1968).
10. Holmes, D.W., *Lift Fluctuations of Aerofoils in Transverse and Streamwise Gusts*, Ph.D. Thesis, University of Cambridge, England (1972).
11. Naumann, H. and Yeh, H., "Lift and Pressure Fluctuations of Cambered Airfoil Under Periodic Gusts and Applications in Turbomachinery", *J. Engrg. Power, ASME*, **95**, p. 165 (1973).
12. Rao, S.S.P., *Some Studies on Turbomachine Blade Dynamics and Response in Incompressible Flow*, Ph.D. Thesis, Indian Institute of Technology, Kharagpur (1980).
13. Rao, S.S.P., Mukhopadhyay, V. and Rao, J.S., "Lift and Moment Fluctuations of a Cambered Aerofoil under Non-Convecting Streamwise Gust", *J. Royal Aero. Society*, p. 83 (1977).
14. Graham, J.M.R., "Similarity Rules for Thin Airfoils in Non-Stationary Subsonic Flows", *J. Fluid Mechanics*, **43**, p. 753 (1970).
15. Amiet, R. and Sears, W.R., "The Aerodynamic Noise of Small Perturbation Subsonic Flows", *J. Fluid Mechanics*, **44**, #2 (1970).
16. Osborne, C., *Compressibility Effects in the Unsteady Interactions Between Blade Rows*, Ph.D. Thesis, Cornell University, Ithaca, NY (1971).
17. Osborne, C., "Unsteady Thin Airfoil Theory for Subsonic Flows", *AIAA Journal*, **11**, p. 205 (1973).
18. Miles, J.M., "On the Compressibility Correction for Subsonic Unsteady Flows", *J. Aero. Sci.*, **17**, p. 181 (1950).
19. Kemp, N.H., "Closed Form Lift and Moment for Osborne's Unsteady Thin Airfoil Theory", *AIAA J.*, **11**, p. 1358 (1973).
20. Amiet, R.K., "Compressibility Effects in Unsteady Thin Airfoil Theory", *AIAA J.*, **12**, p. 252 (1974).

21. Kemp, N.H., "Lift and Moment for Arbitrary Power Law Upwash in Oscillating Subsonic Unsteady Thin Airfoil Theory", *AIAA J.*, **12**, p. 413 (1974).
22. Amiet, R.K., "Airfoil Response to an Incompressible Skewed Gust of Small Spanwise Wave Number", *AIAA J.*, **14**, p. 541 (1976).
23. Timothy, S., Lewis and Sirovich, L., "Approximate and Exact Numerical Computation of Supersonic Flow over an Airfoil", *J. Fluid Mechanics*, 112, p. 265 (1981).
24. Kemp, N.H. and Sears, W.R., "Aerodynamic Interference Between Moving Blade Rows", *J. Aero. Sci.*, **20**, p. 585 (1953).
25. Rao, J.S., "Computer Program for Determining Unsteady Forces of an Elementary Turbomachine Stage", TM 75 WRL M11, Wehle Research Lab, Rochester Institute of Technology, NY (1975).
26. Kemp, N.H. and Sears, W.R., "The Unsteady Forces Due to Viscous Wakes in Turbomachines", *J. Aero. Sci.*, **22**, p. 478 (1955).
27. Lotz, M. and Raabe, J., "Blade Oscillations in One Stage Axial Flow Turbomachinery", *J. Basic Engrg., ASME*, p. 405 (1968).
28. Mani, R., "Compressibility Effects in Kemp-Sears Problem", *Fluid Mechanics and Design of Turbomachinery*, NASA SP-304, *Unsteady Flow and Noise*, p. 513 (1970).
29. Adachi, T. and Fukusada, K., "Study on Interference Between Moving Blade Rows in Axial Flow Blower", *Bull. JSME*, **17**, p. 904 (1974).
30. Rao, S.S.P., Mukhopadhyay, V. and Rao, J.S., "Unsteady Aerodynamic Forces in Cambered Blades of an Elementary Axial Flow Turbomachine Stage", *J. Aero. Soc. of India*, **30**, p. 4 (1978).
31. Sisto F., "Unsteady Aerodynamic Reactions on Airfoils in Cascade", *J. Aero. Sci.*, **22**, p. 297 (1955).
32. Chang, Ch. and Chu, W.H., "Aerodynamic Interference of Cascade Blades in Synchronized Oscillation", *J. Appld. Mech.*, **22**, p. 503 (1955).
33. Sebautik, A. and Sisto, F., "A Survey of Aerodynamic Excitation Problems in Turbomachines", *ASME*, **78**, p. 555 (1956).
34. Whitehead, D.S., "Force and Moment Coefficients for Vibrating Airfoils in Cascade" ARC R&M, 3254 (1960).
35. Whitehead, D.S., "Bending Flutter of Unstalled Cascades at Finite Deflection" ARC R&M, 3386 (1962).
36. Tanabe, K. and Horlock, J.H., "Simple Method for Predicting Performance of Cascades of Low Solidity", *Aero. Qly.*, **18**, p. 225 (1967).
37. Parker, R., "Calculation of Flow Through Cascades of Blades Having Relative Motion and the Generation of Alternating Pressures and Forces Due to Interaction Effects", *Proc. I. Mech. E.*, **182**, #1 (1967-68).
38. Parker, R., "Pressure Fluctuations Due to Interaction Between Blade Rows in Axial Flow Compressors", *Proc. I. Mech. E.*, **183**, p. 153 (1968-69).
39. Partington, A.J., "Experimental Study of the Sources of Blade Vibration Excitation in a Low Pressure Turbine", Westinghouse Engrg. Report EC-384 (1968).
40. Horlock, J.H., "Fluctuating Lift on Airfoils Moving Through Transverse and Chordwise Gusts", *J. Basic Engrg., ASME*, **90**, p. 494 (1968).
41. Gearhart, W.S. et al, "The Quasi Steady Design of A Compressor or Pump Stage for Minimum Fluctuating Lift", *J. Engrg. Power, ASME*, p. 31 (1969).
42. Kaji, S. and Okajaki, T., "Generation of Sound by Rotor Stator Interaction", *J. Sound & Vib.*, **13**, p. 281 (1970).

43. Murata, S. and Tsujimoto, Y., "The Unsteady Forces on Flat Plate Airfoils in Cascade Moving Through Sinusoidal Gusts", ZAMM, **56**, p. 205 (1976).
44. Murata, S. and Tsujimoto, Y., "Unsteady Flows Through Cascades", *Bull. JSME*, **20**, #147 (1977).
45. Henderson, R.E. and Daneshyar, H., "Theoretical Analysis of Fluctuating Lift on the Rotor of An Axial Turbomachine", ARC R&M, 3684 (1970).
46. Henderson, R.E. and Horlock, J.H., "An Approximate Analysis of the Unsteady Lift on Airfoils in Cascade", *J. Engrg. Power, ASME*, p. 233 (1972).
47. Parker, R. and Watson, J.F., "Interaction Effects Between Blade Rows in Turbomachines", *Proc. I. Mech. E.*, **186**, 21/72 (1972).
48. Walker, G.J. and Oliver, A.R., "The Effect of Interaction Between Wakes from Blade Rows in An Axial Flow Compressor on the Noise Generated by Blade Interaction", *J. Engrg. Power, ASME*, **94**, Series A (1972).
49. Rao, J.S., Rao, V.V.R. and Seshadri, V., "Non-steady Forces in Turbomachine Stage", *Vib. in Rotating Machinery, I. Mech. E.*, p. 243 (1984).
50. Bedrosyan, L., Maier, W. and Waggott, J., "Unsteady Turbine Blade Forces — A Practical Approach", *Vib. Blades and Bladed Disk Assemblies, ASME*, p. 105 (1985).
51. Tanaka, M. et al., "Study on Aerodynamic Force Exciting Vibration of Turbine Rotor Blade", *Proc. IFToMM Intl. Conf. Rotor Dynamics*, Tokyo, p. 307 (1986).
52. Rao, V.V.R. and Rao, J.S., "Effect of Downwash on the Unsteady Forces in a Turbomachine Stage", *11th Biennial ASME Conf. Vib. and Noise*, Boston, *Bladed Disk Assemblies*, DE-Vol 6, p. 21 (1987).
53. Preiswerk, E., "Application of the Methods of Gas Dynamics to Water Flows with Free Surface, Flows with No Energy Dissipation", NACA TM 934 (1940).
54. Preiswerk, E., "Application of the Methods of Gas Dynamics to Water Flows with Free Surface, Flows with Momentum Discontinuities (Hydraulic Jumps)", NACA TM 935 (1940).
55. Loh, W.H.T., "Theory of Hydraulic Analogy for Steady and Unsteady Gas Dynamics", *Modern Developments in Gas Dynamics*, Plenum Press, New York (1969).
56. Bryant, R.A.A., "The One-Dimensional and Two-Dimensional Gas Dynamics Analogies", *Australian J. Applied Sci.*, **7**, p. 296 (1956).
57. Harleman, D.R.F. and Ippen, A.T., "The Range of Application of the Hydraulic Analogy in Transonic and Supersonic Aerodynamics", *Anniversary Volume for M.D. Riabouchinsky*, Publication of Scientifiques et Tech. du Ministre de l'Air, Paris (1954).
58. Karman, T., "The Similarity Law of Transonic Flow", *J. Maths and Physics*, **26**, p. 182 (1947).
59. Byrd, J.L. and Williams, J.G., "Static Pressure Distribution Along Inclined Set Back Plate with Attached Jet Using Hydraulic Analogy", *Army Missile Command Report* RG-TR-63-15, Red Stone Arsenal, Alabama (1963).
60. Adams, D.M., "Application of the Hydraulic Analogy to Axisymmetric Non-Ideal Compressible Gas Systems", *J. Space Craft*, **4**, p. 359 (1967).
61. Schorr, C.J., "Pressure Ratio Correction Factor When Utilizing Hydraulic Analogy", *J. Space Crafts and Rockets*, **5**, p. 1119 (1968).
62. Rao, J.S., "Hydraulic Analogy for Compressible Gas Flow in Converging Nozzles", 80 ID 002-1, Stress Technology Inc., Rochester, NY (1980).

63. Loh, W.H.T., *A Study of the Dynamics of the Induction and Exhaust Systems of a Four Stroke Engine by a Hydraulic Analogy*, Sc.D. Thesis, M.I.T. (1946).
64. Bryant, R.A.A., "The Hydraulic Analogy as a Distorted Dissimilar Model", *J. Aero. Sci.*, **23**, p. 282 (1956).
65. Rao, J.S., "Simulation of Compressible Gas Flow Through Converging-Diverging Nozzles by the Use of Hydraulic Analogy", 80 ID 002-2, Stress Technology Inc., Rochester, NY (1981).
66. Gilmore, F.R., Plesset, M.S. and Crossley, H.E., "The Analogy Between Hydraulic Jumps in Liquids and Shock Waves in Gases", *J. Appld. Phy.*, **21**, (1950).
67. Laitone, E.V., "A Study of Transonic Gas Dynamics by the Hydraulic Analogy", *J. Aero. Sci.* (1952).
68. Laitone, E.V. and Neilson, H., "Transonic Flow Past Wedge Profiles by Hydraulic Analogy" *J. Aero. Sci.*, p. 498 (1954).
69. Bryant, R.A.A. and Grant, J.N.G., "Two-Dimensional Bow Shock Wave Detachment Distances", *J. Royal Aero. Soc.*, **61** (1957).
70. Saiva, G., "Hydraulic Analogy Studies of Shock Attachment to Wedges", *Proc. 1st Australian Conf. Hydraulics and Fluid Mechanics* (1962).
71. Klein, E.J., "Interaction of a Shock Wave and a Wedge, An Application of Hydraulic Analogy", *AIAA J.*, **3**, #5 (1965).
72. Rao, J.S., "An Examination of Errors in Hydraulic Analogy for Nozzle flows with Compressive Normal Shock", 80-ID-002-3, Stress Technology Inc., Rochester, NY, (1981).
73. Rao, J.S., Rao, V.V.R. and Seshadri, V., "Hydraulic Analogy for Non-Ideal Compressible Gas Flows", *Proc. IFToMM Intl. Conf. Rotor Dynamic Problems in Power Plants*, Rome, p. 267 (1982).
74. Rao, J.S., "A Study of Nozzle Exit Flows by Hydraulic Analogy", 80-ID-002-5, Stress Technology Inc, Rochester, NY (1981).
75. Rao, J.S., "The Effect of Straight Oblique Shock Waves on Hydraulic Analogy", 80-ID-002-4, Stress Technology Inc, Rochester, NY (1981).
76. Bryant, R.A.A., "Use of Hydraulic Analogy for Inside Problems", *J. Aerospace Sci.*, **25**, #8 (1958).
77. Laitone, E.V., "The Limitations of the Hydraulic Analogy", *J. Aerospace Sci.*, **28**, #8 (1961).
78. Gupta, O.P., "An Analytical Method for Evaluating the Optimum Depth in Hydraulic Analogy Experiments", *AIAA J.*, **3**, #10 (1954).
79. Owczarek, J.A., "On Wave Phenomenon in Turbines", *J. Engrg. Power ASME*, **88**, #3 (1976).
80. Giraud, F.L., *Experimental Studies on Supersonic Cascades Using Water Analogy*, M.S. Thesis, M.I.T. (1968).
81. Rieger, N.F., Wicks, A.L., Crofoot, J.F. and Nowak, J., "Experiments and Theories for Non-Steady Forces on Turbine Blades", Naval Ship Engrg. Centre, Arlington, VA (1978).
82. Rao, J.S., Rao, V.V.R. and Seshadri, V., "Rotating Water Table for Measurement of Non-Steady Forces in a Turbomachine Stage", *Def. Sci. J.*, **33** p. 273 (1983).
83. Sandhu, P.P.S., *Non-steady Force Measurement in an Orpheus Gas Turbine Engine Using Hydraulic Analogy*, M. Tech. Thesis, Indian Institute of Technology, New Delhi (1984).
84. Rao, V.V.R., *Unsteady Blade Forces in a Turbomachine*, Ph.D. Thesis, Indian Institute of Technology, New Delhi (1982).

85. Mylonas, J., "Unsteady 2D Flow Calculation in Turbomachinery Cascades", Proc. Fourth Symposium on Unsteady Aerodynamics and Aeroelasticity of Turbomachines and Propellers, September 6-10, Aachen (1987).
86. Martensen, E., "Berechnung der Druckverteilung an Gitterprofilen in ebener Potentialstromung mit einer Fredholmschen Integralgleichung", *Arch. Rat. Mech. Analysis*, **3**, p. 235 (1959).
87. Jacob, K., "Erweiterung des Martensen — Verfahrens auf Einzel — und Gitterprofile mit eckiger Hinterkante oder sehr kleinem Abrundungsradius", *AVA-Bereicht*, **67 A**, 21 (1967).
88. Leinhart, W., "Berechnung der Instationaren Stromung durch gegeneinander bewegte Schaufelgitter und der Schaufelkraftschwankungen", *VDI-Forschungshaft*, **562**, Dusseldorf (1974).
89. Pfost, H., *Zur Berechnung der Kompressiblen Potentialstromung durch ebene Schaufelgitter fur Unterschallgeschwindigkeiten*, Ph.D. Thesis, Technical University, Munich (1969).
90. Krammer, P., *Potentialtheoretische Berechnung der Instationaren Schaufelkrafte in Turbomaschinen mit Simulation von Nachlaufdellen*, Ph.D. Thesis, Technical University, Munich (1983).
91. Truchenbrodt, E., "Neuere Erkenntnisse uber die Berechnung von Stromungsgrenzschichten mittels einfacher Quadraturformeln", *Ing. Arch.*, **43**, p. 9 (1973).
92. Hubensteiner, M. and Mylonas, J., "Experimental and Theoretical Investigations on Unsteady Cascade Flow in Turbomachines", II Intl. Symposium on Transport Phenomena, *Dynamics and Design of Rotating Machines*, p. 73 (1988).
93. Rai, M.M., "Navier-Stokes Simulations of Rotor-Stator Interaction using Patched and Overlaid Grids", *AIAA*-85-1519.
94. Gundy-Burlet, K.L., Rai, M.M., Stauter, R.C. and Dring, R.P., "Temporally and Spatially Resolved Flow in a Two Stage Axial Compressor. Part 2 — Computational Assessment", *ASME*, 90-GT-299.
95. Stauter, R.C., Dring, R.P. and Carta, F.O., "Temporally and Spatially Resolved Flow in a Two Stage Axial Compressor, Part 1 — Experiment", *ASME*, 90-GT-259.

Chapter 7

Damping

7.1 INTRODUCTION

In the earlier chapters, we defined the blade structural system to determine the free vibration characteristics in bending, torsion and coupled modes. We have also shown that the blades are excited by external forces, principal amongst them being the nozzle passing frequency excitation with several harmonics. When the frequency of any of these harmonics coincide with one of the several natural frequencies, resonance takes place. The machine speeds at which such resonance takes place are called critical speeds and are depicted on a Campbell diagram, see Fig. 1.2, for convenience.

In general a turbomachine blade may experience two types of resonances, one under steady operation and the other under transient conditions of operation. The steady resonant or near resonant conditions of even one blade stage are not healthy for any machine, as it would increase the chances of high cycle fatigue failure, despite proper design practices that may have been adopted. These steady operating conditions are generally avoided as a rule and only under specific circumstances, some stages may be allowed to operate at near resonant conditions with appropriate means for dissipation of the energy in vibratory motion.

Transient conditions of operation exist principally for two different reasons, one due to unsteady flow conditions arising out of steam or gas admission to the stator and the other due to acceleration and deceleration of the rotor during starting and shutting operations of the machine. Under transient conditions of steam or gas admission the blades experience variable magnitudes of excitation force which cause vibratory response primarily at the fundamental mode of vibration. The resulting stresses may be high and lead to low cycle fatigue problems of the blade. These stresses can be minimized by good operating procedures. However, the transient conditions of operation that exist during the machine start-up and shutdown conditions, generate resonant stresses at the corresponding modes of vibration. These stresses are a function of acceleration value and the only way they can be controlled is by dissipation of the vibratory energy at the transient resonant conditions. Thus damping becomes an important aspect of the blade design. Unfortunately, it is also a complex problem and there are no simple procedures by which an effective way can be developed to determine the damping values in such structures.

Some of the earliest studies of the damping properties of materials were made by Rowett [1], who measured the decay of vibrations in steel specimens, and devised a stress power law. Den Hartog [2], Jacobsen [3], Kimball [4], amongst others, studied the response of mechanical systems on the relation between damping and forced vibration amplitude for simple linear systems. The first published studies of the damping properties of blades appear to date from around 1945, which deal with the selection of dampers for the suppression of vibrations in jet engine compressor blades, see Shannon [5].

Hanson [6] described a rig for testing gas turbine and compressor blade damping properties, which consisted of a disc carrying a single blade which could be rotated at speeds up to 15,000 rpm. The blade could be excited by an air jet, or by the impact of a small ball. Blade damping data is given on root friction damping coefficient *vs* rpm, and on material damping *vs* rpm, for several types of blade roots. Fir-tree root blades were found to have the highest root damping rates of amplitude decay, whereas pin-root and wedge-root blades gave consistently smaller damping values. He reported in all instances the damping to increase as rpm was increased. Hanson, Meyer and Manson [7] investigated a proposed compressor blade friction-damping device. Using the same rotating blade apparatus, they demonstrated a 3 : 1 decrease in blade vibration amplitudes when this friction damping device was used on compressor blades in the test rig. Goodman and Klumpp [8] investigated certain slip damping properties of riveted sandwich beams, which had a common interface along their neutral axis (maximum shear plane). It was found that beneficial effects from interface damping ceased beyond a certain interface load value. This is contrary to what Hanson found, where the damping increased with speed. Once the blade gets locked in the root, the slip action disappears and therefore the damping is due to material only. They also found that a highly beneficial reduction of vibrational amplitude could be achieved by friction damping, in the order of a 7 : 1 decrease, where the friction interface conditions were optimized. Beards [9] evaluated the likely effectiveness of root slip damping for suppressing vibrations of compressor blades based on an examination of the results given in the above three papers.

Grady [10] tested several dummy intermediate-pressure blades in disc attachments using a pull-test machine, and used a dynamic shaker to excite the blades to specified force levels. Attachment prestress conditions were changed by varying the root interference fits. It was shown that an optimum root flexibility condition could be developed which would result in dynamic stresses about 1/3 to 1/2 of those in a rigid (welded) joint. Where root stiffness is high, less damping is developed. Considerable variability existed in root flexibility, depending on blade assembly procedures and blade vibration history. He [11] later performed tuning fork vibration tests in air to measure blade material damping under simulated centrifugal load conditions in a tensile test machine. The results showed that the material logarithmic decrement values were highest with high dynamic stresses. A maximum test logarithmic decrement value of 0.033 was obtained. Damping decreased with increase in centrifugal load, and increased with frequency of excitation. Most of the results obtained are in general agreement with results obtained by Lazan [12] for material damping. Wagner [13] conducted a program of damping tests on rotating steam turbine blade groups in a test turbine. The objective of this program was to investigate the magnitude of blade damping, the influence of certain root types and blade sizes on damping and the effectiveness of certain damping devices. During the tests, blades were rapidly rotated through axially-directed water jets which impulsively excited them into vibratory motion. The blade response was obtained using weldable strain gages and a slip-ring assembly.

Rieger and Beck [14] developed a laboratory test rig and investigated the damping in several sizes and types of steam turbine blades, with firtree roots and ball roots inserted in a corresponding segment of turbine disc. The centrifugal load is simulated by thermal expansion of the rig, pulling a pair of identical blades which are joined together by a flexible attachment. Results were given for blade damping at several (non-rotating) simulated high centrifugal attachment load values achieved with the test rig. The variation of logarithmic decrement with vibration amplitude was investigated following impulse excitation. A substantial number of nominally identical blades were tested in each instance to determine the inherent statistical scatter of the results. Peck, Johnson and House [15] continued similar tests on long LP axial entry steam turbine blades and HP tangential entry steam turbine blades. Johnson [16] conducted studies on LP blades under high vibration and dynamic stress conditions. Each of these investigations was undertaken to determine the extent to which attachment slip could be utilized to optimize blade damping. The results showed that the damping in steam turbine blades is primarily material damping, and that both the root attachment conditions and the environmental gas conditions contributed little damping under the test conditions used.

Rao et al [17] developed a test rig using thermal cooling to simulate the centrifugal load and carried out tests in both air and vacuum. They observed the damping ratio to decrease with simulated centrifugal load. They also conducted a finite element study of the dynamic contact problem with a Coulomb friction model at the contact surfaces. Material and gas damping properties are not included in the analysis. A free vibration analysis of a blade with T and straddle T type root junctions was carried out to determine the logarithmic decrement at successive time steps. Their theoretical analysis shows that damping decreases with the increase in centrifugal load due to locking effect but beyond a certain value of centrifugal load, damping is found to increase further due to the presence of lips in the disc which open out and cause slip friction on the blade. With increase in coefficient of friction, the damping is found to increase in a linear fashion. It is also found from the analysis that the damping is very sensitive to the applied force. The straddle T root has given a considerably larger damping ratio in comparison with T root for the same centrifugal load and applied lateral load. Their experimental values have been consistently higher; this is obvious, since the theoretical model assumed only Coulomb damping at the contact surfaces.

Rao, Gupta and Vyas [18] designed and built a spin rig with nozzle passing excitation simulated by electromagnets. Transient excitation of rotating blades is caused by suddenly shutting off the excitation to the rotating blades. This enabled them to measure the damping more accurately than before. They obtained damping ratio as a function of rotational speed and tip displacements for the first four modes. Average values of the damping for these first four modes was plotted as a function of the rotor speed which has shown a distinct threshold speed beyond which the blade gets locked with the root and the damping mechanism is then mainly due to material deformation. Rao and Vyas [19] subsequently developed modal damping envelopes for different critical speeds as a function of tip displacement. This enabled them to perform a forced vibration analysis using a nonlinear damping model and such an analysis has shown dramatic improvement in the theoretical results that agreed more closely with experimental values.

A series of theoretical and experimental studies of damping in compressor blades were made by Jones and Muszynska. They [20] developed a simple two-mass analytical model to represent the vibrational behaviour of a jet engine compressor blade in its fundamental mode, allowing for slip at the blade attachment interface. This work is an attempt to develop a theoretical basis for earlier work by Goodman and Klumpp [8] discussed for slip in vibrating beam type structures, and by Hanson, Meyer and Manson [7], and William and Earles [21], on the effect of slip on the response of simple cantilever beams with internal clappers. Vibration tests were carried out to experimentally verify the analysis using a particular blade geometry. A blade with a simple root geometry was inserted in a heavy fixture containing a matching axial-entry tree head root. The blade was loaded with radial forces to simulate the centrifugal load on the root during operation. Tests were conducted to determine the response amplitude at various frequencies corresponding to several centrifugal loads. Nonlinear equations of motion with slip at the root were developed and solved by the method of matched parameters. The correlation between analysis and experiment shows that it is possible to model the blade response using a simple two degree system with slip at the root level. Their response curves follow the response for a linear model until slip occurs at the loaded interface. Beyond this point the blade vibration amplitude ceased to grow and remained constant over a substantial range of frequency. Certain dimensionless parameters which govern the optimized performance of the model are identified and correlated with the experimental results.

Muszynska and Jones [22] investigated the dynamic response of blades on a rigid disc allowing for interface slip at the attachment interface and blade hysteritic damping. Two types of mass model of a single blade in a disc were used. The first model allowed for root slip and blade hysteresis and the second model allowed a damper concept which incorporated platform slip to be studied. The results are compared with experimental response data for a single blade in a test fixture under low level harmonic excitation. It is shown that the contribution of root slippage to blade damping falls off with increasing rotational speed in the first model but that this could be compensated with a suitably sized damper device of the type used in the second model.

Jones and Muszynska [23] further investigated the influence of slip between components in the blade platform region. The purpose of this study was to determine the extent to which damping in a blade geometry could be optimized by allowing contact between the blade platform and the disc to occur at high speed. Approximate nonlinear equations of motion are derived and solved using a harmonic balance procedure. Solutions are obtained in parametric form for a number of cases, and numerical results for typical cases are included. It was found that the proposed device could provide high levels of slip damping, provided that the relevant stiffnesses are properly selected. Although this paper does not contain any significant experimental results, such testing is proposed together with a spin pit investigation to validate the influence of practical platform dampers.

Muszynska et al [24] investigated the nonlinear response of a set of compressor blades mounted on a rigid disc and interconnected with a dry friction coupling near each blade platform. A set of blades is considered, each blade consisting of a discrete two-mass system connected to adjacent blades by a flexible link, and a dry friction contact. Dry friction contacts to the disc from each blade are also included. A single blade was initially investigated experimentally to identify numerical modal values for the blade models. Analytical results are given for amplitudes and phase angles for the multi-blade set. Several system effects were examined, including blade mistuning and influences from the magnitude and distribution of exciting forces, and phase differences between the exciting forces on adjacent blades. They have shown that the fundamental modes of the blades can be accurately modelled and their parameters precisely identified from tests. An optimum value of the friction parameter (ratio of friction force to driving force amplitude) seems to exist for which response amplitudes are at a minimum for all blades.

Srinivasan, Cutts and Sridhar [25] performed an extensive investigation on compressor fan blade damping. Root damping was found to be negligible at operating speeds for the dovetail root attachment structure used in these tests. The material damping in the titanium alloy blades was also found to be very small. The shroud interface contacts were therefore found to be the most likely source of damping in such fan blade structures, but the authors indicate that damping from shroud or platform rubbing may be difficult to define for consistent results. The logarithmic decrement values for titanium blades were found to range from 0.0006 to 0.0015.

7.2 DAMPING MECHANISMS

Any vibrating structure possesses a certain capacity to dissipate energy, which may be made up of several damping mechanisms. For a turbine blade, in present context, these damping mechnisms include material damping, Coulomb friction damping at interfaces, gas dynamic damping and possibly impactive damping. Of these Coulomb damping can be specified by a simple mathematical model and a solution to the governing differential equations of simple systems can be obtained, e.g., see Rao and Gupta [26]. Material damping characteristics are well established by Lazan [12], but again they are not very amenable for simple mathematical solutions, even for structures that can be described by simple models.

The most convenient form of damping mechanism from the point of view of analysis is viscous damping. Viscous action gives rise to a force which opposes motion, and which is directly proportional to the velocity of the element to which a viscous damper is attached. As it is evident to the reader, the beauty of such a mechanism is, it fits well into the description of simple linear second order differential equation of a system and delivers an elegant solution to the periodic excitation forces. This comes handy, if the modal equations of a structure contain such a viscous damping property. Unfortunately, very few real structures have pure viscous damping, more so with the complex problem of a turbine blade. This is the reason why equivalent viscous damping concept is developed by several researchers, e.g., Jacobsen [3]. This is done by equating the energy loss per cycle of a given structure to the energy loss of an equivalent viscous damper and thus determine the equivalent viscous damping coefficient [26]. This is of course possible, if we have a good definition of energy loss per cycle of the structure. A procedure to define equivalent viscous damping envelopes from experimental observations for turbine blades is given later in this chapter. Some information on individual damping mechanisms is first given.

7.2.1 Material Damping

Material damping has been studied by many investigators. Rowett [1] investigated the torsional damping properties of certain grades of steel shafting, and suggested an early form of the stress-damping law.

$$E_H = J\sigma^n$$

where E_H is energy loss per unit volume, J is a constant of proportionality, σ is the local stress and n is a damping exponent of value between 2 to 3 for most materials. The total energy dissipated by material hysteresis is obtained by integrating the above stress relationship throughout the structure.

Lazan [12] conducted comprehensive studies into the general nature of material damping, and presented damping results data for almost 2000 materials and test conditions, with an extensive review of the subject literature in his treatise on this subject. This work is now the basic reference on material damping. It contains details of all material damping laws applied to a number of materials. These results are presented in chart form, in which specific damping energy is related to dynamic stress. Procedures for determining the amplitude of damped vibrating structures are presented, including an application to the first mode vibrations of a simple turbine blade. Lazan's results show that logarithmic decrement values will increase with dynamic stress, i.e., with vibration amplitude, where material damping is the dominant mechanism.

Material damping coefficients of the above equation for a typical group of steels is given in Table 7.1, see Lazan [12].

TABLE 7.1

Material Damping Coefficients and Exponents [12]

Material & Condition	*Coefficient* J	*Exponent* n	*Stress Range psi*
Stainless steel AISI 403 12 Cr	2.5×10^{-14}	2.900	< 30 ksi
Stainless steel	1.688×10^{-13}	2.359	< 14 ksi
17-7 PH Annealed	2.34×10^{-18}	3.534	> 14 ksi
Stainless steel 17-4 PH Annealed	1.088×10^{-13}	2.124	< 45 ksi
High strength	2.153×10^{-17}	3.343	< 38 ksi
steel 4340 RC-40	1.224×10^{-26}	5.360	> 38 ksi
Titanium 90-6A1-4V	3.382×10^{-12}	1.969	< 16 ksi
Annealed	1.035×10^{-12}	2.091	> 16 ksi

7.2.2 Gas Dynamic Damping

Fluid medium in the stage, steam or gas provides a very small contribution to the energy loss of a vibrating blade, as some work is required to move this medium along with the vibrating blade surface. Obviously this amount is very small and its contribution, if any, can be determined by tests in an environmental chamber and comparing the results with those conducted in vacuum.

However, an important contribution by the fluid medium is the damping that arises from variation of lift force due to changes in the flow parameters because of blade vibration. Essentially this is an aeroelastic problem that opposes the blade motion in general, thus contributing to damping of the structure. In some specific cases, the aeroelastic characteristic is such that it aids the blade motion, thus giving rise to self excited motion. Self excited vibrations arising out of aeroelastic coupling is very commonly found in isolated airfoils giving rise to flutter. In turbomachine blades, such motions are rarely found and thus are not discussed here. Gas dynamic damping arising out of stage interaction is considered here for a simple case, to understand the underlying principle.

Consider an airfoil shown in Fig. 7.1. The airfoil is subjected to uniform upstream flow of velocity V, at an angle of attack δ. The blade chord is $2c$. The lift force can be written as

$$L = \rho_G c V^2 C_n(\delta) \tag{a}$$

where ρ_G is the density of the fluid medium and $C_n(\delta)$ is lift coefficient as a function of the angle of attack. Let this blade be a rotor blade in a

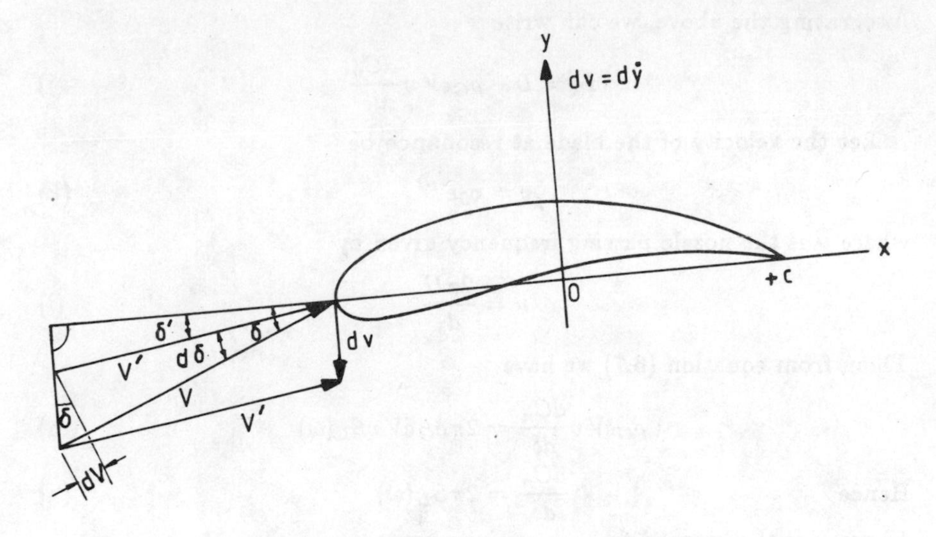

Fig. 7.1 Gas dynamic damping

stage with a peripheral velocity equal to U, executing bending vibrations in flapwise mode. From the mean equilibrium position of the vibrating blade, consider the instant when the blade velocity is incremented to $dv = d\dot{y}$. The new velocity in the upstream is V' with an angle of attack δ' as shown in Fig. 7.1. The change in the angle of attack is

$$d\delta = -\frac{dv \cos\delta}{V} \tag{b}$$

Hence

$$\frac{d\delta}{dv} = -\frac{\cos\delta}{V} \tag{c}$$

For small angles of attack, the above can be approximated as

$$\frac{d\delta}{dv} = -\frac{1}{V} \tag{d}$$

Also, from Fig. 7.1,

$$dV = -dv \sin\delta \tag{e}$$

Again for small angles of attack the above equation gives

$$\frac{dV}{dv} = 0 \tag{f}$$

Now differentiating equation (a) with v, we get

$$\frac{dL}{dv} = \frac{dL}{d\delta}\frac{d\delta}{dv} + \frac{dL}{dV}\frac{dV}{dv} = \rho_G c\left(V^2\frac{dC_n}{d\delta}\frac{d\delta}{dv} + 2C_nV\frac{dV}{dv}\right) \tag{g}$$

Using equations (d) and (f), the above becomes

$$\frac{dL}{dv} = -\rho_G cV\frac{dC_n}{d\delta} \tag{h}$$

Therefore the change in lift force due to a change in velocity of the blade is

$$dL = -\rho_G cV\frac{dC_n}{d\delta}dv \tag{i}$$

Integrating the above, we can write

$$L_t = L - \rho_G c V v \frac{dC_n}{d\delta} \tag{j}$$

Let the velocity of the blade at resonance be

$$v = v_0 e^{i\nu t} \tag{k}$$

where ν is the nozzle passing frequency given by

$$\nu = \frac{2\pi U}{d_s} \tag{l}$$

Then, from equation (6.7) we have

$$\rho_G c V v \frac{dC_n}{d\delta} = 2\pi \rho_G c V v S_L(\omega) \tag{m}$$

Hence

$$\frac{dC_n}{d\delta} = 2\pi S_L(\omega) \tag{n}$$

where ω is the reduced frequency given by

$$\omega = \frac{\nu c}{V} \tag{o}$$

Absolute value of $S_L(\omega)$ is given in Table 7.2, see Kemp [7].

TABLE 7.2

ω	S_{AL}	ω	S_{AL}
0	1.000	1.8	0.294
0.02	0.966	2.0	0.280
0.04	0.931	2.5	0.252
0.06	0.898	3.0	0.230
0.08	0.866	3.5	0.213
0.10	0.837	4.0	0.199
0.20	0.720	4.5	0.188
0.30	0.637	5.0	0.178
0.40	0.574	6.0	0.163
0.50	0.527	7.0	0.150
0.60	0.488	8.0	0.140
0.70	0.457	9.0	0.133
0.80	0.426	10.0	0.124
0.90	0.397	11.0	0.120
1.00	0.390	12.0	0.115
1.20	0.312	13.0	0.104
1.40	0.334	14.0	0.110
1.60	0.311	15.0	0.103

From equations (j) and (n) we find that the lift force decreases for a motion in the same direction and hence the damping is positive. The damping coefficient can be written as

$$C_G = 2\pi \rho_G c V S_L(\omega) \tag{p}$$

Consider the data of Sollman [30] for the estimation of blade damping ratios due to the stage interference with the rotor blade subjected to a simple transverse gust.

Area of cross-section of the blade $= 232 \text{ mm}^2$
Chord length of the blade $= 20 \text{ mm}$
Young's modulus $= 68670 \text{ N/mm}^2$
Length of the blade $= 200 \text{ mm}$
Second moment of area $= 1107 \text{ mm}^4$
$\rho_b = 2.8 \times 10^{-6} \text{ kg/mm}^3$
$\rho_G = 1.293 \times 10^{-9} \text{ kg/mm}^3$

The first two flap-wise modes are

$$p_1 = \frac{3.516}{200^2}\sqrt{\frac{68670 \times 1107 \times 1000}{2.8 \times 10^{-6} \times 232}} = 950.9 \text{ rad/s}$$

$$p_2 = \frac{22.034}{200^2}\sqrt{\frac{68670 \times 1107 \times 1000}{2.8 \times 10^{-6} \times 232}} = 5958.9 \text{ rad/s}$$

The reduced frequencies are

$$\omega_1 = \frac{950.9 \times 20}{100000} = 0.19 \qquad \omega_2 = \frac{5958.9 \times 20}{100000} = 1.19$$

The Sear's function values from Table 7.2 are

$$S(\omega_1) = 0.73 \qquad S(\omega_2) = 0.36$$

The damping coefficients are

$$\begin{aligned} C_{G1} &= 2\pi \times 1.293 \times 10^{-9} \times 20 \times 100000 \times 200 \times 0.73 \\ &= 2.372 \text{ kg/s (Ns/m)} \end{aligned}$$

$$\begin{aligned} C_{G2} &= 2\pi \times 1.293 \times 10^{-9} \times 20 \times 100000 \times 200 \times 0.36 \\ &= 1.17 \text{ kg/s (Ns/m)} \end{aligned}$$

The critical damping values are

$$C_{c1} = 2 \times 2.8 \times 10^{-6} \times 232 \times 200 \times 950.9 = 247.08 \text{ kg/s (Ns/m)}$$
$$C_{c2} = 2 \times 2.8 \times 10^{-6} \times 232 \times 200 \times 5958.9 = 1548.36 \text{ kg/s (Ns/m)}$$

Hence the damping ratios for the first two modes are

$$\xi_1 = 0.0096 \qquad \xi_2 = 0.00076$$

The above calculation is based on a simple lift function assumed for the blade. In reality this function is more complicated. As we see later, the damping ratios above due to gas damping are small compared to the interfacial damping and material damping values and hence, as a first approximation, they can be ignored.

7.2.3 Interface Damping

Interfacial damping, arising out of friction between two surfaces moving relative to each other, is a complex analytical problem, particularly for the type of blade structures with blade root junctions, shroud bands, lacing wires and platform dampers. An attempt has been made recently to assess this damping due to Coulomb friction between the blade root junction surfaces by using contact finite elements, see Rao et al [17]. Brief discussion on this is given below.

Static Contact Problem

When there are two separate bodies such as the blade and the disc with the corresponding root, the contact under static loading should be first established before modelling the dynamic contact problem. In the case of a blade root system, the contact can take place with a separation and friction, which is a nonlinear and irreversible case. The quasistatic conditions corresponding to the effect of centrifugal load on the blade and the disc represent the equilibrium state about which oscillations would occur. Thus it is necessary to solve the static contact problem before arriving at the initial conditions for the forced vibration problem.

The structure is assumed to be an assembly of a large number of finite elements as shown in Fig. 7.2(a). We can use quadratic isoparametric elements for the blade and root structure. These elements are conveniently handled by normalized coordinates. Fig. 7.2(b) shows the parent element and the transformed element.

Figs. 7.3 and 7.4 show two elastic bodies A and B before and after they come into contact. The body B is considered to have fixed boundary conditions and the body A is acted upon by a force vector $\{f\}$. Elastically the problem is same if the contact is brought about by rigid body movement of the support of body B and the force vector $\{f\}$ results in the body A

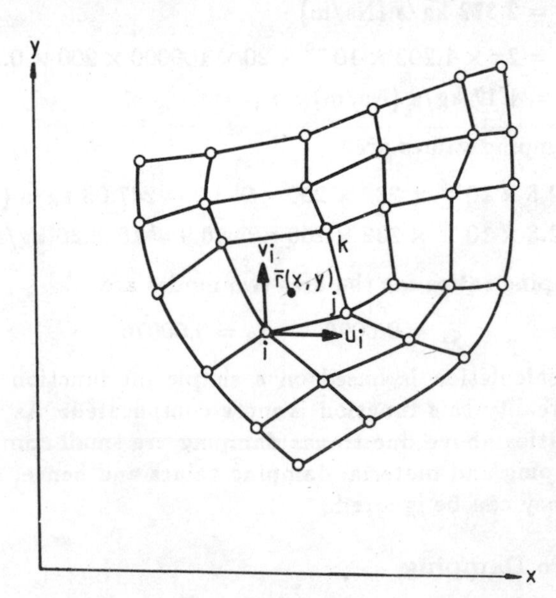

Fig. 7.2(a) Finite element idealization of a plane region

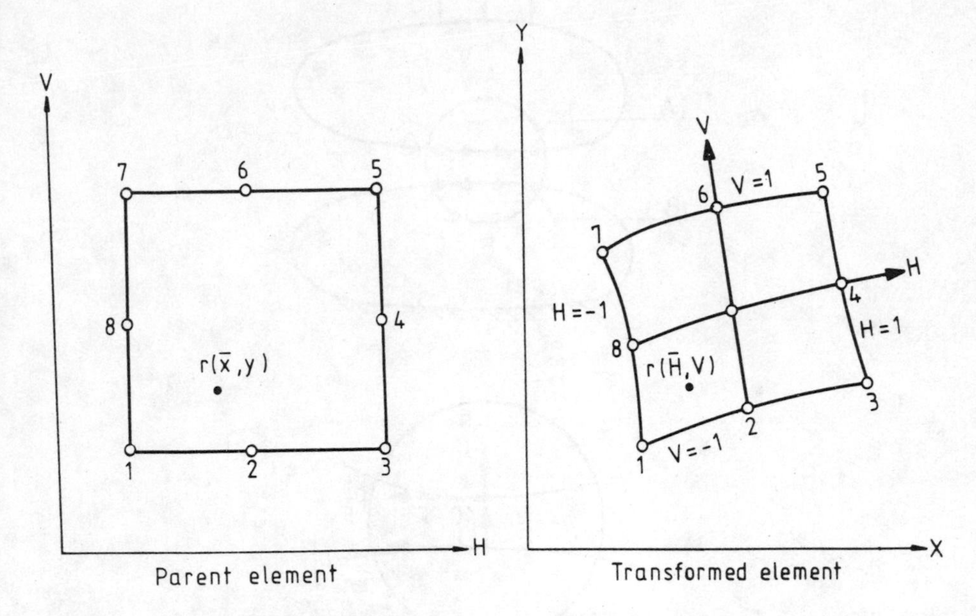

Fig. 7.2(b) Coordinate transformation

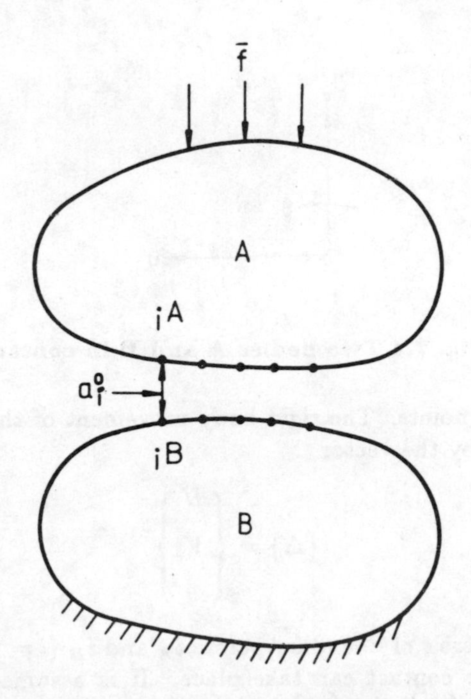

Fig. 7.3 Bodies A and B before contact

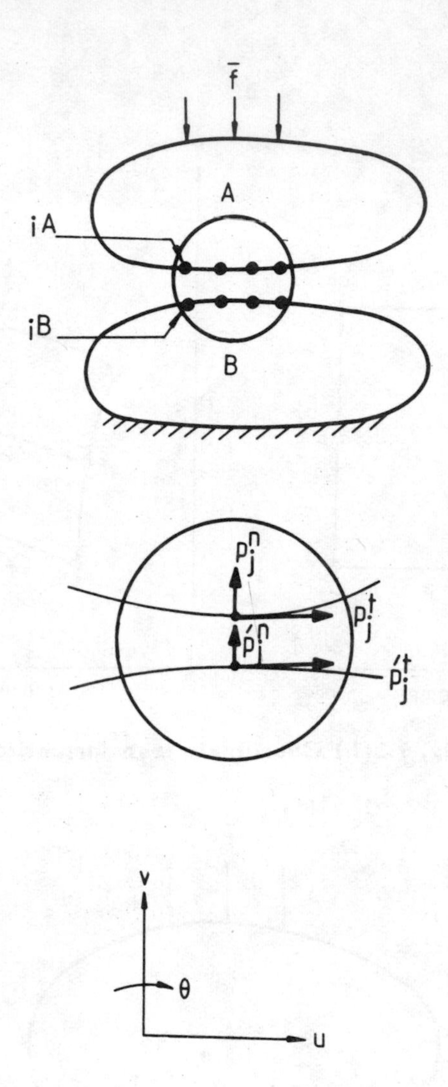

Fig. 7.4 Two bodies A and B in contact

at the boundary points. The rigid body movement of the supports of the body B is given by the vector

$$\{\Delta\} = \begin{Bmatrix} U \\ V \\ \theta \end{Bmatrix} \tag{7.1}$$

Consider the series of boundary points i_A and i_B ($i = 1$ to n) in Fig. 7.3 at which possible contact can take place. It is assumed that the nodal contact forces are $\{p_j\}$ and $\{p'_j\}$ on body A and B respectively. It is to be noted that these vectors have two components, one in the tangential and the other in normal directions as below.

$$\{p_j\} = \left\{ \begin{array}{c} \{p_j^t\} \\ \{p_j^n\} \end{array} \right\}$$

and,

$$\{p'_j\} = \left\{ \begin{array}{c} \{p_j'^t\} \\ \{p_j'^n\} \end{array} \right\} \tag{7.2}$$

If the body B has fixed supports and the only forces acting are the contact forces, the displacements of points i in the body B can be written as

$$\{a_i^B\} = \sum_{j=1}^{n} [C_{ij}^B]\{p'_j\} \tag{7.3}$$

where $[C_{ij}^B]$ above represents a 2×2 submatrix representing flexibility coefficients corresponding to the tangential and normal deflections, respectively at the node i due to the force at j.

The submatrix of flexibility coefficients is obtained by eliminating all the nodes except those where possible contact can take place and inverting the so formed condensed stiffness matrix. If the supports of the body B are given the rigid displacements given in equation (7.1), the total displacement of the point i is given by

$$\{a_j^B\} = \sum_{j=1}^{n} [C_{ij}^B]\{p'_j\} + [\sigma_i]\{\Delta\} \tag{7.4}$$

where,

$$[\sigma_i] = \begin{bmatrix} \cos\phi & \sin\phi & -(y\cos\phi - x\sin\phi) \\ -\sin\phi & \cos\phi & -(x\cos\phi + y\sin\phi) \end{bmatrix} \tag{7.5}$$

is the kinematic transformation matrix, see Fig. 7.5.

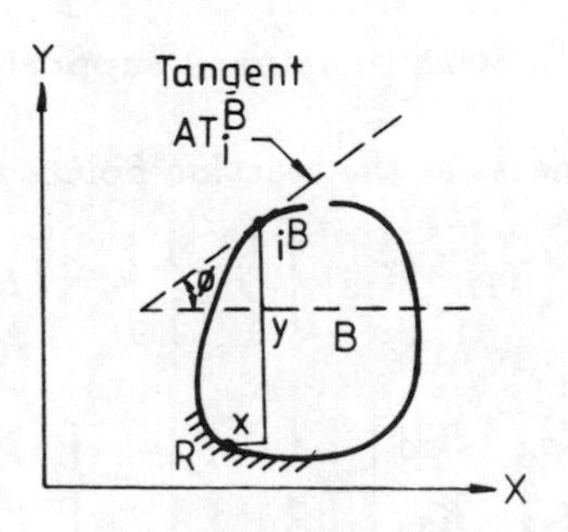

Fig. 7.5 Computation of transformation matrix

As the body A has force boundary condition, it is free to have rigid body movement. Hence a slightly different method is to be used for getting the flexibility coefficients for this. The rigid body movement of A is restricted by assuming two imaginary supports as shown in Fig. 7.6. Displacements are assumed zero in both the directions at one of the points and at the other point in one direction only. Hence three degrees of freedom are restrained. All the nodes except those at the imaginary reaction points and the likely contact points and those with the external forces are eliminated. The condensed system of equations in general can be written as

$$\begin{bmatrix} K_{11} & K_{12} & K_{13} \\ K_{21} & K_{22} & K_{23} \\ K_{31} & K_{32} & K_{33} \end{bmatrix} \begin{Bmatrix} \{a^R\} \\ \{a^C\} \\ \{a^f\} \end{Bmatrix} = \begin{Bmatrix} \{R\} \\ \{p^C\} \\ \{f\} \end{Bmatrix} \tag{7.6}$$

where K are submatrices of the condensed stiffness matrix of body A, $\{a^R\}$ is the vector of displacement of reaction points, $\{a^C\}$ is the vector of displacements at contact points, $\{a^f\}$ is the vector of displacements at points where external forces are acting on A. $\{R\}$ is the vector of reaction forces, $\{p^C\}$ and $\{f\}$ are the vectors of forces at the contact points and at the points where external forces are applied.

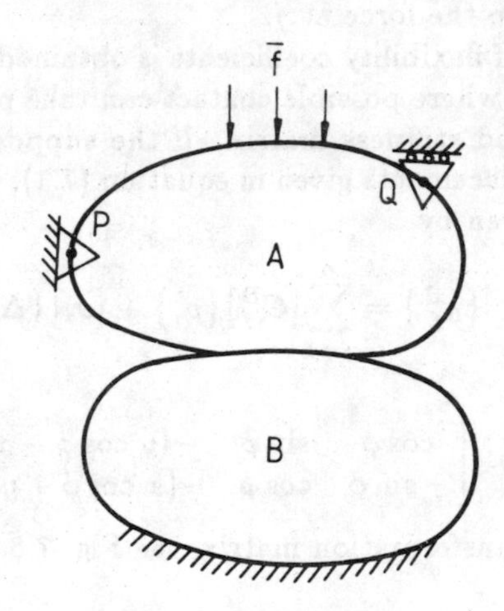

Fig. 7.6 Body A with imaginary support points P and Q

Since the displacements at the reaction points are zero,

$$[K_{12} \quad K_{13}] \begin{Bmatrix} \{a^C\} \\ \{a^f\} \end{Bmatrix} = \{R\} \tag{7.7}$$

and,

$$\begin{bmatrix} K_{22} & K_{23} \\ K_{32} & K_{33} \end{bmatrix} \begin{Bmatrix} \{a^C\} \\ \{a^f\} \end{Bmatrix} = \begin{Bmatrix} \{p^C\} \\ \{f\} \end{Bmatrix} \tag{7.8}$$

From the above two equations we get,

$$[K_{12} \quad K_{13}] \begin{Bmatrix} K_{22} & K_{23} \\ K_{32} & K_{33} \end{Bmatrix}^{-1} \begin{Bmatrix} \{p^C\} \\ \{f\} \end{Bmatrix} = \{R\} = 0 \tag{7.9}$$

since in reality there are no external forces present in this case.

The displacement of points i in the direction normal to the surface are given by the expression

$$\{a_i^A\} = \sum_{j=1}^{n}[C_{ij}^A]\{p_j\} + \sum_{k=1}^{k=e}[C_{i,n+k}]\{f_k\} \tag{7.10}$$

where e is the number of nodes with external forces and $\{f_k\}$ is the vector of external forces at the kth loaded node such that

$$\{f_k\} = \begin{Bmatrix} f_k^t \\ f_k^n \end{Bmatrix} \tag{7.11}$$

where f_k^t and f_k^n are the components in the tangential and normal directions respectively.

It may be noted that

$$[C_{ij}] = \begin{bmatrix} K_{22} & K_{23} \\ K_{32} & K_{33} \end{bmatrix}^{-1} \tag{7.12}$$

This is easily accomplished by inverting the condensed matrix obtained after imposing the boundary conditions for the fictitious nodes. When the two bodies are in contact the following compatibility equation must be satisfied.

$$\{a_i^A\} = \{a_i^B\} + \{a_i^0\} \tag{7.13}$$

where $\{a_i^0\}$ is the vector of clearances. Hence from equations (7.4) and (7.10),

$$\sum_{j=1}^{j=n}[C_{ij}^A]\{p_j\} + \sum_{k=1}^{k=e}[C_{i,n+k}]\{f_k\} = \sum_{j=1}^{j=n}[C_{ij}^B]\{p_j'\} + [\sigma_i]\{\Delta\} + \{a_i^0\} \tag{7.14}$$

At the contact nodes $\{p_j\} = -\{p_j'\}$ and hence

$$\sum_{j=1}^{j=n}([C_{ij}^A] + [C_{ij}^B])\{p_j\} - [\sigma_i]\{\Delta\} = -\sum_{k=1}^{k=e}[C_{i,n+k}]\{f_k\} + \{a_i^0\} \tag{7.15}$$

where n is the number of nodes in contact and e is the number of nodes with external forces.

For the case when the nodes slip relative to each other in the direction parallel to the contact surface, the compatibility condition (7.13) is not valid for the tangential direction. Hence the above equation will not be valid along the tangential direction and the corresponding equations are replaced by the slip condition

$$\{p_j^t\} = \mu\{p_j^n\} \tag{7.16}$$

where μ is the coefficient of dynamic friction and the subscript j represents the node in contact. If m nodes are in contact, the above two equations constitute $2m$ equations in $2m + 3$ unknowns and the three additional equations are the equations of equilibrium suitably represented by (7.9).

This system is linear and the solution is straightforward if the contact zone is known *a priori* and friction is sufficient to prevent slipping. However, in practice the area of contact is not known and some nodes are observed to be slipping. Hence the following steps are to be followed.

1. Assume a set of nodal points to be in contact and solve equation (7.15) for a prescribed load.
2. Delete all those points where nodal forces become negative from the possible contact zone.
3. If the ratio of nodal force in tangential direction to the nodal force in the normal direction is greater than the coefficient of friction, the nodes are slipping, then the portion of equation (7.15) corresponding to the tangential direction is replaced by equation (7.16).
4. The above steps are repeated until all normal forces in the contact zone come out to be positive only and the ratio of tangential force to normal force for all the nodes in contact is either less or equal to the coefficient of friction.

Now consider a quasistatic problem of blade root-disc deformation under centrifugal loading. Rao et al [17] studied a T and straddle T type root interface joints. Fig. 7.7 shows the finite element model of T root model of the 19th stage of a 210 MW steam turbine. Body A has been modeled with 34 elements having 282 degrees of freedom while the body B has 26 elements with 230 degrees of freedom. Eight noded quadratic isoparametric elements, see [27] and [28], are used in the analysis. For simplicity, all the centrifugal load is assumed to act at the tip of the blade. Fig. 7.8 shows the discretization for the straddle T root of the 13th stage IP rotor of a 120 MW machine. Body A has been modeled with 38 elements having 342 degrees of freedom and body B has been modeled with 33 elements having 286 degrees of freedom. In both the cases, the body B has been fixed along the edge AB, while the body A has been imposed with axial force of prescribed value along the surface. Fictitious nodes for restraining rigid body motion have been assumed for body A at two nodes on opposite faces of the blade.

Figure 7.9 shows the enlarged details of pairing of contact nodes for the T root junction. The deformation of two body assembly under the action of axial force is shown in Fig. 7.10. The deformation is symmetric about the central line. For the given mesh, two nodes lift and ten are found to slip when an axial load of 1000 kg is applied. The deformations shown are enlarged 400 times. When the axial force is increased to 2000 kg, the deformed shape indicates that two nodes lift and eight are slipping. The dovetail is nearly locked and the lips of the disc open out, causing slip on the two side surfaces. The displacements shown in Fig. 7.11 are enlarged 300 times.

For the straddle T root, the deformation of the two body assembly under the action of axial load 1000 kg is shown in Fig. 7.12. The deformation is symmetric about the central line. For the given mesh 10 nodes are lifting, eight are slipping, while two are locked.

The nodal displacements thus determined in the quasistatic problem constitute the initial condition for the dynamic problem.

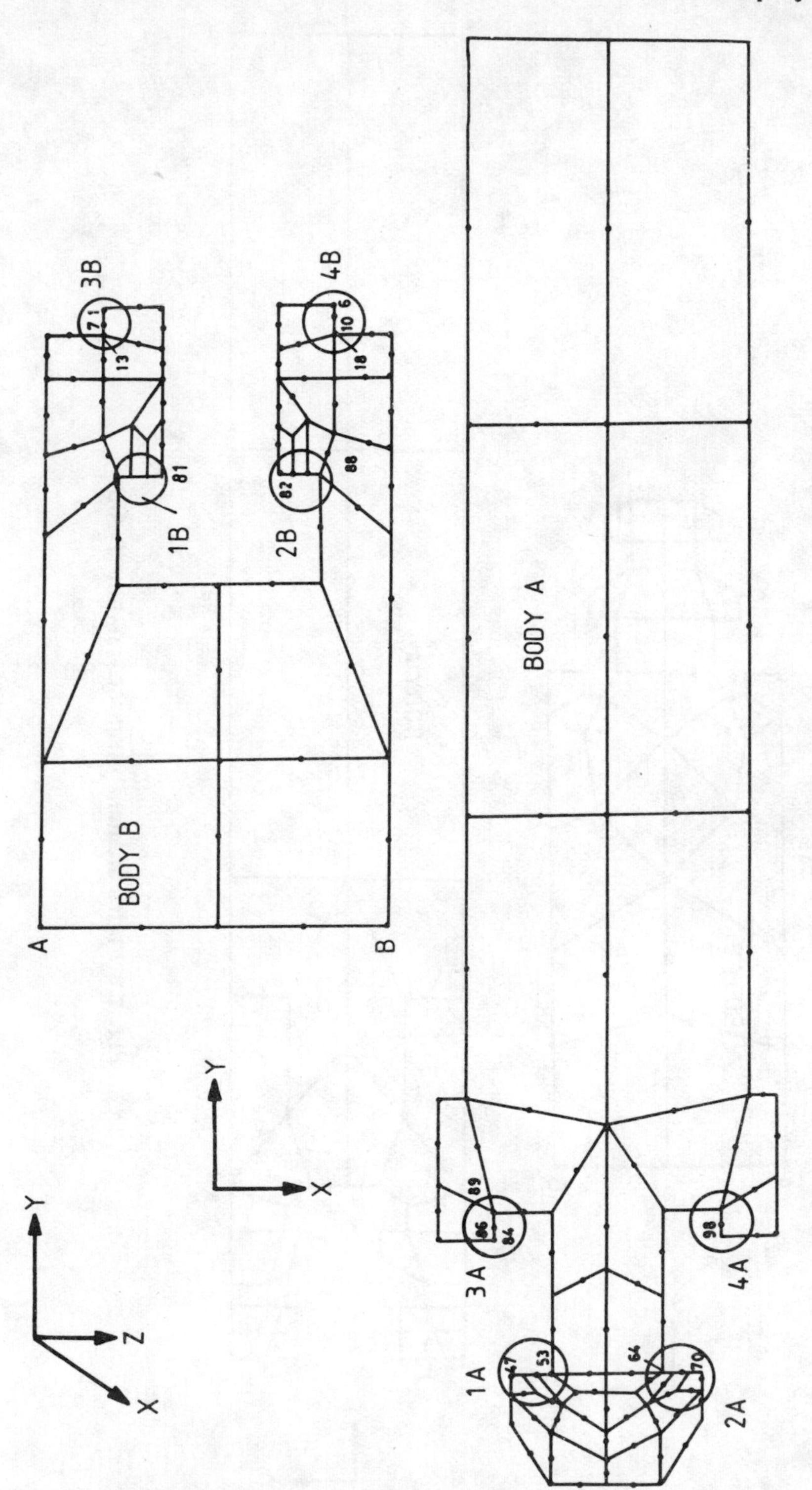

Fig. 7.7 Finite element mesh of T root blade and disc

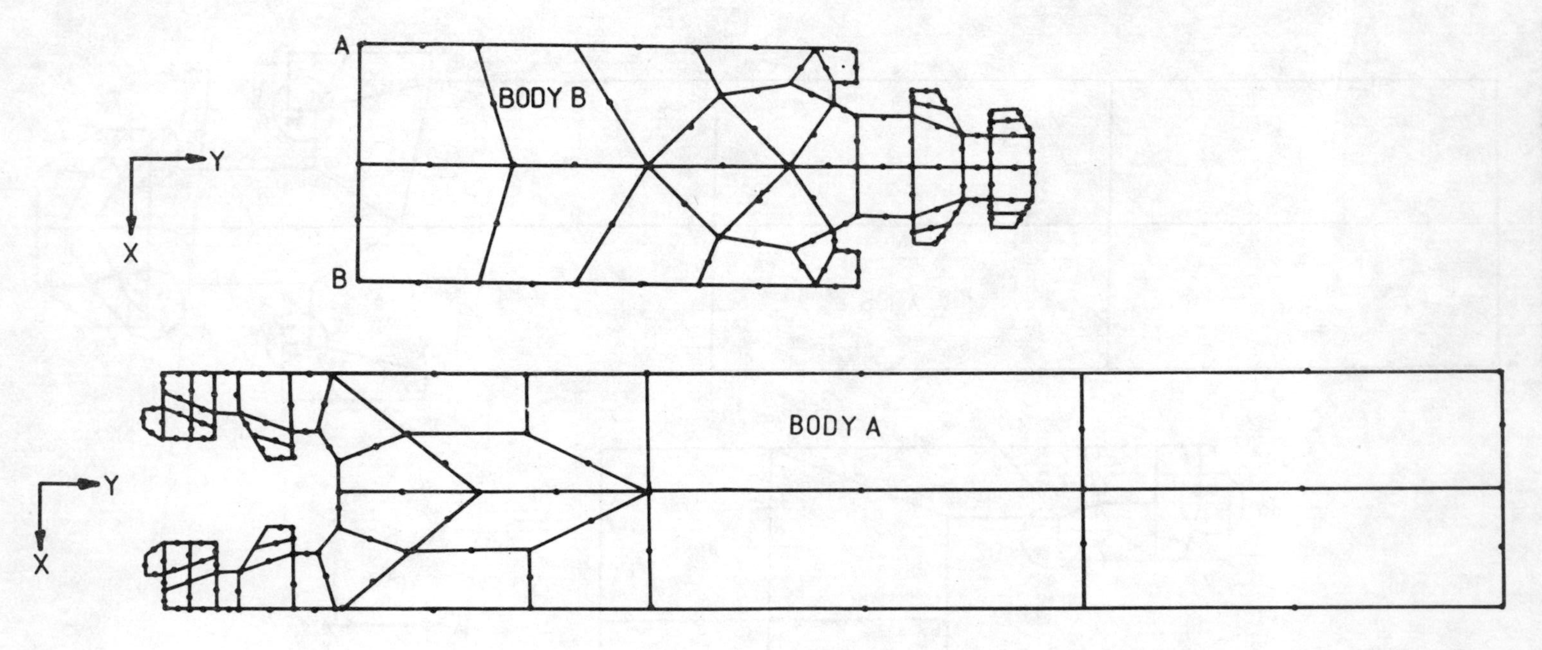

Fig. 7.8 Finite element mesh of straddle T-root

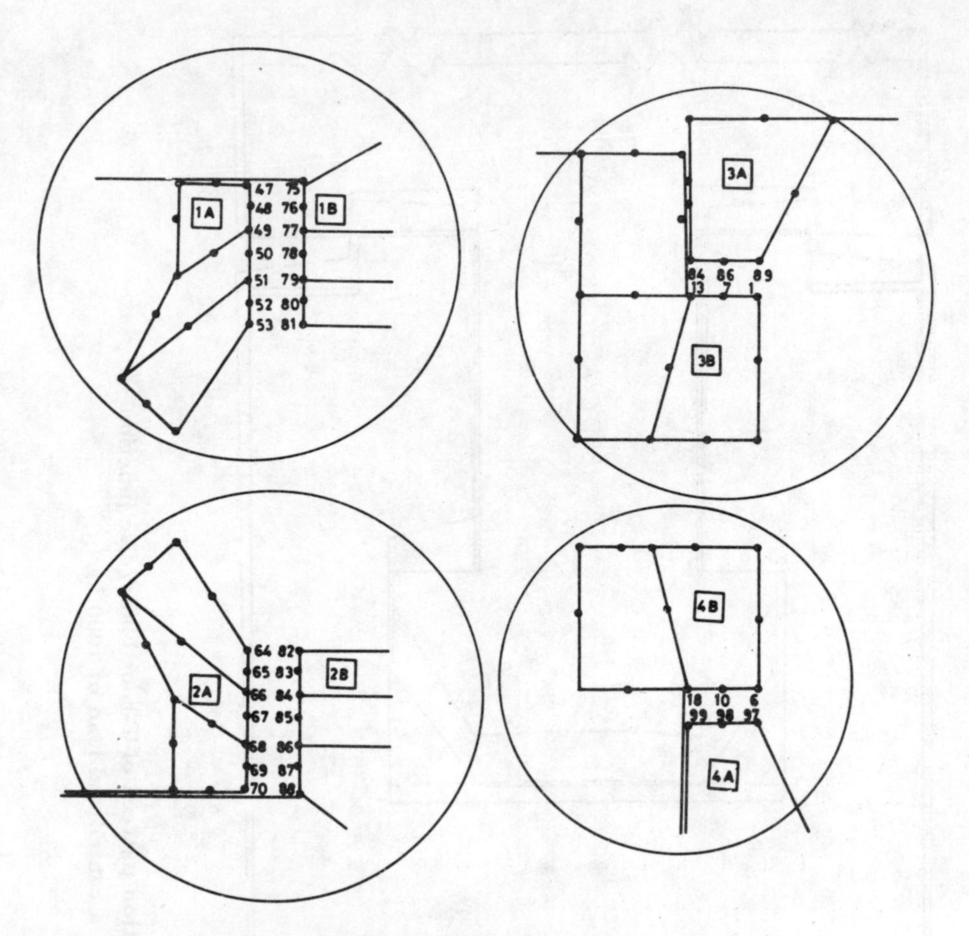

Fig. 7.9 Details of finite element mesh in the neighbourhood of contact region

Dynamic Contact Problem

The system equation for a single blade assembly in the absence of material damping is

$$[M]\{\ddot{a}\} + [K]\{a\} = \{f\} \tag{7.17}$$

The behaviour of the contact zone between A and B is nonlinear on account of opening and slipping which are to be suitably modeled. Thus, partitioning the displacement vector

$$\{a_t\}^A = \begin{Bmatrix} \{a_t^C\} \\ \{a_t^A\} \end{Bmatrix}$$

and,

$$\{a_t\}^B = \begin{Bmatrix} \{a_t^{C'}\} \\ \{a_t^B\} \end{Bmatrix} \tag{7.18}$$

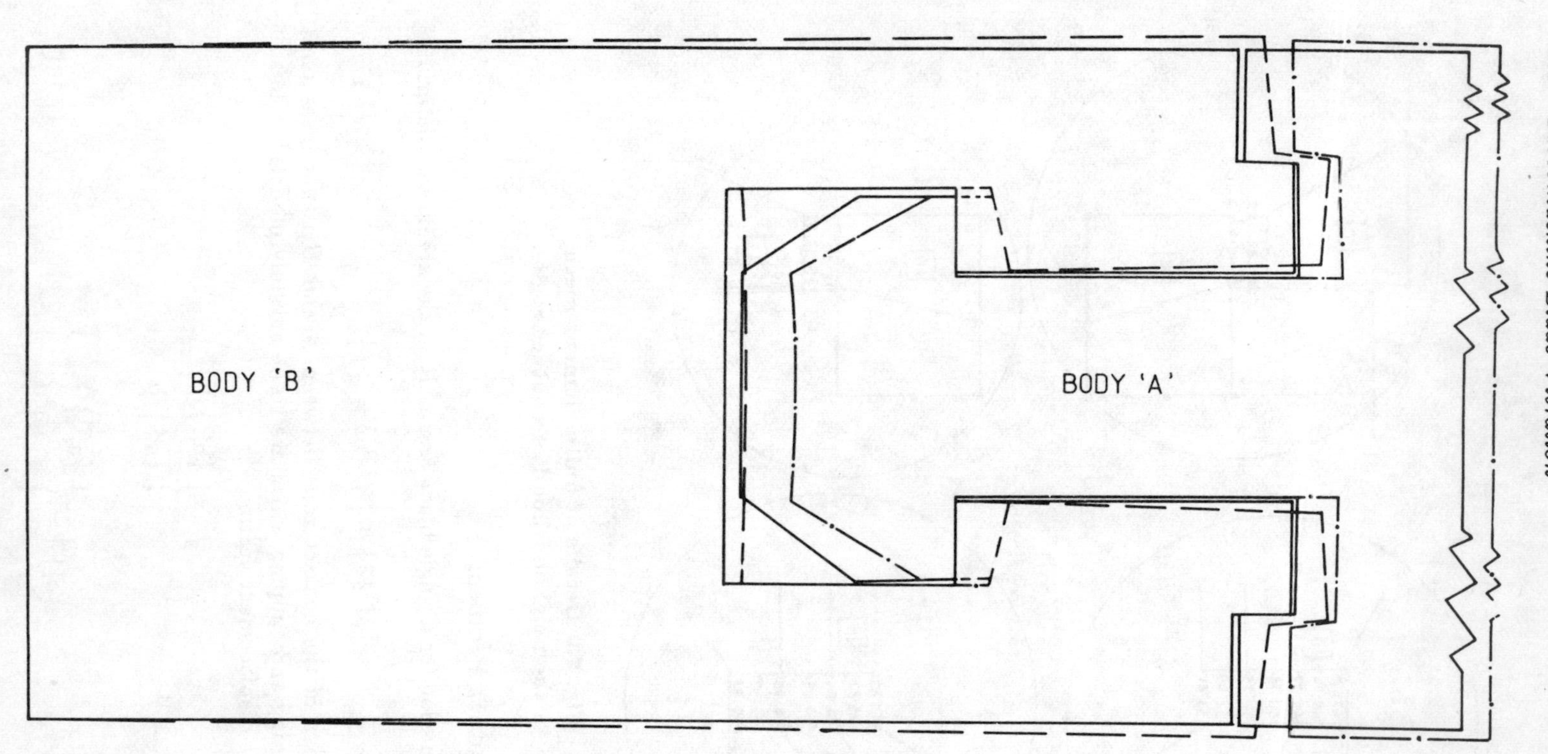

Fig. 7.10 Deformation pattern of T-root blade/disc junction under a centrifugal load of 1000 kg

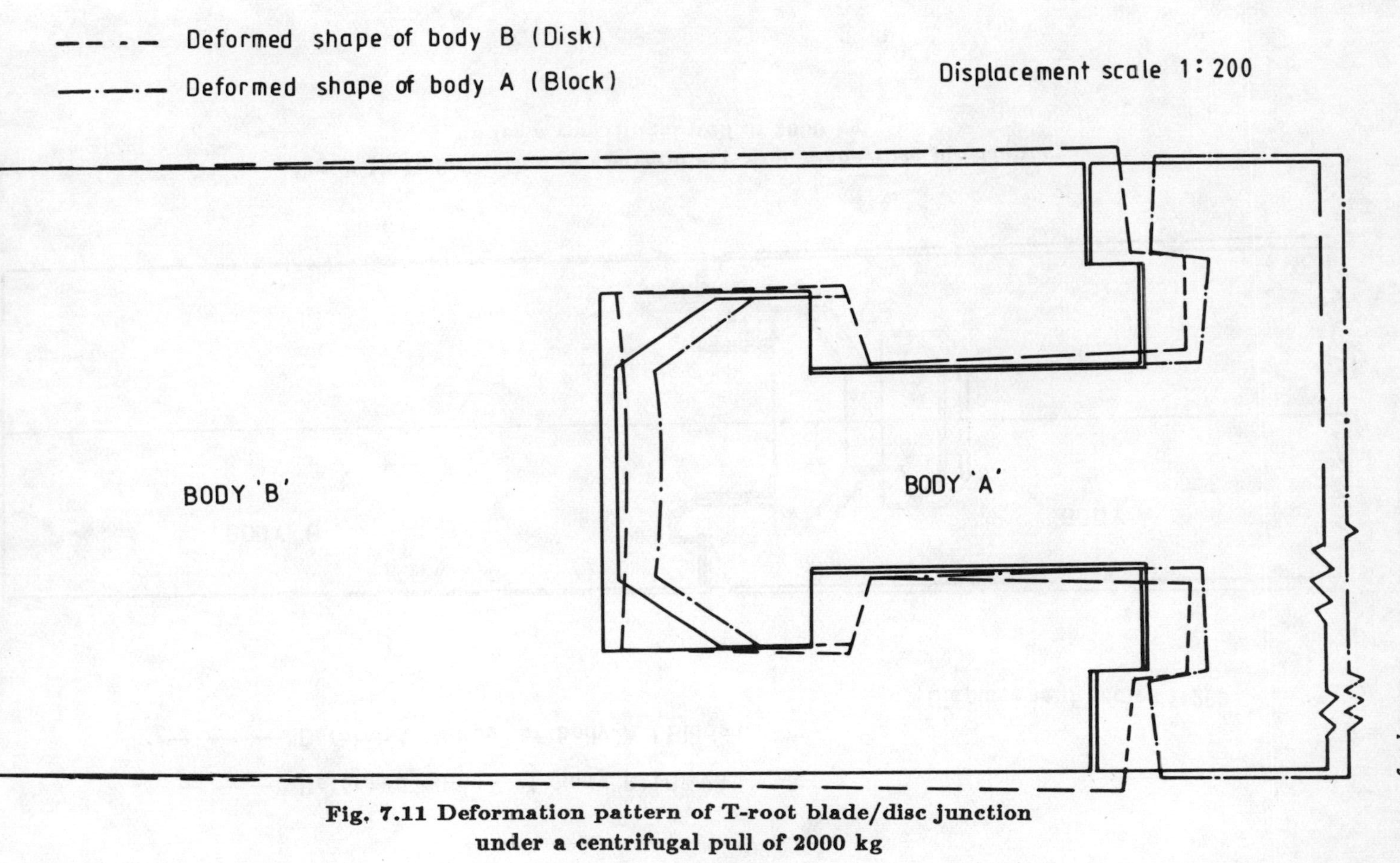

Fig. 7.11 Deformation pattern of T-root blade/disc junction under a centrifugal pull of 2000 kg

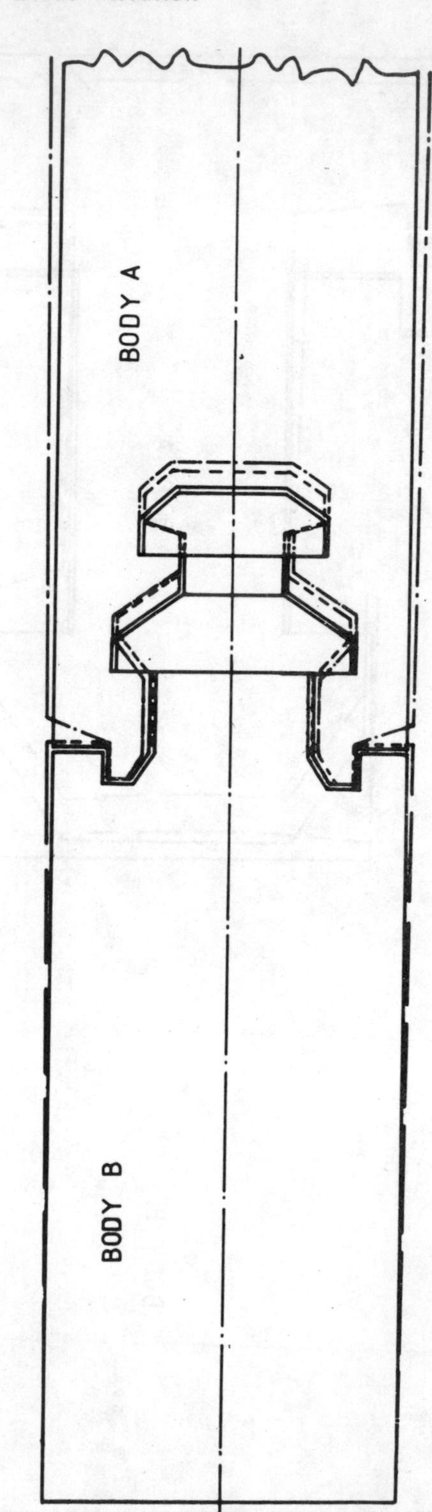

Fig. 7.12 Deformation of a straddle T-root blade/disc junction under a centrifugal pull of 2000 kg

we can write the equations of motion for both the bodies as

$$[M^B]\begin{Bmatrix}\{\ddot{a}_t^{C'}\}\\ \{\ddot{a}_t^B\}\end{Bmatrix}+[K^B]\begin{Bmatrix}\{a_t^{C'}\}\\ \{a_t^B\}\end{Bmatrix}=\begin{Bmatrix}\{f_t^{C'}\}\\ \{f_t^B\}\end{Bmatrix} \tag{7.19}$$

$$[M^A]\begin{Bmatrix}\{\ddot{a}_t^{C}\}\\ \{\ddot{a}_t^A\}\end{Bmatrix}+[K^A]\begin{Bmatrix}\{a_t^{C}\}\\ \{a_t^A\}\end{Bmatrix}=\begin{Bmatrix}\{f_t^{C}\}\\ \{f_t^A\}\end{Bmatrix} \tag{7.20}$$

On account of the nonlinearities, modal methods are not found suitable, instead direct integration algorithms will be useful.

The Wilson-θ method, discussed extensively by Bath and Wilson [29] is an implicit integration method, in which the acceleration varies linearly over the time interval from t to $t+\theta\,\Delta t$ where $\theta \geq 1$ is determined to obtain optimum stability and accuracy characteristics. θ is usually taken as 1.4. Equation (7.17) is written as

$$[M]\{\ddot{a}_{t+\theta\,\Delta t}\}+[K]\{a_{t+\theta\,\Delta t}\}=\{f_{t+\theta\,\Delta t}\} \tag{7.21}$$

where,

$$\{f_{t+\theta\,\Delta t}\}=\{f_t\}+\theta(\{f_{t+\Delta t}\}-\{f_t\})$$

For adopting the Wilson-θ scheme, the following procedure is used.

Initial calculations

1. Form stiffness matrix $[K]$ and mass matrix $[M]$.
2. Initialize $\{a_0\}$, $\{\dot{a}_0\}$, $\{\ddot{a}_0\}$.
3. Select time step Δt.

$$A0=\frac{6}{(\theta\,\Delta t)^2};\quad A2=\frac{6}{\theta\Delta t};\quad A4=\frac{\Delta\theta}{\theta};\quad A5=-\frac{A2}{\theta}$$

$$A6=1-\frac{3}{\theta};\quad A7=\frac{\Delta t}{2};\quad A8=\frac{\Delta t^2}{6} \tag{7.22}$$

4. Form the effective stiffness matrix

$$[\widehat{K}]=[K]+A0[M] \tag{7.23}$$

5. Carry out the reduction of equation to upper triangular form $[\widetilde{K}]$.
6. For each time step, the effective load at time $t+\theta\,\Delta t$ is calculated as

$$\begin{aligned}\{\hat{f}_{t+\theta\,\Delta t}\}&=\{f_t\}+\theta(\{f_{t+\Delta t}\}-\{f_t\})\\&\quad+[M](A0\{a_t\}+A2\{\dot{a}_t\}+2\{\ddot{a}_t\})\end{aligned} \tag{7.24}$$

7. Form the following equation to determine $\{a_{t+\theta\,\Delta t}\}$.

$$[\widetilde{K}]\{a_{t+\theta\,\Delta t}\}=\{\hat{f}_{t+\theta\,\Delta t}\} \tag{7.25}$$

8. Determine accelerations, velocities and displacements at time $t + \Delta t$.

$$\begin{aligned}
\{\ddot{a}_{t+\Delta t}\} &= A4(\{a_{t+\theta\,\Delta t}\} - \{a_t\}) + A5\{\dot{a}_t\} + A6\{\ddot{a}_t\} \\
\{\dot{a}_{t+\Delta t}\} &= \{\dot{a}_t\} + A7(\{\ddot{a}_{t+\Delta t}\} + \{\ddot{a}_t\}) \\
\{a_{t+\Delta t}\} &= \{a_t\} + \Delta t\,\{\dot{a}_t\} + A8(\{\ddot{a}_{t+\Delta t}\} + 2\{\ddot{a}_t\})
\end{aligned} \tag{7.26}$$

$$\{\ddot{a}_{t+\theta\,\Delta t}\} = \frac{6}{\theta^2\,\Delta t^2}(\{a_{t+\theta\,\Delta t}\} - \{a_t\}) - \frac{6}{\theta\,\Delta t}\{\dot{a}_t\} - 2\{\ddot{a}_t\}.$$

To solve the transient dynamic contact problem which is nonlinear, the dynamic equations of motion for the two bodies are separately considered. Each region further consists of a general domain and a contact zone. The dynamic equations are condensed out for the contact zone only for the two bodies A and B. Equations of equilibrium and compatibility along with the constitutive relations for contact behaviour are set up to obtain the solution for the behaviour through a heuristic iterative procedure. Consider equation (7.19) at $t + \theta\,\Delta t$. Then from the last equation of (7.26), we get

$$[P^B]\begin{Bmatrix} \{a^{C'}_{t+\theta\,\Delta t}\} \\ \{a^B_{t+\theta\,\Delta t}\} \end{Bmatrix} = [Q^B]\begin{Bmatrix} \{a^{C'}_t\} + \theta\,\Delta t\,\{\dot{a}^{C'}_t\} + \frac{\theta^2\,\Delta t^2}{3}\{\ddot{a}^{C'}_t\} \\ \{a^B_t\} + \theta\,\Delta t\,\{\dot{a}^B_t\} + \frac{\theta^2\,\Delta t^2}{3}\{\ddot{a}^B_t\} \end{Bmatrix} + \begin{Bmatrix} \{f^{C'}_{t+\theta\,\Delta t}\} \\ \{f^B_{t+\theta\,\Delta t}\} \end{Bmatrix} \tag{7.27}$$

where,

$$[P^B] = \frac{6}{\theta^2\,\Delta t^2}[M^B] + [K^B] = \begin{bmatrix} P'_{CC} & P_{CB} \\ P_{BC} & P_{BB} \end{bmatrix}$$

$$[Q^B] = \frac{6}{\theta^2\,\Delta t^2}[M^B] = \begin{bmatrix} Q'_{CC} & Q_{CB} \\ Q_{BC} & Q_{BB} \end{bmatrix}$$

For body A, similarly, we get

$$[P^A]\begin{Bmatrix} \{a^C_{t+\theta\,\Delta t}\} \\ \{a^A_{t+\theta\,\Delta t}\} \end{Bmatrix} = [Q^A]\begin{Bmatrix} \{a^C_t\} + \theta\,\Delta t\,\{\dot{a}^C_t\} + \frac{\theta^2\,\Delta t^2}{3}\{\ddot{a}^C_t\} \\ \{a^A_t\} + \theta\,\Delta t\,\{\dot{a}^A_t\} + \frac{\theta^2\,\Delta t^2}{3}\{\ddot{a}^A_t\} \end{Bmatrix} + \begin{Bmatrix} \{f^C_{t+\theta\,\Delta t}\} \\ \{f^A_{t+\theta\,\Delta t}\} \end{Bmatrix} \tag{7.28}$$

where,

$$[P^A] = \begin{bmatrix} P_{CC} & P_{CA} \\ P_{AC} & P_{AA} \end{bmatrix}$$

$$[Q^A] = \begin{bmatrix} Q_{CC} & Q_{CA} \\ Q_{AC} & Q_{AA} \end{bmatrix}$$

If the pairing nodes i_A and i_B are in contact, then

$$\begin{aligned} \{a^C_{t+\theta\,\Delta t}\} &= \{a^{C'}_{t+\theta\,\Delta t}\} \\ \{f^C_{t+\theta\,\Delta t}\} &= -\{f^{C'}_{t+\theta\,\Delta t}\} = \{p^C_{t+\theta\,\Delta t}\} \end{aligned} \tag{7.29}$$

Imposing the above conditions to equations (7.27) and (7.28), we get

$$\begin{aligned} &[C^*_B + C^*_A]\{p^C_{t+\theta\,\Delta t}\} = \\ &\quad [C^*_A S^*_1 - C^*_B Q^*_1]\Big\{\{a^C_t\} + \theta\,\Delta t\,\{\dot{a}^C_t\} + \frac{\theta^2\,\Delta t^2}{3}\{\ddot{a}^C_t\}\Big\} \\ &\quad + C^*_A S^*_2\Big\{\{a^A_t\} + \theta\,\Delta t\,\{\dot{a}^A_t\} + \frac{\theta^2\,\Delta t^2}{3}\{\ddot{a}^A_t\}\Big\} \\ &\quad - C^*_B Q^*_2\Big\{\{a^B_t\} + \theta\,\Delta t\,\{\dot{a}^B_t\} + \frac{\theta^2\,\Delta t^2}{3}\{\ddot{a}^B_t\}\Big\} \\ &\quad - C^*_A P_{CA} P^{-1}_{AA}\{f^A_{t+\theta\,\Delta t}\} \end{aligned} \tag{7.30}$$

where,

$$\begin{aligned} C^*_A &= \left[P_{CC} - P_{CA}P^{-1}_{AA}P_{AC}\right]^{-1} & C^*_B &= \left[P'_{CC} - P_{CB}P^{-1}_{BB}P_{BC}\right]^{-1} \\ Q^*_1 &= Q_{CC} - P_{CA}P^{-1}_{AA}Q_{AC} & Q^*_2 &= Q_{CA} - P_{CA}P^{-1}_{AA}Q_{AA} \\ S^*_1 &= Q'_{CC} - P_{CB}P^{-1}_{BB}Q_{BC} & S^*_2 &= Q_{CB} - P_{CB}P^{-1}_{BB}Q_{BC} \end{aligned}$$

The solution of linear equation (7.30) directly yields the unknown contact forces at the nodes. This can be used at any particular time instant to obtain the nodal contact forces, if the nodes in contact are known *a priori*. However, this is usually unknown and some assumptions are necessary. These assumptions are to be further confirmed by the signs of interface forces which have to be compressive. If some of these are found to be tensile, node opening occurs and these nodes have to be excluded from the contact zone, during which further redistribution of contact pressures occur. The contact nodes have to be further checked for slip and the modelling of slip has to be done accurately.

Simulation of the Opening

At a particular time instant, given the right hand side of equation (7.30), the nodal contact forces can be first computed by solving for $\{p^C_{t+\theta\,\Delta t}\}$. If some of the nodes exhibit tensile forces, these cannot be supported by the contact zone. Hence the nodal forces have to be recomputed by deleting these nodes from the list of original contact nodes and the interface forces have to be computed again.

Simulation of Slip

At any particular time instant, after the nodes in contact are firmly established, the contact nodes have to be checked for conditions of slip. Thus for all the nodes, the ratio of the tangential to the normal load force is to be computed. If this ratio exceeds the prescribed coefficient of dynamic friction then the no-slip condition assumed is not valid. Hence the compatibility condition along the tangential direction is to be replaced by equation (7.16), for the particular node under consideration. Then equation (7.30) gives rise to $2m$ equations corresponding to m nodes in contact in $2m$ variables, viz., the unknown contact forces. When certain nodes are found to slip, slip conditions have to be imposed, thereby giving rise to as many equations as there are unknowns and an iterative procedure is to be used to determine the contact forces accurately.

Contact along Inclined Surfaces

If the contact is along inclined surface, the conditions of compatibility and equilibrium have to be imposed along the direction $x'y'$. This is easily accomplished by transforming the stiffness matrix corresponding to these nodes along the directions $x'y'$, see Fig. 7.13. The transformed stiffness matrices are then assembled and condensed in arriving at the final form of equation (7.30), to determine the contact forces along the tangential and normal directions. It is obvious that an efficient computing procedure should be developed to reduce the number of iterations.

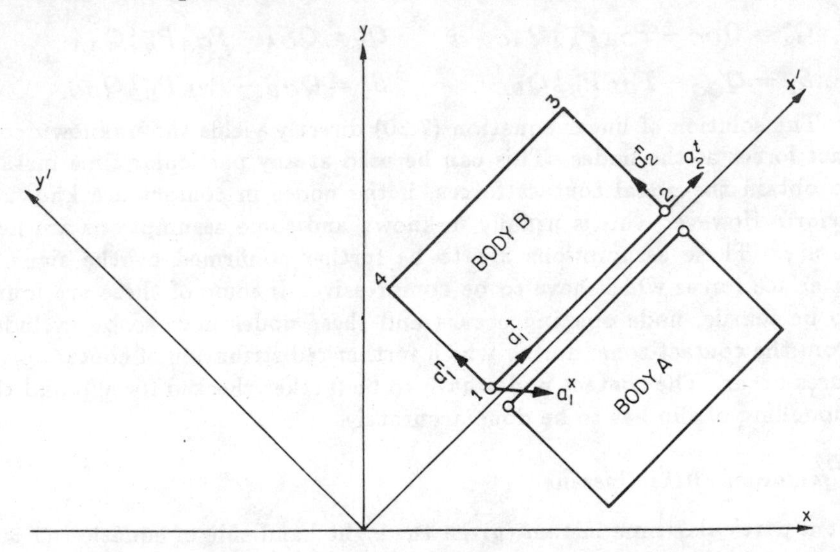

Fig. 7.13 Computation of transformation matrix

Consider the T-root blade example discussed earlier for the static contact problem, shown in Fig. 7.7. The contact zone extends over four differentsurfaces shown in Fig. 7.9. No contact is expected along other surfaces on account of the ample clearances present in the assembly. The contact region on the surface $1A$ and $2A$ is modelled using 6 elements, while the contact on surfaces $3A$ and $4A$ is modelled with 2 elements only.

The displacements obtained in the quasistatic analysis constitute the initial conditions for the dynamic problem. The initial velocities and accelerations are assumed to be zero.

The blade in Fig. 7.7 is considered with unit thickness and Young's modulus $= 2.07 \times 10^{11}$ Pa, Poisson's ratio $= 0.3$ and coefficient of dynamic friction $= 0.4$. The blade has an overall length of 29 cm and width 5 cm. The excitation due to suddenly applied load is as shown in Fig. 7.14. For an axial load of 9820 N and a transverse load of 196 N, the tip displacement is obtained as a function of time. A time step of 0.0002 sec was used so that at least 20 such steps are included in one cycle of response. The natural period was found to be 0.00432 sec. Fig. 7.14 shows that the decay in response is clearly due to Coulomb friction model. The average damping ratio is estimated as 0.0022.

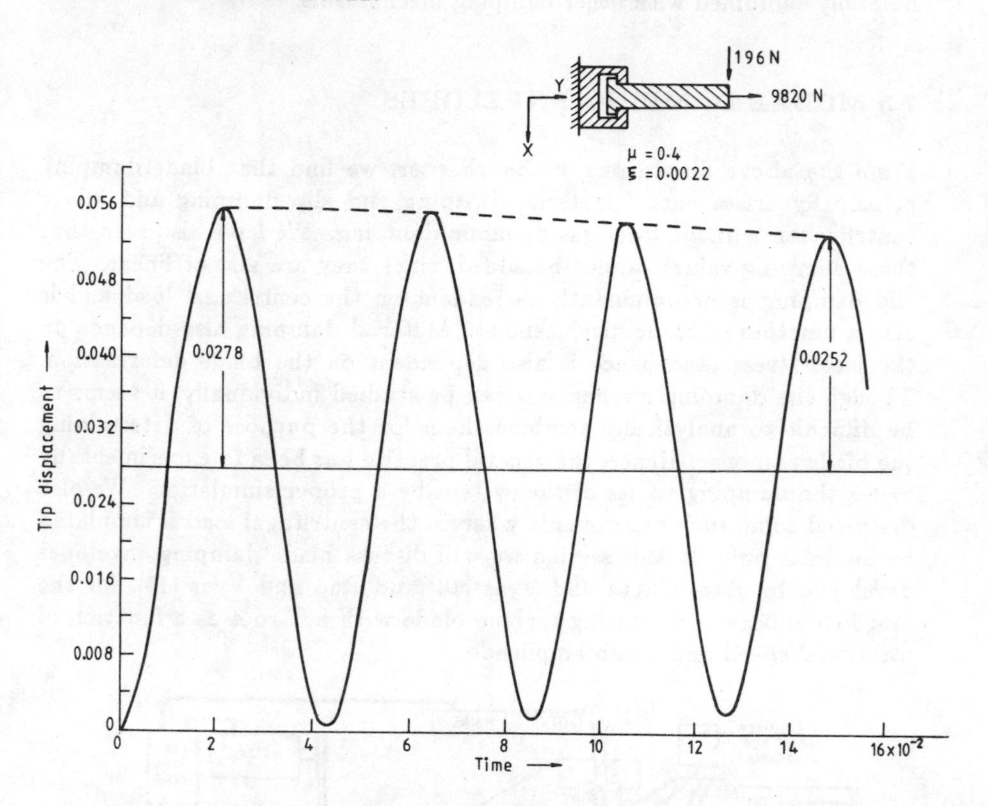

Fig. 7.14 Tip displacement vs time curve

When the axial load is increased to 11780 N, the equivalent viscous damping was found to be 0.00158, showing thereby, that the blade is getting locked in the root of the disc. At 14730 N axial load, very little difference in the successive amplitudes was found, showing that there is practically no slip between the blade and the root.

When the axial load was further increased to 19640 N the value of damping ratio actually increased to 0.0033. This behaviour is attributed to the deformation pattern of the joint assembly as shown in Fig. 7.11. While the surfaces 1A and 2A, which are perpendicular to the centrifugal load and therefore get locked under increased axial pull, the two surfaces 3A and 4A

being parallel to the application of axial pull, now slip and contribute to the additional damping in the system.

From a study of the straddle T-root case under similar conditions of T-root case above it is found that in general the straddle root provides more damping as generally observed in practice. For example, when the axial load is 9820 N, the damping ratio improved from 0.0022 to 0.016.

The finite element analysis is therefore useful in studying the type of roots that may provide optimum damping under certain conditions of operation. However, the problem is complex and time consuming and cannot be easily combined with other damping mechanisms.

7.3 MODAL DAMPING ENVELOPES

From the above discussions in this chapter, we find that blade damping principally arises out of material damping and slip damping and minor contribution is made from gas dynamic damping. We have also seen that these damping values cannot be added, since they are all not linear. The slip damping is predominantly dependent on the centrifugal load and is also a function of blade displacement. Material damping also depends on the local stress, and hence is also dependent on the blade deformation. Though the damping mechanisms can be studied individually, it seems to be difficult to analytically combine them for the purpose of determining the blade response. Hence, the general practice has been to experimentally assess the damping values of the system by a proper simulation. We also discussed some such experiments wherein the centrifugal load is simulated by an axial pull. In this section we will discuss blade damping envelopes developed by Rao, Gupta and Vyas [18] and Rao and Vyas [19], for the first four modes of a rotating turbine blade with a T-root as a function of rotational speed and strain amplitude.

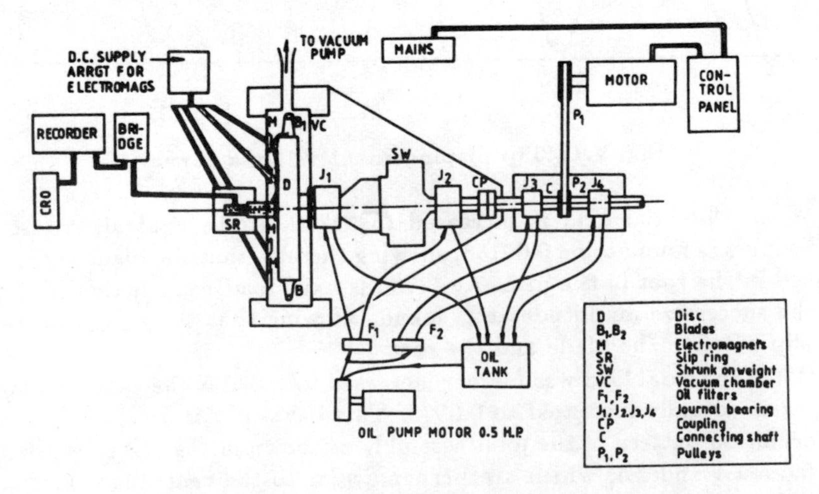

Fig. 7.15 Schematic layout of the system

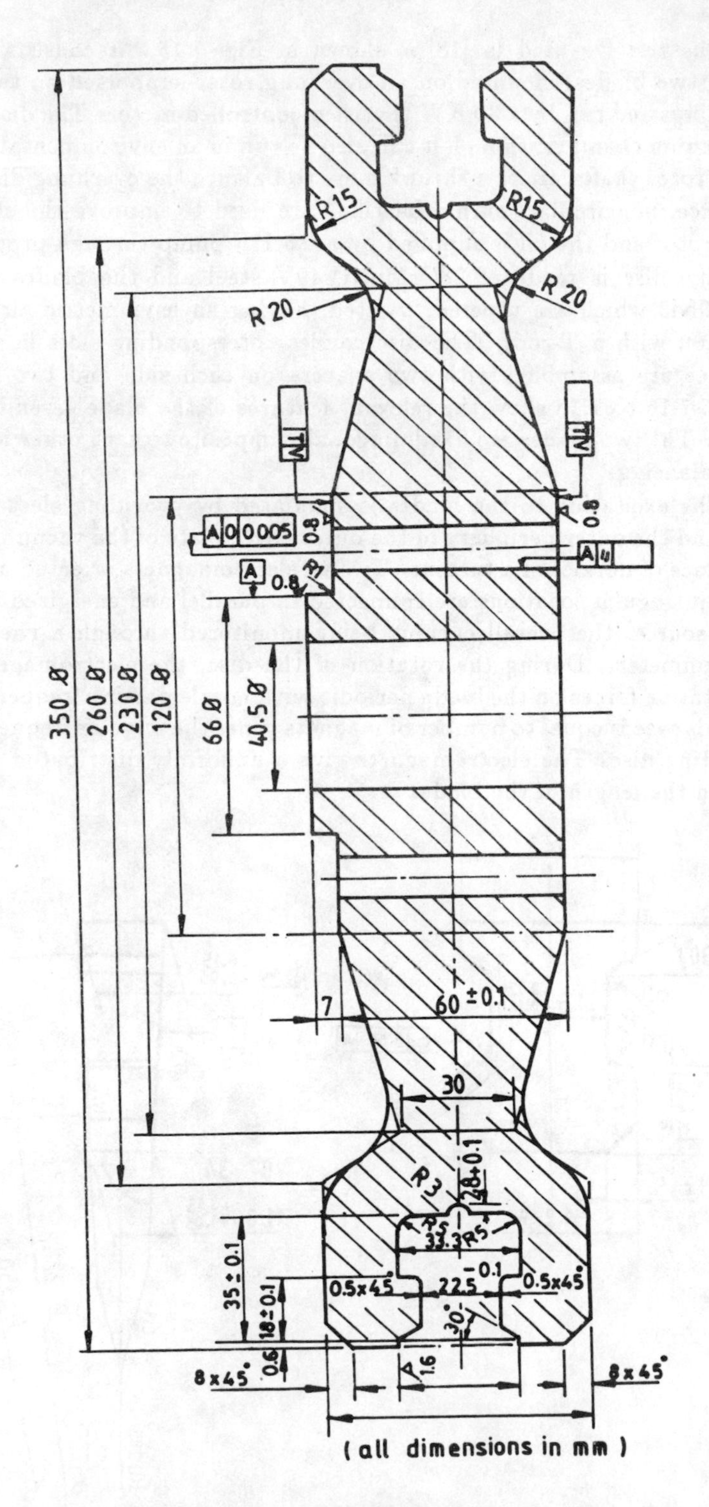

Fig. 7.16 The disc

The test rig used in [18] is shown in Fig. 7.15. It consists of a disc with two blades, mounted on an overhung rotor supported on two journal bearings and run by a 30 KW thyristor controlled motor. The disc is run in a vacuum chamber, though it can also be run in an environmental chamber. The rotor shaft carries a shrunk mass to balance the overhung disc. Multi-surface, noncircular journal bearings are used to improve the stability of the rotor and the oil is supplied by a 0.5 HP pump through proper filters.

The disc is made of 28CrMoNiV49V steel and the blades are made of 40Ni3 which are tapered, twisted, having an asymmetric airfoil cross-section with a T-root. The disc carries corresponding slots in which the blades are assembled with two spacers on each side and two segments. Figs. 7.16 to 7.18 show the relevant features of the blade assembly on the disc. The two blades are fixed diagonally opposite to each other for reasons of balancing.

The excitation to the blades is simulated by providing electromagnets around the outer periphery of the outer cover plate of the vacuum chamber in place of nozzles of a turbine. Twelve electromagnets oriented and placed at equiangular locations are connected in parallel and energized by a 12 V DC source, the overall current being monitored through a rheostat and an ammeter. During the rotation of the disc, the electromagnets cause excitation forces on the blade periodic with nozzle passing frequency, which in this case is equal to number of magnets times the angular frequency of the rotating disc. The electromagnets give a uniformly distributed excitation along the length of the blade.

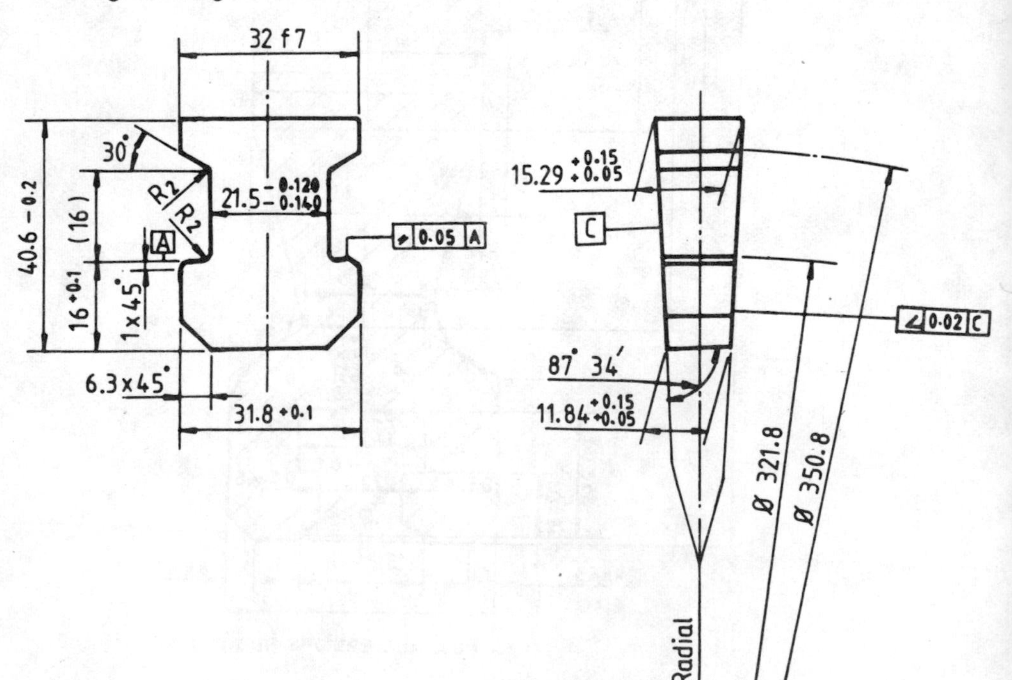

Fig. 7.17 The spacer

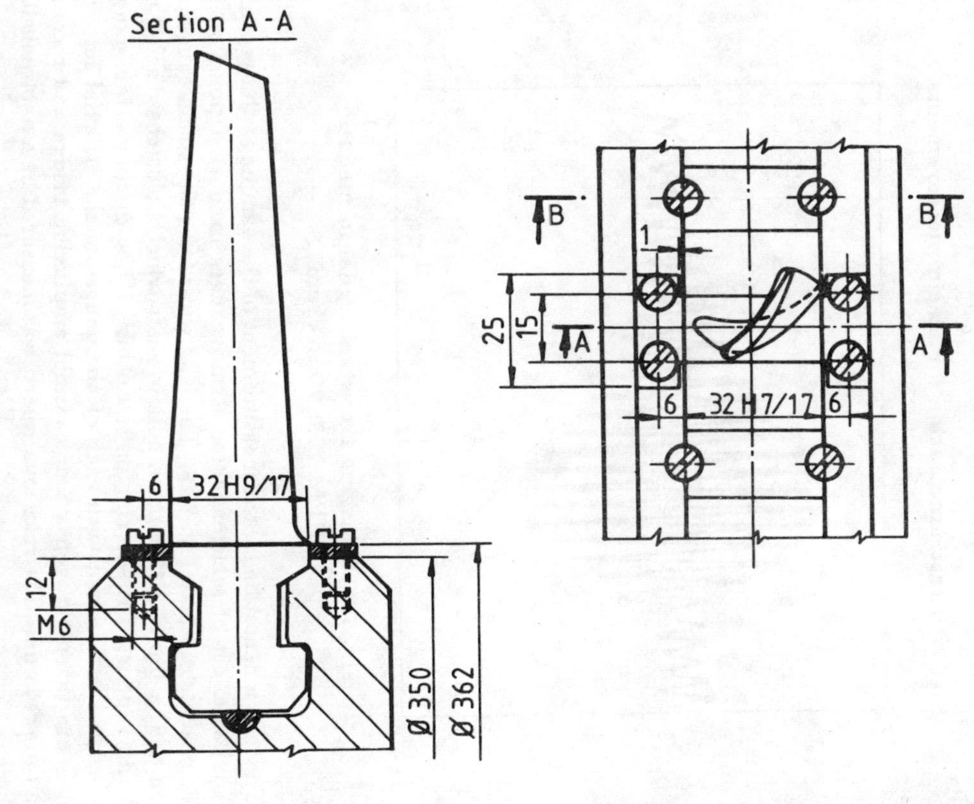

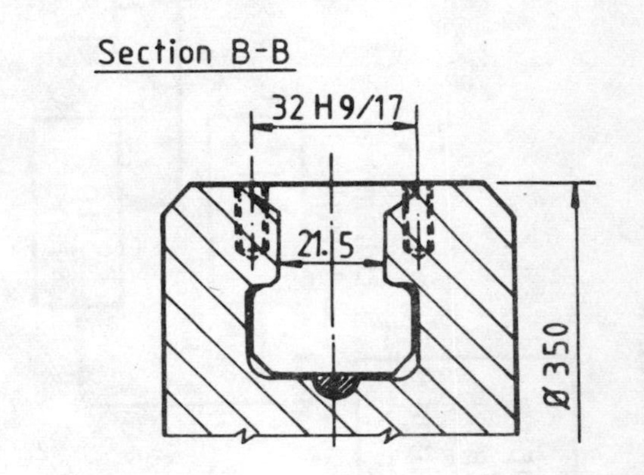

Fig. 7.18 The blade assembly

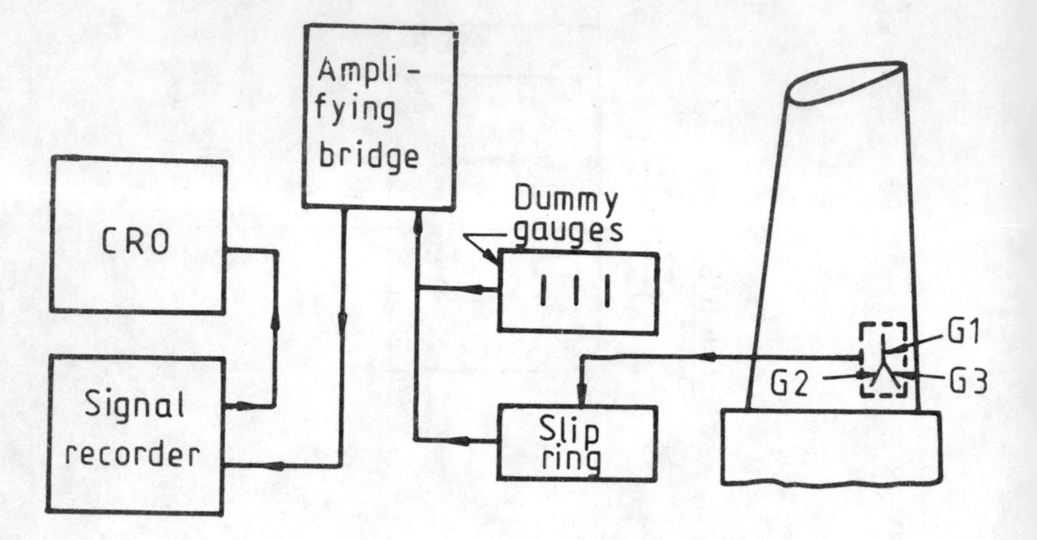

Fig. 7.19 Instrumentation for strain measurements

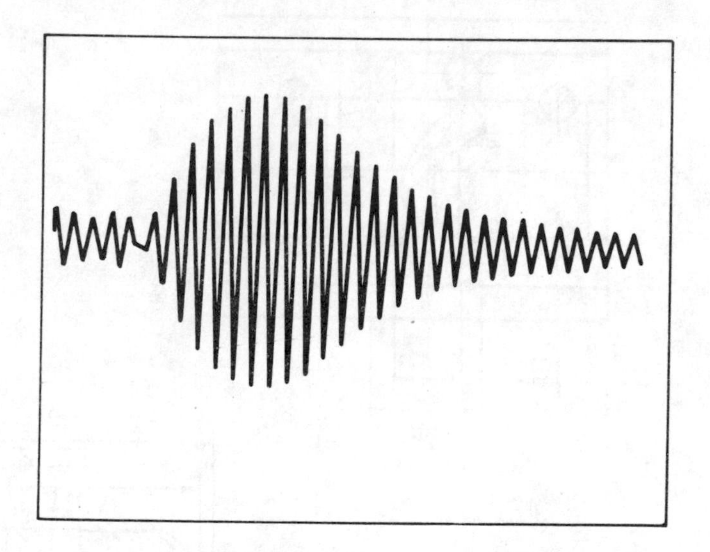

Fig. 7.20 Decaying transient signal in mode I
RPM 700; 20 mV/cm

The instrumentation for the measurement of the damping values consists of a tri-axial set of semiconductor strain gauges fixed at a point on the blade near its root, see Fig. 7.19. The lead wires from the strain gauges are taken to the slip ring through a hollow shaft which holds the disc on one end and supports the slip ring on the other. Using dummy strain gauges, half bridges are formed with each of the gauges in the tri-axial set. The signals are taken through a four channel amplifying bridge and recorded on a tape recorder. The rotational speed was measured by a photosensitive pick-up.

To measure the modal damping, the rotor is simply run at a desired speed with the electromagnetic excitation on. The excitation is suddenly switched off at a convenient time, so that the transients are set up in the blade. The entire process of the initiation and decay of the transient response caused due to the shutting-off the magnetic excitation, is recorded through one of the strain gauges of the tri-axial set. From the decaying part of the overall signal, the response for a particular natural mode is filtered through a tunable filter and played on a dual trace oscilloscope, see Fig. 7.20.

The equivalent viscous damping values are determined for different amplitude values, cycle by cycle, for the decaying part of the signal. Figs. 7.21 to 7.24 show the first four modal damping ratios as a function of strain amplitude at different rotor speeds. Fig. 7.25 shows the average damping ratios for the first four modes as a function of rotor speed only, obtained by using the entire decay signal.

It can be observed that the modal damping values in all the four modes show similar trends of variation with the strain amplitude and speed of rotation. As is well established in literature, the results show that the damping values progressively decrease with the speed of rotation. However an interesting freature revealed by the results is the existence of a threshold speed, at which the modal damping values start decreasing rather rapidly. For the case under study, the threshold speed is about 400 rpm, see Fig. 7.25. It can be observed that below this rpm, the variation in modal damping with the strain amplitude, particularly in mode I, is negligible, indicating a predominantly root damping effect. For rotational speeds 700 rpm and above, the root damping effect appears to be negligible and the modal damping values increase with the strain amplitudes. The strain amplitudes are quite small, since the gauges were located very close to the root. From Fig. 7.25, it is observed that for all rotational speeds, the modal damping values are slightly higher in higher modes. Below the threshold speed of 400 rpm, the modal damping values range from 3 to 5 per cent. This data will be used in the next two chapters to determine the steady and transient response of a turbine blade subjected to nozzle passing excitation.

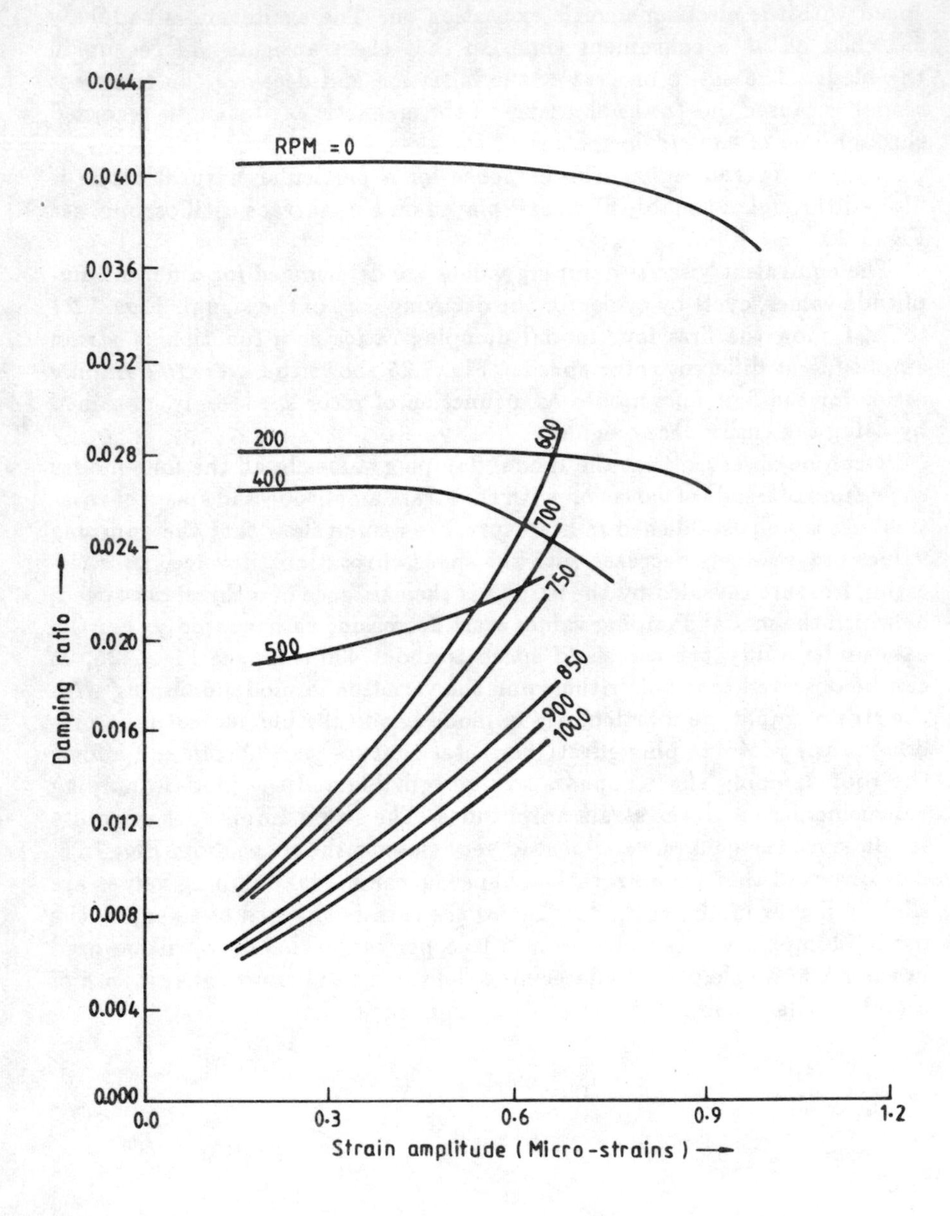

Fig. 7.21 Damping ratio vs strain amplitude — mode I

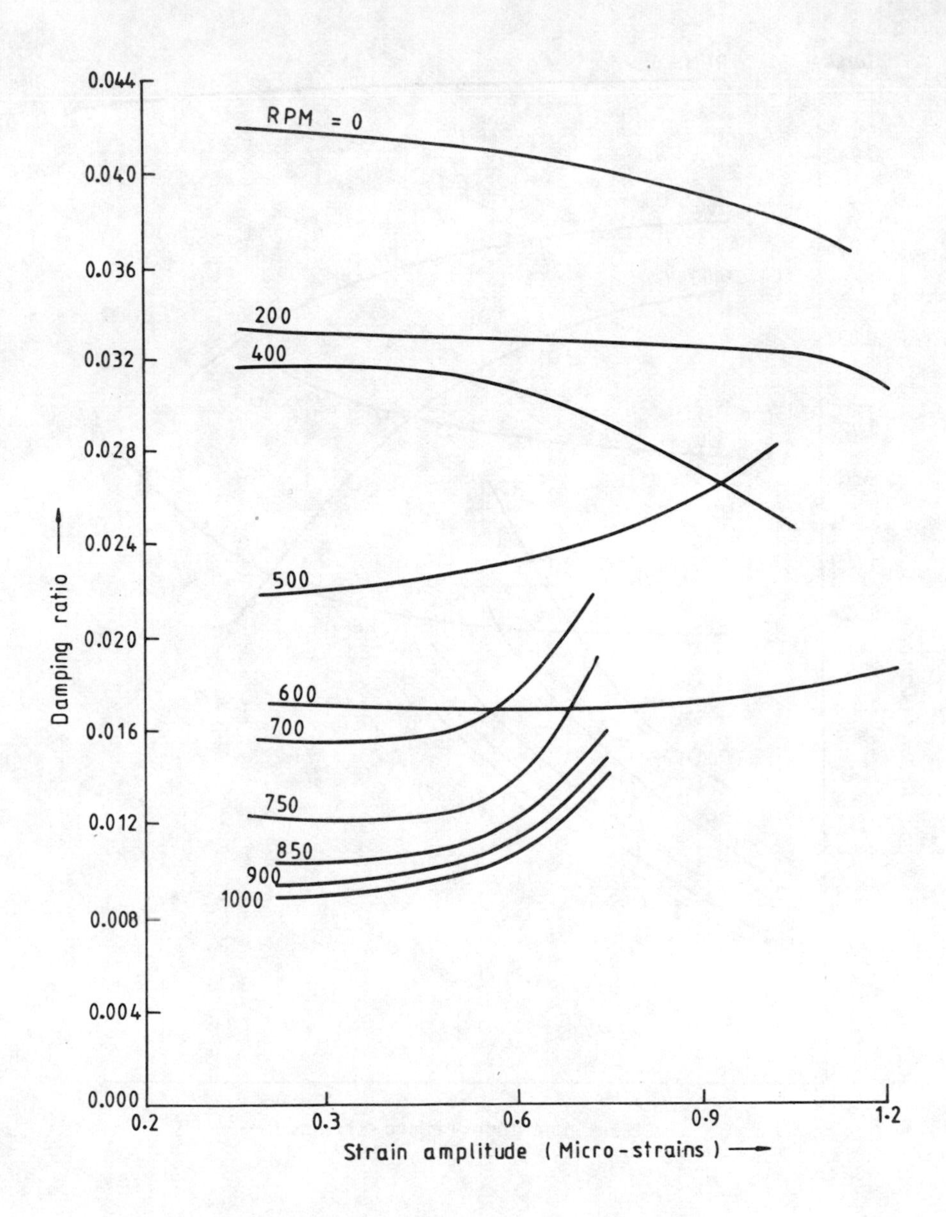

Fig. 7.22 Damping ratio vs strain amplitude — mode II

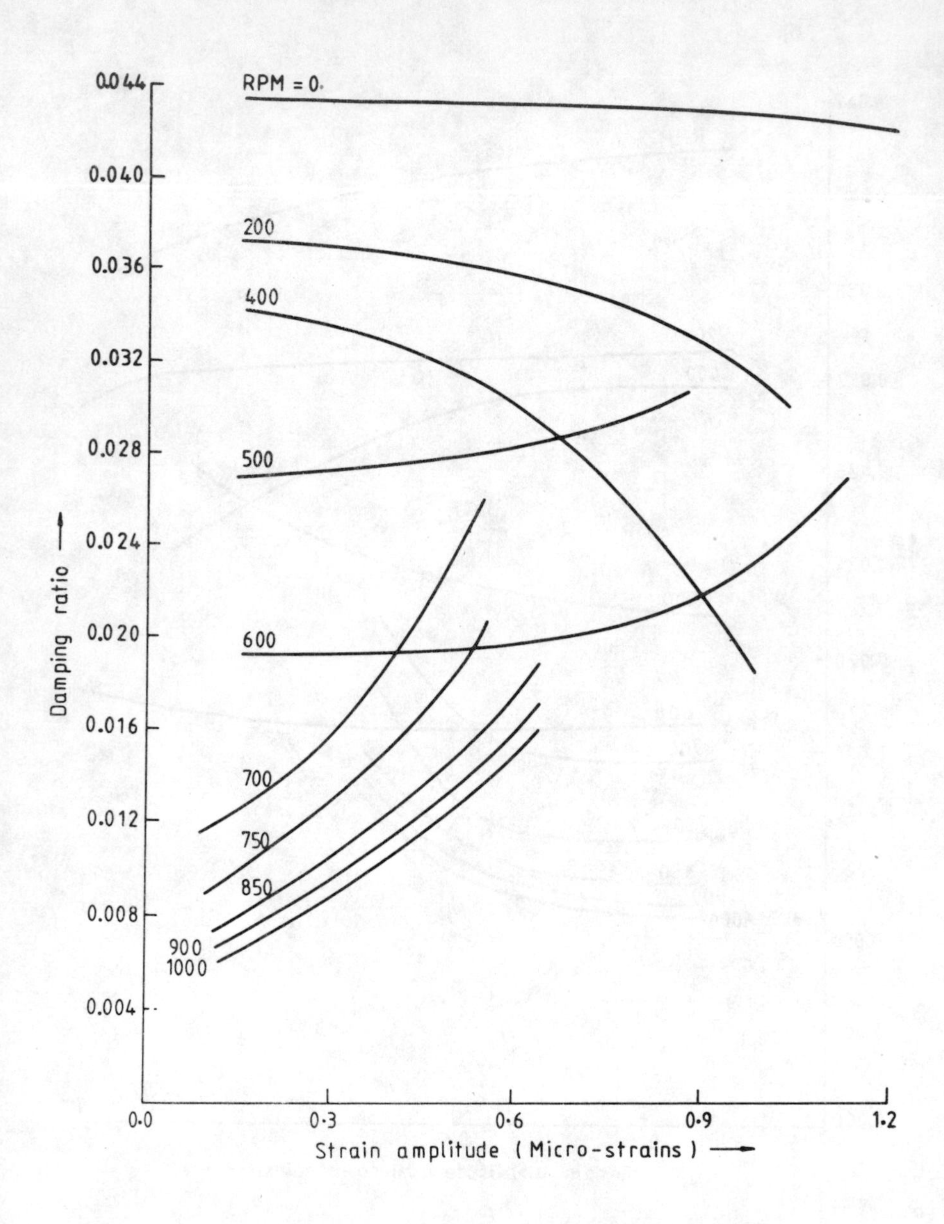

Fig. 7.23 Damping ratio vs strain amplitude — mode III

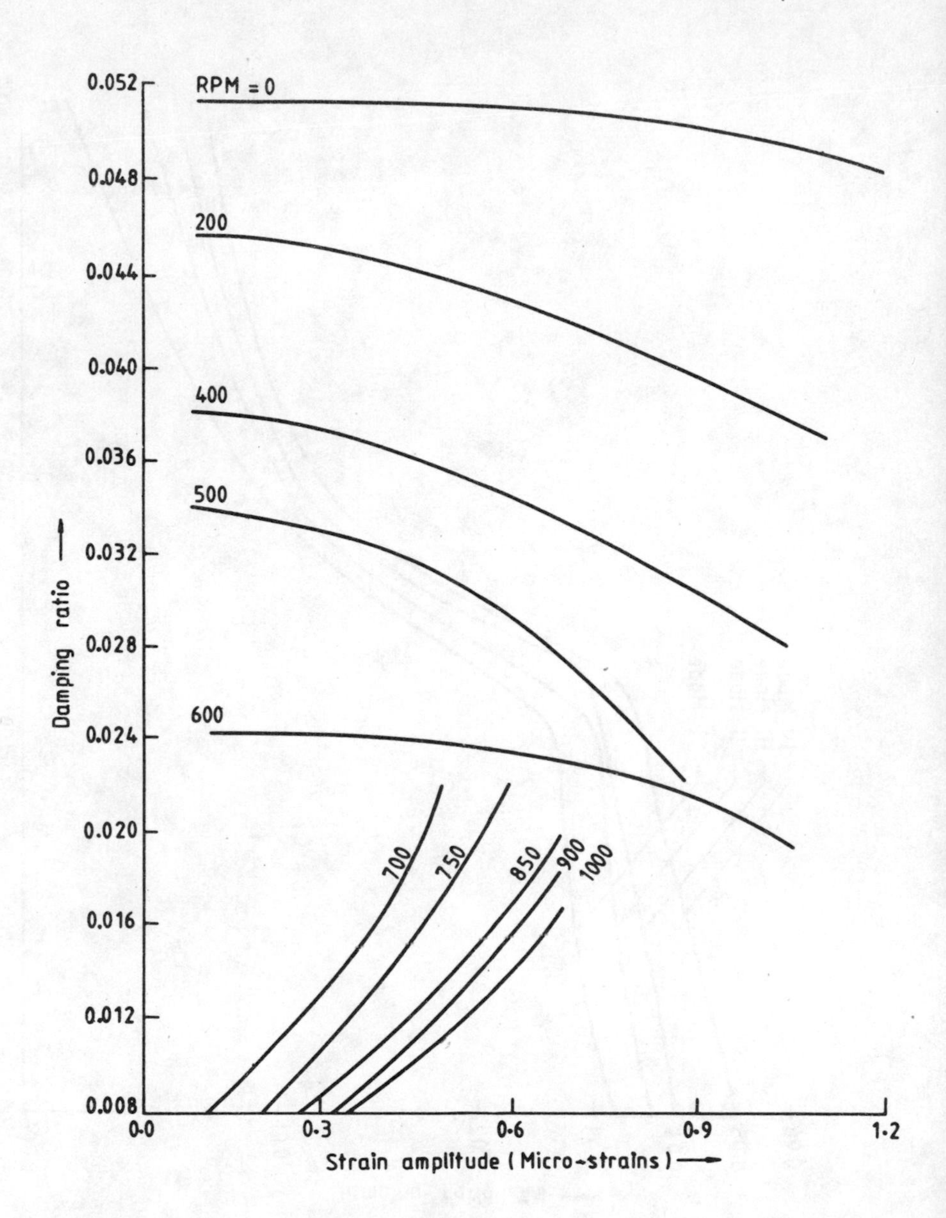

Fig. 7.24 Damping ratio vs strain amplitude — mode IV

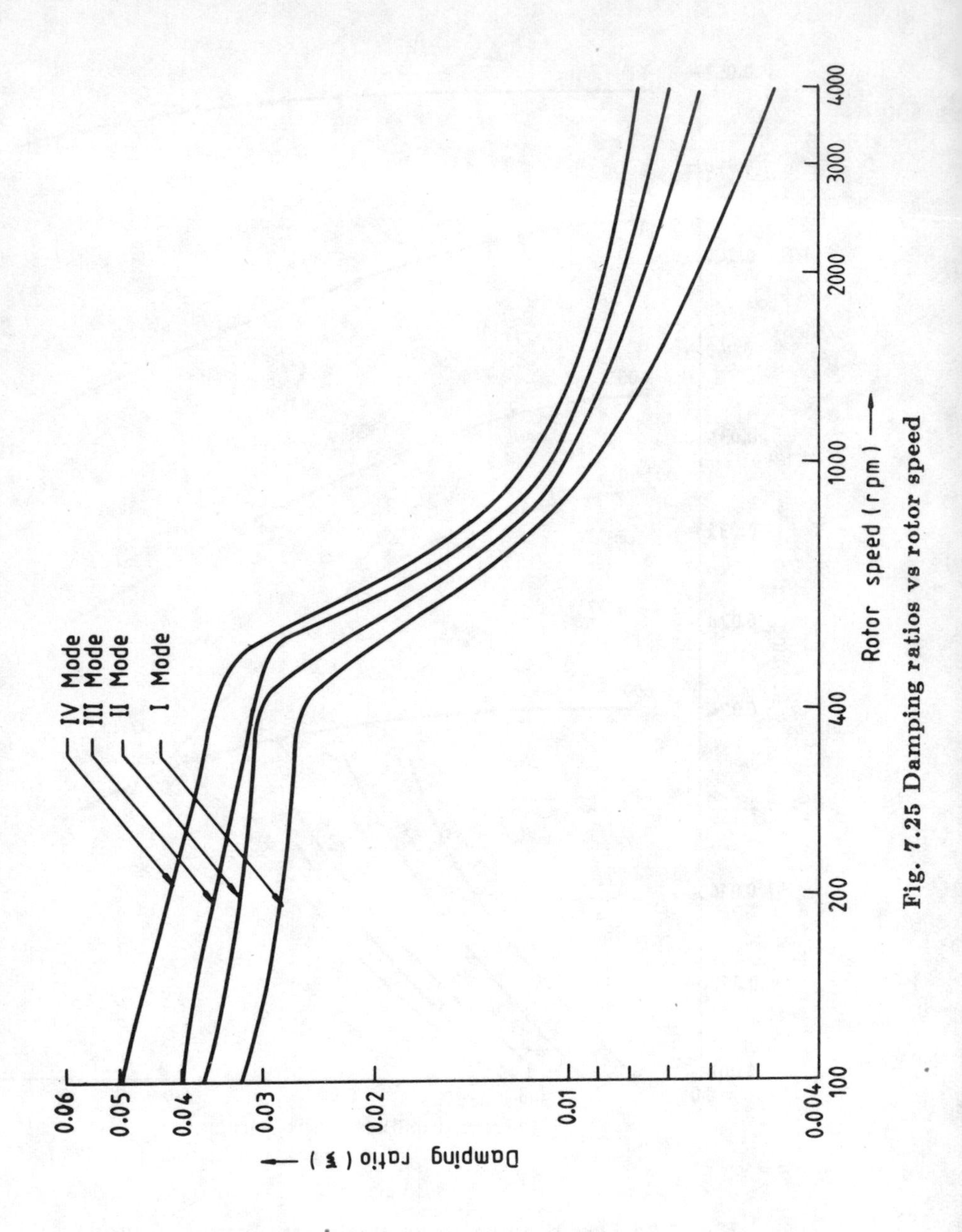

Fig. 7.25 Damping ratios vs rotor speed

REFERENCES

1. Rowett, F.E., "Elastic Hysteresis in Steel", *Proceedings of Royal Society*, **89**, (1914).
2. Den Hartog, J.P., "Forced Vibrations with Combined Viscous and Coulomb Damping" *Phil. Mag.*, **9**, Series **7**, p. 801 (1930).
3. Jacobsen, L.S., "Steady Forced Vibrations as Influenced by Damping", *Trans. ASME*, **50**, p. 169 (1930).
4. Kimball, A.L. "Vibration Problems, Part 5, Friction and Damping in Vibrations", *Trans. ASME*, **63**, p. A-135 (1941).
5. Shannon, J.F., "Investigation of Failures in Jet Engines", Aeronautics Council of Great Britain, Reports and Memoranda, HMSO (1945).
6. Hanson, M.P., "A Vibration Damper for Axial Flow Compressor Blading", *Proc. Soc. Exptl. Stress Analysis XIV*, p. 155 (1955).
7. Hanson, M.P., Meyer, A.J. and Manson, S.S., "A Method for Evaluating Loose Blade Mountings as a Means of Suppressing Turbine and Compressor Blade Vibrations", *Proc. Soc. Exptl. Stress Analysis*, **10** (2) p. 103 (1953).
8. Goodman, L.E. and Klumpp, J.H., "Analysis of Slip Damping with Reference to Turbine Blade Vibration", *Journal of Applied Mech., Trans. ASME*, **23** (3), p. 421 (1956).
9. Beards, J.E., "Damping in Structural Joints", *Shock Vibration Digest*, **11** (8), p. 35 (1979).
10. Grady, R.F., "Investigations of Dovetail Damping Contribution of Propulsion Steam Turbine Buckets", GE Report No. B5-94390 (November 1967).
11. Grady, R.F., "Investigation of Material Damping Properties of Propulsion Turbine Blade Material", GE Report No. B5-94390 (December 1967).
12. Lazan, B.J., *Damping of Materials and Members in Structural Mechanics*, Pergamon Press, New York (1968).
13. Wagner, J.T., "Blade Damping Tests", Westinghouse Engrg. Report, EC-401, No. BSN00024-67-C-5494 (May 1969).
14. Rieger, N.F. and Beck, C.M., "Damping Tests on Steam Turbine Blades", EPRI Project RP-1185 1, Palo Alto, California (1980).
15. Peck, L.C., Johnson, P.J. and House, D.F., "Evaluation of Turbine Blade Root Designs to Minimize Fatigue Failures", EPRI 1185-2, Palo Alto, CA (1981).
16. Johnson, P.J., "Evaluation of Damping in Free Standing Low Pressure Steam Turbine Blades", EPRI Project 016-2, Palo Alto, California (1982).
17. Rao, J.S., Usmani, M.A.W. and Ramakrishnan, C.V., *Interface Damping in Blade Attachment Region*, Proc. 3rd Intl. Conference Rotor Dynamics, Lyon, p. 185, (September 1990).
18. Rao, J.S., Gupta, K. and Vyas, N.S., "Blade Damping Measurement in a Spin Rig with Nozzle Passing Excitation Simulated by Electromagnets", *Shock & Vib. Bull.*, **56**, Part 2, p. 109 (1986).
19. Rao, J.S. and Vyas, N.S., "Resonant Stress Determination of a Turbine Blade with Modal Damping as a Function of Rotor Speed and Vibrational Amplitude", *ASME*, 89-GT-27.
20. Jones, D.I.G. and Muszynska, A., "Vibrations of a Compressor Blade with Slip at the Root", *Shock & Vib. Bull*, **48**, Pt. 2, p. 53 (1978).
21. Williams, E.J. and Earles, S.W.E., "Optimization of the Response of Frictionally Damped Beam Type Structures with Reference to Gas Turbine Compressor Blading", *ASME*, Paper No. 73-Det-108.

22. Muszynska, A. and Jones, D.I.G., "On Discrete Modelisation of Response of Blades with Slip and Hysteretic Damping", *Proc. Fifth World Congress on Theory of Machines and Mechanisms*, p. 641 (1979).
23. Jones, D.I.G. and Muszynska, A., "Design of Turbine Blades for Effective Slip Damping", *Shock & Vib. Bull.*, **49**, Pt. 2 (1979).
24 Muszynska, A., Jones, D.I.G., Lagnese, T. and Whitford, L., "On Nonlinear Response of Multiple Blade Systems", *Shock and Vibration Symposium*, San Diego, California (October 1980).
25. Srinivasan, A.V., Cutts, D.G. and Sridhar, S., "Turbojet Engine Blade Damping", United Technologies Report No. R 81-91441031, Submitted to NASA (Report No. CR 165406) (July 1981).
26. Rao, J.S. and Gupta, K., *Theory and Practice of Mechanical Vibration*, Wiley Eastern, New Delhi (1984).
27. Huebner, K.H., *The Finite Element Method for Engineers*, John Wiley & Sons (1975).
28. Zienkiwicz, O.C., *The Finite Element Method*, Tata McGraw-Hill (1979).
29. Bath, K.J. and Wilson, E.L., *Numerical Methods in Finite Element Analysis*, Prentice Hall of India (1978).
30. Schmidt, R. and Sollman, H., "Schaufelschwingungen Axialer Turbomaschinen", Band B, Technische Universitat, Dresden, Heft 5/1978.

Chapter 8

Forced Vibrations

8.1 INTRODUCTION

Different sources of excitation of a turbine blade are discussed in Chapter 6. The major source of excitation under steady operating conditions is due to stage interference at nozzle passing frequency and its harmonics. Different sources of damping in a blade are discussed in Chapter 7. Under resonant or near resonant conditions of operation, the vibratory amplitudes are controlled by dissipation of energy.

Plunkett [1] studied free and forced vibrations of simple rotating cantilever blades. Schnittger [2] used a simplified analysis in which a single stiff blade with one translational and one pitching mode is studied. A significant contribution to calculate the stresses in a banded group of turbine buckets was made by Prohl [3]. The method was based on the assumption that damping is small and the energy input to the banded group from a prescribed nozzle stimulus is equated to the energy dissipated in damping to determine vibration and stress levels at resonance. The calculations for typical blade groups were presented by Weaver and Prohl [4]. Rao and Carnegie [5, 6] determined the response of a rotating cantilever blade due to a sinusoidal excitation function at the blade tip by the Ritz method; Coriolis components were included and viscous damping is assumed. Jones [7] considered a simple analytical model of a vibrating jet engine compressor blade using modal analysis; he allowed for slip at the root. The analysis predicted the minimum frictional force needed to avoid infinite amplitudes under high level excitation. The analysis was extended to a two mass model; results were nonlinear when slip occurred beyond threshold level at the root [8]. Jones and Muszynska [9] refined the model by assuming that one mass moves in the space between two elastic non-inertial buffers; the motion of the masses is damped by dry friction, fluid pumping in gaps and hysteretic damping of elastic members. An external harmonic force was considered. Approximate nonlinear equations of motion were derived for a blade with low frictional forces at high speeds; a gap introduced closed up only at high speeds [10]. Slip at the root and hysteresis of the blade were considered further [11] and compared with experimental response data. A refined discrete model was given subsequently [12]. Response of the airfoil cross-section of rotating blades due to flow excitation has been studied by Rao et al [13] using the Rayleigh-Ritz method; viscous damping was assumed. Forced and self-excited vibrations of blades in a stage due to flow interference have been considered [14]. MacBain [15] used the finite element method to determine transient response of a cantilever plate subjected to normal impact by a ballistic pendulum. The ANSYS program has been used to determine dynamic stress of a blade group due to flow path excitation [16]. Salama and Petyt [17] used the finite element method to determine the response of a packet of blades to harmonic excitation of different engine orders. Component modal analysis has been used for a typical fan design by Srinivasan et al [18]; viscous damping due to rubbing of the shroud interfaces was included. Fabunmi [19] used a semi-empirical method that utilized modal analysis and was based on experimentally obtained mode shapes; he determined the vibration response of a 23-bladed axial compressor rotor.

Kimberling [20] considered slip damping at circular root geometry and determined the response both analytically and experimentally. The available unsteady force coefficients were used by Matsuura [21] to determine bending response of blades and checked the results with experimental values. Jones [22] gave a simple analysis for the response of a blade with slip at the root and checked his results with experimental observations; Griffin [23] investigated the effect of slipping force and stiffness of a damper on the resonant response both analytically and experimentally. Fu-Sheng [24] considered friction and material damping and optimized Coulomb damping for resonant operating conditions.

Hoyniak and Fleeter [25] made an energy balance between unsteady aerodynamic work and energy dissipated due to aerodynamic damping to predict blade resonant vibrations; the forced response is determined by using experimental unsteady aerodynamic gust data for flat and cambered cascade airfoils. Tardos and Botman [26] also considered aerodynamic interaction of a stage to determine the dynamic response. The coupled bending-torsion vibration response was determined by Rao [27] due to stage interference effects in incompressible flow. Jadvani and Rao [28] used Lagrange's equations and modal analysis to determine forced vibrations of rotating pre-twisted blades. Namura [29] described a test turbine to determine resonant stresses due to nozzle passing and impulse excitation; their measured and calculated results agreed well. An experimental test rig with nozzles simulated by permanent magnets is described by Rao and Jadvani [30] to test pre-twisted blades; both analytical and experimental results were compared and good agreement observed. Based on Lagrange's equations and modal analysis, Jadvani [31] gave a general forced vibration program due to nozzle passing frequency for pre-twisted asymmetric tapered rotating blades. Rao et al [32] used unsteady lift coefficients based on Kemp and Sears model to determine self-excited vibration of turbomachine blades. Sisto and Chang [33] considered the gyroscopic forces on the dynamic behaviour of rotating blades; Coriolis forces were also included in the analysis [34]. The instabilities due to harmonic variation with time of the precessional rate due to whirling and other causes were given by

Sisto, Chang and Sutcu [35]. Sivaneri and Chopra [36] used finite element method based on Hamilton's principle to study the dynamic stability of rotor blades with quasi-steady 2D airfoil theory. Kline [37] used Prohl's energy method to determine the forced vibrations and dynamic stresses of a typical blade group. A Goodman diagram fatigue criterion is applied to evaluate the probability of blade group failure.

Rao and Jadvani [38] used tip excitation to determine the vibratory response of twisted tapered rotating blades; the analytical results were verified by experimental values obtained from a spin rig operating in vacuum. This analysis has been extended to distributed forces along the blade length [39]. Vibratory stresses have been determined experimentally with the nozzle passing excitation simulated by electromagnets [40]; average modal damping was used in the theoretical model obtained by Reissner's energy principle. The agreement between theory and experiment is found to be satisfactory. Rao and Vyas [41] extended this analysis to a nonlinear damping model by considering the modal damping to be a function of rotational speed and tip displacement; the theoretical results improved dramatically, showing a good agreement with experimental values.

Menq and Griffin [42] used finite element analysis to determine the forced response of a frictionally damped beam. The response of such a beam to variable normal loads was obtained later by Menq, Griffin and Bielak [43]. Fu-Sheng and Mosimann [44] gave an analysis for determining resonant stresses in turbine blades with friction damping. Sinha and Griffin [45] studied the effects of static friction on the forced response of frictionally damped turbine blades.

In this chapter, the Lagrangian procedure adopted in Chapter 2 is extended to illustrate the procedure of determining the response of a general beam type of blade due to nozzle passing frequency and its harmonics. For blade groups, Prohl's energy method is described to determine resonant response factors. Reissner's energy method is then illustrated to determine the forced vibration stress response by considering the average modal damping values as well as nonlinear modal damping characteristics of a turbine blade.

8.2 MODAL ANALYSIS OF CONTINUOUS SYSTEMS

The potential and kinetic energy expressions for a general blade using beam theory are derived in Section 2.2. For simplicity in the analysis the damping is assumed to be viscous and proportional (C_v). Thus the energy dissipated is

$$W_d = \frac{C_v}{2}\int_0^L A(z)\left[(\dot{x}+r_y(z)\dot{\theta})^2+(\dot{y}-r_x(z)\dot{\theta})^2\right]dz$$
$$+\frac{C_v}{2}\int_0^L I_p(z)(\dot{\theta})^2dz-\frac{C_v}{2}\omega^2\int_0^L\Big\{(x'+r_y'(z)\theta+r_y(z)\theta')^2$$
$$+(y'-r_x'(z)\theta-r_x(z)\theta')^2\Big\}(RI_1+I_2)dz$$

$$+\frac{C_v}{2}\omega^2\sin^2\phi\int_0^L A(z)(x+r_y(z)\theta)^2dz$$
$$+\frac{C_v}{2}\omega^2\cos^2\phi\int_0^L A(z)(y-r_x(z)\theta)^2dz$$
$$-\frac{C_v}{2}\omega^2\sin 2\phi\int_0^L A(z)(x+r_y(z)\theta)(y-r_x(z)\theta)dz \tag{8.1}$$

The excitation force is defined by components $F_x(z,t)$, $F_y(z,t)$ in x and y directions and the moment $M(z,t)$. The work done by the external forces therefore is given by

$$W_f=\int_0^L(F_x x+F_y y+M\theta)dz \tag{8.2}$$

For simplicity of analysis, the excitation force is assumed to be a point force acting at $z=L$. Hence

$$W_f=F_x(t)x_L+F_y(t)y_L+M(t)\theta_L$$

If there are n_s number of nozzles, the nozzle passing frequency is $\nu=n_s\omega$. The force components are expressed in the form of Fourier series.

$$F_x(t)=\left[a_{0x}+\sum_m a_{mx}\cos m\nu t+\sum_m b_{mx}\sin m\nu t\right]$$
$$F_y(t)=\left[a_{0y}+\sum_m a_{my}\cos m\nu t+\sum_m b_{my}\sin m\nu t\right] \tag{8.3}$$
$$M(t)=\left[a_{0m}+\sum_m a_{mM}\cos m\nu t+\sum_m b_{mM}\sin m\nu t\right]$$

where,

$$a_{0x}=\frac{\nu}{2\pi}\int_0^{\frac{2\pi}{\nu}}F_x(t)\,dt$$
$$a_{mx}=\frac{\nu}{\pi}\int_0^{\frac{2\pi}{\nu}}F_x(t)\cos m\nu t\,dt$$
$$b_{mx}=\frac{\nu}{\pi}\int_0^{\frac{2\pi}{\nu}}F_x(t)\sin m\nu t\,dt$$

and so on.

Substituting the response functions given in Chapter 2, equation (2.30), the expressions for W_d and W_f become

$$W_d=\frac{C_v}{2}\int_0^L A(z)\left[\left(\sum f_i\dot{q}_{1i}+r_y(z)\sum F_i\dot{q}_{3i}\right)^2\right.$$
$$\left.+\left(\sum f_i\dot{q}_{2i}-r_x(z)\sum F_i\dot{q}_{3i}\right)^2\right]dz+\frac{C_v}{2}\int_0^L I_p(z)\left(\sum F_i\dot{q}_{3i}\right)^2dz$$
$$-\frac{C_v}{2}\omega^2\int_0^L\left[\left(\sum f_i'q_{1i}+r_y'(z)\sum F_iq_{3i}+r_y(z)\sum F_i'q_{3i}\right)^2\right.$$

$$+\left(\sum f_i' q_{2i} + r_x'(z)\sum F_i q_{3i} - r_x(z)\sum F_i' q_{3i}\right)^2\Bigg](RI_1 + I_2)dz$$
$$+\frac{C_v}{2}\omega^2 \sin^2\phi \int_0^L A(z)\left(\sum f_i q_{1i} + r_y(z)\sum F_i q_{3i}\right)^2 dz$$
$$+\frac{C_v}{2}\omega^2 \cos^2\phi \int_0^L A(z)\left(\sum f_i q_{2i} - r_x(z)\sum F_i q_{3i}\right)^2 dz$$
$$-\frac{C_v}{2}\omega^2 \sin 2\phi \int_0^L A(z)\left(\sum f_i q_{1i} + r_y(z)\sum F_i q_{3i}\right)\cdot$$
$$\left(\sum f_i q_{2i} - r_x(z)\sum F_i q_{3i}\right)dz \tag{8.4}$$

$$W_f = \left[a_{0x} + \sum_m a_{mx}e^{im\nu t} + \sum_m b_{mx}e^{i\left(\frac{\pi}{2}-m\nu t\right)}\right]\left(\sum f_i q_{1i}\right)_{z=L}$$
$$+\left[a_{0y} + \sum_m a_{my}e^{im\nu t} + \sum_m b_{my}e^{i\left(\frac{\pi}{2}-m\nu t\right)}\right]\left(\sum f_i q_{2i}\right)_{z=L}$$
$$+\left[a_{0m} + \sum_m a_{mM}e^{im\nu t} + \sum_m b_{mM}e^{i\left(\frac{\pi}{2}-m\nu t\right)}\right]\left(\sum F_i q_{3i}\right)_{z=L} \tag{8.5}$$

Using Lagrange's equations, the equation of motion for forced vibrations become

$$[M]\{\ddot{q}\} + [K]\{q\} + [C]\{\dot{q}\}$$
$$= \{Q_0\} + \sum_{m=1}^{6}\{Q_m\}e^{im\nu t} + \sum_{m=1}^{6}\{Q_{m+6}\}e^{i\left(\frac{\pi}{2}-m\nu t\right)} \tag{8.6}$$

The matrices $[M]$ and $[K]$ are defined in section 2.2 and the matrix $[C]$ is given by

$$[C] = \begin{bmatrix} \overline{C}_1 & \overline{C}_2 & \overline{C}_3 \\ \overline{C}_4 & \overline{C}_5 & \overline{C}_6 \\ \overline{C}_7 & \overline{C}_8 & \overline{C}_9 \end{bmatrix} \tag{8.7}$$

where,

$$\overline{C}_{1i,j} = C_v \int_0^L A(z) f_i f_j \, dz$$
$$\overline{C}_{2i,j} = 0$$
$$\overline{C}_{3i,j} = C_v \int_0^L A(z) r_y(z) F_i F_j \, dz$$
$$\overline{C}_{4i,j} = 0$$
$$\overline{C}_{5i,j} = \overline{C}_{1i,j}$$
$$\overline{C}_{6i,j} = -C_v \int_0^L A(z) r_x(z) F_i f_j \, dz$$
$$\overline{C}_{7i,j} = C_v \int_0^L A(z) r_y(z) f_i F_j \, dz$$

$$\overline{C}_{8i,j} = -C_v \int_0^L A(z) r_x(z) f_i F_j \, dz$$

$$\overline{C}_{9i,j} = C_v \int_0^L \left[I_p(z) + A(z)\left(r_x^2(z) + r_y^2(z)\right)\right] F_i F_j \, dz$$

$$j = 1, n; \quad i = 1, n$$

Only six harmonics are considered in (8.3). The forcing vectors are given by

$$\{Q_0\} = \begin{Bmatrix} a_{0x}\left(f_j\right)_{z=L} \\ a_{0y}\left(f_j\right)_{z=L} \\ a_{0M}\left(F_j\right)_{z=L} \end{Bmatrix}$$

$$\{Q_m\} = \begin{Bmatrix} a_{mx}\left(f_j\right)_{z=L} \\ a_{my}\left(f_j\right)_{z=L} \\ a_{mM}\left(F_j\right)_{z=L} \end{Bmatrix} \tag{8.8}$$

$$\{Q_{m+6}\} = \begin{Bmatrix} b_{mx}\left(f_j\right)_{z=L} \\ b_{my}\left(f_j\right)_{z=L} \\ b_{mM}\left(F_j\right)_{z=L} \end{Bmatrix}$$

Making use of modal analysis for forced vibration response

$$\{q\} = [U]\{\eta\} \tag{8.9}$$

where $[U]$ is the modal matrix obtained from the free vibration analysis. Substituting equation (8.9) in equation (8.6)

$$[M][U]\{\ddot{\eta}\} + [K][U]\{\eta\} + [C][U]\{\dot{\eta}\} = \{Q_0\} + \sum_{m=1}^{6}\{Q_m\}e^{im\nu t} + \sum_{m=1}^{6}\{Q_{m+6}\}e^{i\left(\frac{\pi}{2} - m\nu t\right)} \tag{8.10}$$

Pre-mulitplying equation (8.10) by $[U]^T$, and using orthogonality properties,

$$[\underline{M}]\{\ddot{\eta}\} + [\underline{K}]\{\eta\} + [\underline{C}]\{\dot{\eta}\} = \{Q_{0N}\} + \sum_{m=1}^{6}\{Q_{mN}\}e^{im\nu t} + \{Q_{(m+6)N}\}e^{i\left(\frac{\pi}{2} - m\nu t\right)} \tag{8.11}$$

where,

$$\begin{aligned} [\underline{M}] &= [U]^T\,[M]\,[U] \\ [\underline{K}] &= [U]^T\,[K]\,[U] \\ [\underline{C}] &= [U]^T\,[C]\,[U] \end{aligned} \tag{8.12}$$

and

$$\begin{aligned}\{Q_{0N}\} &= [U]^T \{Q_0\} \\ \{Q_{mN}\} &= [U]^T \{Q_m\} \\ \{Q_{(m+6)N}\} &= [U]^T \{Q_{m+6}\}\end{aligned} \tag{8.13}$$

From the above, the following uncoupled equations are obtained.

$$M_k(\ddot{\eta}_k + 2\varsigma\omega_k\dot{\eta}_k + \omega_k^2\eta_k) = Q_{0N_k} + \sum_{m=1}^{6} Q_{mN_k}e^{im\nu t} + \sum_{m=1}^{6} Q_{(m+6)N_k}e^{i(\frac{\pi}{2}-m\nu t)} \tag{8.14}$$

where,

$$\omega_k^2 = \frac{K_k}{M_k} \quad \text{and} \quad \varsigma = \frac{C_k}{2M_k\omega_k} \tag{8.15}$$

The solution η_k due to each term of the forcing function in equation (8.14) is

$$\begin{aligned}\eta_{0_k} &= \frac{\frac{Q_{0N_k}}{M_k}}{\omega_k^2} \\ \eta_{m_k} &= \frac{\frac{Q_{mN_k}}{M_k}e^{im\nu t}}{[\omega_k^2 - (m\nu)^2] + 2im\nu\omega_k\varsigma} \\ \eta_{(m+6)_k} &= \frac{\frac{Q_{(m+6)N_k}}{M_k}e^{i(\frac{\pi}{2}-m\nu t)}}{[\omega_k^2 - (m\nu)^2] + 2im\nu\omega_k\varsigma} \\ m &= 1,6; \quad k = 1,3n\end{aligned} \tag{8.16}$$

With the help of equations (8.9) and (8.16), the expressions for bending and torsional deflections can be obtained for zero frequency component of the forcing function and for first to sixth harmonic components of the forcing function.

Based on the above analysis, the computer program discussed in Chapter 2 to determine the natural frequencies and mode shapes of a turbine blade can be extended for finding bending-bending and torsional deflections due to forced vibration. The excitation force for each nozzle can be in the form of digital data as a function of rotor angle with the blade approaching the nozzle and then receding from the nozzle. A Fourier series analysis can be built in the program for carrying out the modal analysis and obtain the bending-bending and torsional displacement amplitudes at selected intervals along the length of the blade for a given R.P.M. of the rotor.

Rao and Jadvani [28, 30] tested analytically and experimentally the following pre-twisted blade:

$I_{xx} = 0.010513$ cm^4		$E = 2.07 \times 10^{11}$ Pa
$I_{yy} = 1.521776$ cm^4		$G = 0.827 \times 10^{11}$ Pa
$\phi = 90°$		$L = 16.85$ cm
$\gamma = 58°$		$A = 1.232$ cm^2
$C = 34.81$ Nm^2/rad		$R = 16.2$ cm
$\rho = 7860$ kg/m^3		$n_s = 12$

In the test rig, the nozzles were simulated by permanent magnets and the excitation force components due to a passing magnet acting at the blade tip are given in figures 8.1 – 8.3. The theoretical and experimental values of the rotor speeds where resonance occurs (intersection points on the Campbell diagram) are given in Table 8.1, which shows an excellent agreement between them.

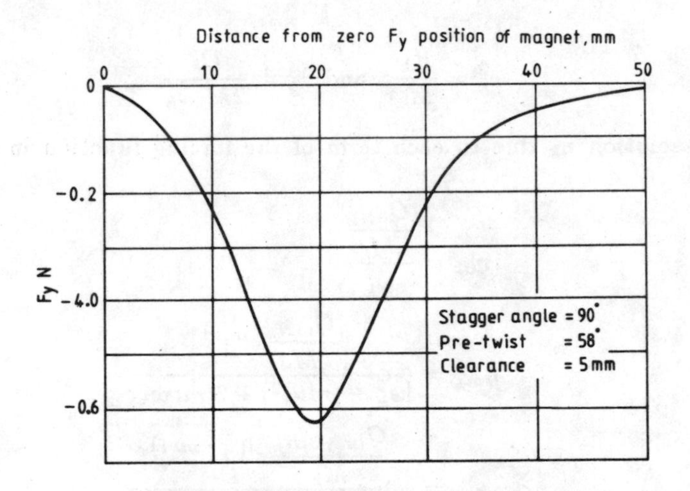

Fig. 8.1 Magnetic attractive force F_y

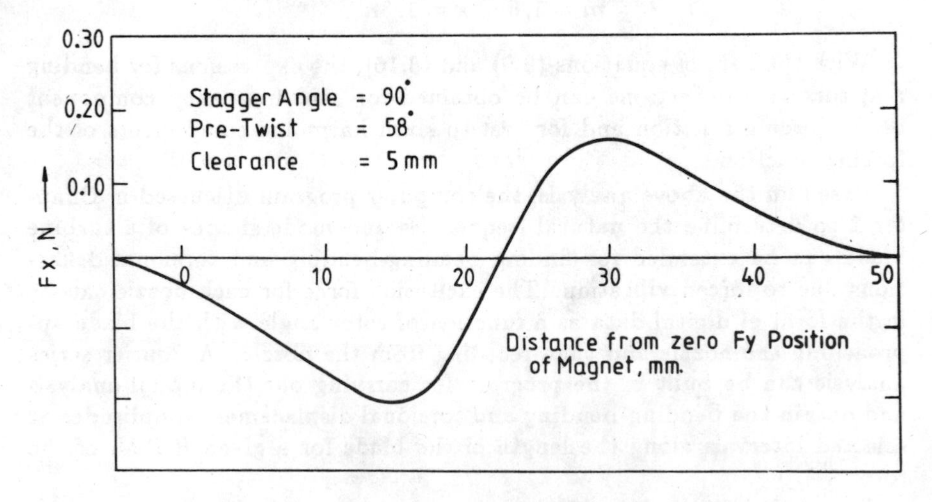

Fig. 8.2 Magnetic attractive force F_x

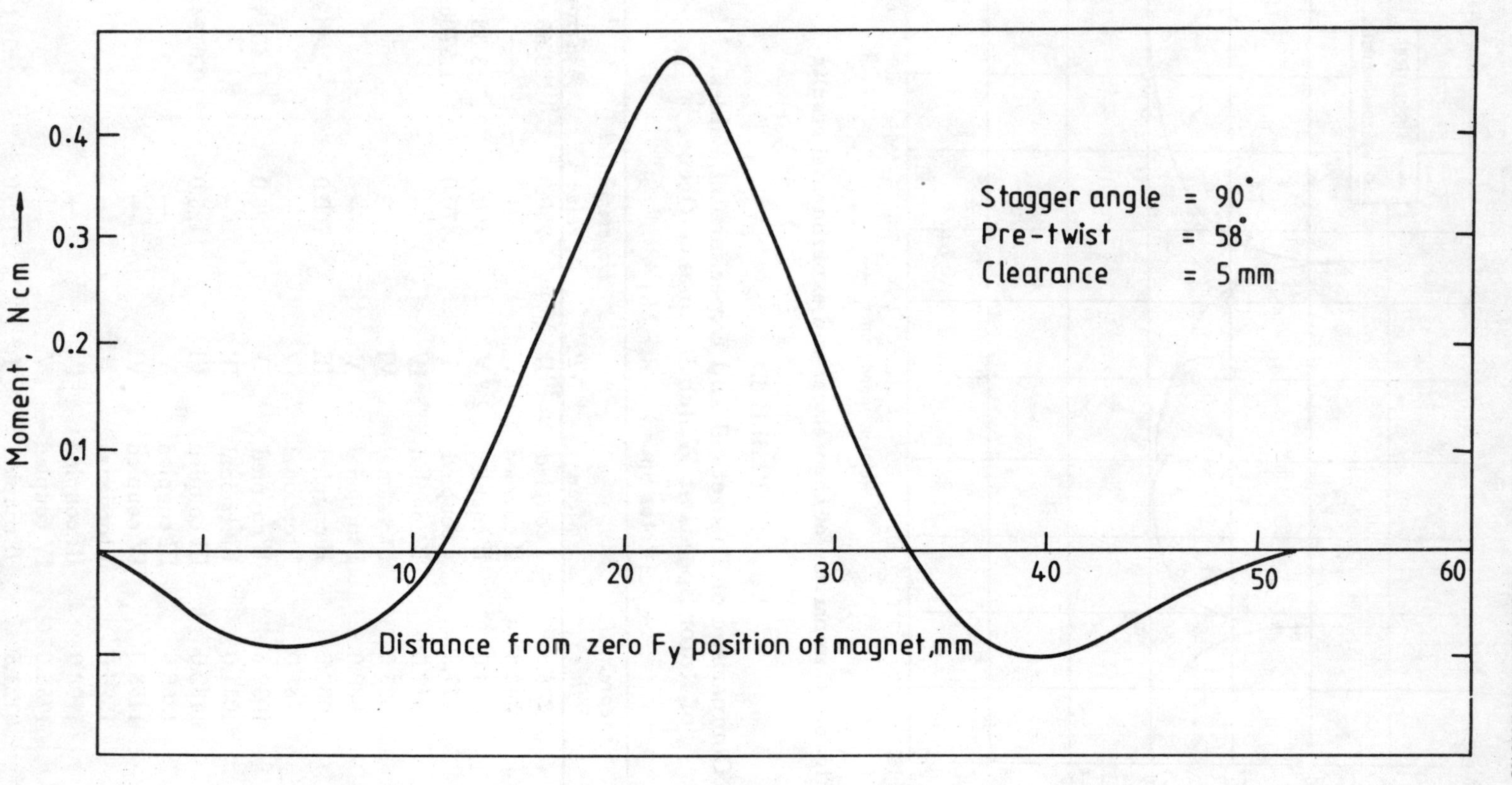

Fig. 8.3 Moment M

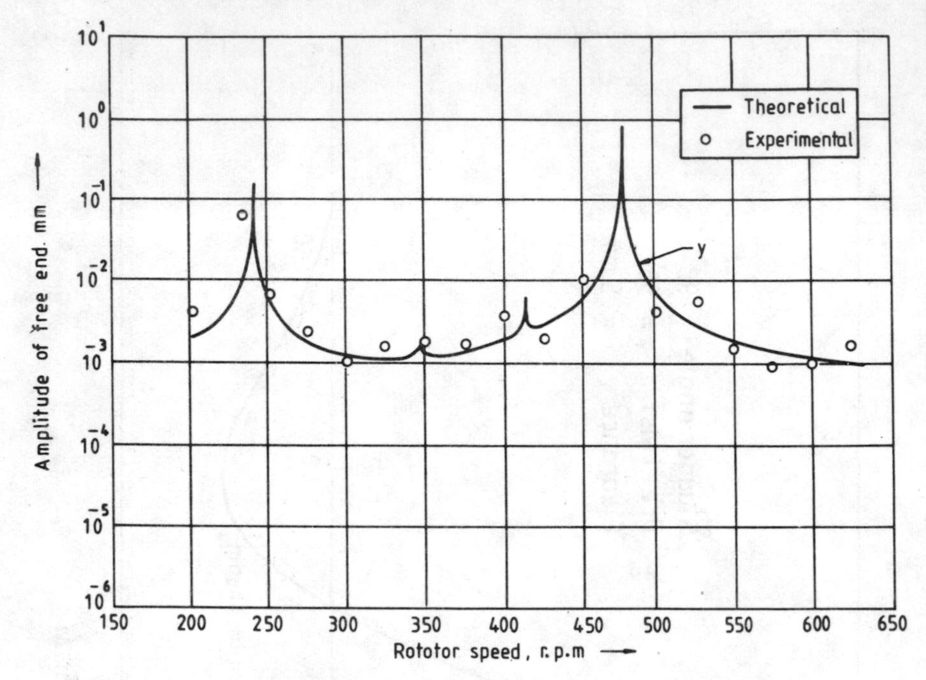

Fig. 8.4 Comparison of theoretical and experimental results

TABLE 8.1

Comparison of Theoretical and Experimental Values of Rotor Speeds at Which Resonance Occurs

(Rotor Speed in R.P.M.)

No.	*Theoretical Value*	*Mode*	*NPF*	*Experimental Value*	*% Error*
1	238.5	I coupled	II	230.0	3.56
2	343.3	II coupled	VI	—	
3	414.0	II coupled	V	400.0	3.38
4	477.5	I coupled	I	470.0	1.5706
5	519.0	II coupled	IV	—	
6	525.0	I torsional	VI	—	
7	630.0	I torsional	V	—	
8	690.0	II coupled	III	700.0	−1.449
9	788.0	I torsional	IV	—	
10	1039.6	II coupled	II	1025.0	1.4015
11	1051.0	I torsional	III	—	
12	1132.0	III coupled	VI	1125.0	0.6184
13	1366.0	III coupled	V	—	
14	1458.0	IV coupled	VI	—	
15	1576.0	I torsional	II	—	
16	1708.0	III coupled	IV	—	
17	1755.0	IV coupled	V	—	
18	2121.5	II coupled	I	2125.0	−0.1649

TABLE 8.2

Digital Form of Data for Nozzle Excitation for a Typical Turbine Blade

(Number of Nozzles = 24 Nozzle Pitch Circle Radius = 27.3 cm)

No.	*Forcing Function* $F_y - N$	*Forcing Equation* $F_x - N$	*Moment* $M - N$ cm
1	0.00	0.00	0.00
2	-0.025	-0.02	-3.928
3	-0.069	-0.049	-7.856
4	-0.128	-0.074	-9.329
5	-0.216	-0.098	-8.838
6	-0.344	-0.133	-7.365
7	-0.491	-0.157	-3.928
8	-0.579	-0.191	-4.419
9	-0.609	-0.206	15.712
10	-0.530	-0.182	27.496
11	-0.412	-0.088	39.280
12	-0.314	0.000	46.645
13	-0.216	0.088	37.316
14	-0.137	0.137	24.550
15	-0.093	0.157	14.730
16	-0.069	0.142	3.928
17	-0.045	0.114	-4.615
18	-0.034	0.088	-8.838
19	-0.020	0.059	-9.820
20	-0.015	0.039	-8.347
21	-0.006	0.020	-5.892
22	0.00	0.006	-3.437
23	0.00	0.00	0.00
..	...	...	...
..	...	...	...
39	0.00	0.00	0.00

Using the tip calibration under static conditions, the fundamental mode response at 230 and 470 rpm is compared with undamped theoretical values in figure 8.4. The overall agreement of the results is excellent. Appropriate modal damping is to be used to analytically determine the peak amplitude observed experimentally.

Case study of a steam turbine blade given in figure 2.40 and Table 2.6 is next considered. The excitation data at 40 equal intervals of the blade tip travel across each of 24 nozzles is given in Table 8.2.

Figure 8.5 shows the Campbell diagram for the blade, with the first three coupled modes. The rotor speeds at which resonance occurs are given in Table 8.3 below.

TABLE 8.3

Rotor Speeds At Which Resonance Occurs

(Rotor Speed in RPM)

NPF Order	*I Coupled Mode*	*II Coupled Mode*	*III Coupled Mode*
I	2808.0	—	—
II	1400.0	—	—
III	933.33	2597.6	—
IV	699.9	1947.56	—
V	559.9	1557.8	2118.25
VI	466.54	1298.08	2541.93

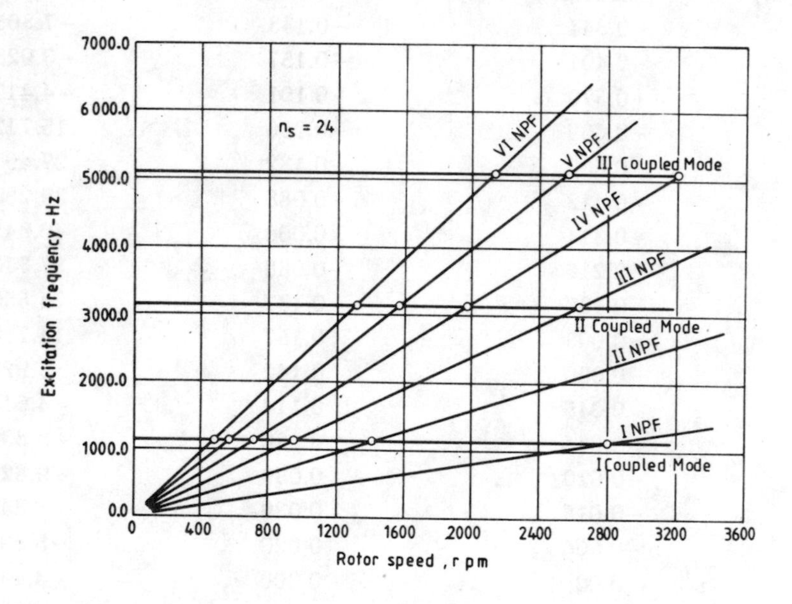

Fig. 8.5 Campbell diagram for typical turbine blade

Figures 8.6 and 8.7 show the variation of bending-bending and torsion displacement amplitudes at the free end of the blade with rotor RPM. The range of speed considered is from 1200 RPM to 2900 RPM. It can be seen from these figures that the amplitudes reach their peak values at the points when the nozzle passing frequency lines intersect the curves of natural frequencies as shown in the Campbell diagram. It can also be observed from the nature of these figures that the intensity of resonance depends on the order of the nozzle passing frequency. At a rotor speed of 2808 RPM, the resonance is quite pronounced due to the first nozzle passing frequency and its effect exists over a large band width of rotor speed. As against this, the resonance at a rotor speed of 2118 RPM due to fifth nozzle passing frequency is quite weak, occupying a very small band width of the rotor speed. Also at rotor speeds of 1400 RPM and 2808 RPM, the coupling between bending-bending and torsion motion is quite pronounced as all the

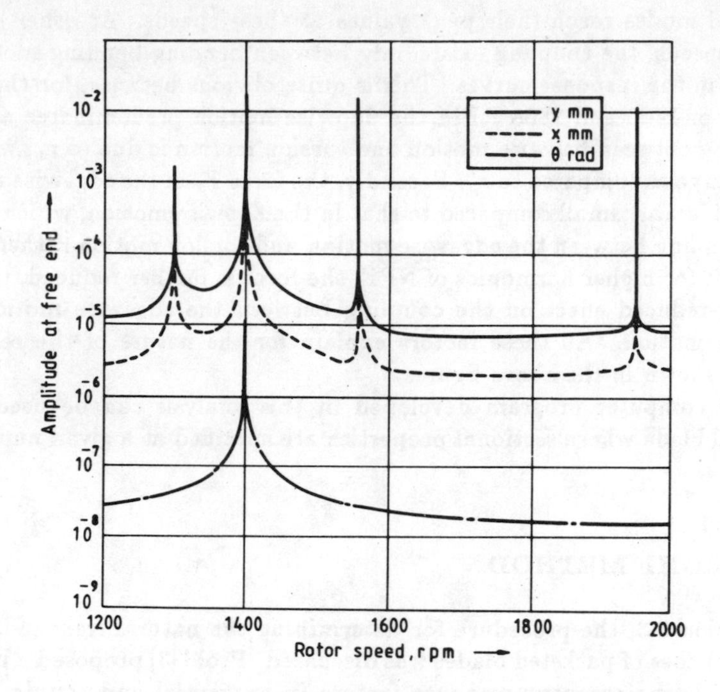

Fig. 8.6 Response of typical turbine blade speed range (1200 – 2000 RPM)

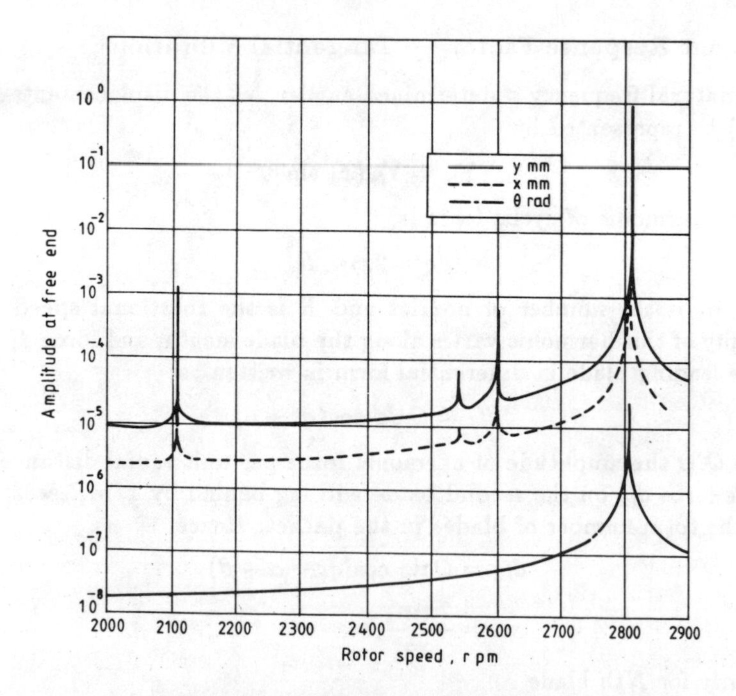

Fig. 8.7 Response of typical turbine blade speed range (2000–2900 RPM)

coupled modes reach their peak values at these speeds. At other critical rotor speeds, the coupling exists only between bending-bending motion as shown in the response curves. This is quite obvious because, for the rotor speeds of 1400 and 2808 RPM the flapwise motion predominates and the coupling between flapwise motion and torsion motion is due to r_x, which is quite large as compared to r_y. Secondly, the force F_x in the edgewise motion is considerably small compared to that in the flapwise motion, which makes the coupling between the edgewise motion and torsion motion rather weak. Thirdly, for higher harmonics of NPF, the force is further reduced, thereby having reduced effect on the coupling between the edgewise motion and torsion motion. All these factors explain for the nature of the response curves shown in the above figures.

The computer program developed in this analysis can be used for a general blade whose sectional properties are specified at a given number of stations.

8.3 PROHL METHOD

In section 5.3, the procedure for determining the natural frequencies and mode shapes of packeted blades was discussed. Prohl [3] proposed a method of calculating resonant response factors in tangential and coupled axial-torsional modes, based on the assumption that the unsteady energy input to the system is totally dissipated by damping of the blade packet.

Resonant Response Factor — Tangential Vibration

For a natural frequency p determined earlier, let the displacements (mode shape) be represented by

$$Y_N = Y_N(x) \sin pt \tag{8.17}$$

The ith harmonic of cyclic force is

$$q = 2\pi i n_s R \tag{8.18}$$

where n_s is the number of nozzles and R is the rotational speed in Hz. Intensity of this harmonic varies along the blade length, and force f_1 acting on the leading blade in differential form is written as

$$df_1 = Q\,dx \cos(qt + \beta) \tag{8.19}$$

where Q is the amplitude of harmonic force per unit radial distance.

The force df_2 on the second blade will lag behind by $1/mR$ secs, where m is the total number of blades in the packet. Hence

$$\begin{aligned} df_2 &= Q\,dx \cos(qt - \alpha + \beta) \\ \alpha &= \frac{2\pi i n_s}{m} \end{aligned} \tag{8.20}$$

Similarly for Nth blade

$$df_N = Q\,dx \cos\{qt - (N-1)\alpha + \beta\} \tag{8.21}$$

Hence the total energy supplied to blade N will be

$$E_{sN} = \int_{x=0}^{D+L} \int_{pt=0}^{2\pi} Y_N(x) Q \, dx \cos\{qt - (N-1)\alpha + \beta\} \cos pt \, d(pt) \tag{8.22}$$

Since $p = q$ for resonance,

$$E_{sN} = \pi \int_0^{D+L} Y_N(x) Q \cos\{\beta - (N-1)\alpha\} dx \tag{8.23}$$

$$E_s = \pi \sum_{N=1}^{s} \sum_{n=0}^{r} \cos\{\beta_0 + \beta_n - (N-1)\alpha\} Y_{Nn} \, dQ_n \tag{8.24}$$

The above equation can be written as

$$E_s = \pi Y_m Q_0 S\{C \cos\beta_0 + D \sin\beta_0\} \tag{8.25}$$

where,

$$S = \text{stimulus coefficient}$$

$$Q_0 = \text{driving force}$$

$$C = \sum_{N=1}^{s} A_N \cos(N-1)\alpha + B_N \sin(N-1)\alpha$$

$$D = \sum_{N=1}^{s} A_N \sin(N-1)\alpha - B_N \cos(N-1)\alpha$$

$$A_N = \sum_{n=0}^{r} \cos\beta_n \left\{\frac{Y}{Y_m}\right\}_{Nn} \left\{\frac{dQ}{Q_0}\right\}_n$$

$$B_N = \sum_{n=0}^{r} \sin\beta_n \left\{\frac{Y}{Y_m}\right\}_{Nn} \left\{\frac{dQ}{Q_0}\right\}_n$$

Maximum value of energy supplied is

$$E_{s\,\max} = \pi Y_m Q_0 S \sqrt{C^2 + D^2} \tag{8.26}$$

The dissipated energy of the blade packet is

$$E_d = \delta p^2 Y_m^2 G_t \tag{8.27}$$

where,

$$\delta = \text{logarithmic decrement}$$

$$G_t = \sum_{N=1}^{s} \sum_{n=0}^{r} m_n \left\{\frac{Y}{Y_m}\right\}_{Nn}^2$$

From equations (8.26) and (8.27) we get Y_v, the values of Y_m as

$$Y_v = \frac{\pi}{\delta} S \frac{Q_0 L}{2} \frac{2}{Lp^2} \frac{\sqrt{C^2 + D^2}}{G_t} \tag{8.28}$$

The tangential bending moment at the root due to resonance is therefore given by

$$M_{vt} = \frac{\pi}{\delta} S M_s K_t \tag{8.29}$$

where,

$$M_s = \text{steam bending moment at root} = \frac{1}{2} Q_0 L$$

$$K_t = \frac{M_m}{Y_m} \frac{2}{Lp^2} \frac{\sqrt{C^2 + D^2}}{G_t}$$

M_m/Y_m is obtained from frequency and mode shape calculations. From equation (8.29), the maximum stress at the root can be calculated. K_t is called resonant response factor in tangential vibration.

Resonant Response Factor — Coupled Motion

A. Axial Excitation

On similar lines as above, it can be shown that

$$M_{va} = \frac{\pi}{\delta} S M_s K_a \tag{8.30}$$

where,

$$K_a = \frac{M_m}{Y_m} \frac{2}{Lp^2} \frac{\sqrt{C^2 + D^2}}{G_a}$$

and,

$$G_a = \sum_{N=1}^{s} \sum_{n=0}^{r} \left[m_n \left\{ \frac{Y}{Y_m} \right\}_{Nn}^2 + J_n \left\{ \frac{\phi}{Y_m} \right\}_{Nn}^2 \right]$$

with corresponding axial displacement and excitation values.

B. Torsional Excitation

Y/Y_m is to be replaced by ϕ/ϕ_m and dQ/Q_0 by dT/T_0 for this case. As before

$$T_v = \frac{\pi}{\delta} S T_s K_r \tag{8.31}$$

where,

$$K_r = \frac{T_m}{\phi_m} \frac{1}{p^2} \frac{\sqrt{C^2 + D^2}}{G_r}$$

and,

$$G_r = \sum_{N=1}^{s} \sum_{n=0}^{r} \left[J_n \left\{ \frac{\phi}{\phi_m} \right\}_{Nn}^2 + m_n \left\{ \frac{Y}{\phi_m} \right\}_{Nn}^2 \right]$$

In all the above calculations for foced vibration response, uniform stimulus is assumed, i.e.,

$$\left\{ \frac{dQ}{Q_0} \right\}_n = \frac{1}{r+1} \tag{8.32}$$

Also, since the system is assumed to be in resonance, β_n is taken as 90 degrees and $A_N = 0$ in equation (8.25).

The results of Weaver and Prohl [4] for the data given in section 5.3, for the resonant response factors are given in figures 5.8 and 5.9. K is plotted as a function of the ratio of the number of nozzles k per 360 deg to the number of buckets m per 360 deg times the harmonic order n of the excitation. K is periodic function of nk/m and repeats for each unit interval. It is of interest to note that at certain nk/m, K becomes zero. This indicates that with proper selection of nozzles and bucket pitching, it is possible to reduce considerably the resonant vibration stress in a particular mode, even though operating precisely at the speed required to excite this natural frequency.

8.4 REISSNER'S VARIATIONAL PRINCIPLE

It is generally known that the potential energy methods, as followed in section 8.2, lead to good displacement fields but poor stress approximations. Usage of complementary energy principle conversely leads to good stress fields but with poor displacement fields. The Galerkin method, failing to ensure that all the shape functions used satisfy all the boundary conditions, may lead to inaccurate results. The discrete model approaches, on the other hand, suffer from the drawback of discretization process of the distributed mass and elasticity, thus yielding lower bound results. They also depend on appropriate stress-strain-displacement relations that have to be adopted in the analysis, and are generally prone to inaccurate determination of stresses, particularly under dynamic conditions.

Several difficulties encountered in the above procedures are eliminated by usage of Reissner's functional, which involves both the potential energy and complementary energy considerations and is known to yield simultaneously good stress and displacement fields, see Dym and Shames [46]. Subrahmanyam [47] applied Reissner method for the free vibration analysis of uniform symmetric blading, pre-twisted blading, rotating pre-twisted tapered blading. He used beam theory to derive the kinetic energy term and Reissner's dynamic functional, and the frequency equation through Ritz process. Vyas [48] compared solutions of simple beams by the exact method, Lagrange procedure and Reissner's method for both bending and torsional vibrations and showed that the natural frequencies, mode shapes, free as well as forced vibrational displacement and stress fields obtained by Reissner's method are better than those obtained by Lagrange procedure and also gave rapid convergence. This procedure is illustrated below to determine the forced vibration response of a turbine blade subjected to nozzle passing excitation.

The Reissner's functional is given by

$$I_R = \iiint_v \left[\tau_{ij} e_{ij} - U^*(\tau_{ij})\right] dv - \iiint_v B_i u_i \, dv - \iint_s T_i u_i \, ds \qquad (8.33)$$

where $\tau_{ij}e_{ij}$ is the potential energy density, $U^*(\tau_{ij})$ is the complementary energy density function, B_i and T_i are the body and traction force terms respectively.

Referring to Figs. 2.1 and 2.2, the displacement field is defined by equations (2.2) and (2.3). The corresponding strains are defined in equations (2.23), (2.25) and (2.26). The respective non-zero stress components are

$$\begin{aligned} \tau_{xz} &= -KGy_1\theta' \\ \tau_{yz} &= -KGx_1\theta' \\ \tau_{zz} &= E(-x_1x'' - y_1y'') \end{aligned} \tag{8.34}$$

where K is shear correction factor. The shear force effects are neglected here and the bending moments are

$$\begin{aligned} M_x &= \int_A \tau_{zz}y_1\, dA = -E(I_{x_1y_1}x_1'' + I_{x_1x_1}y_1'') \\ M_y &= \int_A \tau_{zz}x_1\, dA = -E(I_{y_1y_1}x_1'' + I_{x_1y_1}y_1'') \end{aligned} \tag{8.35}$$

The strain energy can be obtained as

$$\iiint_v \tau_{ij}e_{ij}\, dv = \int_0^l (-M_yx'' - M_xy'' + T_\theta\theta')dz \tag{8.36}$$

Evaluating x_1'' and y_1'' from equation (8.35) and substituting them in equation (8.34), we can obtain the complementary energy as

$$\begin{aligned} \iiint_v U^*(\tau_{ij})\, dv &= \iiint_v \left\{ \frac{\tau_{zz}^2}{2E} + \frac{\tau_{zx}^2 + \tau_{zy}^2}{2KG} \right\} dv \\ &= \int_0^l \left[\frac{M_x^2 I_{y_1y_1} + M_y^2 I_{x_1x_1} - 2M_xM_yI_{x_1y_1}}{2E(-I_{x_1y_1}^2 + I_{y_1y_1}I_{x_1x_1})} + \frac{T_\theta^2}{2C} \right] dz \end{aligned} \tag{8.37}$$

The body forces can be assumed negligible. The excitation forces are given by equation (8.3), which can now be generalized as functions of blade axial coordinate z. Hence the Reissner's functional can be written as

$$\begin{aligned} I_R = \int_0^l \Bigg[&-M_yx'' - M_xy'' + T_\theta\theta' \\ &- \frac{M_x^2 I_{y_1y_1} + M_y^2 I_{x_1x_1} - 2M_xM_yI_{x_1y_1}}{2E(-I_{x_1y_1}^2 + I_{y_1y_1}I_{x_1x_1})} + \frac{T_\theta^2}{2C} \Bigg] dz \\ - \int_0^l \Bigg[&\left\{ a_{0x}(z) + \sum_m a_{mx}(z) \cos m\nu t + \sum_m b_{mx}(z) \sin m\nu t \right\} x(z) \\ &+ \left\{ a_{0y}(z) + \sum_m a_{my}(z) \cos m\nu t + \sum_m b_{my}(z) \sin m\nu t \right\} y(z) \\ &+ \left\{ a_{0M}(z) + \sum_m a_{mM}(z) \cos m\nu t + \sum_m b_{mM}(z) \sin m\nu t \right\} \theta(z) \Bigg] dz \end{aligned} \tag{8.38}$$

The kinetic energy expression is given by equation (2.22). Ritz's criterion can now be applied to the dynamic Reissner's functional given by

$$L = T - I_R \tag{8.39}$$

In addition to the shape functions given in equation (2.30), the following shape functions are taken.

$$\begin{aligned} M_x &= \sum_{i=1}^{n} h_i(Z) q_{4i} \\ M_y &= \sum_{i=1}^{n} h_i(Z) q_{5i} \\ T_\theta &= \sum_{i=1}^{n} h_i(Z) q_{6i} \end{aligned} \tag{8.40}$$

where,

$$h_i = \frac{(1-Z)^i}{i+1}$$

satisfying the boundary condition

$$M_x = M_y = T_\theta = 0 \quad \text{at } Z = 1$$

After substituting the shape functions in Reissner's dynamic functional given in equation (8.39), applying Ritz's criterion

$$\frac{\partial L}{\partial q_{1i}} = \frac{\partial L}{\partial q_{2i}} = \cdots = \frac{\partial L}{\partial q_{6i}} = 0$$

and assuming proportional damping, we get

$$\begin{aligned} &[M]\{\ddot{q}\} + [C]\{\dot{q}\} + [K]\{q\} \\ &\quad = \{Q_0\} + \sum_m \{Q_m\} \cos m\nu t + \sum_m \{Q_{m+6}\} \sin m\nu t \end{aligned} \tag{8.41}$$

where,

$$\{q\}^T = \{q_{1i}\ q_{2i} \cdots q_{6i}\} \tag{8.42}$$

$[M]$ and $[K]$ are mass and stiffness matrices with 6×6 sub-matrices and excitation vectors $\{Q\}$ contain sub-vectors, whose elements can be generated (similar to equations (2.35) by appropriate integrals).

The solution to equation (8.41) can be obtained by modal analysis given in equations (8.9) through (8.16).

The response due to mth harmonic for kth mode is given by

$$\begin{aligned} H_{m_k} &= \eta_{m_k} + \eta_{(m+6)_k} \\ &= \frac{\sqrt{\dfrac{Q_{mN_k}^2 + Q_{(m+6)N_k}^2}{M_k}}}{\sqrt{\{p_k^2 - (m\nu)^2\}^2 + 4\varsigma_k^2 p_k^2 (m\nu)^2}} \cos(m\nu t - \psi_{m_k} - \delta_{m_k}) \end{aligned} \tag{8.43}$$

where,

$$\psi_{m_k} = \tan^{-1} \frac{2\zeta_k p_k m\nu}{p_k^2 - (m\nu)^2}$$

and

$$\delta_{m_k} = \tan^{-1} \frac{Q_{(m+6)N_k}}{Q_{mN_k}} \tag{8.44}$$

With the help of the first equation of (8.16), (8.43), (8.9) and (2.30) and (8.40), the deflections and moments of the blade for the force harmonics can be determined. For computation of stresses, we can use the following formulae.

$$\begin{aligned}
M_X &= M_x \cos\gamma_z - M_y \sin\gamma_z \\
M_Y &= M_y \cos\gamma_z + M_x \sin\gamma_z \\
\sigma_z &= \frac{M_Y \underset{\cdot}{x}}{I_{yy}} + \frac{M_X \underset{\cdot}{y}}{I_{xx}} \\
\tau_{zx} &= \frac{T_\theta \underset{\cdot}{y}}{I_p} \\
\tau_{zy} &= \frac{T_\theta \underset{\cdot}{x}}{I_p}
\end{aligned} \tag{8.45}$$

The principal stresses are

$$\begin{aligned}
\sigma_1 &= 0 \\
\sigma_{2,3} &= \frac{\sigma_z}{2} \pm \sqrt{\left(\frac{\sigma_z}{2}\right)^2 + \tau_{yz}^2 + \tau_{zx}^2}
\end{aligned} \tag{8.46}$$

The basic mean stress (zeroth harmonic) and alternating stress components are determined by using the corresponding principal stresses by using the following formula.

$$\sigma_{mv}, \sigma_{alt} = \sqrt{\frac{1}{2}[\sigma_2^2 + \sigma_3^2 - 2\sigma_2\sigma_3]} \tag{8.47}$$

A general computer program was developed by Rao, Gupta and Vyas [40] to determine natural frequencies, mode shapes, displacement and stress fields. They also used the experimental rig described in Chapter 7, to determine the response of the blade shown in Fig. 7.18. The geometrical and material properties of this blade are given in Table 8.4.

The electromagnets used to simulate the nozzle passing excitation were calibrated in a simple test to determine the excitation forces per pitch length of each nozzle given to the rotating blade. They are given in Fig. 8.8. Table 8.5 gives the Fourier component amplitudes of the forcing functions given in this figure.

Fig. 7.25 gives the average modal damping ratios obtained experimentally as a function of rotor speed, for the first four modes. Using these damping values and excitation forces defined in Fig. 8.8 with the Fourier components in Table 8.5, the results obtained for the blade in Table 8.4 are discussed here.

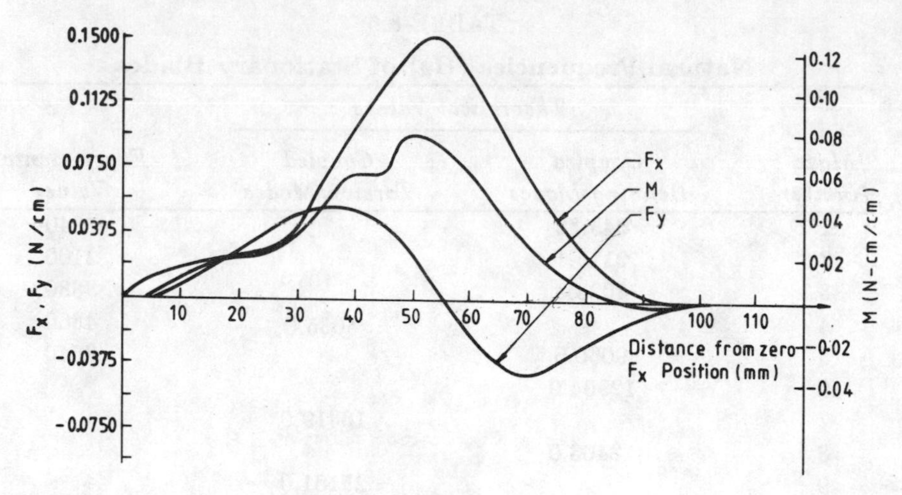

Fig. 8.8 Forcing functions

TABLE 8.4

Geometrical and Material Properties of the Blade

$L = 11.10$ cm $\quad R = 18.0$ cm $\quad \phi = 90°$

$E = 2.0 \times 10^{11}$ Pa $\quad G = 0.75 \times 10^{11}$ Pa $\quad \rho = 8150$ kg/m^3

	Section					
Property	1	2	3	4	5	6
Distance from Root cm	0.0	2.5	5.0	7.5	10.0	11.0
A cm^2	2.127	1.738	1.647	1.263	1.008	0.736
I_{xx} cm^4	.3447	.3651	.4927	.5054	.4356	.3258
I_{yy} cm^4	.9489	.6717	.5174	.3097	.1543	.0858
I_{xy} cm^4	.4705	.4158	.4146	.2943	.2243	.1572
$C \times 10^{-3}$ Nm^2/rad	.9702	.7776	.7576	.6113	.4424	.3087
r_x cm	−.213	−.094	−.082	−.178	−.282	−.303
r_y cm	.075	.135	.073	−.113	−.222	−.453

TABLE 8.5

Fourier Component Amplitudes (N/cm)/(N cm/cm)

$a_{0x} = 0.0622 \quad a_{0y} = -.0064 \quad a_{0M} = 0.0280$

m	a_{mx}	b_{mx}	a_{my}	b_{my}	a_{mM}	b_{mM}
1	−.0642	−.0169	.0153	−.0392	−.0269	−.0173
2	.0110	.0140	−.0063	.0095	.0037	.0089
3	−.0024	.0005	.0024	.0005	−.0006	.0004
4	−.0027	.0013	.0012	.0009	−.0010	.0011
5	−.0001	−.0020	.0007	−.0012	.0001	−.0012
6	−.0005	.0028	−.0018	.0015	−.0007	.0017

TABLE 8.6

Natural Frequencies (Hz) of Stationary Blades

Mode Number	*Theoretical Values* *Coupled Bending Modes*	*Coupled Torsion Modes*	*Experimental Values*
1	643.86		640
2	2115.1		2100
3	4023.3		3880
4		5035.0	4860
5	9060.6		8640
6	12904.0		
7		16319.0	
8	18463.0		
9		25161.0	
10	33532.0		
11		35195.0	

TABLE 8.7

Resonant Rotor Speeds (RPM)

NPF Order	*Mode I*	*Mode II*	*Mode III*
I	3250.0		
II	1614.6		
III	1074.5	3530.0	
IV	805.0	2646.1	
V	644.0	2116.0	4035.0
VI	536.7	1763.3	3358.3

TABLE 8.8

Resonant Stress Components (MPa) at Critical Blade Sections

RPM	σ_z	τ_{xz}	τ_{yz}	σ_a	σ_m (*At Root*)
537	0.177	0.0	.0002	0.177	.2907
644	0.110	0.0	.0001	0.110	.2906
805	0.762	0.0	.0012	0.762	.2904
1075	0.647	0.0	.0010	0.647	.2897
1615	3.851	0.0	.0059	3.851	.2881
1763	0.160	0.0006	.0041	0.160	.2975
2116	0.119	0.0005	.0030	0.119	.2859
2646	0.137	0.0005	.0039	0.137	.2932
3250	23.589	0.0	.0378	23.589	.2795
3358	0.396	0.0	.0002	0.396	.2788
3530	0.049	0.0002	.0016	0.049	.2754

Table 8.6 gives comparison of the theoretical and experimental values of the natural frequencies of the stationary blade. It can be observed that the first two theoretical values match very well with the first two experimental values. Higher natural frequencies match within 3 to 6 per cent. Fig. 8.9 is the Campbell diagram for the blade with the six nozzle passing harmonics interacting with the natural modes to give resonant rotor speeds up to 4000 RPM listed in Table 8.7.

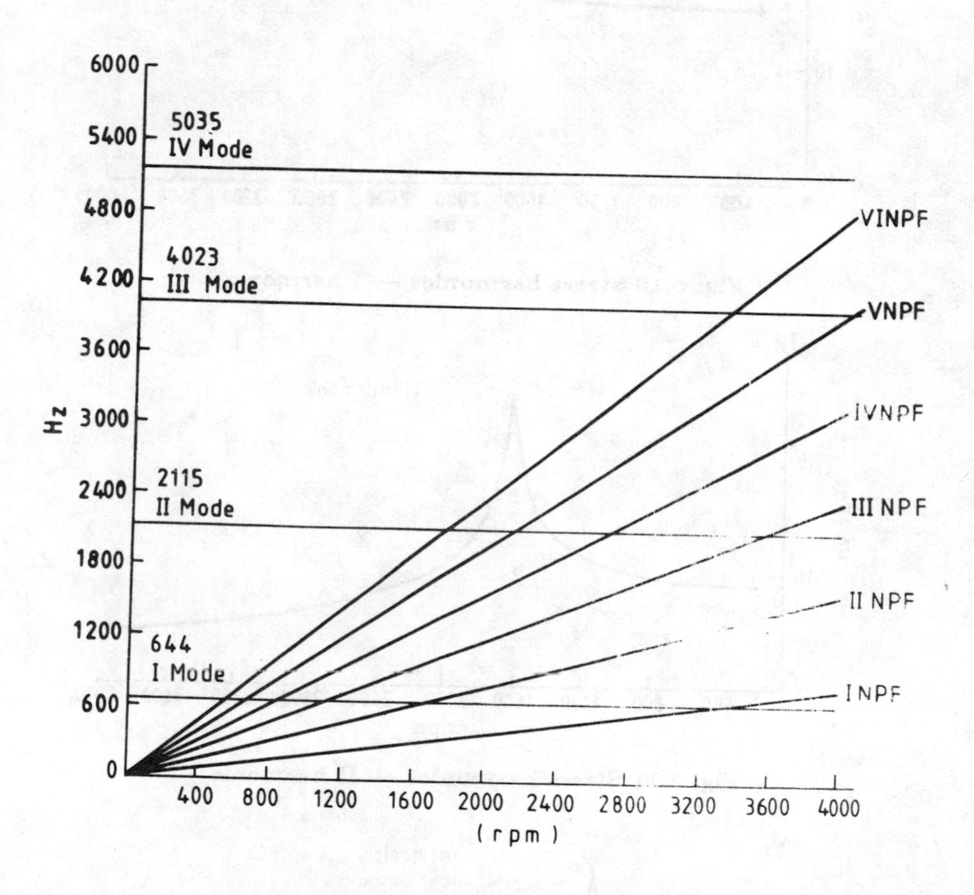

Fig. 8.9 Campbell Diagram

The blade stress fields have been computed for various rotor speeds upto 4000 RPM including resonances. The stresses are determined at ten equidistant stations along the blade length. Figs. 8.10 to 8.15 show the plots of the basic stress harmonics at the blade root as a function of rotor speed. In Fig. 8.16, representative resonant stresses are plotted as functions of blade length to identify critical blade sections. Table 8.8 gives the resonant stress components at critical blade sections. It can be observed that the stress fields on the blade are functions of the order of the nozzle passing harmonic, the amplitude of the forcing functions components, the interacting blade natural mode and the blade geometry.

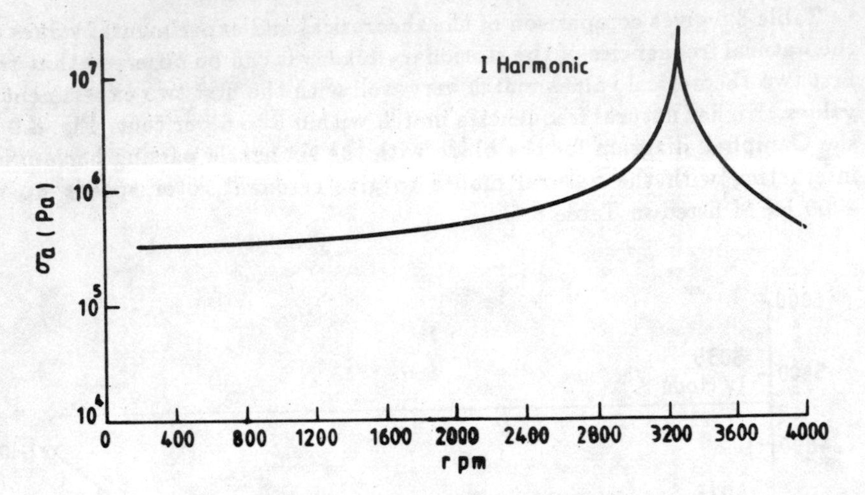

Fig. 8.10 Stress harmonics — I harmonic

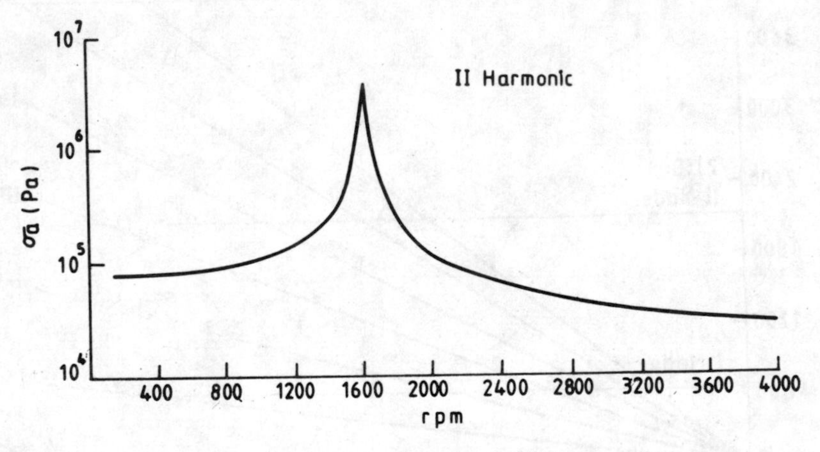

Fig. 8.11 Stress harmonics — II harmonic

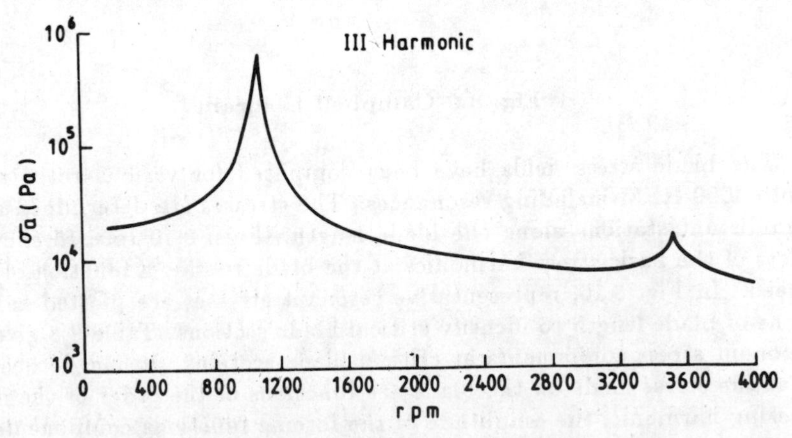

Fig. 8.12 Stress harmonics — III harmonic

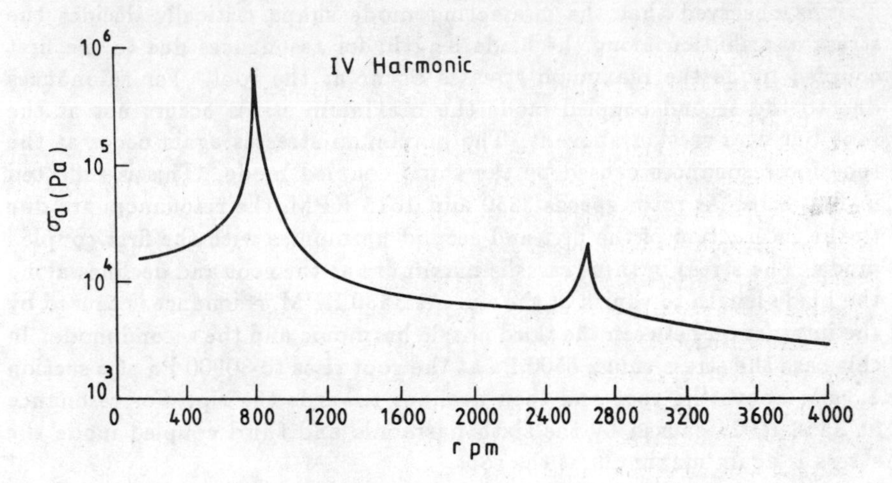

Fig. 8.13 Stress harmonics — IV harmonic

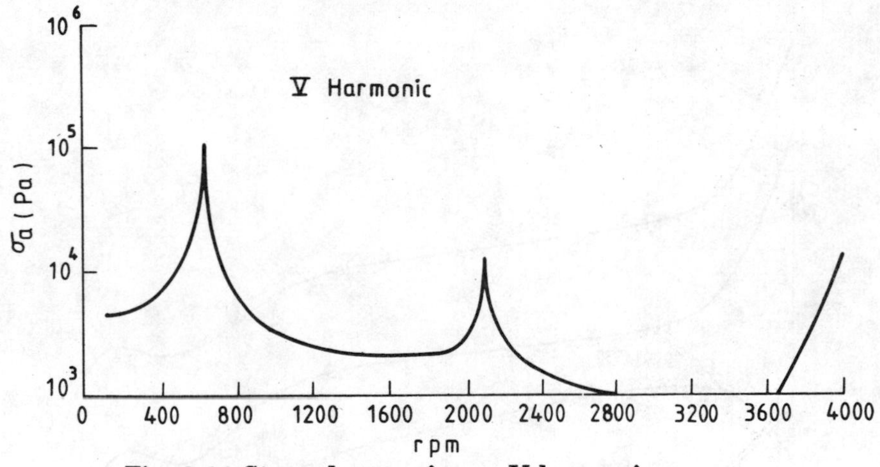

Fig. 8.14 Stress harmonics — V harmonic

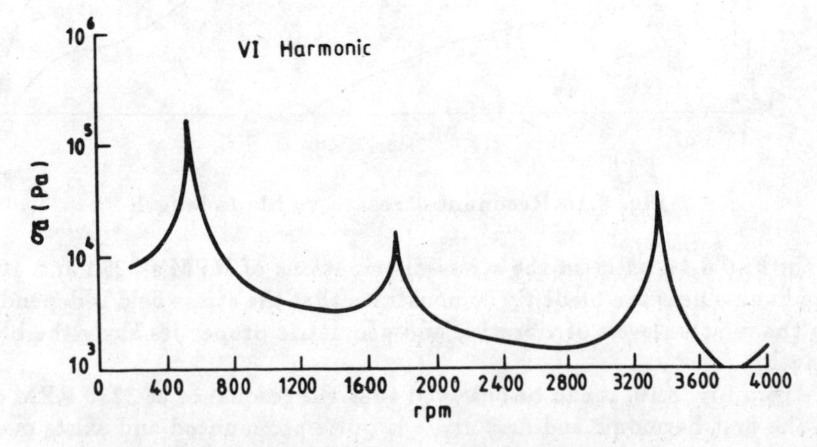

Fig. 8.15 Stress harmonics — VI harmonic

It is observed that the interacting mode shape critically decides the stress distribution along the blade length; for resonances due to the first coupled mode the maximum stresses occur at the root. For resonances due to the second coupled mode the maximum stress occurs not at the root but at a section above it. The maximum stresses again occur at the root for resonances caused by the third coupled mode. This is reflected in Fig. 8.16. At rotor speeds 3250 and 1615 RPM, the resonances are due to the interaction of the first and second harmonics with the first coupled mode. The stress in these cases is maximum at the root and declines along the blade length to vanish at the tip. At 3530 RPM, resonance is caused by the interaction between the third nozzle harmonic and the second mode. In this case the stress value, 8500 Pa at the root rises to 30900 Pa at a section 1.3 cm above the root and then declines towards the tip. For resonance at 3358 RPM caused by the sixth harmonic and third coupled mode the stress is again maximum at the root.

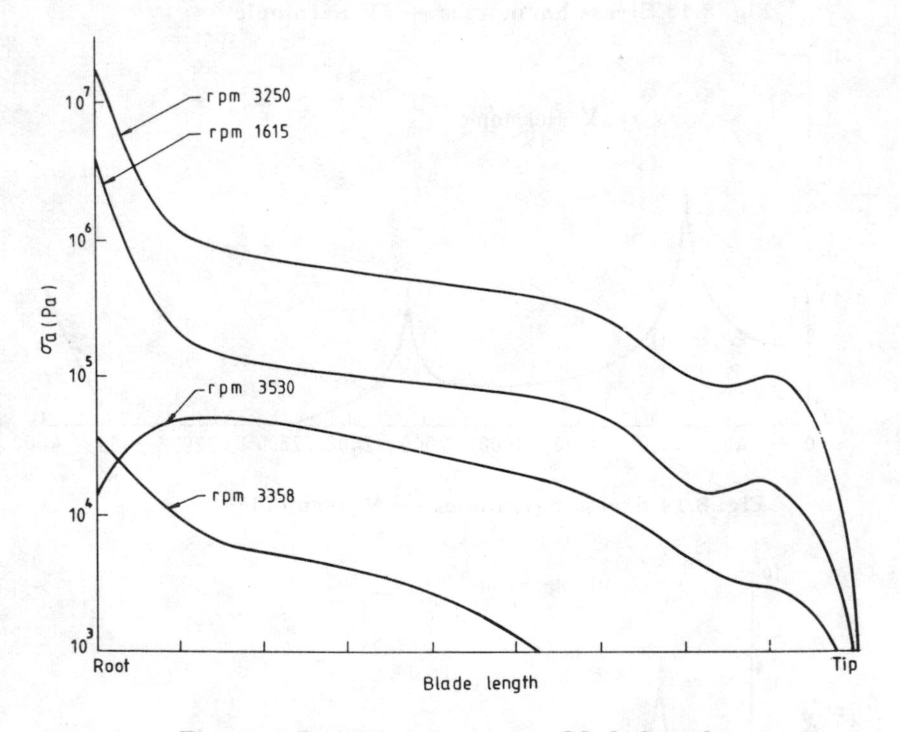

Fig. 8.16 Resonant stresses vs blade length

In Fig. 8.16 itself, in the stress distributions of RPM's 3250 and 1615, the humps near the blade tip demonstrate that the stress field is dependent on the relative levels of vibration and geometric properties along the blade length.

From Fig. 8.10, it can be observed that the resonance at 3250 RPM due to the first harmonic and first mode is quite pronounced and exists over a large range of rotor speeds, while that at 537 RPM, due to sixth harmonic and first mode, is quite weak. It can also be seen that the non-resonant stress value due to the first harmonic is higher than the resonant stress value due to the sixth harmonic at 537 RPM.

In Table 8.8, it can be noticed that for resonances at 537, 1763 and 3358 RPM caused by the interaction of the sixth harmonic with first, second and third natural modes respectively, the resonant stress values are respectively lower, the first coupled mode being more pronounced than the second than the third.

In Fig. 8.12, the resonant stress value at 1075 RPM due to the third harmonic, is 0.6465 MPa, which is lower than the stress value of 0.7620 MPa, due to the fourth harmonic at 805 RPM. This can be attributed to the fact that the amplitudes of the fourth order harmonic components of the forcing functions are larger than those of the third, see Table 8.5.

Comparisons between the stress measured experimentally on the blade and the theoretical stresses are illustrated in Figs. 8.17 to 8.22. The experimental stresses are computed from the measured strains using the formulae for the 120 rosette.

$$\sigma_{1,2} = \frac{E}{3(1-\mu)}\left[(e_1+e_2+e_3) \pm \sqrt{2(e_1-e_2)^2 + 2(e_2-e_3)^2 + 2(e_3-e_1)^2}\right]$$

$$\sigma_a = \sqrt{\sigma_1^2 + \sigma_2^2 - \sigma_1\sigma_2}$$

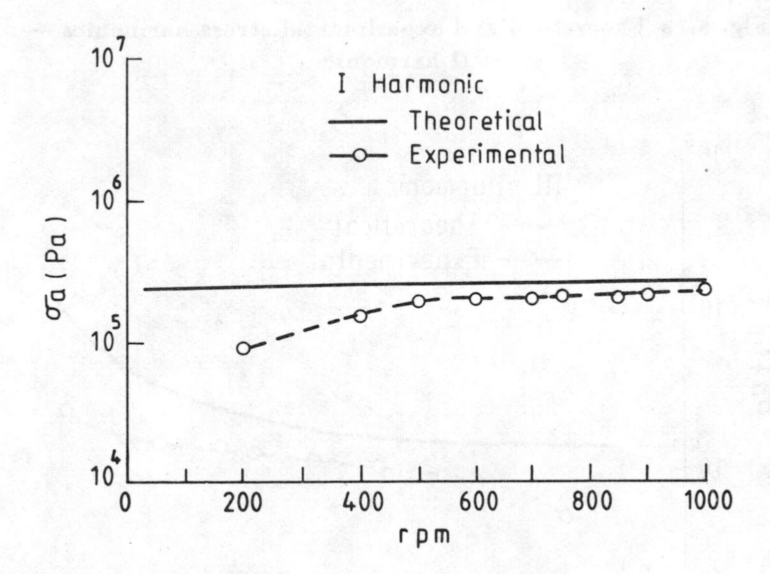

Fig. 8.17 Theoretical and experimental stress harmonics — I harmonic

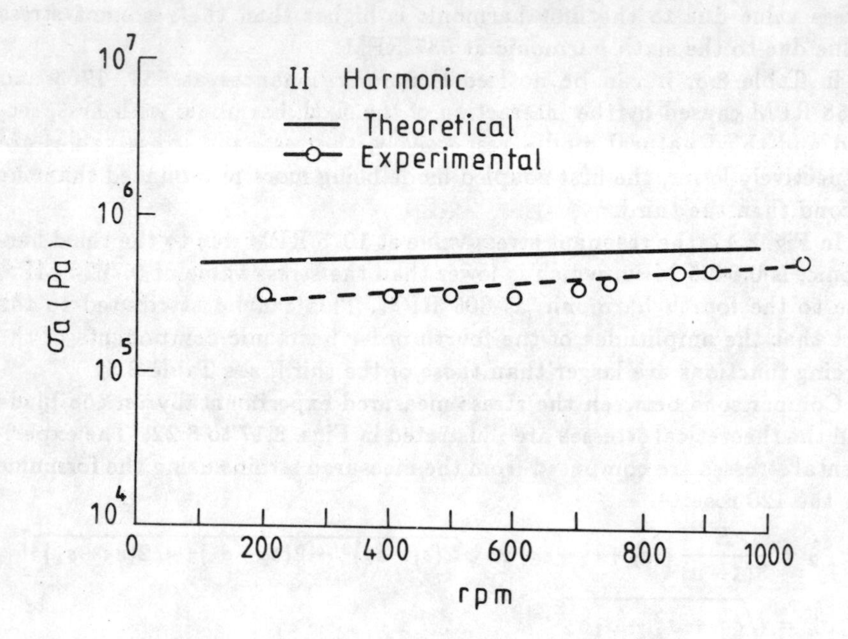

Fig. 8.18 Theoretical and experimental stress harmonics — II harmonic

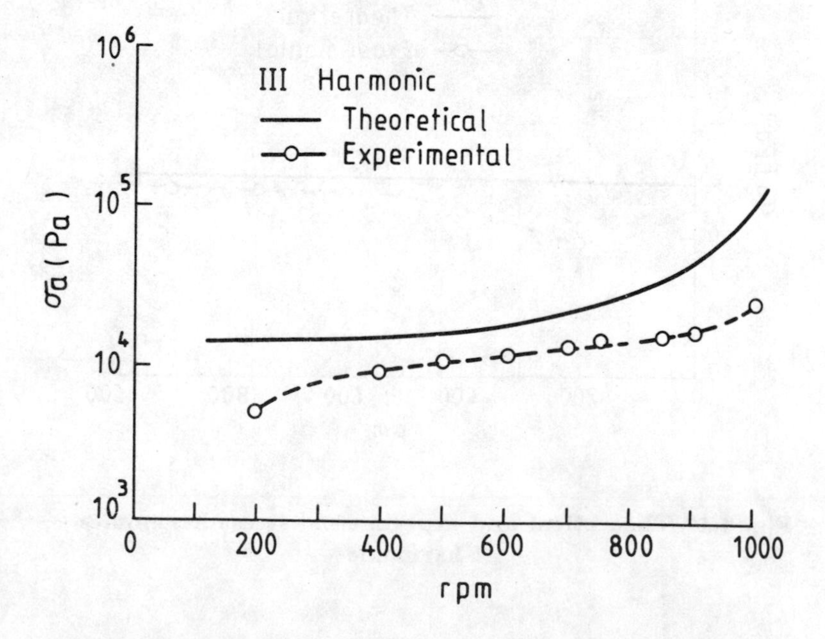

Fig. 8.19 Theoretical and experimental stress harmonics — III harmonic

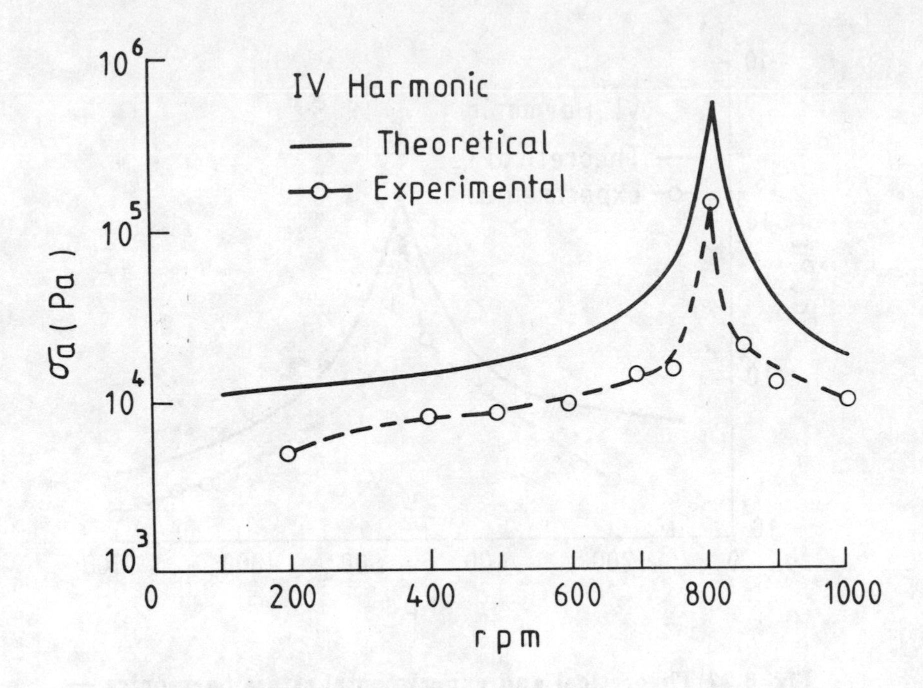

Fig. 8.20 Theoretical and experimental stress harmonics — IV harmonic

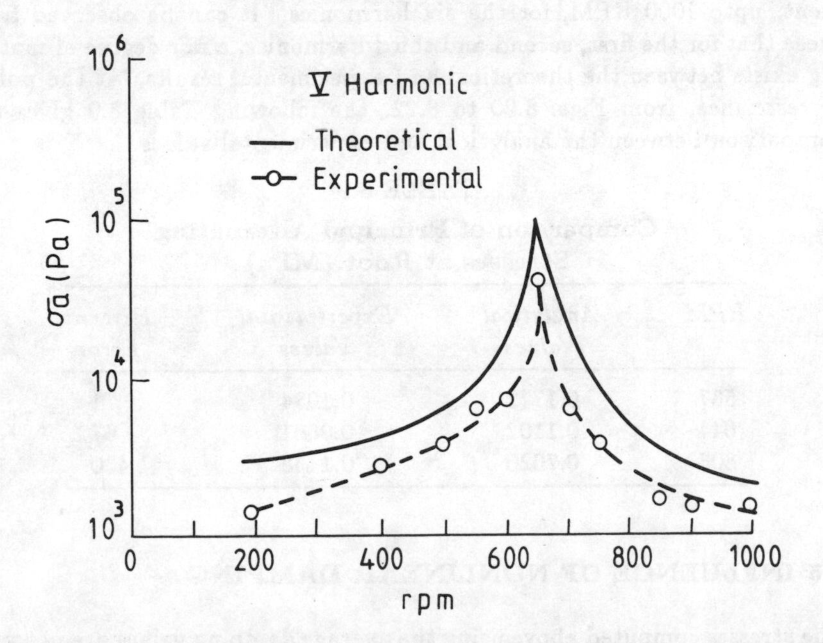

Fig. 8.21 Theoretical and experimental stress harmonics — V harmonic

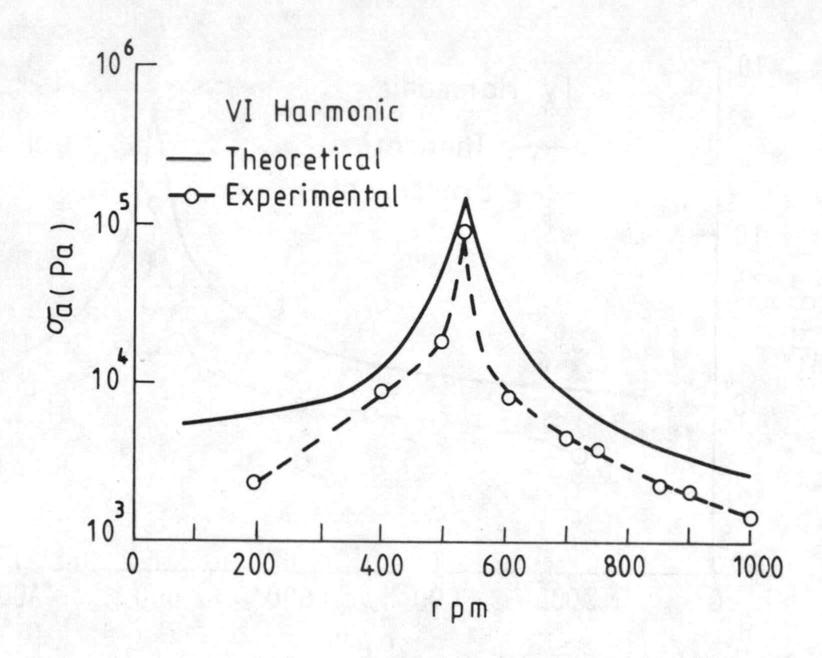

Fig. 8.22 Theoretical and experimental stress harmonics — VI harmonic

The plots show the basic alternating stresses at the point of measurement, upto 1000 RPM, for the six harmonics. It can be observed from these that for the first, second and third harmonics, a fair degree of matching exists between the theoretical and experimental results. At the points of resonance, from Figs. 8.20 to 8.22, the following Table 8.9 gives the comparison between the analytical and experimental values.

TABLE 8.9

Comparison of Principal Alternating Stresses at Root (MPa)

RPM	*Analytical Values*	*Experimental Values*	*Percent Error*
537	0.1772	0.1034	71
644	0.1102	0.0661	67
805	0.7620	0.1438	430

8.5 INFLUENCE OF NONLINEAR DAMPING

The stresses computed above using the average damping values are not very good, particularly at the higher speed 805 RPM. This is because the modal damping is predominantly dependent on the vibrational amplitude, which has been neglected. The analysis can be improved by using a nonlinear damping model for the modal analysis.

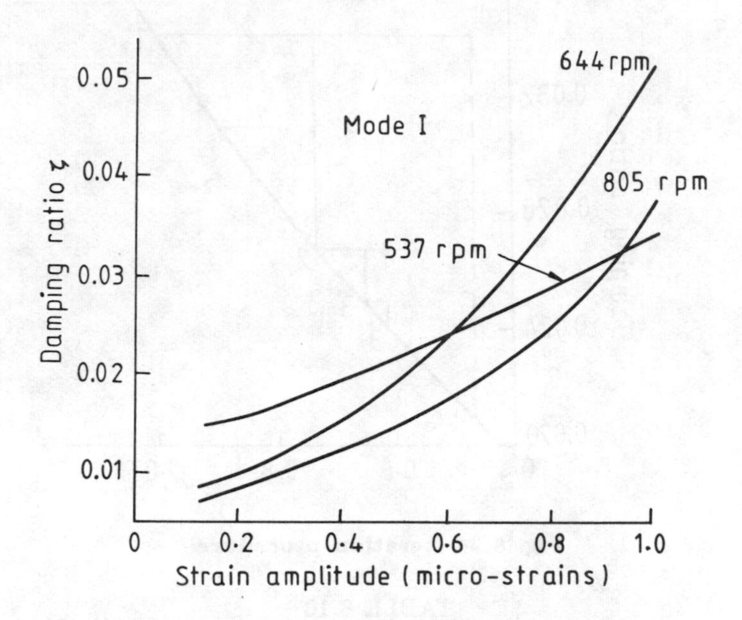

Fig. 8.23 Damping ratios vs strain amplitude

Fig. 8.23 shows the variation of damping ratios at resonant speeds of 537 RPM, 644 RPM and 805 RPM for the I mode. It may be noted here that the experimental investigations were restricted for rotor speeds upto only 1000 RPM. The values in this figure have been calculated using cycle by cycle decrement of strain amplitude. The resonances at 537 RPM, 644 RPM and 805 RPM are caused by the interaction of the first blade mode with the sixth, fifth and fourth orders of the nozzle passing frequency respectively.

The computer program described above is modified to generate modal damping surfaces in the rotor speed-strain amplitude plane, from the previously recorded experimental data, using curve fitting subroutines with sixth degree polynomials. The strain amplitude axis can be readily proportioned to stress amplitudes for all principal components of stress, using stress-strain relationships, since the blade material is isotropic and root effects can be assumed to remain approximately constant.

Starting with an assumed value of strain and corresponding modal damping ratio, point 1, Fig. 8.24, the principal component of stress and resultant strain is estimated, point 1′. The average of the assumed and estimated values of strain and the corresponding modal damping ratio are taken as the starting values for the next iteration. The process is repeated till two successive values of stress are obtained within a specified limit of

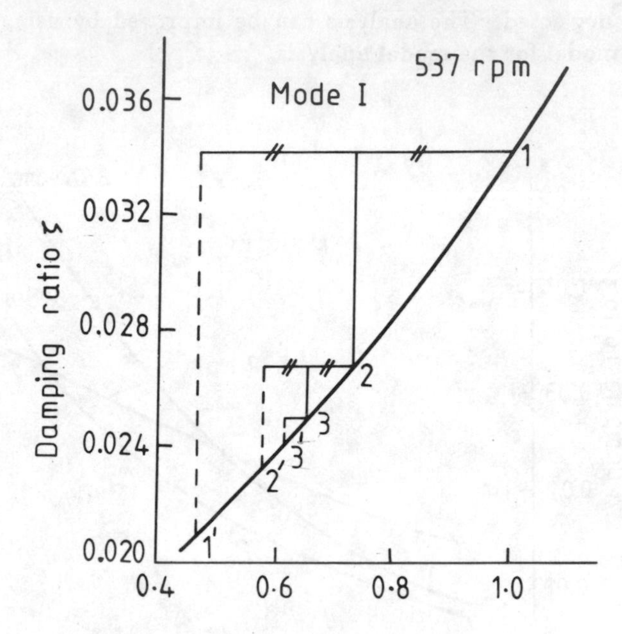

Fig. 8.24 Iteration procedure

TABLE 8.10

Iteration Values

Starting Values		*Resultant Values*	
Micro Strain	*Damping Ratio*	*Stress MPa*	*Micro Strain*
a. 537 RPM			
1.00000	0.03400	0.09098	0.45490
0.72747	0.02879	0.11487	0.57435
0.65091	0.02499	0.12295	0.61475
0.63284	0.02458	0.12496	0.62480
0.62882	0.02449	0.12541	0.62705
0.62794	0.02447	0.12551	0.62755
b. 644 RPM			
1.00000	0.05000	0.03311	0.16555
0.58278	0.02325	0.06988	0.34840
0.48557	0.01819	0.08871	0.44355
0.45456	0.01776	0.09082	0.45410
c. 805 RPM			
1.00000	0.03700	0.17128	0.85830
0.92814	0.03271	0.19316	0.98580
0.94696	0.03386	0.18672	0.93360
0.94028	0.03345	0.18898	0.94490
0.94259	0.03359	0.18819	0.94095
0.94178	0.03354	0.18847	0.94235

accuracy. The iterative steps and final principal stresses obtained within an accuracy of 0.1%, for resonant rotor speeds of 587 RPM, 644 RPM and 805 RPM are listed in Table 8.10.

A comparison with the corresponding values in Table 8.9 clearly shows the critical importance of incorporating the variation of damping ratio with vibrational amplitude for resonant calculations. A comparison between the theoretical stresses obtained by average damping method and the method above is given in Table 8.11 along with the experimental results.

TABLE 8.11
Comparison of Analytical and Experimental Stresses

Rotor Speed RPM	*Exptl. Stress MPa*	*Analytical Stress*			
		Avg. Damping	*% Error*	*Nonlinear Damping*	*% Error*
537	0.1034	0.1772	71	0.1255	21
644	0.0661	0.1102	87	0.0908	37
805	0.1438	0.7620	430	0.1885	31

It can be seen that the theoretical values obtained by nonlinear damping model have significantly reduced the gap between experimental and analytical results, particularly for the case of 805 RPM. This reinforces the critical aspect of the dependence of modal damping not only on the speed but also on vibrational amplitude.

REFERENCES

1. Plunkett, R., "Free and Forced Vibration of Rotating Blades", *J. Aero. Sci.*, **18**, p. 278 (1951).
2. Schnittger, J.R., "The Stress Problem of Vibrating Compressor Blades", *J. of Appl. Mechs., ASME*, p. 57 (1955).
3. Prohl, M.A., "A Method for Calculating Vibration Frequency and Stress of a Banded Group of Turbine Buckets", *Trans. ASME*, **80**, p. 169 (1958).
4. Weaver, F.L. and Prohl, M.A., "High Frequency Vibration of Steam Turbine Buckets", *Trans. ASME*, p. 181 (1958).
5. Rao, J.S. and Carnegie, W., "Nonlinear Vibration of Rotating Cantilever Beams", *J. Roy. Aero. Soc.*, **74**, p. 161 (1970).
6. Rao, J.S. and Carnegie, W., "Nonlinear Vibrations of Rotating Cantilever Beams Treated by Ritz Averaging Procedure", *J. Roy. Aero. Soc.*, **76**, p. 566 (1972).
7. Jones, D.I.G., "Effect of Slip on Response of a Vibrating Compressor Blade", *ASME*, Paper No. 77-WA/GT-3.
8. Jones, D.I.G. and Muszynska, A., "Vibration of Compressor Blade with Slip at the Root", *Shock Vib. Bull.*, **48**, Pt. 2, p. 53 (1978).
9. Jones, D.I.G. and Muszynska, A., "Harmonic Response of a Damped Two Degree of Freedom System with Gaps", *Nonlin. Vib. Problems*, **19**, p. 153 (1978).

10. Jones, D.I.G. and Muszynska, A., "Design of Turbine Blades for Effective Slip Damping at High Rotational Speeds", *Proc. 49th Shock Vib. Symp.* (1978).
11. Jones, D.I.G. and Muszynska, A., "Nonlinear Modelisation of Nonconservative Blade Vibration Response", 8th Intl. Conf. Nonlinear Oscillation, Prague (Sept. 11-16, 1978).
12. Muszynska, A. and Jones, D.I.G., "On the Discrete Modelization of Response of Blades with Slip and Hysteretic Damping", *Proc. 5th World Congr. IFToMM*, Montreal, ASME Publ., p. 641 (July 1979).
13. Rao, J.S. et al, *Unsteady Forces — Blade Response*, Project Rept. 1978-79, R. and D. Board, Ministry of Defence, India (1979).
14. Mukhopadhyay, V., Rao, S.S.P. and Rao, J.S., "Turbomachinery Blade Vibration due to Self Excitation and Neighboring Cascade Effect", 8th Natl. Conf. on Fluid Mech. and Fluid Power, India, p. 63 (Nov. 1978).
15. MacBain, J.C., *Transient Response of a Cantilevered Plate to Impact Using Holographic Interferometry and Finite Element Techniques*, Air Force Aero Propulsion Lab, Wright Patterson AFB, OH, Rept. No. AFAPL TR-76 56 (Aug. 1976).
16. Rieger, N.F. and Nowak, W.J., "Analysis of Fatigue Stresses in Steam Turbine Blade Groups", EPRI Seminar on Steam Turbine Availability, Palo Alto, CA (Jan. 1977).
17. Salama, A.L. and Petyt, M., "Dynamic Response of Packets of Blades by the Finite Element Method", *J. Mech. Des., Trans. ASME*, **100** (4), p. 660 (1978).
18. Srinivasan, A.V., Lionberger, S.R. and Brown, K.W., "Dynamic Analysis of an Assembly of Shrouded Blades Using Component Modes", *J. Mech. Des., Trans. ASME*, **100** (3), p. 520 (1978).
19. Fabunmi, J.A., "Forced Vibrations of a Single Stage Axial Compressor Rotor", *ASME*, Paper No. 79-GT-108.
20. Kimberling, M.C., *Slip Damping of Turbine Blades*, AFIT/GAE/AA. 78 D 11 (Dec. 1978).
21. Matsuura, T., "Blade Bending Vibration Induced by Wakes", *J. Mech. Engrg. Sci.*, **21** (5), p. 361 (1979).
22. Jones, D.I.G., *Vibrations of a Compressor Blade with Slip at the Root*, AFWAL-TR-80-4003, (1980).
23. Griffin, J.H., "Friction Damping of Resonant Stresses in Gas Turbine Engine Airfoils", *J. Engrg. for Power, ASME*, **102**, p. 329 (1980).
24. Fu-Sheng, C., Wassel, M. and Mosimann, J., "Combined Friction and Material Damping of Resonant Steam Turbine Blade Response", ASME 81-DET-136.
25. Hoyniak, D. and Fleeter, S., "Prediction of Aerodynamically Induced Vibrations in Turbomachinery", Winter Annual Meeting of ASME, p. 1 (Nov. 1981).
26. Tardos, R.N. and Botman, M., "Dynamic Response of Blades and Vanes to Wakes in Axial Turbomachinery", ASME 81-DET-33.
27. Rao, S.S.P., *Some Studies on Turbomachine Blade Dynamics and Response in Incompressible Flow*, Ph.D. Thesis, Ind. Inst. of Tech., Kharagpur (1980).
28. Jadvani, H.M. and Rao, J.S., "Forced Vibrations of Rotating Pre-Twisted Blades", *Proc. IFToMM Int. Conf. Rotordynamic Problems in Power Plants*, p. 259 (Sept. 28–Oct. 1, 1982).

29. Namura, K., "Tangential Vibration of Integral Turbine Blades due to Partial Admission", *Fluid Structure Interactions in Turbomachinery*, Winter Annual Meeting of ASME, p. 25 (Nov. 1981).
30. Rao, J.S., Jadvani, H.M. and Reddy, P.V., "Analytical and Experimental Investigation of Rotating Blade Response due to NPF Excitation", *Shock and Vib. Bull.*, **53**, Pt. 4, p. 85 (1983).
31. Jadvani, H.M., *Vibration of Turbine Blades*, Ph.D. Thesis, Ind. Inst. of Technology, Delhi (1982).
32. Rao, S.S.P., Mukhopadhyay, V. and Rao, J.S., "Self Excited Vibration of Turbomachine Blades", *Proc. 5th Int. Symp. Airbreathing Engines*, Bangalore, India (Feb. 16–21, 1981).
33. Sisto, F. and Chang, A.T., "The Influence of Gyroscopic Forces on the Dynamic Behaviour of Rotating Blades", *Proc. 5th Int. Symp. Airbreathing Engines*, Bangalore, India (Feb. 16–21, 1981).
34. Sisto, F., Chang, A.T. and Sutcu, M., "The Influence of Coriolis Forces on Gyroscopic Motion of Spinning Blades", 27th Int. Gas Turbine Conf., London (Apr. 18–22, 1982).
35. Sisto, F., Chang, A.T. and Sutcu, M., "Blade Excitation Due to Gyroscopic Forces", *Proc. IFToMM Int. Conf. Rotordynamic Problems in Power Plants*, p. 279 (Sep. 28–Oct. 1, 1982).
36. Sivaneri, N.T. and Chopra, I., "Dynamic Stability of a Rotor Blade Using Finite Element Analysis", *AIAA J.*, **20** (5), p. 716 (1982).
37. Kline, P.J., *Three Dimensional Resonant Vibrations and Stresses in Turbine Blade Groups*, M.S. Thesis, Rochester Inst. of Tech., Rochester (1981).
38. Rao, J.S. and Jadvani, H.M., "Free and Forced Vibration of Turbine Blades", *Vibrations of Bladed Disc Assemblies*, ASME Publication, p. 11 (1983).
39. Rao, J.S. and Vyas, N.S., "Response of Steam Turbine Blades Subjected to Distributed Harmonic Nozzle Excitation", *Proc. 3rd Intl. Modal Analysis Conf.*, p. 618 (1985).
40. Rao, J.S., Gupta, K. and Vyas, N.S., "Analytical and Experimental Investigations on Vibratory Stresses of A Rotating Steam Turbine Blade under NPF Excitation", *Proc. IFToMM Intl. Conf. Rotor Dynamics*, Tokyo, p. 289 (1986).
41. Rao, J.S. and Vyas, N.S., "Resonant Stress Determination of a Turbine Blade with Modal Damping as a Function of Rotor Speed and Vibrational Amplitude", ASME 89-GT-27, Gas Turbine and Aero Engine Congress and Exposition, Toronto, Canada (June 4–8, 1989).
42. Menq, C.H. and Griffin, J.H., "A Comparison of Transient and Steady State Finite Element Analyses of the Forced Response of a Frictionally Damped Beam", *J. Vib. Acoust. Stress Rel. Des., Trans. ASME,* **107**, p. 19 (1985).
43. Menq, C.H., Griffin, J.H. and Bielak, J., "The Influence of a Variable Normal Load on the Forced Vibration of a Frictionally Damped Structure", *J. Engrg. Gas Turbines Power, Trans. ASME*, **108**, p. 300 (1986).
44. Fu Sheng, C. and Mosimann, J.G., "Analysis of Friction Damped Resonant Stresses in Turbine Blades", *Vibrations of Blades and Bladed Disc Assemblies*, ASME Publication, p. 45 (1983).
45. Sinha, A. and Griffin, J.H., "Effects of Static Friction on the Forced Response of Frictionally Damped Turbine Blades", *J. Engrg. Gas Turbines Power, Trans. ASME,* **106**, p. 65 (1984).

Chapter 9

Transient Vibrations

9.1 INTRODUCTION

In Chapter 8, we have seen how to conduct a forced vibration analysis of rotating turbomachine blade subjected to steady excitation forces at nozzle passing frequency, with modal damping represented by either average equivalent viscous damping or nonlinear modal envelopes as a function of the amplitude of vibration.

Transient conditions of operation exist principally for two different reasons, one due to unsteady flow conditions arising out of steam or gas admission to the stator and the other due to acceleration and deceleration of the rotor during starting and shutting operations of the machine. Under transient conditions of steam or gas admission, the blades experience variable magnitudes of excitation force, which cause vibratory response primarily at the fundamental mode of vibration. The resulting stresses may be high and lead to low cycle fatigue problems of the blade. These stresses can be minimized by good operating procedures. However, the transient conditions of operation that exist during the machine start-up and shut-down conditions, when the rotor crosses the criticals on the Campbell diagram, see Fig. 8.9 and Figs. 8.10 to 8.15, generate resonant stresses at the corresponding modes of vibration. The stresses generated at such criticals are actually a function of the acceleration value and cannot be determined by the method of Chapter 8.

The study of determination of transient stresses has been a recent development. Irretier [1] used beam theory and modal approximation technique to present a numerical model for an untwisted blade to obtain transient vibration response due to partial admission during a change in rotational speed. He extended this analysis [2] to include the run-up conditions and later [3] considered nozzle excitation resonances also.

Vyas, Gupta and Rao [4] determined the transient response of a turbine blade during step-up and step-down operations, during which the blade may be excessively stressed while passing through resonant rotor speeds. They first studied the response of a stationary cantilever beam to a sinusoidal force with varying excitation frequency using Duhamel's integral. The response is obtained for damped and undamped systems in terms of non-dimensional parameters for different frequency sweep rates. The analysis is then extended to the case of a turbine blade, considered in Chapter 8. Reissner's dynamic functional for a tapered twisted asymmetric aerofoil cross-section blade mounted at a stagger angle on a disc under rotational acceleration was derived. Average modal damping values are assumed in the analysis. Results are presented in terms of stress levels for different harmonics of nozzle excitation and rotor acceleration rates for the start-up and shut-down operations. This chapter presents the analysis outlined in [4].

9.2 TRANSIENT RESPONSE OF A CANTILEVER

To establish the type of response of a system due to variable frequency excitation, let us first consider a simple case of a cantilever beam shown in

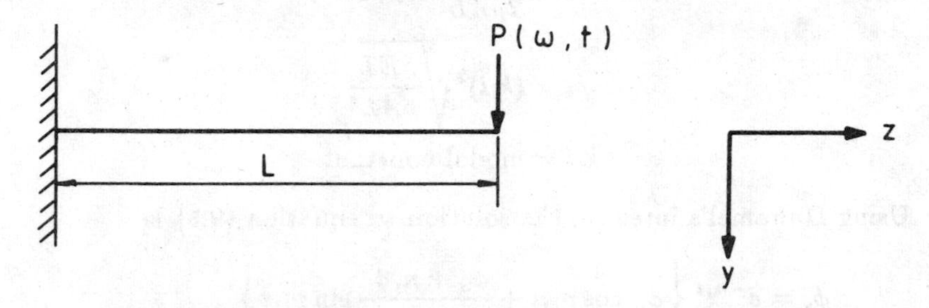

Fig. 9.1 Cantilever beam excited by a sinusoidal force with varying excitation frequency

Fig. 9.1. Let the beam be excited by a pulsating force $P(\omega, t)$ at the free end. Let the frequency ω be time dependent, such that

$$\omega = \omega_0 + \alpha t \tag{9.1}$$

where,

$$\omega_0 = \text{initial frequency}$$
$$\alpha = \text{sweep rate}$$
$$\omega = \text{instantaneous frequency}$$

The excitation force, for sine pulsation, is

$$P = P_0 \sin\left(\omega_0 t + \frac{1}{2}\alpha t^2\right) \tag{9.2}$$

The equation of motion of a uniform beam is

$$EI\frac{\partial^4 y}{\partial z^4} + m\frac{\partial^2 y}{\partial t^2} = P_0 \sin\left(\omega_0 t + \frac{1}{2}\alpha t^2\right) \tag{9.3}$$

The response of the beam can be written as

$$y = \sum_j \phi_j(t) Z_j(z) \tag{9.4}$$

where,

$$\phi_j(t) = \text{time function}$$
$$Z_j(z) = \text{normal mode}$$

Following modal analysis [5], and including proportional viscous damping, we can obtain

$$\ddot{\phi}_i + 2n_i\dot{\phi}_i + p_i^2\phi = F_0 \sin\left(\omega_0 t + \frac{1}{2}\alpha t^2\right) \tag{9.5}$$

where,

$$F_0 = \frac{P_0\{Z_i\}_{z=L}}{\rho AL}$$
$$n_i = \frac{C_v}{2\rho AL} = \zeta_i p_i$$
$$p_i = (k_i l)^2\sqrt{\frac{EI}{\rho AL^4}}$$
$$k_i l = \text{modal constant}$$

Using Duhamel's integral, the solution to equation (9.5) is

$$\phi_i = e^{-n_i t}\left\{\phi_0 \cos p_{1i}t + \frac{\dot{\phi}_0 + n_i\phi_0}{p_{1i}} \sin p_{1i}t\right\}$$
$$+ \frac{F_0}{p_{1i}}\int_0^t e^{-n_i(t-\tau)} \sin p_{1i}(t-\tau) \sin\left(\omega_0\tau + \frac{1}{2}\alpha\tau^2\right) d\tau \tag{9.6}$$

where,

$$p_{1i} = \sqrt{p_i^2 - n_i^2}$$

The first two terms in equation (9.6), which form the complementary function, take into account of the initial conditions, and the third term is the particular integral. Equation (9.6) can be written as

$$\phi_i = e^{-n_i t}\left\{\phi_0 \cos p_{1i}t + \frac{\dot{\phi}_0 + n_i\phi_0}{p_{1i}} \sin p_{1i}t\right\} + \frac{F_0}{p_{1i}}A_s(\sin p_{1i}t + \lambda_s) \tag{9.6a}$$

with

$$A_s = \frac{1}{2}\sqrt{(I_1 + I_2)^2 + (-I_3 + I_4)^2}$$
$$\lambda_s = \tan^{-1}\frac{-I_3 + I_4}{I_1 + I_2}$$
$$I_1 = \int_0^t e^{-n_i(t-\tau)} \sin\left[\left(\omega_0\tau + \frac{1}{2}\alpha\tau^2\right) - p_{1i}\tau\right] d(p_{1i}\tau)$$

$$I_2 = \int_0^t e^{-n_i(t-\tau)} \sin\left[\left(\omega_0\tau + \frac{1}{2}\alpha\tau^2\right) + p_{1i}\tau\right] d(p_{1i}\tau)$$

$$I_3 = \int_0^t e^{-n_i(t-\tau)} \cos\left[\left(\omega_0\tau + \frac{1}{2}\alpha\tau^2\right) - p_{1i}\tau\right] d(p_{1i}\tau)$$

$$I_4 = \int_0^t e^{-n_i(t-\tau)} \cos\left[\left(\omega_0\tau + \frac{1}{2}\alpha\tau^2\right) + p_{1i}\tau\right] d(p_{1i}\tau)$$

Similarly, for a cosine excitation

$$P_0 \cos\left(\omega_0 t + \frac{1}{2}\alpha t^2\right)$$

the response is

$$\phi_i = e^{-n_i t}\left\{\phi_0 \cos p_{1i}t + \frac{\dot{\phi}_0 + n_i\phi_0}{p_{1i}} \sin p_{1i}t\right\} + \frac{F_0}{p_{1i}} A_c(\sin p_{1i}t + \lambda_c) \tag{9.7}$$

where,

$$A_c = \frac{1}{2}\sqrt{(-I_1 + I_2)^2 + (I_3 + I_4)^2}$$

$$\lambda_c = \tan^{-1}\frac{I_3 + I_4}{-I_1 + I_2}$$

The response of the cantilever beam can be determined from equation (9.6) for $\omega_0 = 0$ and zero initial conditions. The following nondimensional parameters can be used.

$\dfrac{\phi_i}{\delta_{st}}$ = dynamic magnification factor

$\delta_{st} = \dfrac{F_0}{p_{1i}^2}$

$\dfrac{\alpha t}{p_{1i}}$ = nondimensional instantaneous velocity

$\dfrac{\alpha}{p_{1i}^2}$ = nondimensional sweep rate

$\dfrac{n_i}{p_i} = \varsigma_i$ = nondimensional damping

In Fig. 9.2, the amplitude of the cyclic variation of the Dynamic Magnification factor is plotted against nondimensional instantaneous velocity for different values of nondimensional sweep rates with no damping. It is well known that when the excitation frequency is constant, resonance occurs when $\omega = p_{1i}$. In the absence of damping, infinite amplitude of vibration would occur, when $\omega = p_{1i}$. However, it takes time for these amplitudes to build up. As observed in the figure, the amplitude of vibration builds up as the instantaneous frequency αt approaches p_{1i}. At $\alpha t = p_{1i}$, infinitely large amplitudes would have built up, had the system been allowed to dwell at this point. The amplitude of vibration starts

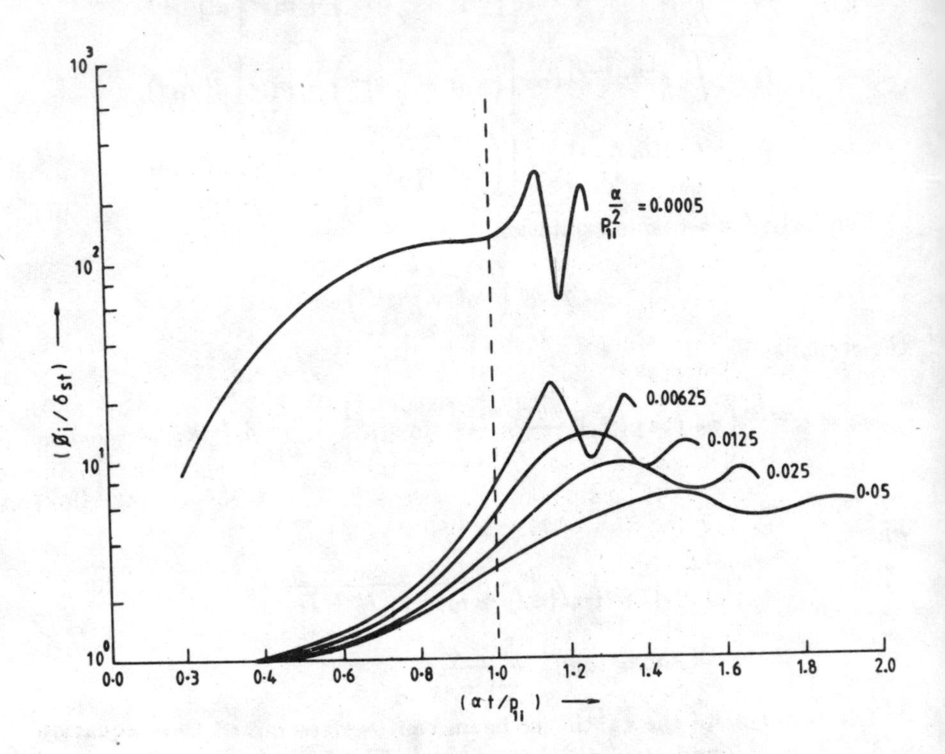

Fig. 9.2 Undamped response ($n/p_i = 0.00$) (dynamic magnification vs non-dimensional instantaneous excitation frequency)

building up at this point, but once past $\alpha t = p_{1i}$, the post resonance effect comes into play. Under these influences a large but finite amplitude of vibration is obtained, not at $\alpha t = p_{1i}$, but at an instance shifted from this position, e.g., at $\alpha t/p_{1i} = 1.112$ for $\alpha/p_{1i}^2 = 0.0005$. It can be observed that for larger values of nondimensional sweep rates, the peak amplitudes get reduced and are shifted farther from linear resonance case. This is so because, for larger sweep rates, the transit through resonance being faster, the energy supplied is utilized more in accelerating the system than in building up the amplitudes of vibration.

Fig. 9.3 illustrates the response for similar sweep rates and similar time instance as in Fig. 9.2, but with a damping ratio $\varsigma = 0.1$. With a finite amount of damping in the system, a portion of the energy supplied is dissipated, leaving less energy for building up of the vibration amplitudes. As can be observed in Fig. 9.3, the resonant peaks are lower than those in Fig. 9.2, significant among them being the lower sweep rate $\alpha/p_{1i}^2 = 0.0005$. We can now make a similar analysis for the case of a turbine blade subjected to rotor accelerations, by extending the theory in section 8.4.

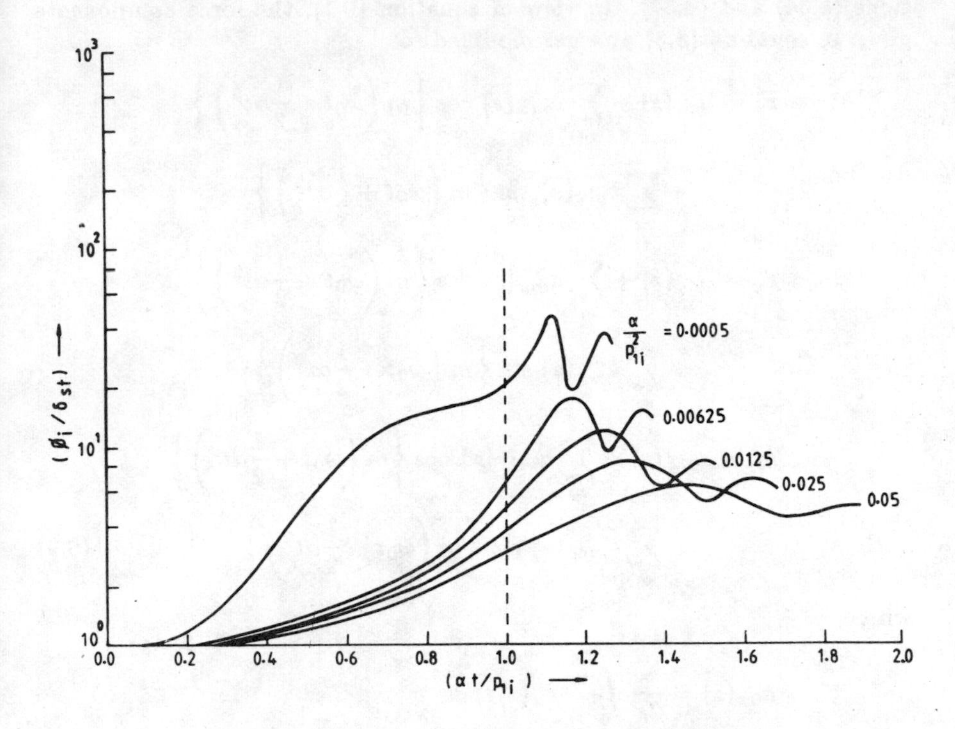

Fig. 9.3 Undamped response ($n/p_i = 0.01$) (dynamic magnification vs non-dimensional instantaneous excitation frequency)

9.3 RESPONSE OF ACCELERATING TURBINE BLADE

Referring to Chapter 2, the kinetic energy expression is given by equation (2.22) in which ω is substituted from equation (9.1).

$$\begin{aligned}T = &\frac{\rho}{2}\int_0^L A\left[(\dot{x} + r_y\dot{\theta})^2 + (\dot{y} - r_x\dot{\theta})^2\right]dz + \frac{1}{2}\int_0^L I_{cg}\dot{\theta}^2\,dz \\ &+ \frac{\rho}{2}(\omega_0 + \alpha t)^2\Big[-\int_0^L \Big\{(x' + r_y'\theta + r_y\theta')^2 \\ &+ (y' - r_x'\theta - r_x\theta')^2\Big\}(RI_1 + I_2)dz \\ &+ \sin^2\phi \int_0^L A(x + r_y\theta)^2 dz + \cos^2\phi \int_0^L A(y - r_x\theta)^2 dz \\ &- \sin 2\phi \int_0^L A(x + r_y\theta)(y - r_x\theta)\,dz\Big]\end{aligned} \tag{9.8}$$

The strain and complemetary energy terms remain the same as in equations (8.36) and (8.37). In view of equation (9.1), the force components given in equation (8.3) now get modified to

$$F_x = a_{0x}(z) + \sum_m a_{mx}(z) \cos\left\{m\left(\omega_0 t + \frac{1}{2}\alpha t^2\right)\right\}$$
$$+ \sum_m b_{mx}(z) \sin\left\{m\left(\omega_0 t + \frac{1}{2}\alpha t^2\right)\right\}$$
$$F_y = a_{0y}(z) + \sum_m a_{my}(z) \cos\left\{m\left(\omega_0 t + \frac{1}{2}\alpha t^2\right)\right\}$$
$$+ \sum_m b_{my}(z) \sin\left\{m\left(\omega_0 t + \frac{1}{2}\alpha t^2\right)\right\}$$
$$M = a_{0M}(z) + \sum_m a_{mM}(z) \cos\left\{m\left(\omega_0 t + \frac{1}{2}\alpha t^2\right)\right\}$$
$$+ \sum_m b_{mM}(z) \sin\left\{m\left(\omega_0 t + \frac{1}{2}\alpha t^2\right)\right\} \tag{9.9}$$

where,

$$a_{0x}(z) = \frac{\nu}{2\pi}\int_0^{\frac{2\pi}{\nu}} F_x(z,t)\,dt$$
$$a_{mx}(z) = \frac{\nu}{\pi}\int_0^{\frac{2\pi}{\nu}} F_x(z,t)\cos\left\{m\left(\omega_0 t + \frac{1}{2}\alpha t^2\right)\right\}dt$$
$$b_{mx}(z) = \frac{\nu}{\pi}\int_0^{\frac{2\pi}{\nu}} F_x(z,t)\sin\left\{m\left(\omega_0 t + \frac{1}{2}\alpha t^2\right)\right\}dt$$
$$\text{etc.,} \tag{9.10}$$

where the nozzle passing frequency ν is

$$\nu = n_s\omega = n_s(\omega_0 + \alpha t) \tag{9.11}$$

Following the procedure outlined in section 8.4, we get

$$M_k(\ddot{\eta}_k + 2\zeta_k p_k \dot{\eta}_k + p_k^2 \eta_k) = Q_{0N_k} + \sum_m Q_{mN_k} \cos\left\{m\left(\omega_0 t + \frac{1}{2}\alpha t^2\right)\right\}$$
$$+ \sum_m Q_{(m+6)N_k} \sin\left\{m\left(\omega_0 t + \frac{1}{2}\alpha t^2\right)\right\} \tag{9.12}$$

The solution to the above can be obtained by using Duhamel's integral, see reference [5], as

$$\begin{aligned} \eta_{0_k} &= I_0 \\ \eta_{m_k} &= A_c \cos(p_{1k}t + \lambda_c) \\ \eta_{(m+6)_k} &= A_s \cos(p_{1k} + \lambda_s) \end{aligned} \tag{9.13}$$

where,

$$A_c = \frac{1}{2}\sqrt{(-I_{1c} + I_{2c})^2 + (I_{3c} + I_{4c})^2}$$
$$A_s = \frac{1}{2}\sqrt{(I_{1s} + I_{2s})^2 + (-I_{3s} + I_{4s})^2}$$
$$\lambda_c = \tan^{-1}\frac{I_{3c} + I_{4c}}{-I_{1c} + I_{2c}}$$
$$\lambda_s = \tan^{-1}\frac{-I_{3s} + I_{4s}}{I_{1s} + I_{2s}} \tag{9.14}$$

and

$$I_0 = \int_0^t \frac{Q_{0N_k}}{p_{1k}M_k} e^{-n_k(t-\tau)} \sin p_k(t-\tau)\, d\tau$$
$$I_{1c} = \int_0^t \frac{Q_{mN_k}}{p_{1k}M_k} e^{-n_k(t-\tau)} \sin\left\{m\left(\omega_0\tau + \frac{1}{2}\alpha\tau^2\right) - p_{1k}\tau\right\} d\tau$$
$$I_{2c} = \int_0^t \frac{Q_{mN_k}}{p_{1k}M_k} e^{-n_k(t-\tau)} \sin\left\{m\left(\omega_0\tau + \frac{1}{2}\alpha\tau^2\right) + p_{1k}\tau\right\} d\tau$$
$$I_{3c} = \int_0^t \frac{Q_{mN_k}}{p_{1k}M_k} e^{-n_k(t-\tau)} \cos\left\{m\left(\omega_0\tau + \frac{1}{2}\alpha\tau^2\right) - p_{1k}\tau\right\} d\tau$$
$$I_{4c} = \int_0^t \frac{Q_{mN_k}}{p_{1k}M_k} e^{-n_k(t-\tau)} \cos\left\{m\left(\omega_0\tau + \frac{1}{2}\alpha\tau^2\right) + p_{1k}\tau\right\} d\tau$$
$$I_{1s} = \int_0^t \frac{Q_{(m+6)N_k}}{p_{1k}M_k} e^{-n_k(t-\tau)} \sin\left\{m\left(\omega_0\tau + \frac{1}{2}\alpha\tau^2\right) - p_{1k}\tau\right\} d\tau$$
$$I_{2s} = \int_0^t \frac{Q_{(m+6)N_k}}{p_{1k}M_k} e^{-n_k(t-\tau)} \sin\left\{m\left(\omega_0\tau + \frac{1}{2}\alpha\tau^2\right) + p_{1k}\tau\right\} d\tau$$
$$I_{3s} = \int_0^t \frac{Q_{(m+6)N_k}}{p_{1k}M_k} e^{-n_k(t-\tau)} \cos\left\{m\left(\omega_0\tau + \frac{1}{2}\alpha\tau^2\right) - p_{1k}\tau\right\} d\tau$$
$$I_{4s} = \int_0^t \frac{Q_{(m+6)N_k}}{p_{1k}M_k} e^{-n_k(t-\tau)} \cos\left\{m\left(\omega_0\tau + \frac{1}{2}\alpha\tau^2\right) + p_{1k}\tau\right\} d\tau$$
$$n_k = \zeta_k p_k \tag{9.15}$$

For the data of the blade in section 8.4 and damping given in Fig. 7.25, the response is obtained for the start-up of the rotor from 0 to 4000 rpm and shut-down at two different acceleration rates ±800 and ±1600 rpm/min. The response is obtained in terms of stress field at the root as a function of instantaneous rotor speed. Table 9.1 gives the stress levels for instantaneous rotor critical speeds. Figs. 9.4 and 9.5 illustrate graphically the response during start-up and shut-down operations. It can be noted from Table 9.1 and Fig. 9.4 that the stress levels are lower for an accelerating blade as compared with those obtained for constant speed operation. Also in compliance with the analysis of Figs. 9.2 and 9.3, the stress levels for sweep rates 800 rpm/min are higher than those for sweep rates 1600

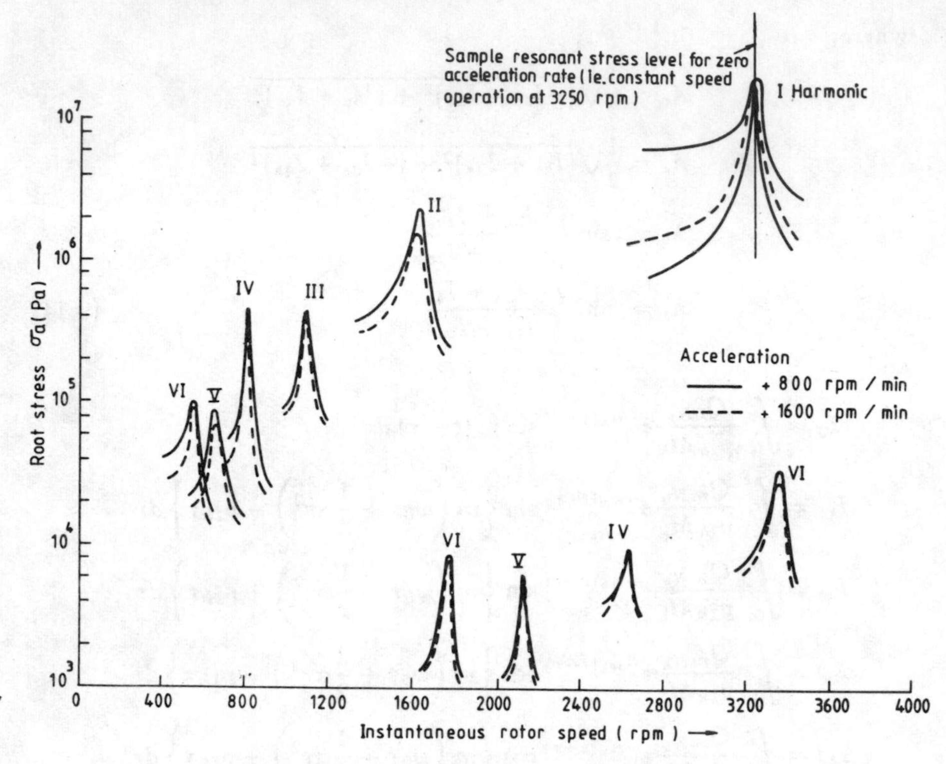

Fig. 9.4 Stress response during step-up

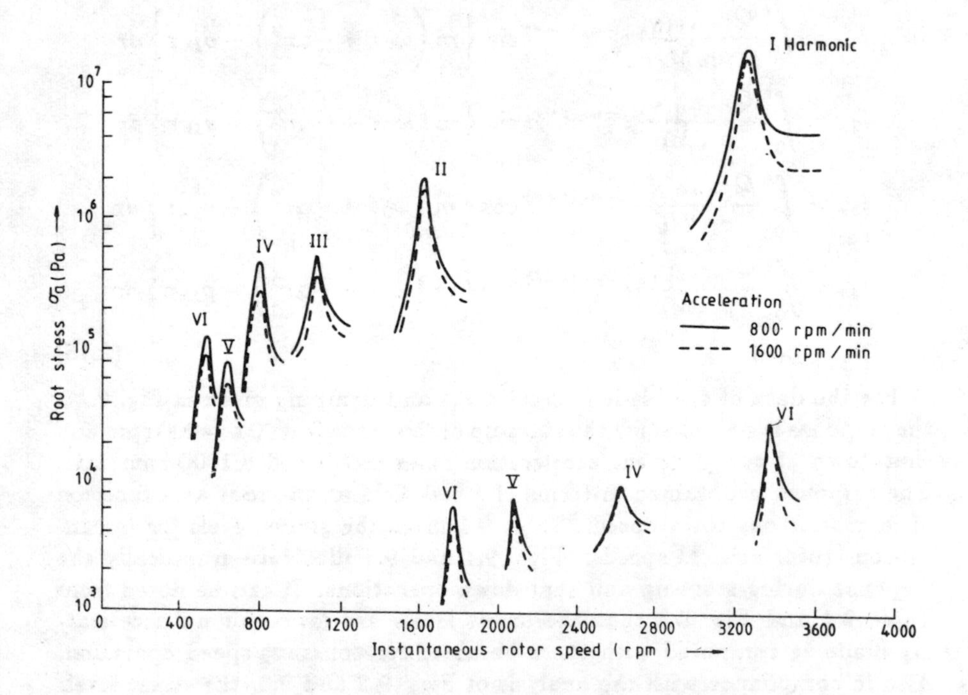

Fig. 9.5 Stress response during step-down

rpm/min. Hence faster rate of transit through resonances during step-up/down operations of a turbine is advisable to incur less harmful stresses on the blade. We will use this information once again in Chapter 11, while studying the life estimation procedures.

TABLE 9.1

Transient Resonant Stress σ_s (MPa) Harmonics

Rotor Speed	*Accel. Values in RPM per Minute*				
RPM	0	800	1600	−800	−1600
537	0.177	0.096	0.092	0.126	0.084
644	0.110	0.097	0.070	0.079	0.053
805	0.762	0.471	0.312	0.454	0.276
1075	0.647	0.451	0.423	0.525	0.377
1615	3.851	2.337	1.625	2.117	1.688
1763	0.160	0.009	0.008	0.007	0.004
2116	0.119	0.007	0.006	0.008	0.007
2646	0.137	0.010	0.009	0.010	0.099
3250	23.589	20.800	18.550	20.790	18.480
3358	0.040	0.038	0.037	0.036	0.035

In section 8.5, we considered the influence of nonlinear damping on resonant response and found it plays a significant role in limiting the peak amplitudes. The influence of nonlinear damping on transient response was studied by Rao and Vyas [6] using similar iteration procedure outlined in section 8.5. For determining the transient resonant stresses, it may be important to consider such an analysis.

REFERENCES

1. Irretier, H., "Numerical Analysis of Transient Responses in Blade Dynamics", *Vibrations in Rotating Machinery*, I. Mech. E., p. 255 (1984).
2. Irretier, H., "Computer Simulation of the Run-up of a Turbine Blade Subjected to Partial Admission", *Proc. 10th ASME Conf. Mech. Vib. Noise*, p. 1 (1985).
3. Irretier, H., "Transient Vibrations of Turbine Blades due to Passage Through Partial Admission and Nozzle Excitation Resonances", *Proc. IFToMM Intl. Conf. Rotor Dynamics*, Tokyo, p. 301 (1986).
4. Vyas, N.S., Gupta, K. and Rao, J.S., "Transient Response of Turbine Blade", *Proc. 7th World Cong. IFToMM*, Sevilla, Spain, p. 697 (1987).
5. Rao, J.S. and Gupta, K., *Theory and Practice of Mechanical Vibrations*, John Wiley & Sons (1984).
6. Rao, J.S. and Vyas, N.S., "Transient Stress Response of a Turbine Blade under Nonlinear Damping Effects", *ASME*, 90-GT-269.

Chapter 10

Coupled Blade-Disc Vibrations

10.1 INTRODUCTION

In modern aircraft gas turbines, catastrophic failures have occurred when bladed-disc assemblies have failed owing to large amplitudes of vibrations. To avert such failures it is essential to accurately predict the natural frequencies and mode shapes of the assembly in order to avoid strong resonances with anticipated excitation frequencies in the operating speed range of the rotor.

As we have seen in earlier chapters, considerable research has been undertaken with the object of developing methods for vibration analysis of the complex shaped blading of modern turbomachinery. However, the importance of analyzing the complete bladed-disc assembly including the shroud, in order to study turbine blade vibration is not well established, the main reason being that interblade coupling through the disc contributes significantly to the nature of the basic vibration properties. The frequencies of actual engine discs are controlled by the geometry of the disc and the method of attaching it to the drive shaft. In the lower order modes, the frequencies are mainly influenced by the thickness of the disc hub and the radius of attachment, while the higher order modes are more influenced by the rim thickness and the geometry of the blade attachment. Thus, for a realistic situation, it becomes important to consider the non-uniformity in the thickness of the disc (disc rim and hub), the blading and the twist of the blading. Experimentally obtained data for natural frequencies and mode shapes and also for resonant vibration response levels can only be fully interpreted by refering to such an analysis.

Since early twentieth century, attempts are being made to study vibration characteristics of bladed-disc assemblies. Stodola [1], Sezawa [2] and Smith [3] established the resulting differential equations, but due to algebraic complexity involved in solving the differential equations of motion of such assemblies, exact solutions could not be obtained. Since the classical work of Campbell [4] and Stodola [1], the vibrations of a rotating disc has been well understood. These vibrations were found to comprise of wave patterns involving integer numbers of nodal diameters and nodal circles; these patterns rotating forward and backward in the disc. The detailed investigation by Campbell [4] on full scale turbine discs, combined with evidence from the disc failures, indicated the serious nature of the standing wave. His approach, which was based on simple mathematical identity, assumes that in a rotating disc, the forward travelling wave is heavily damped by windage, thus leaving the backward moving wave virtually in command. Forward and backward moving waves are stated in relation to the rotation of the disc; this approach is contrary to the dynamic theory, since it presupposes that one half of normal wave can behave independently of the other. The vibration characteristics of blades alone have been studied extensively by several authors, by continuum models as well as discretized models; three recent review articles for this may be referred [5, 6, 7].

A set of blades mounted on a rotor exhibit more complex vibration characteristics than do single or packeted cantilever blades. Armstrong [8] is one of the earliest workers to discuss in detail the coupling between the turbine disc and the blade for vibration analysis. He used receptance technique [9] to couple the disc and blade motions. His analysis [10] has been made with the assumption that all the blades are identical. By this means the receptance coupling procedure is applied to just two components only, the disc and the blading. In addition to that, the method also presupposes the form of modes of vibration of the assembly and is effectively restricted to systems with a large number of blades. His work can be outlined as the oscillations in a coupled blade-disc mode characterised by diametral and circular modes. The blade, being constrained in the disc at the rim, will vibrate in bending motion at antinodes, in torsional motion at nodes and in combined bending-torsion mode elsewhere. The circular modes may lie in the disc, but will more often be located in the blades too. He also carried out extensive tests on assemblies comprising of uniform thickness discs and uniform untwisted blades. He [11] described different blade vibration techniques as were in practice in 1966.

The blade mistuning case was studied by Whitehead [12] and was differentiated from flutter, in which the effect of mistuning is always favourable. He investigated the effect of blade mistuning on the vibrations of blades induced by wakes in the incident gas flow which rotate relative to the blades in question. Wagner [13] also considered the case of detuned blades.

Ewins [14] after Armstrong made significant contributions to bladed-disc analysis by receptance technique while considering multi-stage discs and the effect of detuning. Dye and Henry [15] discussed the aspect of detuning and its importance in the failures of 'rough' blades, (blades involved in unexplained failures are termed as rough blades). They examined two situations, since the blades of a particular stage must in practice have slightly varying natural frequencies and the rotor disc on which the blades are mounted will be to a certain extent flexible and will transmit forces between any blade and its neighbours. Their intention was to determine if these two situations could simulate rough blade conditions. Ewins [16] has later shown the effects of detuning upon the forced vibration of bladed-discs. He used a simplified model to study the implications of a real situation, i.e., blades being non-identical owing to manufacturing process, from consideration of the vibration levels which these components might suffer in practice. He described the dynamic characteristics of a system having pairs of close natural modes, when it is forced to vibrate under typical operating conditions with excitation being derived from presence of obstruction in flow. Ewins [17] examined the condition of resonance coincidence in bladed-discs and also the possibility that minor dimensional variations between one unit and the other could result in their exhibiting significantly different vibration characteristics. It was shown that the blade imperfections cause vibration levels 10 per cent in excess of those of tuned blades. He [18] later reviewed the whole situation of the vibrations of a bladed turbine wheel, and fully investigated the case of mistuning of 24 bladed-disc assembly. The finite element method was used by Kirkhope and Wilson [19], and the effects of rotation and temperature loading on vibrations of turbine assemblies were investigated. Ewins [20] summarized all his previous works and presented an analysis to predict the natural frequencies and mode shapes of uniform bladed-discs. He also concluded that a detuned system is susceptible to many more resonances than is an equivalent tuned system.

Cottney and Ewins [21] made an investigation aimed at improving the efficiency of shrouded bladed-disc assemblies. The assembly was analysed by studying each component individually and then combining them together with a receptance coupling technique by matching forces and displacements at each connecting point. A study was also undertaken to establish which of the six coordinates at each junction need to be included in this process. The results clearly indicated the necessity of including axial translation, tangential rotation and radial rotation, whereas others may be omitted with a little loss of accuracy.

Kirkhope and Wilson [22] used finite element method for the bladed-disc assemblies. Their theory is based on dynamic stiffness concept rather than the receptance concept. This method ultimately results in an algebraic eigenvalue problem for the modes and frequencies and offers significant computational advantages. Ewins and Sadasiva Rao [23] studied the effect of blade damping on the free response of the bladed-discs; only material damping was considered. The effect of medium and heavy damping on the forced vibration response of mistuned assemblies was reported. They [24], in a subsequent study considered the effect of aerodynamic damping. The aerodynamic damping was modelled by the inclusion of discrete viscous dampers between each blade tip and ground. The study of vibration of bladed-discs with a continuous part-span-shroud was reported subsequently [25].

Salama et al [26] analyzed the bladed-disc by a wave propagation technique and a matrix difference method. Thomas and Sabuncu [27] studied the vibration characteristics of an asymmetric cross sectional rotating bladed-disc using a finite element analysis with a wave propagation technique. Irretier [28] has solved the differential equation of motion of bladed-disc assemblies using Runge-Kutta integration and a shooting method.

As discussed in Chapter 5, significant work on packeted blades mounted on a rigid disc has been proposed by Prohl [29]. Ewins and Imregun [30] have included the effect of flexibility of the disc on the vibration of packeted blades by using receptance technique via a substructuring process. Blade-disc inertial coupling in tuned and severely mistuned bladed-discs have been studied analytically and experimentally by Crawley [31]. Later [32], he included the stagger angle dependence in this analysis. Ferrari et al [33] presented results on the predictions of natural frequencies and mode shapes of bladed rotating axisymmetric structures of jet engines. Karadag [34] modelled bladed-discs with thin and thick 18 degrees of freedom beam and plate elements with conforming sector disc elements. He determined shear centre effects on practical bladed-discs.

Shiga [35–37] studied the characteristics of natural frequencies of bound blades of the entire disc periphery of a steam turbine. Component mode synthesis has been used by Irretier [38] to study mistuned bladed-disc assemblies. A mode localization phenomenon associated with mistuning has been given by Valero and Bendicksen [39]. Griffin and Hoosac [40] made statistical investigation of blade mistuning. Aerodynamic and structural interaction have been studied by Basu and Griffin [41] among all blades and neighbouring ones. Optical data acquisition system has been described for a 56 bladed rotor by Lawrence and Meyn [42]. Nishiwaki et al [43] described similar tests for impellers. Rao et al [44, 45] conducted tests on an Orpheus engine bladed-disc assembly. Dominic [46] tested the bladed-disc assembly of the high pressure fuel turbo-pump of the main engine of the space shuttle in a spin pit.

MacBain and Stange [47] used holographic methods to determine the vibratory nature of mistuned rotating bladed-discs. Kuo [48] used NASTRAN based finite element models to demonstrate how the coupling between blade and disc subsystems would alter the entire bladed-disc assemblies. Elchuri [49, 50] developed added capabilities to NASTRAN to study the normal mode, aeroelastic and forced response of turbine and advanced turbo-propeller bladed-discs. Swaminadham et al [51] introduced a near automated computer procedure to generate a geometric model of a complex bladed-disc assembly using the PATRAN-G graphics program and the finite element data compatible to any general purpose code, particularly to COSMIC/NASTRAN to determine normal modes. They also conducted laser holographic experiments to identify and record system mode shapes of a complex shaped compressor disc with curved and swept blades.

Forced vibration response of mistuned bladed-discs has been studied by Afolabi [52, 53] and by Ewins and Han [54]. Kubota et al [55] considered stationary distributed force excitation on a rotating bladed-disc by both analytical and experimental methods. A closed form expression for the maximum forced resonant response of mistuned bladed-discs has been derived by MacBain and Whaley [56]. Experimental work and test results have been reported by several authors [57–62]. Whaley and Macbain [63] studied forced vibration of a bladed-disc as a function of mistuning for aeroelastic coupling in subsonic flow. Kaza and Kielb [64] used a geometric nonlinear theory and Hamilton's principle to derive aeroelastic equations of mistuned bladed-discs; they considered both structural and aerodynamic coupling in determining flutter response [65]. Bendicksen [66] considered a mistuned fan rotor and showed that a critical reduced frequency exists below which mistuning alone cannot stabilize a rotor. Griffin and Sinha [67] considered the interaction between mistuning and friction in the forced response of bladed-disc assemblies.

Statistics of the forced response of mistuned bladed-disc assemblies have been considered by Sinha [68]. Muszynska and Jones [69, 70] developed simple discrete models for a bladed-disc, taking into account friction and structural damping, mistuning and several types of excitation. The use of friction dampers for torsional vibration flutter of bladed-discs has been considered by Sinha et al [71]. Sinha and Griffin [72] used a lumped parameter model of a rotor stage with negative viscous damping to obtain steady state response of 3, 4 or 5 bladed-discs with friction dampers. The stability of limit cycles was discussed later [73]. Menq, Griffin and Bielak [74], considered a nonlinear friction constraint at the shroud interface in the forced vibration analysis of a fan stage.

In this chapter we will use the method of Receptance Coupling to study the vibration characteristics of bladed-disc assemblies. Both Armstrong's and Ewins' analyses will be illustrated. The tests on an Orpheus aircraft engine stage will be briefly discussed, and the results are compared with theoretical values. NASTRAN finite element program results for a bladed-disc assembly are also presented along with experimental values.

10.2 DISC RECEPTANCES

Receptance coupling method is essentially the application of dynamic equilibrium and compatibility between two connected components. Such a method, when properly developed, becomes very useful in studying the dynamic characteristics of structures involving different components that are governed by displacement fields of different nature, like plates and beams.

When a harmonic force $Fe^{i\omega t}$ acts at some point of a dynamical system, the system takes up a steady motion with the same frequency ω, such that the point of application of force has the displacement

$$x = Xe^{i\omega t} \tag{10.1}$$

If the equations of motion are linear, we can correlate the displacement with the applied force through the following relation.

$$x = \alpha F e^{i\omega t} \tag{10.2}$$

where α depends on the nature of the system and frequency ω, but not on the amplitude F of the force. The quantity α is termed the direct receptance. If on the other hand x is the displacement at some other point of the system other than that at which the force is applied, then equation (10.2) defines a cross receptance. First we will outline the procedure to determine the disc receptances, see references [9] and [75].

10.2.1 Equations of Motion of Circular Disc

The equation of motion of a circular plate vibrating freely at a frequency ω with transverse displacement w, can be written as

$$\nabla^4 w(r,\theta) - \beta^4 w(r,\theta) = 0 \tag{10.3}$$

where,

$$\nabla^2 = \frac{\partial^2}{\partial r^2} + \frac{1}{r}\frac{\partial}{\partial r} + \frac{1}{r^2}\frac{\partial^2}{\partial \theta^2}$$

$$\beta^4 = \frac{\omega^2 \mu}{D}$$

μ is the mass per unit area of the plate, and

$$D = \frac{2Eh^3}{3(1-\nu^2)}$$

with h as the half thickness of the disc and ν is Poisson's ratio.

The solution of above equation is given by

$$\begin{aligned} w = &\left\{A_{1n} J_n(\beta r) + A_{3n} Y_n(\beta r) + B_{1n} I_n(\beta r) + B_{3n} K_n(\beta r)\right\} \sin n\theta \\ &+ \left\{A_{2n} J_n(\beta r) + A_{4n} Y_n(\beta r) + B_{2n} I_n(\beta r) + B_{4n} K_n(\beta r)\right\} \cos n\theta \end{aligned} \tag{10.4}$$

Since Bessel functions Y and K are infinite when their arguments are zero, we can write the above equation for a circular plate with finite deformation as

$$w(r,\theta) = W_n \left\{J_n(\beta r) + \mu_n I_n(\beta r)\right\} \cos(n\theta - \epsilon_n) \tag{10.5}$$

where,

$$\cos(n\epsilon_n) = \frac{A_{1n}}{W_n} = \frac{B_{1n}}{\mu_n W_n}$$

$$\sin(n\epsilon_n) = \frac{A_{2n}}{W_n} = \frac{B_{2n}}{\mu_n W_n}$$

The frequency equation is obtained as

$$\begin{aligned} \frac{b^2}{2n}\Big[&\left\{b^4 + (1-\nu)^2 n^2 (n^2-1)\right\}(J_{n-1} I_{n+1} + J_{n+1} I_{n-1}) \\ &- 2b^2(1-\nu) n \left\{(n-1) I_{n-1} J_{n-1} + (n+1) I_{n+1} J_{n+1}\right\}\Big] = 0 \end{aligned} \tag{10.6}$$

where,

$$b = \beta a$$

and $J_{n\pm1}$ and $I_{n\pm1}$ are $J_{n\pm1}(b)$ and $I_{n\pm1}(b)$ respectively, and a is the radius of the disc.

The above equation is not valid for zero nodal diameters, i.e., $n = 0$. In this case,

$$b(J_0 I_1 + I_0 J_1) - 2(1-\nu) I_1 J_1 = 0 \tag{10.7}$$

If the roots of equations (10.6) and (10.7) for n nodal diameters and s nodal circles are denoted by b_{ns}, $n = 0, 1, 2$, $s = 0, 1, 2$, the natural frequencies are given by

$$\omega_{ns} = b_{ns}^2 \sqrt{\frac{2Eh^3}{3\rho(1-\nu^2)a^4}} \tag{10.8}$$

The corresponding mode shapes are

$$\phi_{ns}(r,\theta) = \left\{ J_n\left(b_{ns}\frac{r}{a}\right) + \mu_{ns} I_n\left(b_{ns}\frac{r}{a}\right) \right\} \cos(n\theta - \epsilon_n) \tag{10.9}$$

where,

$$\mu_{ns} = \frac{b^3 J_n' + (1-\nu)n^2(bJ_n' - J_n)}{b^3 I_n' - (1-\nu)n^2(bI_n' - I_n)} \quad \text{for } n \neq 0 \tag{10.10}$$

$$\mu_{ns} = \frac{b^2 J_n + (1-\nu)bJ_n'}{b^2 I_n - (1-\nu)bI_n'} = \frac{b^3 J_n'}{b^3 I_n'} \quad \text{for } n = 0 \tag{10.11}$$

where J' and I' are derivatives of J and I with respect to r.

Equation (10.9) can be written as

$$\phi_{ns}(r,\theta) = f_{ns}^{(r)} \cos(n\theta - \epsilon_n) \tag{10.12}$$

where,

$$f_{ns}^{(r)} = J_n\left(b_{ns}\frac{r}{a}\right) + \mu_{ns} I_n\left(b_{ns}\frac{r}{a}\right) \tag{10.13}$$

The disc receptances can be represented in either closed form or series form.

10.2.2 Closed Form Disc Receptances

These receptances are obtained by considering a circumferential shear force per unit length, $P_n \cos n\theta\, e^{i\omega t}$, with zero bending moment; and a circumferential bending moment per unit length $M_n \cos n\theta\, e^{i\omega t}$ with zero shear applied at the rim of the disc. Using these boundary conditions, instead of free boundary conditions, the following receptances can be obtained.

Direct receptance between force and displacement

$$_N\alpha_w = \frac{a^2}{D}\frac{N_n}{D_n} \tag{10.14}$$

Cross receptance between force and slope

$$_N\alpha_\phi = \frac{a}{D}\frac{N_n'}{D_n} \tag{10.15}$$

Cross receptance between moment and displacement

$$_C\alpha_w = \frac{a}{D}\frac{N_n'}{D_n} \tag{10.16}$$

Direct receptance between moment and slope

$$_C\alpha_\phi = \frac{1}{D}\frac{N_n''}{D_n} \tag{10.17}$$

where,

$$D_n = \{b^4 + n^2(n^2-1)(1-\nu)^2\}\{I_{n-1}J_{n+1} + I_{n+1}J_{n-1}\}$$
$$- 2b^2 n(1-\nu)\{(n-1)I_{n-1}J_{n-1} + (n+1)I_{n+1}J_{n+1}\} \tag{10.18}$$

$$N_n = (1-\nu)\{I_{n-1}J_{n+1} + J_{n-1}I_{n+1}\} - 4nI_nJ_n \tag{10.19}$$

$$N_n' = n^2(1-\nu)\{I_{n-1}J_{n+1} + I_{n+1}J_{n-1}\}$$
$$- b^2\{I_{n-1}J_{n-1} + I_{n+1}J_{n+1}\} \tag{10.20}$$

$$N_n'' = n^2(1-\nu)\{J_{n+1}I_{n-1} + J_{n-1}I_{n+1}\}$$
$$- b^2 n\{I_{n-1} + I_{n+1}\}\{J_{n-1} - J_{n+1}\} \tag{10.21}$$

For zero nodal diameters, i.e., $n = 0$,

$$D_0 = b^3\left[b\{I_0J_1 + I_1J_0\} - 2(1-\nu)I_1J_1\right] \tag{10.22}$$
$$N_0 = (1-\nu)\left[I_0J_1 + I_1J_0\right] - 2bI_0J_0 \tag{10.23}$$
$$N_0' = b^2\left[I_0J_1 - J_0I_1\right] \tag{10.24}$$
$$N_0'' = 2b^3I_1J_1 \tag{10.25}$$

The closed form receptances take a lot of computer time, since Bessel functions are to be evaluated. Hence series form receptances are desirable, which are given in the next section.

10.2.3 Series Form Disc Receptances

Let a normal force $P = P_k \cos k\theta\, e^{i\omega t}$ be applied at the rim. The response in the (n, s) normal mode may be determined by considering the generalized force,

$$Q_{ns} = \int_0^{2\pi} P_k \cos k\theta \left[f_{ns}^{(r)}\right]_{r=a} \cos n\theta\, d\theta \tag{10.26}$$

Evaluating the above integral, we get

$$Q_{ns} = \begin{cases} 0 & k \neq n \\ \pi P_k & k = n \neq 0 \\ 2\pi P_k & k = n = 0 \end{cases} \tag{10.27}$$

The magnitude of the principal coordinate is

$$q_{ns} = \alpha_{ns}Q_{ns} = \frac{Q_{ns}}{a_{ns}(\omega_{ns}^2 - \omega^2)} \tag{10.28}$$

where,

$$a_{ns} = \begin{Bmatrix} 2 \\ 1 \end{Bmatrix} \frac{1}{2} \int_0^{2\pi}\int_0^a r\{f_{ns}(r)\cos(n\theta - \epsilon_{ns})\}^2 dr\, d\theta \quad \begin{Bmatrix} n = 0 \\ n > 0 \end{Bmatrix} \tag{10.29}$$

The displacement of the rim in the (n, s) normal mode is

$$W_{ns}(r,\theta) = \begin{Bmatrix} 2 \\ 1 \end{Bmatrix} \frac{\pi P_n \cos n\theta}{a_{ns}(\omega_n^2 - \omega^2)}$$
$$= W_{ns} \cos n\theta \quad \begin{Bmatrix} n = 0 \\ n > 0 \end{Bmatrix} \tag{10.30}$$

Similarly for the slope at the rim, we have

$$\begin{Bmatrix} \dfrac{\partial W_{ns}}{\partial r} \end{Bmatrix}_{r=a} = \begin{Bmatrix} 2 \\ 1 \end{Bmatrix} \frac{\begin{Bmatrix} \dfrac{\partial}{\partial r} f_{ns}^{(r)} \end{Bmatrix}_{r=a} \pi P_n}{a_{ns}(\omega_{ns}^2 - \omega^2)} \cos n\theta \quad \begin{Bmatrix} n = 0 \\ n > 0 \end{Bmatrix}$$
$$= W'_{ns} \cos n\theta \tag{10.31}$$

If we define the edge receptances for the (n, s) mode as

$$\alpha_{ns} = \frac{W_{ns}}{P_n}$$
$$\alpha'_{ns} = \frac{W'_{ns}}{P_n} \tag{10.32}$$

then the so-called closed form receptances are given by

$$\alpha^{(n)} = \sum_{s=0}^{\infty} \alpha_{ns} = \begin{Bmatrix} 2 \\ 1 \end{Bmatrix} \sum_{s=0}^{\infty} \frac{\pi}{a_{ns}(\omega_{ns}^2 - \omega^2)} \quad \begin{Bmatrix} n = 0 \\ n > 0 \end{Bmatrix} \tag{10.33}$$

and

$$\alpha'^{(n)} = \sum_{s=0}^{\infty} \alpha'_{ns} = \begin{Bmatrix} 2 \\ 1 \end{Bmatrix} \sum_{s=0}^{\infty} \frac{\pi \begin{Bmatrix} \dfrac{\partial}{\partial r} f_{ns}^{(r)} \end{Bmatrix}_{r=a}}{a_{ns}(\omega_{ns}^2 - \omega^2)} \quad \begin{Bmatrix} n = 0 \\ n > 0 \end{Bmatrix} \tag{10.34}$$

If a couple is applied at the rim, then the slope-circumferential couple receptance is given by

$$\alpha''^{(n)} = \sum_{s=0}^{\infty} \alpha''_{ns} = \begin{Bmatrix} 2 \\ 1 \end{Bmatrix} \sum_{s=0}^{\infty} \frac{\pi \begin{Bmatrix} \dfrac{\partial}{\partial r} f_{ns}^{(r)} \end{Bmatrix}_{r=a}^2}{a_{ns}(\omega_{ns}^2 - \omega^2)} \quad \begin{Bmatrix} n = 0 \\ n > 0 \end{Bmatrix} \tag{10.35}$$

The series form of receptances can thus be calculated from the above equations, for a pre-determined number of terms (q), in the circular directions. They are

$${}_N\alpha_w = \sum_{s=1}^{q} \frac{1}{2T_{ns}\left(1 - \dfrac{\omega^2}{\omega_{ns}^2}\right)} \tag{10.36}$$

$${}_N\alpha_\phi = \sum_{s=1}^{q} \frac{f'_{ns}}{2T_{ns}\left(1 - \dfrac{\omega^2}{\omega_{ns}^2}\right)} \tag{10.37}$$

$$c\alpha_w = N\alpha_\phi \tag{10.38}$$

$$c\alpha_\phi = \sum_{s=1}^{q} \frac{f'^2_{ns}}{2T_{ns}\left(1 - \dfrac{\omega^2}{\omega^2_{ns}}\right)} \tag{10.39}$$

where T_{ns} is the kinetic energy of the disc. If the disc is made of J rings, then

$$T_{ns} = \sum_{k=1}^{J} \frac{\rho\pi r_k \delta r_k h_k}{2} \{f^{(r_k)}_{ks}\omega_{ns}\}^2 \tag{10.40}$$

The above expression is obtained by summing up the kinetic energy of J number of elemental rings of radius r_k and width δr_k; ρ is the mass density of the disc material. This method of determining the receptances has the advantage of not only saving computer time, but can also be used for nonuniform discs.

10.2.4 Point Receptances for the Disc

The diametral receptances $\alpha(n)$, $\alpha'(n)$ and $\alpha''(n)$ may be used to derive the more general point receptances α_{jk}, α'_{jk} and α''_{jk}. These receptances relate the response at point j due to a point excitation applied at k. A point force or couple on the rim may be expressed as a Fourier series of harmonically distributed forces, such as those used to derive $\alpha(n)$. Thus a point force F_0 at $\theta = \theta_k$ is equivalent to

$$F(\theta) = \frac{F_0}{\pi}\left\{\frac{1}{2} + \sum_{m=1}^{\infty} \cos m(\theta - \theta_k)\right\} \tag{10.41}$$

Hence we may express the point receptances in terms of diametral receptances as

$$\alpha_{jk} = \frac{1}{\pi}\left\{\frac{1}{2}\alpha(0) + \sum_{m=1}^{\infty} \alpha(m) \cos m(\theta_j - \theta_k)\right\} \tag{10.42}$$

$$\alpha'_{jk} = \frac{1}{\pi}\left\{\frac{1}{2}\alpha'(0) + \sum_{m=1}^{\infty} \alpha'(m) \cos m(\theta_j - \theta_k)\right\} \tag{10.43}$$

$$\alpha''_{jk} = \frac{1}{\pi}\left\{\frac{1}{2}\alpha''(0) + \sum_{m=1}^{\infty} \alpha''(m) \cos m(\theta_j - \theta_k)\right\} \tag{10.44}$$

10.3 BLADE RECEPTANCES

According to Bernoulli-Euler bending theory, the differential equation of motion of a vibrating beam is

$$\frac{\partial^4 w}{\partial l^4} + \frac{A\rho}{EI}\frac{\partial^2 w}{\partial t^2} = 0 \tag{10.45}$$

Assuming

$$w = W(l)e^{i\omega t} \tag{10.46}$$

we get

$$w = \left\{A\cos(\lambda l) + B\sin(\lambda l) + C\cosh(\lambda l) + D\sinh(\lambda l)\right\}e^{i\omega t} \tag{10.47}$$

where

$$\lambda^4 = \frac{\omega^2 A\rho}{EI} \tag{10.48}$$

The frequency equation for a free beam can be obtained by

$$\cos(\lambda L)\cosh(\lambda L) = 1 \tag{10.49}$$

In the above l is the distance measured along the beam axis and L is the beam length.

The first five roots of the above equation are (in addition to the zero root), 4.730, 7.853, 10.996, 14,137 and 17.278. For higher roots, the following approximation is valid.

$$\lambda_r L = (2r+1)\frac{\pi}{2} \quad r > 5 \tag{10.50}$$

The principal modes can be obtained by writing the characteristic equation as

$$\phi_r(l) = \cosh(\lambda_r l) + \cos(\lambda_r l) - \sigma_r\left\{\sinh(\lambda_r l) + \sin(\lambda_r l)\right\} \tag{10.51}$$

where,

$$\sigma_r = \frac{\cosh(\lambda_r L) - \cos(\lambda_r L)}{\sinh(\lambda_r L) - \sin(\lambda_r L)}$$

The blade receptances can also be represented in either closed form or series form.

10.3.1 Closed Form Blade Receptances

The closed form receptances can be derived by applying a force or couple $Fe^{i\omega t}$, $Me^{i\omega t}$ and determining the appropriate displacement or slope.

The end receptances are the ratios of displacements to the force, both measured at the root of the blade, i.e., $l = 0$. They are:

Direct receptance between force and displacement

$$_N\beta_w = -\frac{F_5}{EI\lambda^3 F_3} \tag{10.52}$$

Cross receptance between force and slope

$$_N\beta_\phi = -\frac{F_1}{EI\lambda^2 F_3} \tag{10.53}$$

Cross receptance between moment and displacement

$$_C\beta_w = -\frac{F_1}{EI\lambda^2 F_3} \tag{10.54}$$

Direct receptance between moment and slope

$$ {}_C\beta_\phi = -\frac{F_6}{EI\lambda F_3} \tag{10.55} $$

Similarly, the receptances between the blade root and the blade tip are

$$ {}_N\beta'_w = \frac{F_8}{EI\lambda^3 F_3} \tag{10.56} $$

$$ {}_N\beta'_\phi = -\frac{F_{10}}{EI\lambda^2 F_3} \tag{10.57} $$

$$ {}_C\beta'_w = -\frac{F_{10}}{EI\lambda^2 F_3} \tag{10.58} $$

$$ {}_C\beta'_\phi = \frac{F_7}{EI\lambda F_3} \tag{10.59} $$

The receptances at the blade tip ($l = L$) are

$$ {}_N\beta''_w = -\frac{F_5}{EI\lambda^3 F_3} \tag{10.60} $$

$$ {}_N\beta''_\phi = \frac{F_1}{EI\lambda^2 F_3} \tag{10.61} $$

$$ {}_C\beta''_w = {}_N\beta''_\phi \tag{10.62} $$

$$ {}_C\beta''_\phi = \frac{F_6}{EI\lambda F_3} \tag{10.63} $$

where,

$$ \begin{aligned} F_1 &= \sin\lambda L \sinh\lambda L \\ F_3 &= \cos\lambda L \cosh\lambda L - 1 \\ F_5 &= \cos\lambda L \sinh\lambda L - \sin\lambda L \cosh\lambda L \\ F_6 &= \cos\lambda L \sinh\lambda L + \sin\lambda L \cosh\lambda L \\ F_7 &= \sin\lambda L + \sinh\lambda L \\ F_8 &= \sin\lambda L - \sinh\lambda L \\ F_{10} &= \cos\lambda L - \cosh\lambda L \end{aligned} \tag{10.64} $$

10.3.2 Series Form Blade Receptances

We can use generalized coordinates to describe the motion, for each mode as

$$ p_r = \beta_r P_r \tag{10.65} $$

where p are the generalized coordinates, P generalized forces and β are the corresponding receptances given by

$$ \beta_r = \frac{1}{a_r(\omega_r^2 - \omega^2)} \tag{10.66} $$

It may be noted that, since these quantities provide means for finding the variations of P_r, which are produced by some specified applied force, their values must be dependent upon the choice of the meaning of unit value for each of the principal coordinates; the magnitudes of the coefficients a_r are in fact dependent upon this choice.

Suppose it is required to find the displacement z of some point Z of a system. The displacement z will be linear function of the p's. We can then write a small increment of z as

$$\delta z = \frac{\partial z}{\partial p_1}\delta p_1 + \frac{\partial z}{\partial p_2}\delta p_2 + \cdots + \frac{\partial z}{\partial p_n}\delta p_n \tag{10.67}$$

where the partial derivatives are constants. Upon integrating, we get

$$z = \frac{\partial z}{\partial p_1}p_1 + \frac{\partial z}{\partial p_2}p_2 + \cdots + \frac{\partial z}{\partial p_n}p_n \tag{10.68}$$

The value of displacement z, due to application of force $Fe^{i\omega t}$, at some other point Y, will therefore be

$$z = \frac{\partial z}{\partial p_1}\left(\beta_1 F\frac{\partial y}{\partial p_1}\right)e^{i\omega t} + \frac{\partial z}{\partial p_2}\left(\beta_2 F\frac{\partial y}{\partial p_2}\right)e^{i\omega t} + \cdots + \frac{\partial z}{\partial p_n}\left(\beta_n F\frac{\partial y}{\partial p_n}\right)e^{i\omega t} \tag{10.69}$$

The above can be written as

$$z = \beta_{zy}Fe^{i\omega t} \tag{10.70}$$

where the cross receptance is given by

$$\beta_{zy} = \beta_{yz} = \beta_1\left(\frac{\partial y}{\partial p_1}\frac{\partial z}{\partial p_1}\right) + \beta_2\left(\frac{\partial y}{\partial p_2}\frac{\partial z}{\partial p_2}\right) + \cdots + \beta_n\left(\frac{\partial y}{\partial p_n}\frac{\partial z}{\partial p_n}\right) \tag{10.71}$$

If Z and Y are same locations, the above expression becomes direct receptance, i.e.,

$$\beta_{yy} = \beta_1\left(\frac{\partial y}{\partial p_1}\right)^2 + \beta_2\left(\frac{\partial y}{\partial p_2}\right)^2 + \cdots + \beta_n\left(\frac{\partial y}{\partial p_n}\right)^2 \tag{10.72}$$

Equation (10.71) can be rewritten as

$$\beta_{zy} = \beta_{yz} = \sum_{r=1}^{\infty}\beta_r\frac{\partial y}{\partial p_r}\frac{\partial z}{\partial p_r} \tag{10.73}$$

In the case of a blade, y and z may be either displacements or slopes. Due to a harmonic excitation force, the deflection, w, at any section is of the form

$$w(x,t) = \sum_{r=1}^{\infty}p_r(t)\phi_r(x) \tag{10.74}$$

Hence from the above two equations, if y and z are identified with $w(x,t)$ and $w(l,t)$ respectively, then

$$\beta_{xl} = \beta_{lx} = \sum_{r=1}^{\infty}\beta_r\phi_r(x)\phi_r(l) \tag{10.75}$$

Now, for the zero frequency mode, for a free-free beam of rectangular cross-section with breadth b and depth d, we get

$$ {}_N\beta_w = -\frac{4}{\rho b d L \omega^2} \tag{10.76} $$

$$ {}_N\beta_\phi = -\frac{6}{\rho b d L^2 \omega^2} \tag{10.77} $$

$$ {}_C\beta_w = {}_N\beta_\phi \tag{10.78} $$

$$ {}_C\beta_\phi = -\frac{12}{\rho b d L^3 \omega^2} \tag{10.79} $$

For the first five modes,

$$ {}_N\beta_w = \sum_{n=1}^{5} \frac{1}{\omega_n^2 - \omega^2} \frac{4}{\rho b d L} \tag{10.80} $$

$$ {}_N\beta_\phi = \sum_{n=1}^{5} \frac{1}{\omega_n^2 - \omega^2} \frac{4}{\rho b d L} \left(\frac{\lambda_n L \sigma_n}{2L} \right) \tag{10.81} $$

$$ {}_C\beta_w = {}_N\beta_\phi \tag{10.82} $$

$$ {}_C\beta_\phi = \sum_{n=1}^{5} \frac{1}{\omega_n^2 - \omega^2} \frac{4}{\rho b d L} \left(\frac{\lambda_n L \sigma_n}{2L} \right)^2 \tag{10.83} $$

For the sixth and higher modes, the receptances can be approximated to

$$ {}_N\beta_w = \frac{64 L^3}{E I \pi^4} 0.000097 \tag{10.84} $$

$$ {}_N\beta_\phi = -\frac{32 L^2}{E I \pi^3} 0.001725 \tag{10.85} $$

$$ {}_C\beta_w = {}_N\beta_\phi \tag{10.86} $$

$$ {}_C\beta_\phi = \frac{16 L}{E I \pi^2} 0.04157 \tag{10.87} $$

The corresponding receptances given in equations (10.76) to (10.87) should be added to give the total receptances.

10.4 ARMSTRONG'S ANALYSIS FOR TUNED SYSTEMS

Consider first a disc represented by B with one blade represented by C, as shown in Fig. 10.1. Only two degrees of freedom are considered. Schematically, the disc and blade are shown as subsystems in Fig. 10.2. Let N and C represent the force and couple, with response x and θ. The displacement field can be written, using the corresponding receptances as,

$$ X_B = {}_N\alpha_w N_B + {}_C\alpha_w C_B \tag{10.88} $$

$$ \theta_B = {}_N\alpha_\phi N_B + {}_C\alpha_\phi C_B \tag{10.89} $$

$$X_C = {}_N\beta_w N_C + {}_C\beta_w C_C \tag{10.90}$$
$$\theta_C = {}_N\beta_\phi N_C + {}_C\beta_\phi C_C \tag{10.91}$$

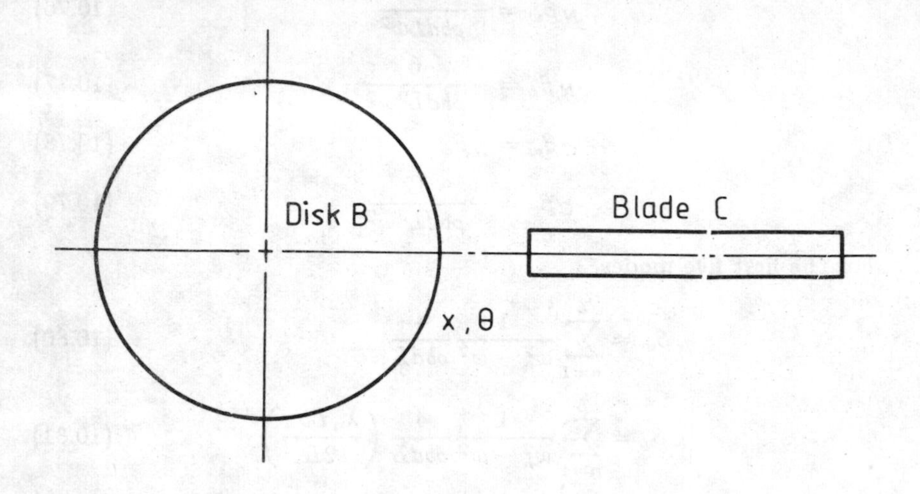

Fig. 10.1 Blade disc — Two degrees of freedom

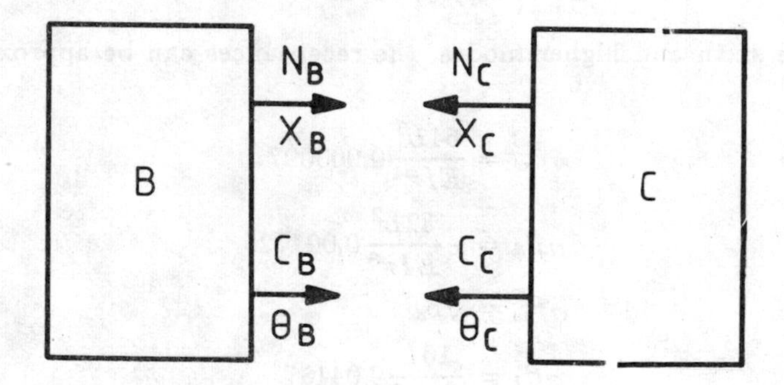

Fig. 10.2 Disc and blade as sub-systems

Let N_1, C_1 be the forces and couples between the blade and disc. Applying conditions of equilibrium and compatibility, we get

$$N_1 = N_B + N_C \tag{10.92}$$
$$C_1 = C_B + C_C \tag{10.93}$$
$$X = X_B = X_C \tag{10.94}$$
$$\theta = \theta_B = \theta_C \tag{10.95}$$

If the applied forces between the disc and blade are zero, then

$$N_B = -N_C \tag{10.96}$$
$$C_B = -C_C \tag{10.97}$$

Using equations (10.94) to (10.97) in equations (10.88) to (10.91), we get the following frequency equation.

$$\begin{bmatrix} ({}_N\alpha_w + {}_N\beta_w) & ({}_C\alpha_w + {}_C\beta_w) \\ ({}_N\alpha_\phi + {}_N\beta_\phi) & ({}_C\alpha_\phi + {}_C\beta_\phi) \end{bmatrix} \begin{Bmatrix} N_B \\ C_B \end{Bmatrix} = 0 \tag{10.98}$$

10.4.1 Bladed-Disc Assembly

Let there be M identical blades (tuned system) and assume that the assembly is vibrating in a mode with n nodal diameters. It is also assumed that the reaction forces and couples between the disc and blades are each distributed around the rim of the disc as shown in Fig. 10.3. The force on jth blade is given by

$$f_j = -2\epsilon A_0 \cos\frac{2\pi nj}{M} \tag{10.99}$$

where A_0 is the amplitude of force distribution and 2ϵ is the angular width of the blade. We can similarly write an expression for the couple C.

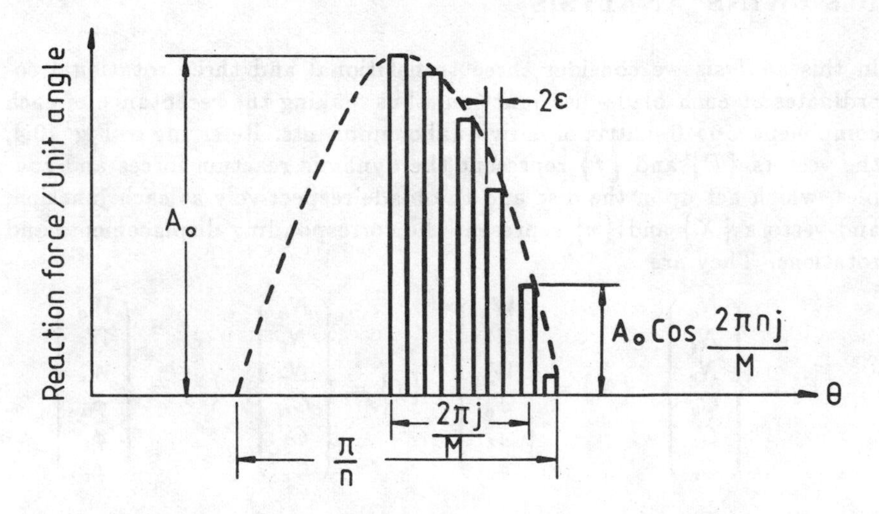

Fig. 10.3 Distribution of reaction forces around disc rim

For the disc we consider the distributed forces acting on the rim. The discrete function in Fig. 10.3 is

$$F(\theta) = \begin{cases} 0 & \left\{\dfrac{2\pi m}{M} + \epsilon\right\} < \theta < \left\{\dfrac{2\pi(m+1)}{M} - \epsilon\right\} \\[2ex] A_0 \cos\dfrac{2\pi mn}{M} & \left\{\dfrac{2\pi m}{M} - \epsilon\right\} < \theta < \left\{\dfrac{2\pi m}{M} + \epsilon\right\} \end{cases}$$

$$m = 1, 2, \ldots, M \tag{10.100}$$

For the disc, we can consider the force as distributed forces acting on its rim, and for $n < M$, we can show

$$F_D(\theta) = \epsilon M A_0 \cos n\theta \tag{10.101}$$

For the location of jth blade, this force is

$$F_D(\theta) = \epsilon M A_0 \cos \frac{2\pi nj}{M} \tag{10.102}$$

Combining equations (10.99) and (10.102), it can be seen that the disc force is half M times that of the blade forces. That is, the disc receptances should be modified as

$$_N\alpha_w \Big|_{\text{modified}} = \frac{1}{2} M _N\alpha_w \tag{10.103}$$

Consequently, the frequency determinant in equation (10.98) becomes

$$\Delta = \begin{bmatrix} \left(\frac{1}{2} M _N\alpha_w + _N\beta_w\right) & \left(\frac{1}{2} M _C\alpha_w + _C\beta_w\right) \\ \left(\frac{1}{2} M _N\alpha_\phi + _N\beta_\phi\right) & \left(\frac{1}{2} M _C\alpha_\phi + _C\beta_\phi\right) \end{bmatrix} \tag{10.104}$$

10.5 EWINS' ANALYSIS

In this analysis we consider three translational and three rotational co-ordinates at each blade-disc junction, thus making the receptance of each component a 6×6 matrix of individual components. Referring to Fig. 10.4, the vectors $\{F\}$ and $\{f\}$ represent the dynamic reaction forces and couples, which act upon the disc and the blade respectively at each junction, and vectors $\{X\}$ and $\{x\}$ represent the corresponding displacements and rotations. They are

$$\{F\} = \begin{Bmatrix} N_a \\ N_t \\ N_r \\ C_a \\ C_t \\ C_r \end{Bmatrix}; \; \{X\} = \begin{Bmatrix} W_a \\ W_t \\ W_r \\ \phi_a \\ \phi_t \\ \phi_r \end{Bmatrix}; \; \{f\} = \begin{Bmatrix} N_a \\ N_t \\ N_r \\ C_a \\ C_t \\ C_r \end{Bmatrix}; \; \{x\} = \begin{Bmatrix} W_a \\ W_t \\ W_r \\ \phi_a \\ \phi_t \\ \phi_r \end{Bmatrix}$$

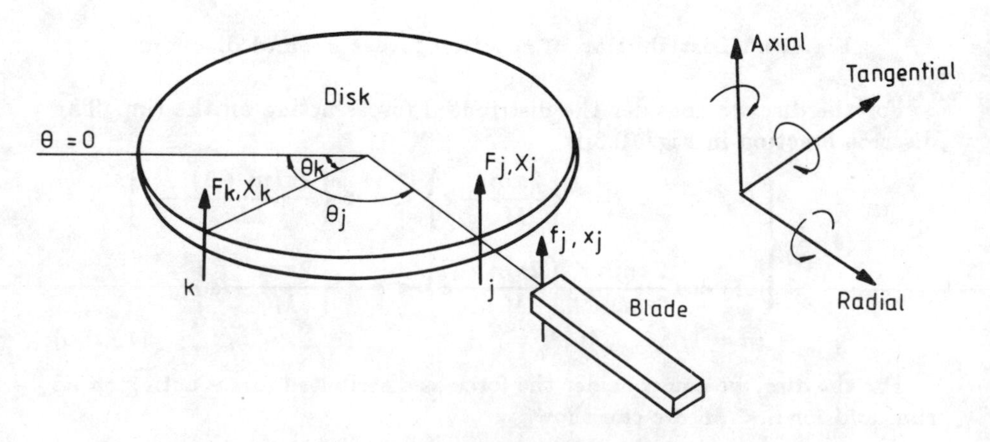

Fig. 10.4 Disc-blade assembly

The displacements of the rim of the disc at point j, X_j will be the sum of the response of each of the M system of forces, F_k, and may be written as

$$\{X\}_j = \sum_{k=1}^{M} [\alpha]_{jk} \{F\}_k \tag{10.105}$$

where $[\alpha]_{jk}$ is point receptance matrix relating to response at j due to force at k.

Consider now the blade attached at point j. We have

$$\{x\}_j = [\beta]_j \{f\}_j \tag{10.106}$$

Assuming the blades to be rigidly fixed to the disc, we have

$$\{X\}_j = \{x\}_j ; \quad j = 1, 2, \ldots, M \tag{10.107}$$

and

$$\{F\}_j = -\{f\}_j ; \quad j = 1, 2, \ldots, M \tag{10.108}$$

From equations (10.105), (10.106) and (10.108), equation (10.107) gives

$$\sum_{k=1}^{M} [\alpha]_{jk} \{F\}_k + [\beta]_j \{F\}_j = 0; \quad j = 1, 2, \ldots, M \tag{10.109}$$

The above leads to the following frequency determinant

$$[\Delta]_{6M \times 6M} \{F\}_{6M \times 1} = 0 \tag{10.110}$$

Once a natural frequency is found by satisfying the determinant $[\Delta] = 0$, we can arbitrarily set $F_1 = N_a^1 = 1$ unit and determine the rest of quantities in $\{F\}$, by solving the $(6M - 1)$ simultaneous equations, and mode shape can be obtained by using equation (10.106).

10.5.1 Blade Mistuning Analysis

When the system is mistuned by allowing small dimensional variations between one blade and next, some of the modes split into distinct but similar pairs. In general a random mistuning of blades will cause all the so-called double modes to split in this way and it is possible to relate the natural frequency split of each mode to the specific detuning arrangement. It is most convenient to select the $1F$ blade cantilever frequency to describe the state of mistuning. It can also be done by specifying the diametral components of the mistuning function $f(\theta)$. Individual blade frequencies f_1, f_2, ..., are given by

$$f_i = f(\theta)\big|_{\theta=\theta_i} ; \quad \theta_i = \frac{2\pi i}{M} \tag{10.111}$$

where $f(\theta)$ is of the form

$$f(\theta) = f_{\text{mean}} \left\{ 1 + a_0 + \sum_{j=1}^{M} a_j \cos(n_j \theta + \phi_j) \right\} \tag{10.112}$$

10.5.2 Response to External Excitation

Let any blade rotating through the pressure field experience a fluctuating force proportional to $\cos n\theta = \cos n\Omega t$, where Ω is the rotational speed. This is nth engine order excitation. Let

$$P_\theta = P_0 \cos n\theta \tag{10.113}$$

be the variation in steady pressure relative to an origin fixed in the machine. The position of jth blade at time t relative to the same origin is

$$\theta_j(t) = \Omega t + \frac{2\pi j}{M} \tag{10.114}$$

The force on that blade is given by

$$P_j = P_0 \cos n \left(\Omega t + \frac{2\pi j}{M} \right) \tag{10.115}$$

In a complex form, the above can be written as

$$P_j = P_0 e^{i\left(\frac{2\pi n j}{M}\right)} e^{i(n\Omega t)} \tag{10.115a}$$

Equation (10.106) can be modified to include the excitation force as

$$\{x\}_j = [\beta_j]\{f\}_j + [\beta_j^*]\{P\}_j; \quad \{P\}_j = \begin{Bmatrix} P_j \\ 0 \\ 0 \\ \vdots \\ 0 \end{Bmatrix} \tag{10.116}$$

where $[\beta^*]$ is the receptance relating to displacement of the root to a unit load at the point where the excitation force acts. If we consider this force to be applied at the root of the blade, the above equation can be written as

$$\{x\}_j = [\beta_j]\{f\}_j + [\beta_j]\{P\}_j \tag{10.117}$$

Using equations (10.107) and (10.108), we get

$$\sum_{k=1}^{M} [\alpha_{jk}]\{F\}_k + [\beta_j]\{F\}_j = [\beta_j]\{P\}_j; \quad j = 1, 2, \ldots, M \tag{10.118}$$

For a specified value of the force, the above equation can be transformed as

$$[\Delta]\{F\} = \{P\} e^{i(n\Omega t)} \tag{10.119}$$

where

$$[\Delta] = \sum_{k=1}^{M} [\alpha_{jk} + \beta_j]_{j=1,2,\ldots,M} \tag{10.120}$$

From the above the response $\{F\}$ can be determined.

10.5.3 Receptance of Damped Blades

(a) *Material Damping*

It can be effectively introduced in the analysis by making the modulus of the blade material complex, i.e.,

$$\overline{E}_b = E_b(1 + i\eta_b)$$

where η_b is the loss factor of the blade material in direct strain.

(b) *Viscous Damping*

The receptance of a viscous damper as represented by a dash pot is $iC\omega$, where C is the damping coefficient and ω is the frequency.

(c) *Hysteretic Damping*

Hysteretic damping at the root of the blade representing friction can also be taken into the analysis by adding the receptance of the damper which is ih, where h is the damping coefficient.

10.5.4 Examples

A computer program was developed by Rao et al [45] based on the analysis of Armstrong and Ewins. This program had four options; option 0, based on Armstrong's analysis, equation (10.104); option 1 to determine the natural frequencies of mistuned bladed-discs using equation (10.110); option 2 to determine the mode shapes of option 1; and option 3 to determine the forced response, using equation (10.119).

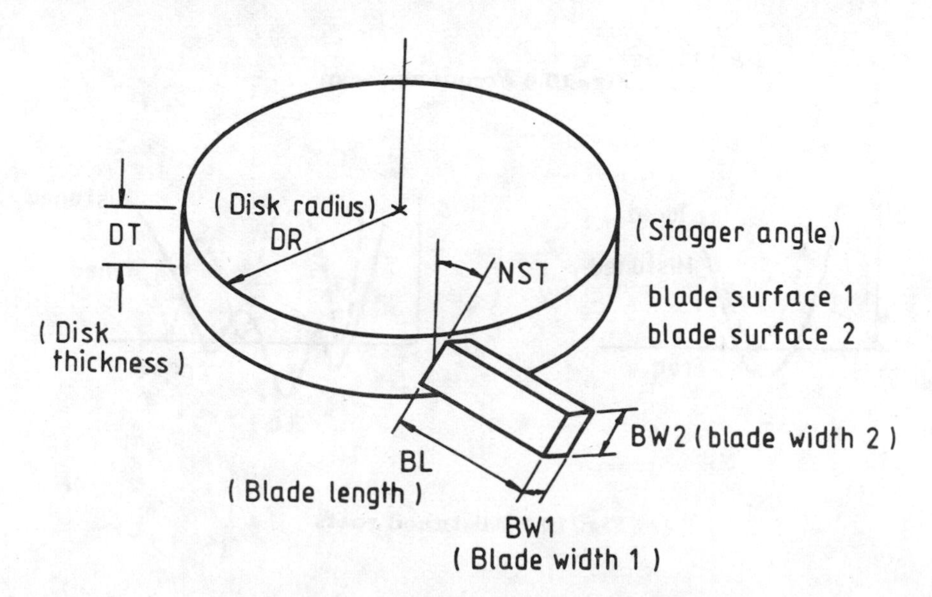

Fig. 10.5 Specification of the bladed-disc

Fig. 10.5 shows the bladed-disc specifications this program can accept. Equivalent blade widths of a rectangular cross-section are to be determined for the actual blade, so that the blade receptances can be computed.

For option 0, in any one frequency scan, the only natural frequencies to be found will be those of modes with a specified number of nodal diameters. As a result, such frequencies will be fairly well spread so that a relatively coarse frequency step may be used. The frequency search in this case is somewhat simple and an accuracy within 0.0001% can be achieved.

In option 1, the natural frequencies of a tuned system are determined from the general analysis; in contrast to option 0, this one detects all in a given frequency range, irrespective of the number of nodal diameters. A typical plot of frequency determinant vs frequency is shown in Fig. 10.6, from which it is clear that the root finding is likely to be a difficult process. The frequency determinant is likely to be either very large or very small. Roots are to be detected either by a change in sign or by a drop of several orders in the magnitude of the frequency determinant.

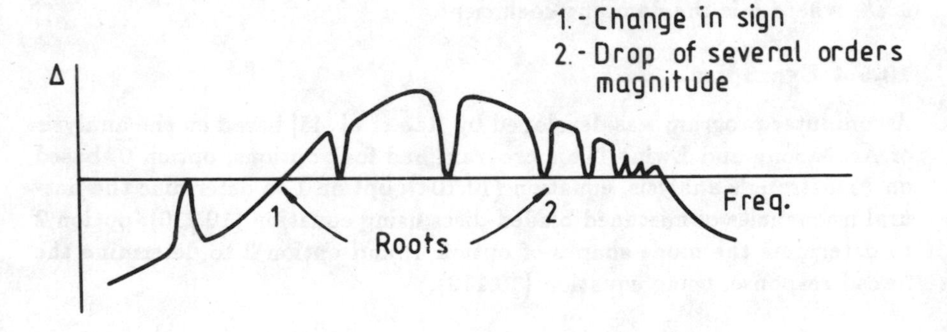

Fig. 10.6 Frequency scan

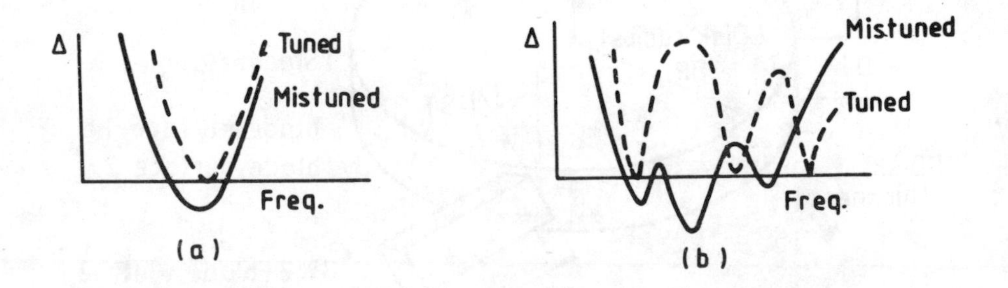

Fig. 10.7 Mistuned roots

When the system is mistuned, some of the roots change in a systematic way, while others are less well ordered, as shown in Fig. 10.7. In either case, the tuned system natural frequencies are the most convenient points to start the search scan process.

Because of the many closely spaced natural frequencies, the response curves in general are very complex, often producing separate resonant responses at any of the natural frequencies. Thus it is important to know all these natural frequencies for a given bladed-disc before attempting to compute response curves. For a tuned system, an m engine order excitation will only excite a mode with m nodal diameters. For the same system mistuned, many modes may be excited as shown in Fig. 10.8.

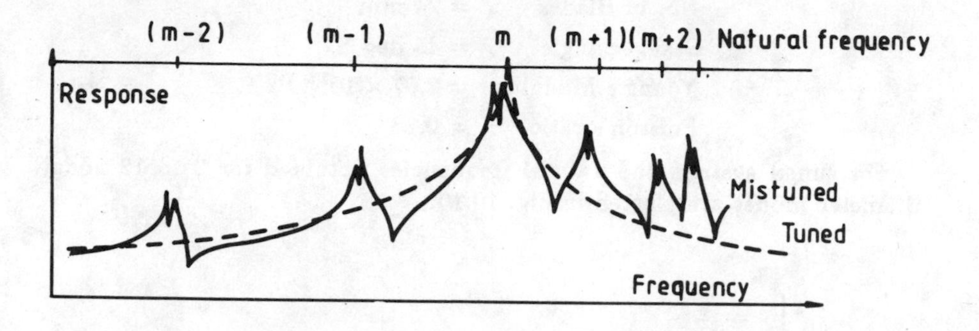

Fig. 10.8 Mistuned bladed-disc response

When damping is not considered, each peak will in theory produce infinite response. It is clear that each peak will have a different strength, which would be reflected in variations in peak values if damping were included. An estimate of relative strength of each peak may be obtained by calculating the response at two frequencies on either side of the natural frequency as shown in Fig. 10.9. The response values at these two points

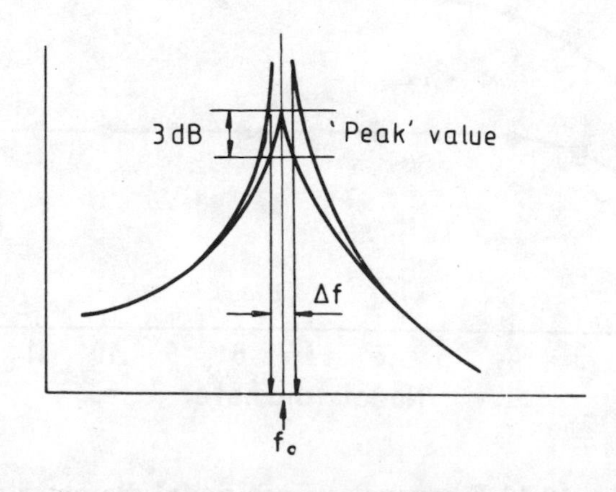

Fig. 10.9 Peak strength

should be approximately equal, if this technique is to provide a useful indication, which it will only do for the case of small damping. For a loss factor of 0.001, the frequency band width Δf should be 0.001 times the mean frequency, f_0.

Consider the bladed-disc assembly with the following data:

Blade length	= 80.26 mm
Width 1	= 3.05 mm
Width 2	= 10.78 mm
Disc thickness	= 5.59 mm
Disc radius	= 127 mm
No. of Blades	= 24 mm
Stagger angle	= 45 deg
Young's Modulus	$= 2.07 \times 10^{11}$ Pa
Poisson's ratio	= 0.287

For tuned system, the natural frequencies obtained for 2 to 12 nodal diameter modes are plotted in Fig. 10.10.

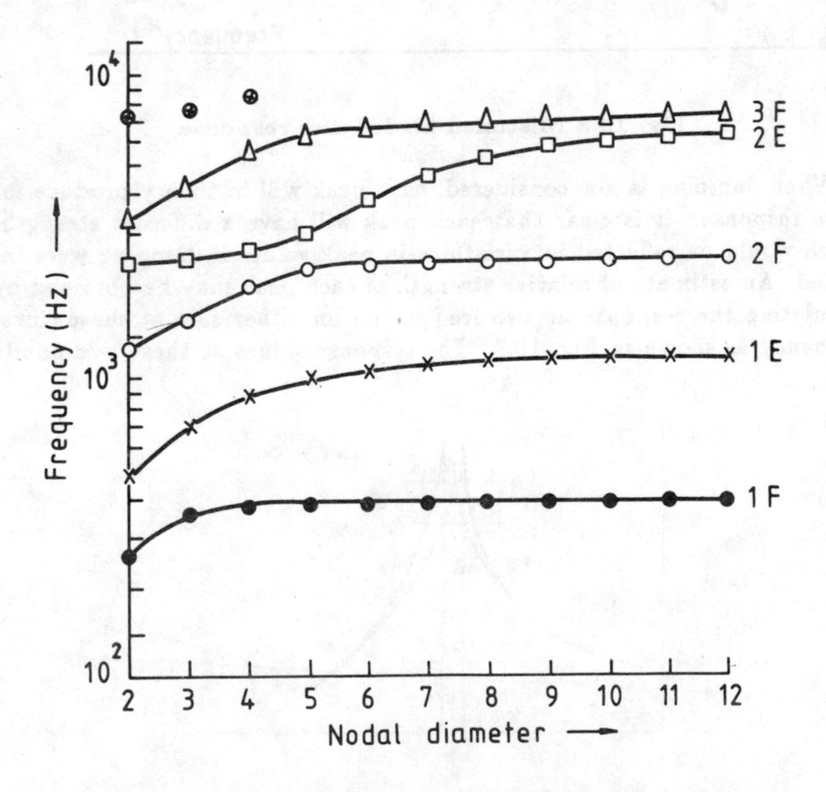

Fig. 10.10 Frequency versus nodal diameter plot for 24 bladed disc assembly

For the mistuned pattern assumed in Table 10.1, the natural frequencies obtained for the same blade are given in Table 10.2.

TABLE 10.1

Mistuned Pattern of 24 Bladed-Disc

Blade No.	*Mistune Ratio*	*Blade No.*	*Mistune Ratio*
1	1.00625	13	0.98625
2	0.99250	14	0.97500
3	0.99750	15	0.99875
4	0.99500	16	1.00375
5	1.00625	17	1.03125
6	0.98125	18	1.04500
7	1.01375	19	1.00375
8	1.00250	20	1.00250
9	0.98875	21	0.99250
10	0.99000	22	0.97500
11	1.00125	23	0.99875
12	0.98375	24	0.99625

TABLE 10.2

Natural Frequencies of 24 Bladed-Disc

No. of Nodal Diameters	*Frequencies in Hz*		
	Tuned System	*Mistuned System*	
2	262.848	262.193	263.663
3	354.368	352.990	353.940
4	373.228	370.785	371.584
5	379.790	376.722	378.082
6	382.927	379.347	383.785
7	384.683	383.188	390.057
8	385.751	382.234	384.319
9	386.422	384.432	388.710
10	386.836	379.924	386.777
11	387.061	386.136	386.390
12	387.133	400.024	—

The bladed-disc assembly above is analyzed to study the material damping, for three different loss factors 0.001, 0.01 and 0.05. The damped response of this mistuned system under 6th engine order excitation is shown in Fig. 10.11.

10.6 MULTI-STAGE DISC-BLADE-SHROUD ANALYSIS

In the previous section, we considered a uniform disc and uniform blades without any twist. In practice, both the blades and the disc are not uniform. The analysis in this section can account for a three-stage concentric

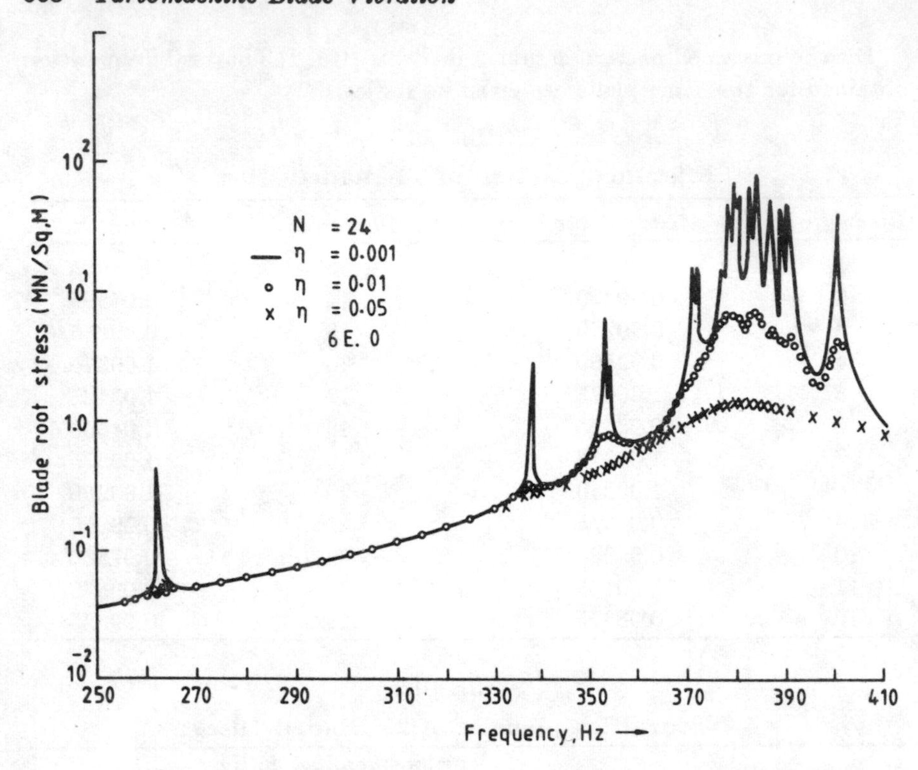

Fig. 10.11 Frequency versus blade root stress plot for different values η for 24 bladed disc assembly

annuli representing the hub, the main body and the rim of the total disc. The blade is also modelled with two or three elements to account for nonuniformity of the blade and can have different stagger angles to account for the twist. The analysis is also extended to account for the shroud of the blades and follows closely the work of Cottney [76].

10.6.1 Receptances of Multi-Stage Disc

Consideration of the internal equilibrium of a disc produces separate expressions relating the flexural deformations to three different types of internal forces, viz., the shear force, bending moment and the twisting moment. These forces will be in equilibrium with externally applied axial forces, tangential and radial torques, respectively applied at the rim of the disc. Since the shear deformation inside the disc is ignored, it can be shown that except in the immediate vicinity of the boundary, the disc will not know any difference between an externally applied axial force and a radial torque provided these forces vary with statically equivalent magnitude, i.e.,

$$Q_1(n) = -\frac{1}{r}\frac{\partial}{\partial \theta}Q_5(n) \tag{10.121}$$

where $Q_1(n)$, and $Q_5(n)$ are the axial force and the radial torque respectively, see Fig. 10.12.

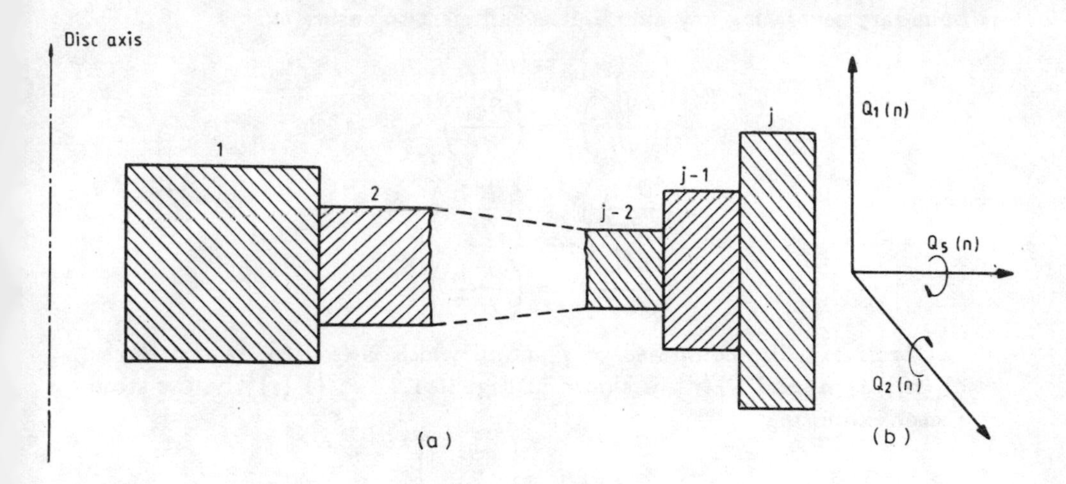

Fig. 10.12 (a) Configuration of disc annuli; (b) rim forces

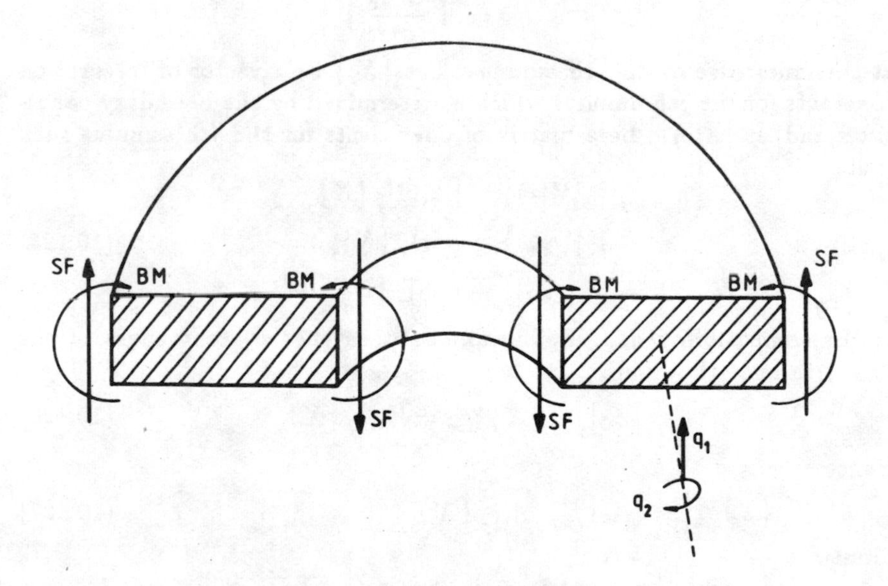

Fig. 10.13 Shear force and bending moment acting on the disc annulus

Therefore, if both the axial forces and the radial torques are applied to a disc, we can combine these forces into a single equivalent shear force, see Fig. 10.13.

$$S_f = Q_1(n) - \frac{1}{r}\frac{\partial}{\partial \theta}Q_5(n) \tag{10.122}$$

In order to satisfy the compatibility and equilibrium across the internal boundary separating any annuli, it is sufficient to ensure that

$$\begin{aligned}
(q_1)_j &= (q_1)_{j-1} \\
\left(\frac{\partial q_1}{\partial r}\right)_j &= \left(\frac{\partial q_1}{\partial r}\right)_{j-1} \\
\left(\frac{\partial^2 q_1}{\partial r^2}\right)_j &= \left(\frac{\partial^2 q_1}{\partial r^2}\right)_{j-1} \\
\left(\frac{\partial^3 q_1}{\partial r^3}\right)_j &= \left(\frac{\partial^3 q_1}{\partial r^3}\right)_{j-1}
\end{aligned} \tag{10.123}$$

Consider a disc composed of j annuli which is excited by rim forces $Q_1(n)$, $Q_2(n)$ and $Q_5(n)$ as shown in Fig. 10.12. Let $\{V(r)\}$ be the state vector expressing

$$\{V(r)\} = \left\{ \begin{array}{c} q_1 \\ \dfrac{\partial q_1}{\partial r} \\ \dfrac{\partial^2 q_1}{\partial r^2} \\ \dfrac{\partial^3 q_1}{\partial r^3} \end{array} \right\} \tag{10.124}$$

at the inner edge of the jth annulus. Let $\{X_j\}$ be a vector of integration constants for the jth annulus which is determined by the boundary conditions and let $[A(r)]_j$ be a matrix of coefficients for the jth annulus such that

$$\begin{aligned}
\{V(r)\} &= [A(r)]_j \{X\}_j \\
\{V(r_{j,i})\} &= [A_i]_j \{X\}_j \\
\{V(r_{j,0})\} &= [A_0]_j \{X\}_j
\end{aligned} \tag{10.125}$$

For compatibility and equilibrium between the adjacent edges of the $(j-1)$th and jth annuli

$$[A_0]_{j-1} \{X\}_{j-1} = [A_i]_j \{X\}_j \tag{10.126}$$

Hence

$$\{X\}_{j-1} = [A_0]_{j-1}^{-1} [A_i]_j \{X\}_j \tag{10.127}$$

Similarly

$$\{X\}_1 = [A_0]_1^{-1} [A_i]_2 [A_0]_2^{-1} [A_i]_3 \ldots [A_i]_j \{X\}_j \tag{10.128}$$

Since by assumption there are no forces at the inner edge of the innermost annulus, we get from equation (10.125)

$$\left\{ \begin{array}{c} 0 \\ 0 \end{array} \right\} = \begin{bmatrix} 0 & 0 & 1 & 0 \\ 0 & 0 & 0 & 1 \end{bmatrix} [A_i]_1 \{X\}_1 \tag{10.129}$$

Substituting (10.128) in the above equation and combining it with the last of equations (10.125) for the outermost jth element, we get

$$\begin{bmatrix} \begin{bmatrix} 0 & 0 & 1 & 0 \\ 0 & 0 & 0 & 1 \end{bmatrix} & [A_i]_1 & [A_0]_1^{-1} & [A_i]_2 & \cdots & [A_i]_j \\ \begin{bmatrix} 0 & 0 & 1 & 0 \\ 0 & 0 & 0 & 1 \end{bmatrix} & [A_0]_j & & & & \end{bmatrix} \{X\}_j$$

$$= \begin{Bmatrix} 0 \\ 0 \\ S_f(r_{j,0}) \\ B_m(r_{j,0}) \end{Bmatrix} \qquad (10.130)$$

$$S_f(r_{j,0}) = Q_1(n) - \frac{1}{r}\frac{\partial}{\partial\theta}Q_5(n)$$

and $$B_m(r_{j,0}) = -Q_2(n)$$

To find the natural frequencies of the disc, we make the external forces zero in equation (10.130) and set the following equation.

$$[B]\{X\}_j = 0 \qquad (10.131)$$

The $[A(r)]$ matrix is given by

$$\begin{bmatrix} J_n & Y_n & I_n & K_n \\ KJ_n' & KY_n' & KI_n' & KK_n' \\ \frac{D}{r^2}(J_n' bu_3 - J_n u_1) & \frac{D}{r^2}(Y_n' bu_3 - Y_n u_1) & \frac{D}{r^2}(I_n' bu_4 - I_n u_1) & \frac{D}{r^2}(K_n' bu_4 - K_n u_1) \\ \frac{D}{r}(J_n u_4 + J_n' u_2) & \frac{D}{r}(Y_n u_4 + Y_n' u_2) & \frac{D}{r}(I_n u_3 + I_n' u_2) & \frac{D}{r}(K_n u_3 + K_n' u_2) \end{bmatrix}$$

(10.132)

where,

$$\begin{aligned} b &= \beta r \\ u_1 &= b^2(1-\nu) \\ u_2 &= -b(1-\nu) \\ u_3 &= u_1 + b^2 \\ u_4 &= u_1 - b^2 \end{aligned}$$

and $J_n'(b)$ etc., are the derivatives with respect to the argument.

10.6.2 Closed Form Receptances

The state vector at the disc rim under the above forces can be obtained from equations (10.130) and (10.125) as

$$\{V(r_{j,0})\} = [A_0]_j \{X\}_j = [A_0]_j [B]^{-1} \begin{Bmatrix} 0 \\ 0 \\ S_f(r_{j,0}) \\ B_m(r_{j,0}) \end{Bmatrix} \qquad (10.133)$$

If an axial force $Q_1(n)$ of magnitude unity is only applied, then the shear force $S_f(r_{j,0})$ will also be unity. Hence

$$\begin{Bmatrix} q_1(n) \\ \dfrac{\partial}{\partial r} q_1(n) \end{Bmatrix} = \begin{bmatrix} 1 & 0 & 0 & 0 \\ 0 & 1 & 0 & 0 \end{bmatrix} [A_0]_j [B]^{-1} \begin{Bmatrix} 0 \\ 0 \\ 1 \\ 0 \end{Bmatrix} \tag{10.134}$$

Alternatively, if a unit tangential torque is applied, then the bending moment $B_m(r_{j,0})$ will be unity. We may now write down the receptance matrix for directions 1 and 2.

$$\begin{bmatrix} \alpha_{11}(n) & \alpha_{12}(n) \\ \alpha_{21}(n) & \alpha_{22}(n) \end{bmatrix} = \begin{bmatrix} 1 & 0 & 0 & 0 \\ 0 & -1 & 0 & 0 \end{bmatrix} [A_0]_j [B]^{-1} \begin{bmatrix} 0 & 0 \\ 0 & 0 \\ 1 & 0 \\ 0 & -1 \end{bmatrix} \tag{10.135}$$

The receptance to a rim radial torque may be evaluated as

$$\alpha_{15}(n) = \frac{n}{a} \alpha_{11}(n)$$
$$\alpha_{25}(n) = \frac{n}{a} \alpha_{21}(n)$$
$$\alpha_{55}(n) = \left(\frac{n}{a}\right)^2 \alpha_{11}(n) \tag{10.136}$$

The remaining receptance components relate to the in-plane motion of the disc and will be assumed to be zero except for 1 and 0 nodal diameter modes, where rigid body motions are possible. These receptances are

$$[\alpha(1)] = \begin{bmatrix} \alpha_{11}^{(1)} & \alpha_{12}^{(1)} & 0 & 0 & \dfrac{\alpha_{11}^{(1)}}{a} & 0 \\ & \alpha_{22}^{(1)} & 0 & 0 & \dfrac{\alpha_{12}^{(1)}}{a} & 0 \\ & & R_1 & 0 & 0 & -R_1 \\ & \text{sym} & & 0 & 0 & 0 \\ & & & & 0 & 0 \\ & & & & & R_1 \end{bmatrix} \tag{10.137}$$

$$[\alpha(0)] = \begin{bmatrix} \alpha_{11}^{(0)} & \alpha_{12}^{(0)} & 0 & 0 & 0 & 0 \\ & \alpha_{22}^{(0)} & 0 & 0 & 0 & 0 \\ & & \alpha^2 R_0 & \alpha^2 R_0 & 0 & 0 \\ & \text{sym} & & R_0 & 0 & 0 \\ & & & & 0 & 0 \\ & & & & & 0 \end{bmatrix} \tag{10.138}$$

where,

$$R_1 = -\frac{\pi}{M_D\omega^2}$$
$$M_D = \text{Mass of the disc}$$
$$R_0 = -\frac{2\pi}{M_4\omega^2}$$
$$M_4 = \text{Polar mass moment of inertia}$$

10.6.3 Series Form Receptances

We will now derive the disc flexural receptances in the form of an infinite series, each term of which represents one natural mode of the disc as a free body. The procedure will be to formulate the receptances in principal coordinates and then transform to local coordinates. Let p_{ns}, P_{ns} and α^{p}_{ns} be the displacement, force and receptance in the principal coordinates of a disc resonating in a freely supported mode with n nodal diameters and s nodal circles. The relation between the principal and local coordinates is

$$\{q(n)\} = \sum_{s=0}^{\infty} p_{ns}\{\phi(a)_{ns}\} \tag{10.139}$$

where $\{\phi(r)_{ns}\}$ is the modal displacement vector, of which the axial component at the rim is made unity, for all possible values of n and s, i.e., $\phi_1(a)_{ns} = 1$.

Initially, we must know a sufficiently large number of terms in the receptance series for each annulus. This implies that we must know the natural frequency and inertia coefficient and also the displacement and slope at the inner and outer edge of each annulus for a reasonably large number of freely supported modes. Let these quantities respectively for the jth annulus be

$$\omega_{ns},\ a_{ns},\ \phi_1(r_{ji})_{ns},\ \phi_2(r_{ji})_{ns},\ \phi_1(r_{jO})_{ns},\ \phi_2(r_{jO})_{ns}$$

In the present case, natural frequencies of complete disc ω_{ns} will only be approximate because annular receptances of each consist of truncated series.

Next, to calculate the disc inertia coefficient, we must solve for the mode shape of the complete disc in terms of both the displacements and forces at junction between the connected annuli. Suppose, when the disc is vibrating in its (n, s)th mode with known amplitudes, we find the forces in directions 1 and 2 on the jth annulus to be

$$\begin{Bmatrix} Q_1(n, r_{ji}) \\ Q_2(n, r_{ji}) \end{Bmatrix} \quad \text{at the inner edge}$$

$$\begin{Bmatrix} Q_1(n, r_{jO}) \\ Q_2(n, r_{jO}) \end{Bmatrix} \quad \text{at the outer edge}$$

We then convert this loading into principal coordinates as follows. Impose a virtual increase δp_{ns} on the (n, s)th principal coordinate of the annulus.

Thus the virtual work equation for the jth annulus is

$$\begin{aligned}\delta W &= Q_1(n, r_{ji})\,\delta q_1(n, r_{ji}) + Q_2(n, r_{ji})\,\delta q_2(n, r_{ji}) \\ &\quad + Q_1(n, r_{j0})\,\delta q_1(n, r_{j0}) + Q_2(n, r_{j0})\,\delta q_2(n, r_{j0})\end{aligned} \tag{10.140}$$

The force on this annulus in principal coordinates is

$$\begin{aligned}P_{ns} &= Q_1(n, r_{ji})\phi_1(r_{ji})_{ns} + Q_2(n, r_{ji})\phi_2(r_{ji})_{ns} \\ &\quad + Q_1(n, r_{j0})\phi_1(r_{j0})_{ns} + Q_2(n, r_{j0})\phi_2(r_{j0})_{ns}\end{aligned} \tag{10.141}$$

Now the deflection of the annulus in the (n, s)th principal coordinate is

$$p_{ns} = P_{ns}\alpha^p_{ns} = \frac{P_{ns}}{a_{ns}(\omega^2_{ns} - \omega^2)} \tag{10.142}$$

Knowing the principal displacements, we can compute the kinetic energy in each mode and, by the following summation, we obtain the kinetic energy of the annulus

$$T_{ns} = \frac{1}{2}\sum_{s=0}^{\infty} a_{ns}\dot{p}^2_{ns} \tag{10.143}$$

By extending this sum over all the annuli, we obtain the total kinetic energy of the disc in the (n, s)th mode.

$$T_{ns} = \frac{1}{2}\sum_{j=1}^{J}\sum_{s=0}^{\infty} a_{ns}\dot{p}^2_{ns} \tag{10.144}$$

Hence we can compute the inertia coefficient of the disc in the (n, s)th mode as

$$a_{ns} = \frac{2T_{ns}}{\dot{p}^2_{ns}} \tag{10.145}$$

Having now calculated the natural frequency and inertia coefficient, we can easily calculate the approximate receptance for the (n, s)th mode using the following.

$$\alpha^p_{ns} = \frac{1}{a_{ns}(\omega^2_{ns} - \omega^2)} \tag{10.146}$$

Transformation of receptances from principal to local coordinates involves the following steps:

(a) Apply a force in the component in global coordinates and then express this as a system of forces in local coordinates.

(b) Form the product of the forces in local coordinates with the relevant local receptances and hence obtain the response in local coordinates.

(c) Interpret this response in terms of displacements in global coordinates.

(d) The global receptance is the quotient of the final response over the initial force.

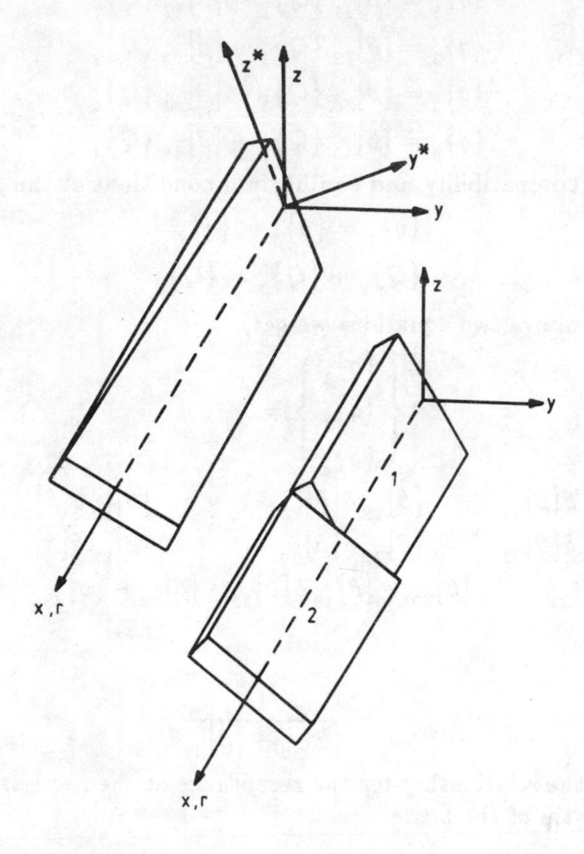

Fig. 10.14a Pre-twisted blade modelled with two straight beam elements

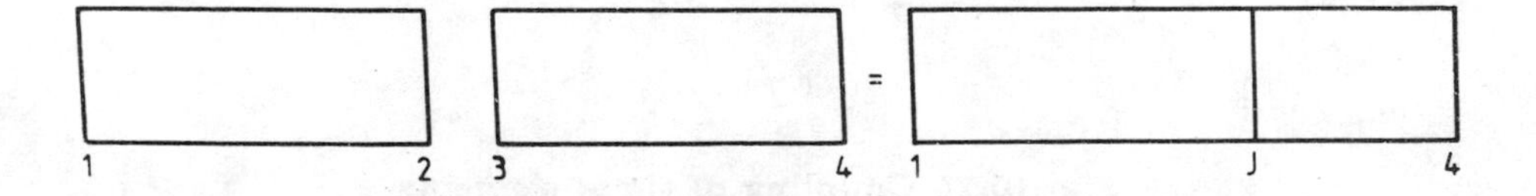

Fig. 10.14b Coupling of blade elements

10.6.4 Receptances of Multi-Stage Blades

It is necessary to obtain a general relationship connecting the receptances of the blade at the root to those at any intermediate point along the length of the blade. If the blade is assumed to be split into two, with end points as 1, 2, 3 and 4, as shown in Figs. 10.14a and 10.14b, we have

$$\begin{aligned}\{q\}_1 &= [\beta]_{11}\{Q\}_1 + [\beta]_{12}\{Q\}_2 \\ \{q\}_2 &= [\beta]_{12}\{Q\}_1 + [\beta]_{22}\{Q\}_2 \\ \{q\}_3 &= [\beta]_{33}\{Q\}_3 + [\beta]_{34}\{Q\}_4 \\ \{q\}_4 &= [\beta]_{43}\{Q\}_3 + [\beta]_{44}\{Q\}_4\end{aligned} \tag{10.147}$$

From the compatibility and equilibrium conditions at the point j,

$$\begin{aligned}\{q\}_2 = \{q\}_3 &= \{q\}_j \\ \{Q\}_2 + \{Q\}_3 &= \{Q\}_j\end{aligned} \tag{10.148}$$

From the above two equations we get,

$$\begin{Bmatrix}\{q\}_1 \\ \{q\}_j \\ \{q\}_4\end{Bmatrix} =$$

$$\begin{bmatrix}[\beta]_{11} - [\beta]_{12}\delta[\beta]_{21} & [\beta]_{12}\delta[\beta]_{33} & [\beta]_{12}\delta[\beta]_{34} \\ [\beta]_{21} - [\beta]_{22}\delta[\beta]_{21} & [\beta]_{22}\delta[\beta]_{33} & [\beta]_{22}\delta[\beta]_{34} \\ [\beta]_{43}\delta[\beta]_{21} & [\beta]_{43} - [\beta]_{43}\delta[\beta]_{33} & [\beta]_{44} - [\beta]_{43}\delta[\beta]_{34}\end{bmatrix} \begin{Bmatrix}\{Q\}_1 \\ \{Q\}_j \\ \{Q\}_4\end{Bmatrix} \tag{10.149}$$

where,

$$\delta = \frac{1}{[\beta]_{22} + [\beta]_{33}} \tag{10.150}$$

which gives the relationship for the receptance at the root, at any position j and at the tip of the blade.

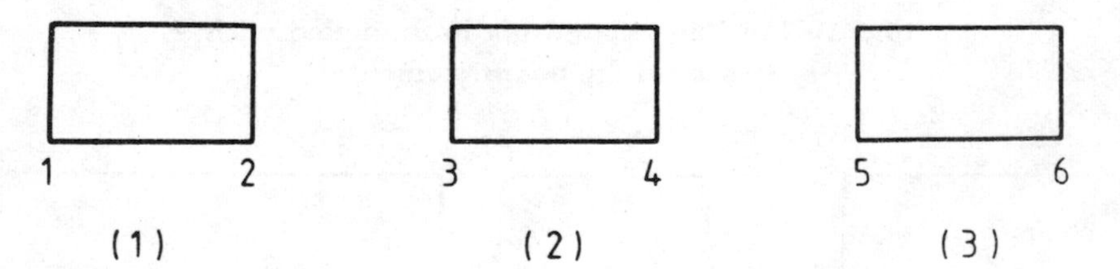

Fig. 10.15 Coupling of three elements

Let there be 3 elements of the blade as shown in Fig. 10.15. Let $[\beta_1]$, $[\beta_2]$ and $[\beta_3]$ be the receptance matrices of elements 1, 2 and 3 respectively of 12×12 size corresponding to all 6 degrees of freedom at both the ends of each element. If the coupled receptance matrix for elements 1 and 2 be denoted as $[\beta_{12}]$, then

$$[\beta_{12}]^{-1} = [\beta_1]^{-1} + [\beta_2]^{-1} \tag{10.151}$$

Similarly the coupled receptance matrix of the newly formed element 12 and element 3, will be given by

$$[\beta_{123}]^{-1} = [\beta_{12}]^{-1} + [\beta_3]^{-1} \tag{10.152}$$

10.6.5 Shroud Receptances

Shrouding commonly employed in gas turbines can be classified as non-interlocking and interlocking types. Some of the interlocking shrouds are shown in Fig. 10.16. In the non-interlocking type, the purpose of the shroud is simply to minimize gas leakage over the blade tips and there is no contact between the adjacent segments, even when the blade vibrates. In Fig. 10.16a, the segments are designed with zig-zag edges such that when the blades are assembled in the disc, they knit together and a preload is established at each interface. In service, centrifugal untwist and thermal expansion increase the preload so that effectively a continuous hoop is formed.

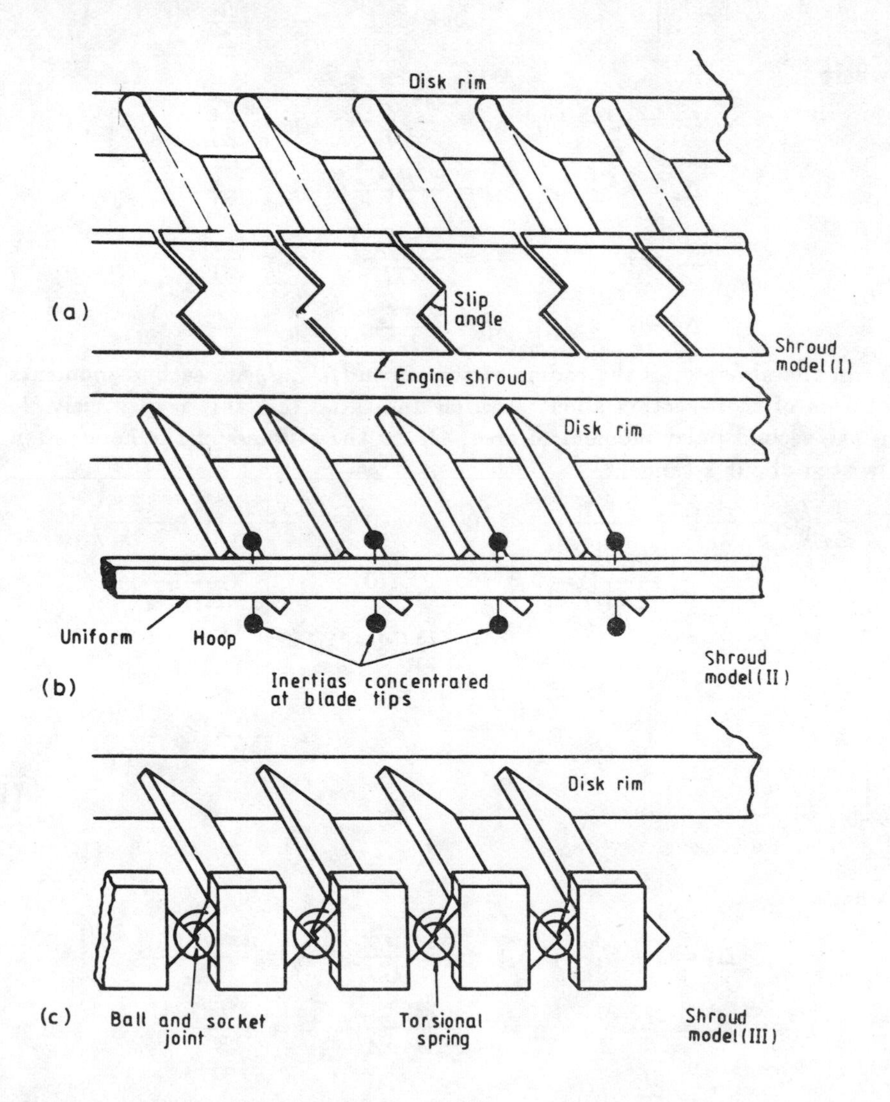

Fig. 10.16 Representation of interlocking shroud

Receptances derived by Cottney [76] are

$$[\gamma(n)]_{n\geq 2} = \begin{bmatrix} \frac{\gamma_0}{\Delta_0} & 0 & 0 & 0 & \frac{n\gamma_0}{r_s\Delta_0} & 0 \\ & \frac{\gamma_t}{\Delta_t} & 0 & 0 & 0 & 0 \\ & & \frac{\gamma_l}{\Delta_l} & 0 & 0 & 0 \\ & \text{sym} & & \frac{n^2\gamma_i}{r_s^2\Delta_i} & 0 & \frac{n\gamma_i}{r_s\Delta_i} \\ & & & & \frac{n^2\gamma_0}{r_s^2\Delta_0} & 0 \\ & & & & & \frac{\gamma_i}{\Delta_i} \end{bmatrix} \quad (10.153)$$

where,

$$\Delta_0 = n^4 - \lambda_0; \quad \lambda_0 = \frac{\omega^2\rho A r_s^4}{EI_0}; \quad \gamma_0 = \frac{r_s^3}{EI_0}$$

$$\Delta_i = n^4 - \lambda_i; \quad \lambda_i = \frac{\omega^2\rho A r_s^4}{EI_i}; \quad \gamma_i = \frac{r_s^3}{EI_i}$$

$$\Delta_t = n^2 - \lambda_t; \quad \lambda_t = \frac{\omega^2\rho I_2 r_s^2}{GI_t}; \quad \gamma_t = \frac{r_s}{GI_t}$$

$$\Delta_l = n^2 - \lambda_l; \quad \lambda_l = \frac{\omega^2\rho r_s^2}{E}; \quad \gamma_l = \frac{r_s}{EA}$$

In the above r_s is the radius of the shroud, I_0, I_i are second moments of area of cross-section about a radial and axial bisectors respectively, I_2 is the second polar moment of area, GI_t is the stiffness of the hoop when twisted about a tangent.

$$[\gamma(1)] = \begin{bmatrix} \frac{\gamma_t}{\Delta_t}\left(C_t - \frac{1}{\lambda_t}\right) & \frac{\gamma_t}{\Delta_t r_s \lambda_t} & 0 & 0 & \frac{\gamma_t}{r_s\Delta_t}\left(C_t - \frac{1}{\lambda_t}\right) & 0 \\ & \frac{\gamma_t}{r_s^2\Delta_t}(1-\lambda_t) & 0 & 0 & \frac{\gamma_t}{\Delta_t r_s^2 \lambda_t} & 0 \\ & & 0 & \frac{\gamma_l}{\Delta_l}(1-\lambda_l) & 0 & 0 \\ & \text{sym} & & 0 & 0 & 0 \\ & & & & \frac{\gamma_t}{r_s^2\Delta_t}\left(C_t - \frac{1}{\lambda_t}\right) & 0 \\ & & & & & \frac{\gamma_l}{\Delta_l}(1-\lambda_l) \end{bmatrix} \quad (10.154)$$

where,

$$\Delta_t = 1 + C_t - C_t\lambda_t; \quad \gamma_t = \frac{r_s^3}{GI_t}; \quad \lambda_t = \frac{\omega^2\rho A r_s^2}{GI_t}$$

$$\Delta_l = 1 - \frac{\lambda_l}{2}; \quad \gamma_l = -\frac{1}{2r_s\omega^2\rho A}; \quad \lambda_l = \frac{\omega^2\rho r_s^2}{E}$$

$$C_t = \frac{I_2}{r_s^2 A}$$

$$[\gamma(0)] = \begin{bmatrix} \frac{\gamma_0}{\Delta_0} & 0 & 0 & 0 & 0 & 0 \\ & \frac{\gamma_t}{\Delta_t} & 0 & 0 & 0 & 0 \\ & & \frac{\gamma_l}{\Delta_l} & \frac{\gamma_l}{r_s \Delta_l} & 0 & 0 \\ & \text{sym} & & \frac{\gamma_l}{r_s^2 \Delta_l} & 0 & 0 \\ & & & & 0 & 0 \\ & & & & & 0 \end{bmatrix} \qquad (10.155)$$

where the notations in the above are similar to that of the case $n \geq 2$, except

$$\Delta_t = \frac{1}{1 - \lambda_t}; \quad \lambda_t = \frac{r_s^2 \rho \omega^2 I_2}{E I_0}$$

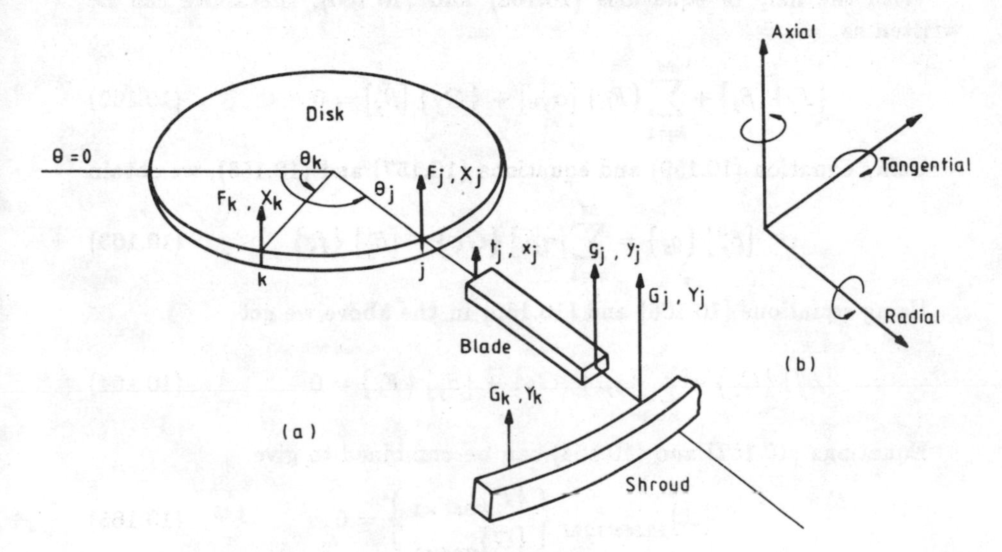

Fig. 10.17 (a) Disc-blade-shroud interaction
(b) Axes referred to point j

10.6.6 Disc-Blade-Shroud Interaction

Fig. 10.17 shows the effect of addition of shrouding. The force displacement relationship for the disc remains same as in equation (10.105). Force displacement relationship for the root of the jth blade in equation (10.106) is modified by considering the force vector $\{q_j\}$ and the corresponding receptance matrix $[\beta_j']$. Then

$$\{X_j\} = [\beta_j]\{f_j\} + [\beta_j']\{g_j\}, \quad j = 1, 2, \ldots, M \qquad (10.156)$$

At the point of attachment to the shroud for the jth blade, we have

$$\{Y_j\} = [\beta_j']\{f_j\} + [\beta_j'']\{g_j\}, \quad j = 1, 2, \ldots, M \qquad (10.157)$$

where $[\beta_j'']$ is the corresponding receptance matrix for forces $\{g_j\}$. Force-displacement relation for the shrouding is similar to that of the disc, i.e.,

$$\{y_j\} = \sum_{k=1}^{M} [\gamma_{jk}] \{G_k\}, \quad j = 1, 2, \ldots, M \tag{10.158}$$

where $[\gamma_{jk}]$ is the point receptance for the shroud. For the shroud junction, we have

$$\{Y_j\} = \{y_j\} \tag{10.159}$$

$$\{G_j\} + \{g_j\} = 0, \quad j = 1, 2, \ldots, M \tag{10.160}$$

Using equation (10.105) and equation (10.156), we obtain

$$\{f_j\} [\beta_j] = \sum_{k=1}^{M} \{F_k\} [\alpha_{jk}] - \{g_j\} [\beta_j'] \tag{10.161}$$

With the help of equations (10.108) and (10.160), the above can be written as

$$\{F_j\} [\beta_j] + \sum_{k=1}^{M} \{F_k\} [\alpha_{jk}] + \{G_j\} [\beta_j'] = 0 \tag{10.162}$$

Using equation (10.159) and equations (10.157) and (10.158), we obtain

$$[\beta_j''] \{g_j\} = \sum_{k=1}^{M} [\gamma_{jk}] \{G_k\} - [\beta_j'] \{f_j\} \tag{10.163}$$

Using equations (10.108) and (10.160) in the above we get

$$[\beta_j''] \{G_j\} + \sum_{k=1}^{M} [\gamma_{jk}] \{G_k\} + [\beta_j'] \{F_j\} = 0 \tag{10.164}$$

Equations (10.162) and (10.164) can be combined to give

$$[\Delta]_{12M \times 12M} \left\{ \begin{matrix} \{F\}_{6M \times 1} \\ \{G\}_{6M \times 1} \end{matrix} \right\} = 0 \tag{10.165}$$

from which the frequencies and mode shapes can be determined.

10.6.7 Examples

The following examples are studied by Rao et al [45].

Example 1

A uniform disc with 119 uniform blades is considered. 6 coordinates are used at the root of each blade for coupling with the disc.

The disc is 254 mm in radius, and 25.4 mm thick. The density of the material is 7800 kg/cu m, $E = 2.07 \times 10^{11}$ Pa and the Poisson's ratio is 0.287.

Each blade has a length of 203.2 mm with a thickness of 3.18 mm and width of 25.4 mm mounted at 15 deg stagger angle. The density and Young's modulus are same as that of the disc.

The natural frequencies obtained are shown in Fig. 10.18.

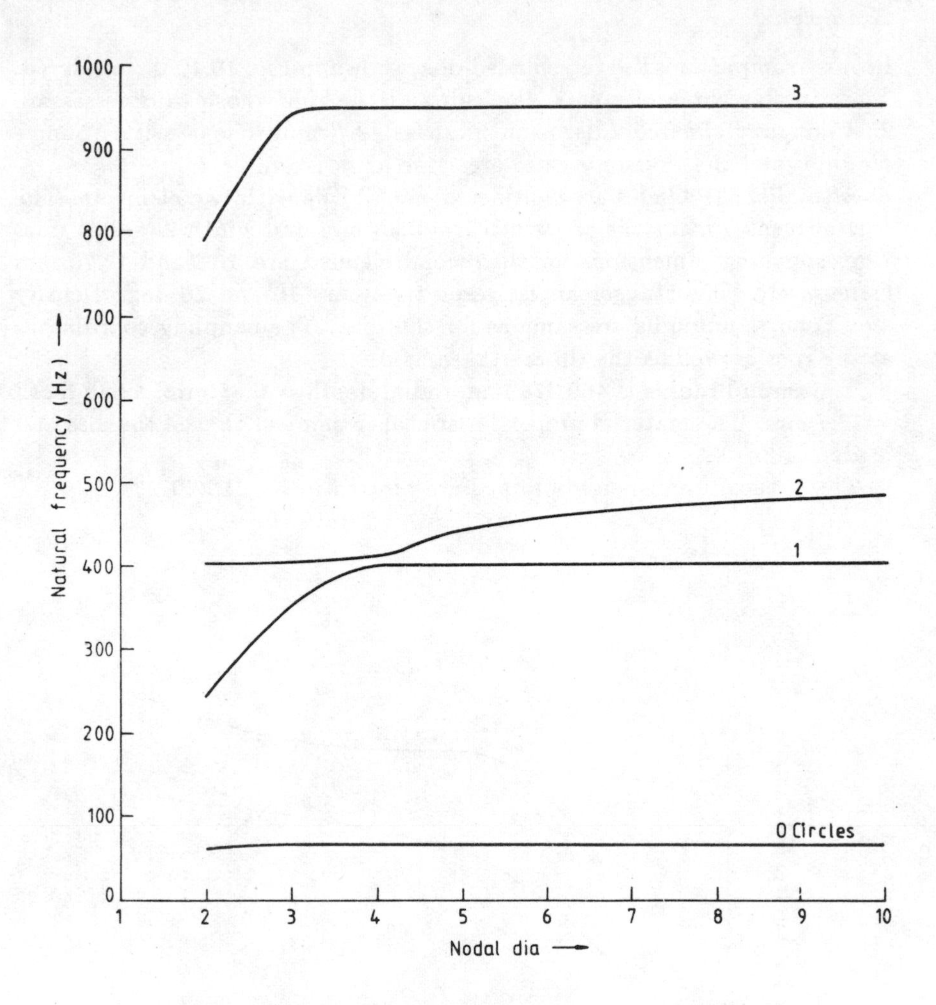

Fig. 10.18 Natural frequency vs nodal diameter plot for Example 1

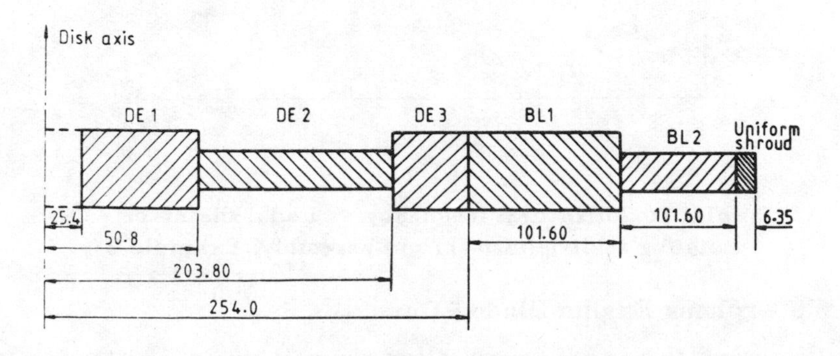

Fig. 10.19 Geometry of disc-blade-shroud assembly for Example 2

Example 2

In this example a multi-stage bladed-disc, shown in Fig. 10.19 is considered. The disc has three elements, the hub and the rim whose thicknesses are 25.4 mm, and the main disc is of thickness 12.7 mm. The density, Young's modulus and the Poisson's ratio are same as in Example 1.

A total of 119 blades are considered, each blade with two elements. The first element dimensions are width 1 = 6.35 mm and width 2 = 25.4 mm. Corresponding dimensions for the second element are 3.18 and 12.70 mm respectively. The stagger angles respectively are 10 and 20 deg. Density and Young's modulus are same as for the disc. The coupling coordinates at the root as well as the tip are taken as 6.

The shroud radius is 460.375 mm, radial depth = 6.35 mm, axial depth = 12.7 mm. The material properties are taken same as that of the disc and blade.

The natural frequencies obtained are plotted in Fig. 10.20.

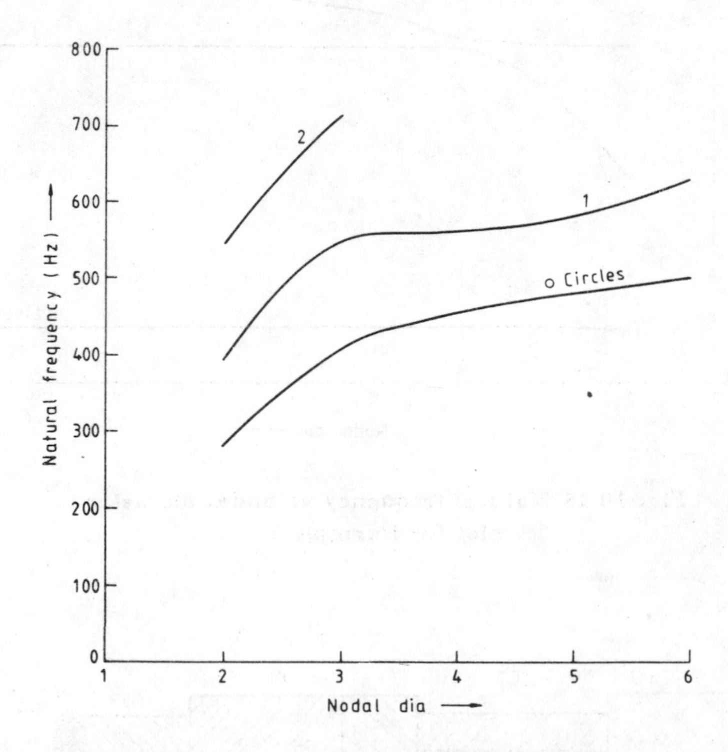

Fig. 10.20 Natural frequency vs nodal diameter plots for a blade-disc-shroud assembly, Example 2

10.6.8 Orpheus Engine Bladed-Discs

In this section, theoretical as well as experimental studies of a gas turbine bladed-disc assembly and third stage of compressor bladed-disc assembly of an Orpheus aircraft engine [45] are reported.

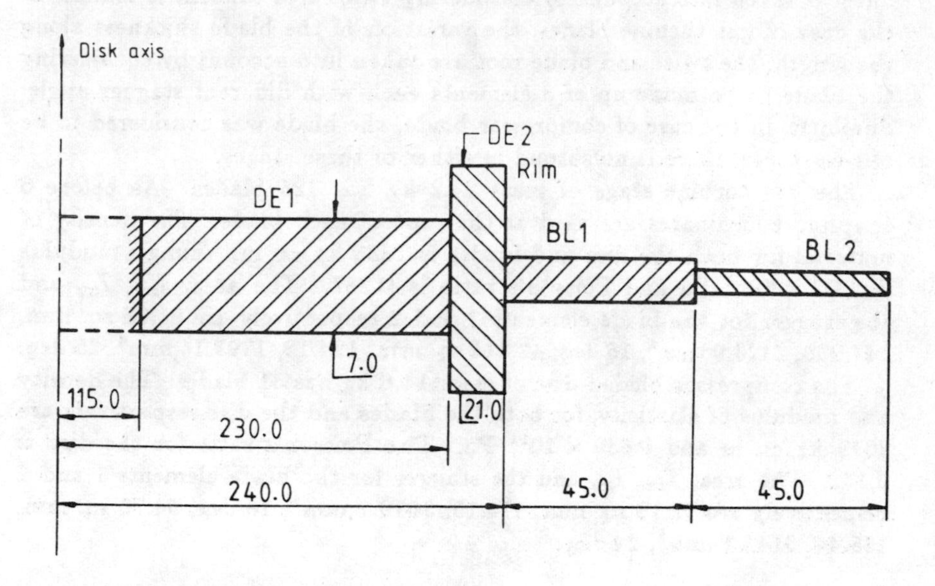

Fig. 10.21 Geometry of an Orpheus engine gas turbine bladed-disc assembly

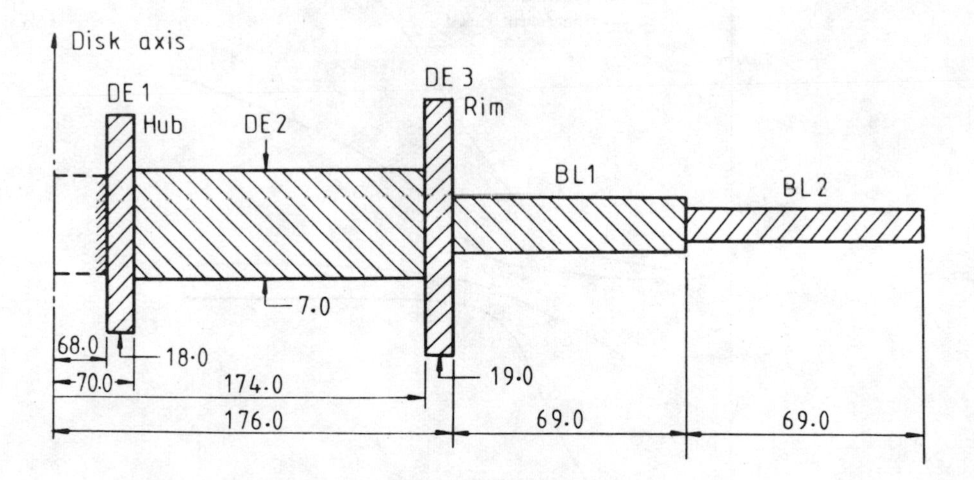

Fig. 10.22 Geometry of an Orpheus engine third stage compressor bladed-disc assembly

The geometry of the gas turbine stage is shown in Fig. 10.21 and for the compressor stage in Fig. 10.22. For both the turbine and compressor stages, non-uniformity in the thickness of the disc (disc, disc rim and disc hub), is taken into account by considering two/three concentric annuli. In the case of gas turbine blades, the variation of the blade thickness along the length, the twist and blade root are taken into account by considering the blade to be made up of 3 elements each with different stagger angle. Similarly, in the case of compressor blade, the blade was considered to be of two stages; there is no shroud in either of these stages.

The gas turbine stage of mass 24.2 kg has 125 blades. As before 6 coupling coordinates are used at the root of each blade. The density of material for both the disc and blades is 7833 kg/cu m, Young's modulus is 2.07×10^{11} Pa and Poisson's ratio is 0.287. The area, I_{xx}, I_{yy} and the stagger for the blade elements 1 and 2 respectively are 61.08 sq. mm, 147.909, 2123.9 mm^4, 15 deg; 42.315 sq. mm, 124.13, 1793.14 mm^4, 25 deg.

The compressor bladed-disc of mass 14.6 kg has 31 blades. The density and modulus of elasticity, for both the blades and the disc respectively are 2699 kg/cu m and 0.689×10^{11} Pa. The Poisson's ratio for the disc is 0.332. The area, I_{xx}, I_{yy} and the stagger for the blade elements 1 and 2 respectively are 72.18 sq. mm, 174.15, 3619.2 mm^4, 18 deg; 54.76 sq. mm, 155.44, 3142.1 mm^4, 24 deg.

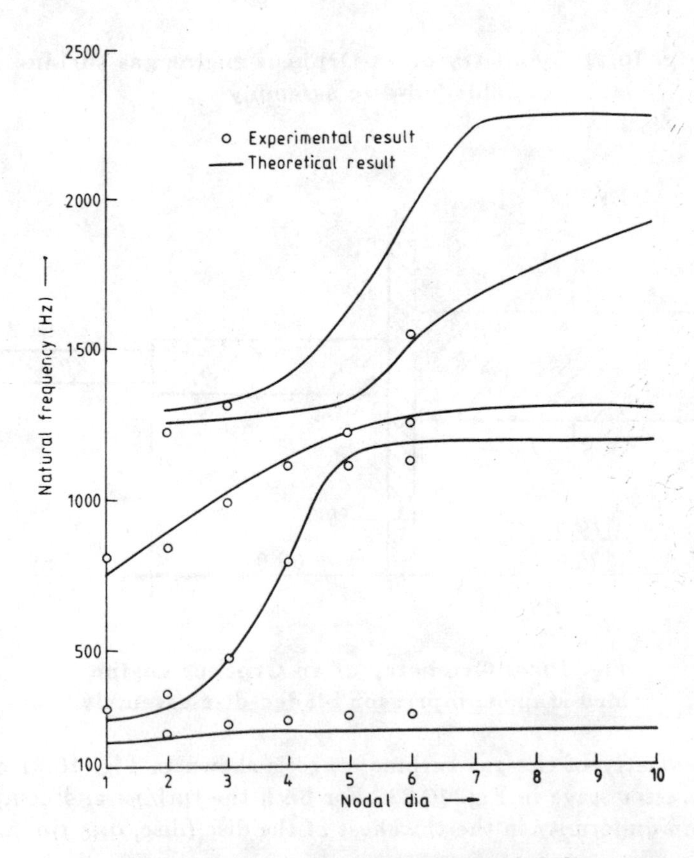

Fig. 10.23 Natural frequency vs nodal diameter plots for gas turbine assembly

An MBIS, automatic servo-controlled permanent magnet exciter controlled by a 2600 series computer (Digital Vibration System) is used to excite the bladed-disc assemblies. A feedback MBIS accelerometer mounted at the centre of the disc controls the system with the desired abort limits to the required level of excitation. Another B&K accelerometer mounted on the disc, whose position can be changed, is used to monitor the response through a B&K 2511 vibration meter. The signal is also monitored through a real time HP 3582A spectrum analyser to obtain the natural frequencies. Both the turbine and compressor stages are mounted through 12 and 10 bolt fixtures respectively, as provided in the original assemblies. After scanning the natural frequencies, the shape of each mode is identified by probing with a hand held accelerometer.

The analytical and experimental results are given in Figs. 10.23 and 10.24. The computer program gives a close prediction of the experimental results.

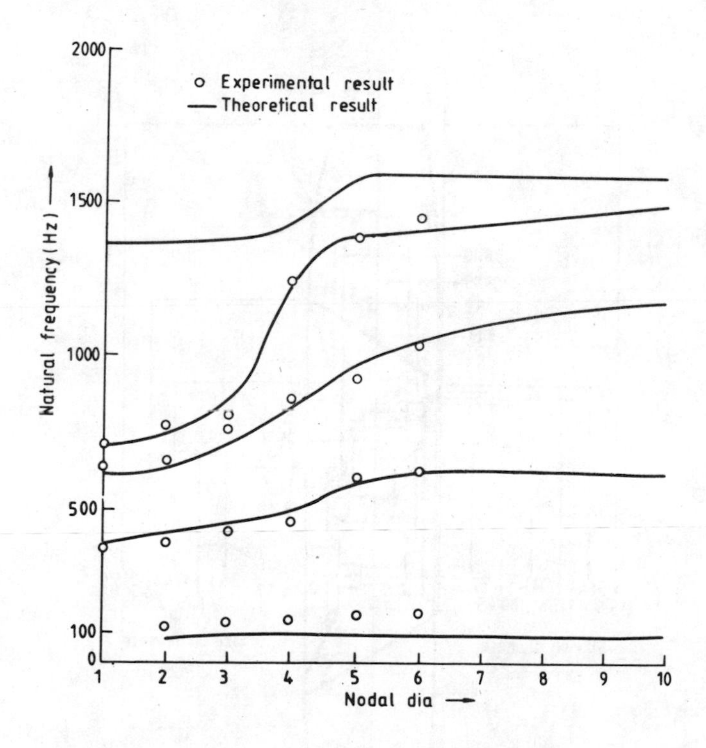

Fig. 10.24 Natural frequency vs nodal diameter plots for compressor assembly

10.6.9 Finite Element Analysis

While the receptance technique is a very useful tool, it does suffer from the number of elements that can be adopted for the disc as well as blades. It also requires development of a good computer code that can be used by any designer; even then it requires some interpretation during the computer run. In certain cases, the bladed-discs are more complicated in their geometry and may be difficult to accurately model by receptance methods. Under such circumstances, particularly when a quick analysis is desirable, standard finite element methods become very handy. Swaminadham et al [51], used NASTRAN program to determine the natural frequencies of a complex bladed-disc. Salient points of this work are discussed here.

The assembly considered has a complex geometry with the blades and disc integrally machined from a solid titanium block. The disc has variable thickness and flares into a conical hub to maximize efficiency of the compressible flow. Twenty blades integral with the disc have curved camber, are severely twisted from root to tip, are swept back, have their stacking axis shifted from the radial line and have both chordwise and spanwise variable thickness. The bladed-disc system is bolted to a shaft-flange attachment and a sectional view with dimensions is shown in Fig. 10.25.

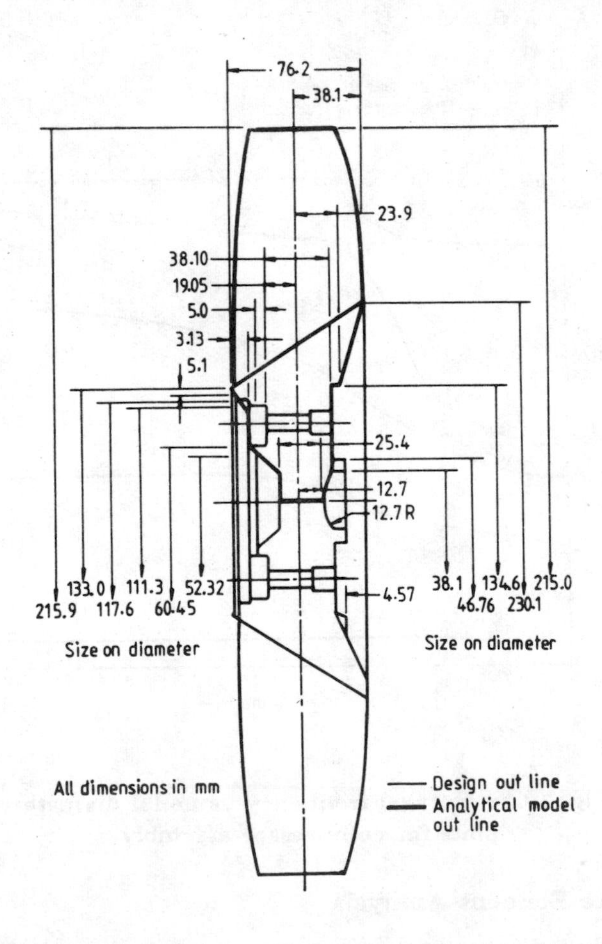

Fig. 10.25 Sectional view of bladed-disc

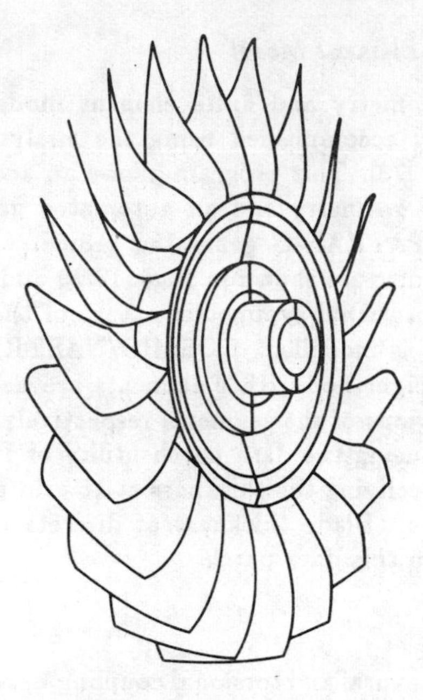

Fig. 10.26 PATRAN generated geometric model of whole bladed-disc

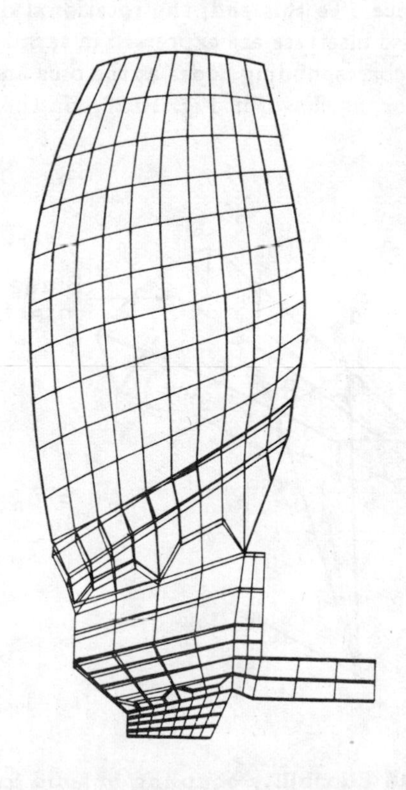

Fig. 10.27 Finite element model of 18° sector of bladed-disc

Geometry and Finite Element Model

The generation of geometry and finite element model is a tedious task and can be efficiently accomplished using the analytical solid modeling program PATRAN-G [78]. This program allows an accurate simulation of the highly curved 3D geometry and an automated generation of a finite element mesh. The PATRAN-G generated geometry and finite element models of the bladed-disc are shown in Figs. 10.26 and 10.27, respectively. Use is made of the rotational symmetric nature of the problem and only an 18 degree segment is modelled. COSMIC/NASTRAN four node plate element QUAD and eight node HEX elements are used to discretize the blade and the disc regions of the segment, respectively. The varying blade thickness is specified using the data patch utility of PATRAN-G. A data patch is defined by specifying the thickness at 16 grid points of the surface representing the blade. Blade thickness at discrete element centroids is then interpolated from this data patch.

NASTRAN Analysis

Compatibility in the flexural and torsional coupling between the QUAD and HEX elements is enforced through the specification of multipoint constraint relations among the degrees of freedom of the two types of elements at the blade-disc interface. To this end, the rotational displacements of the nodes at the blade-disc interface are expressed in terms of the translational displacements of the corresponding nodes at the plus and minus nine degree sides of the disc sector, as shown in Fig. 10.28, via the following relation.

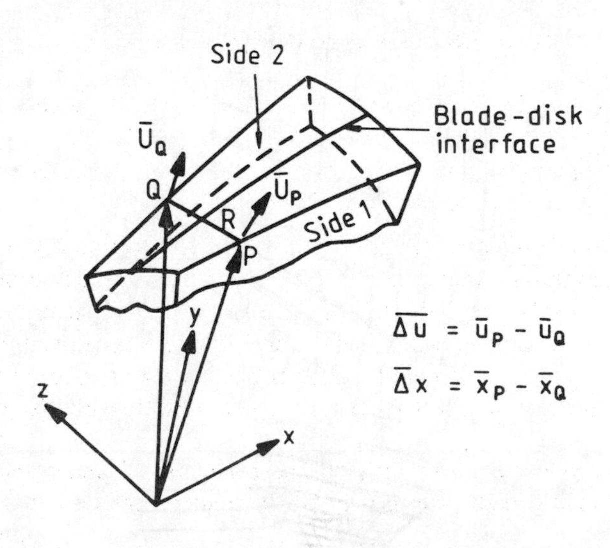

Fig. 10.28 Flexibility coupling of solid and plate elements at disc and blade interface

$$\overline{\theta} = \left(\frac{1}{\overline{\Delta X}}\hat{i}\right) x\overline{\Delta U} \tag{10.166}$$

where $\overline{\theta}$ is the vector of the three orthogonal rotational displacements of the node on the common interface, $\hat{i}$ is the unit vector, and $\overline{\Delta U}$, $\overline{\Delta X}$ are the vectors of relative translational displacements and spatial positions of the corresponding nodes on the $\pm 9°$ sides of the disc segment. Referring to Fig. 10.28,

$$\overline{\Delta X} = \overline{X}_P - \overline{X}_Q \tag{10.167}$$

$$\overline{\Delta U} = \overline{U}_P - \overline{U}_Q \tag{10.168}$$

where $\overline{X}$, $\overline{U}$ are respectively the position vector and displacement vector of the nodes P and Q on the two sides of the disc segment.

The finite element equations for the eigen problem of a cyclic symmetric segment n are given as

$$([K] - \omega^2[M])\{U\}^n = 0 \tag{10.169}$$

together with the inter-segment compatibility conditions

$$\{U\}^{n+1}_{\text{side 2}} = \{U\}^n_{\text{side 1}} \qquad n = 1, 2, \ldots, N \tag{10.170}$$

where $[K]$ and $[M]$ are respectively the stiffness and mass matrices, and $\{U\}^n$ is the vector of nodal displacements of grid points of the segment n, N is the total number of segments, and ω is the modal vibration frequency. Owing to the rotational symmetry, the displacement field $\{U\}^n$ can be expressed as

$$\{U\}^n = \{\overline{U}\}^0 + \sum_{k=1}^{k_L} \{\overline{U}\}^{kc} \cos(n-1)ka + \{\overline{U}\}^{ks} \sin(n-1)ka + (-1)^{\frac{N}{2}-1}\{\overline{U}\}^{\frac{N}{2}} \tag{10.171}$$

where,

$$a = \frac{2\pi}{N}$$

$$k_L = \begin{cases} \dfrac{N-1}{2} & \text{if } N \text{ is odd} \\ \dfrac{N-2}{2} & \text{if } N \text{ is even} \end{cases}$$

The quantities $\{\overline{U}\}^0$, $\{\overline{U}\}^{kc}$, $\{\overline{U}\}^{ks}$ and $\{\overline{U}\}^{\frac{N}{2}}$ are the circumferential harmonic components. Substitution of equation (10.171) in (10.170) leads to the compatibility conditions relating the harmonic components as

$$\begin{aligned}
\{\overline{U}\}^0_2 &= \{\overline{U}\}^0_1 \quad \text{for } k = 0 \\
\{\overline{U}\}^{kc}_2 &= \{\overline{U}\}^{kc}_1 \cos ka + \{\overline{U}\}^{ks}_1 \sin ka \quad \text{for } k = 1, 2, \ldots, k_L \\
\{\overline{U}\}^{ks}_2 &= \{\overline{U}\}^{kc}_1 \sin ka + \{\overline{U}\}^{ks}_1 \cos ka \quad \text{for } k = 1, 2, \ldots, k_L \\
\{\overline{U}\}^{\frac{N}{2}}_2 &= -\{\overline{U}\}^{\frac{N}{2}}_1 \quad \text{for } k = \frac{N}{2}
\end{aligned} \tag{10.172}$$

In view of the above constraints, let $\{\overline{U}\}^K$ be a vector of independent degrees of freedom that include all the interior grid point displacements and the harmonic components of boundary 1, $\{\overline{U}\}^{kc}$ and $\{\overline{U}\}^{ks}$. Then

$$\begin{aligned}\{\overline{U}\}^{kc} &= [G]_{ck}\{\overline{U}\}^K \\ \{\overline{U}\}^{ks} &= [G]_{sk}\{\overline{U}\}^K\end{aligned} \tag{10.173}$$

where the transformation matrices $[G]_{ck}$ and $[G]_{sk}$ are defined using equation (10.172).

Equation (10.169) can now be written as

$$\left([\overline{K}]^K - \omega^2[\overline{M}]^K\right)\{\overline{U}\}^K = 0 \tag{10.174}$$

where,

$$\begin{aligned}[\overline{K}]^K &= [G]^T_{ck}[K][G]_{ck} + [G]^T_{sk}[K][G]_{sk} \\ [\overline{M}]^K &= [G]^T_{ck}[M][G]_{ck} + [G]^T_{sk}[M][G]_{sk}\end{aligned} \tag{10.175}$$

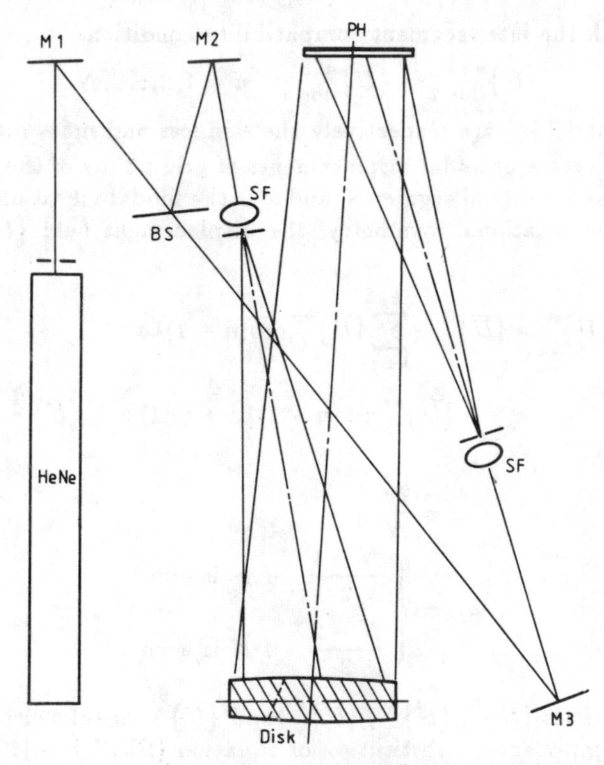

Fig. 10.29 Optical setup for mode shape recording

NASTRAN solves the eigen problem given by the above equation for a given value of the harmonic index k, and outputs the corresponding harmonic component displacement.

Laser holographic tests, see Fig. 10.29 for the optical setup, were conducted to record the mode shapes for comparison with those of the analytical model. Since the compressor disc is rigidly fixed to a shaft by a coupling flange for test purposes the shaft section of the assembly was clamped rigidly by a heavy steel block bolted to the holographic table. Two piezo-electric shakers are mounted on the base of the test assembly to excite the disc in its normal modes. The assembly was excited and the frequency range of interest was scanned to identify the bladed disc modes in real time.

At the natural frequencies thus recorded, time average holographic tests were performed to produce permanent record of the mode shapes. The time average method of recording utilizes standing waves of the object to its advantage. The object's motion is almost non-existent at nodes and the time averaged irradiance distribution at the recording plane is maximum at the nodes, while it goes to minimum at the antinodes. Upon reconstruction, the holographic plates reproduce virtual images of the bladed disc, covered with black and white fringes, whose spacing corresponds to the local amplitude of vibration. Lines of maximum intensity of light indicate the nodal lines in this method.

TABLE 10.3

Compressor Bladed-Disc Assembly Natural Frequencies Hz

Analytical	*Experimental*
391	549
771	823
1275	—
1778	1728

Several trials were attempted to tune the system so that the fundamental frequency of the system duplicated the experimental value. Table 10.3 above gives the best results obtained for 0C modes. There is relatively poor agreement for the lower modes; the accuracy of the analytical values improve considerably with the higher modes.

REFERENCES

1. Stodola, A., *Steam and Gas Turbines*, McGraw-Hill Book Co. (1927).
2. Sezawa, K., "Vibration of Turbine Blades with Shrouding", *Phil. Mag.*, p. 164 (1933).
3. Smith, D.M., "Vibrations of Turbine Blades in Packets", *Proc. 7th Int. Congress Applied Mech.*, London, p. 178 (1949).
4. Campbell, W., "Prediction of Steam Turbine Disk Wheels from Axial Vibration", *ASME*, **46**, No. 1920 (1924).

5. Rao, J.S., "Turbomachine Blade Vibration", *The Shock Vibration Digest*, **19**, No. 5, p. 3 (1987).
6. Lalanne, M., "Vibration Problems in Jet Engines", *The Shock Vibration Digest*, **17**, p. 19 (1985).
7. Leissa, A.W., "Vibrations of Turbine Engine Blades by Shell Analysis", *The Shock Vibration Digest*, **12**, No. 11, p. 3 (1980).
8. Armstrong, E.K., *An Investigation into the Coupling Between the Turbine Discs and Blade Vibrations*, Ph.D. Thesis, University of Cambridge (1955).
9. Bishop, R.E.D. and Johnson, D.C., *The Mechanics of Vibration*, Cambridge University Press (1960).
10. Armstrong, E.K., Christie, P.I. and Hague, W.M., "Natural Frequencies of Bladed-Discs", *Proc. Inst. Mech. Engrs.*, **180**, pt. 31, p. 110 (1965).
11. Armstrong, E.K., "Recent Blade Vibration Techniques", *ASME*, Paper 66-WA/GT-14 (1966).
12. Whitehead, D.S., "Effect of Mistuning on the Vibration of Turbomachine Blades Induced by Wakes", *J. Mech. Engrng. Sci.* (1966).
13. Wagner, J.T., "Coupling of Turbomachine Blades Through the Rotor", *J. Engrng. Power*, ASME, p. 4 (1967).
14. Ewins, D.J., *The Effects of Detuning upon the Vibration of Bladed-Discs*, Ph.D. Thesis, University of Cambridge (1966).
15. Dye, R.C.F. and Henry, T.A., "Vibration Amplitudes of Compressor Blades Resulting from Scatter in Blade Natural Frequencies", *J. Engrng. Power*, ASME, **91**, p. 3 (1969).
16. Ewins, D.J., "The Effects of Detuning upon the Forced Vibrations of Bladed-Discs", *J. Sound and Vib.*, **9**, p. 66 (1969).
17. Ewins, D.J., "A Study of Resonance Coincidence in Bladed-Discs", *J. Mech. Engrng. Sci.*, **12**, No. 5, p. 305 (1970).
18. Ewins, D.J., "A Study of the Vibration of a Bladed Turbine Wheel", *Conf. Vib. Rotating Systems, Inst. Mech. Engr. Proc.* (1972).
19. Kirkhope, J. and Wilson, G.J., "Analysis of Coupled Blade-Disc Vibration in Axial Flow Turbines and Fans", *Proc. 12th AIAA/ASME Conf. Struct. Dyn.* (1971).
20. Ewins, D.J., "Vibration Characteristics of Bladed-Disc Assemblies", *J. Mech. Engrng. Sci.*, **15**, No. 3, p. 165 (1973).
21. Cottney, D.J. and Ewins, D.J., "Towards the Efficient Vibration Analysis of Shrouded Bladed-Disc Assemblies", *J. Engrng. Indus*, ASME, p. 1054 (1974).
22. Kirkhope, J. and Wilson, G.J., "A Finite Element Analysis for the Vibration Modes of A Bladed-Disc" *J. Sound Vib.*, **49**, p. 469 (1976).
23. Ewins, D.J. and Rao, Y.V.K.S., "Effect of Blade Damping on the Forced Vibration Response of Bladed-Discs", SRC Report No. 1, Mech. Engrng. Dept., Imperial College, London (June 1975).
24. Ewins, D.J. and Rao, Y.V.K.S., "The Effect of Aerodynamic Damping on the Forced Response of Bladed-Discs", SRC Report No. 2, Mech. Engrng. Dept., Imperial College, London (May 1976).
25. Ewins, D.J. and Rao, Y.V.K.S., "Vibrations of Bladed-Discs with a Continuous Part-Span Shroud", SRC Report No. 3, Mech. Engrng. Dept., Imperial College, London (September 1976).
26. Salama, A.M., Petyt, M. and Mota Soares, C.A., "Dynamic Analysis of Bladed-Discs by Wave Propagation and Matrix Difference Techniques", *ASME Winter Annual Meeting*, p. 45 (1976).

27. Thomas, J. and Sabuncu, M., "Vibration Characteristics of Asymmetric Cross-Section Bladed-Disc Under Rotation", ASME 79-DET-94 (1979).
28. Irretier, H., "Coupled Vibration of Blades in Bending-Bending Torsion and Discs in Out-of-Plane and In-Plane Motion", ASME 79-DET-90.
29. Prohl, M.A., "A Method for Calculating Vibration Frequency and Stress of a Banded Group of Turbine Buckets", *J. Appl. Mechanics*, ASME, p. 169 (1958).
30. Ewins, D.J. and Imregun, M., "Vibration Modes of Packeted Bladed-Discs", *Vibrations of Bladed-Disc Assemblies*, ASME, p. 25 (1983).
31. Crawley, E.F., "In-Plane Inertial Coupling in Tuned and Severely Mistuned Bladed-Discs", *J. Engrg. Gas Turbines Power*, ASME, **105**, p. 585 (1983).
32. Crawley, E.F. and Mokadam, D.R., "Stagger Angle Dependence of Inertial and Elastic Coupling in Bladed-Discs", *J. Vib. Acoust Stress Rel Des, Trans. ASME*, **106**, p. 181 (1984).
33. Ferrari, G., Henry, R., Lalanne, M. and Trompette, P., "Frequencies and Mode Shapes of Rotating Axisymmetric Structures, Application to Jet Engines", *Vibrations of Blades and Bladed-Disc Assemblies*, ASME, p. 1 (1983).
34. Karadag, V., "Finite Element Dynamic Analysis of Blade Shear Centre Effects on Practical Bladed-Discs", *J. Sound and Vib.*, **94**, p. 183 (1984).
35. Shiga, M., "Characteristics of Natural Frequencies of Steam Turbine Blades (1st Report: Relationship on Vibration Between Grouped Blades and Disc)", *Bull. JSME*, **27**, p. 802 (1984).
36. Shiga, M., "Characteristics of Natural Frequencies of Steam Turbine Blades (2nd Report: Influence of Disc on Vibration of Grouped Blades)", *Bull. JSME*, **27**, p. 1203 (1984).
37. Shiga, M., "Characteristics of Natural Frequencies of Steam Turbine Blades (3rd Report: Vibration of Bound Blades on Entire Disc Circumference)", *Bull. JSME*, **28**, p. 1196 (1985).
38. Irretier, H., "Spectral Analysis of Mistuned Bladed-Disc Assemblies by Component Mode Synthesis", *Vibrations of Blades and Bladed-Disc Assemblies*, ASME, p. 115 (1983).
39. Valero, N.A. and Bendicksen, O.O., "Vibration Characteristics of Mistuned Shrouded Blade Assemblies", *J. Engrng. Gas Turbines Power*, ASME, **108**, p. 293 (1986).
40. Griffin, J.H. and Hoosac, T.M., "Model Development and Statistical Investigation of Turbine Blade Mistuning", *J. Vib. Acoust. Stress Rel. Des.*, ASME, **106**, p. 204 (1984).
41. Basu, P. and Griffin, J.H., "The Effect of Limiting Aerodynamic and Structural Coupling in Models of Mistuned Bladed-Disc Vibration", *Vibrations of Bladed-Disc Assemblies*, ASME, p. 31 (1985).
42. Lawrence, T.J. and Meyn, E.H., "The Use of an Optical Data Acquisition System for Bladed-Disc Vibration Analysis", *Vibrations of Blades and Bladed-Disc Assemblies*, ASME, p. 65 (1985).
43. Nishiwaki, N., Kimura, K. and Shiratori, S., "Impeller Blade Vibration in Turbochargers", *Proc. IFToMM Intl. Conf. Rotor Dynamics*, Tokyo, p. 313 (1986).
44. Rao, J.S., Shah, C.B., Ganesh, Ch.L. and Rao, Y.V.K.S., "Vibration Characteristics of Aircraft Engine Blade-Disc Assembly", *Defence Sci. J.*, **36**, p. 9 (1986).
45. Rao, J.S. et al, *Response of Bladed-Disc Assemblies*, Part 1 of the final report, Project 285, Aero R&D Board, Ministry of Defence, India (1984).

46. Dominic, R.J., "Spin Pit Test of Bladed-Disc with Blade Platform Friction Dampers", *Shock Vib. Bull.*, **55**, Pt. 1, p. 71 (1985).
47. MacBain, J.D. and Stange, W.A., "An Investigation of Dual Mode Phenomena in a Mistuned Disc", ASME 81-DET-133 (1981).
48. Kuo, P.S., "On the Formulation of Coupled/Uncoupled Dynamic Analyses of Bladed-Disc Assemblies", ASME 81-DET-126 (1981).
49. Elchuri, V., Smith, G.C.C., Gallo, A. Michael and Dale, B.J., *NASTRAN Level 16 Theoretical User's Programmer's and Demonstration Manuals — Updates for Aeroelastic Analysis of Bladed-Discs*, NASTRAN Manuals, NASA CR's 159273-15276 (March 1980).
50. Elchuri, V., Smith, G.C.C. and Gallo, A. Michael., "NASTRAN Forced Vibration Analysis of Rotating Cyclic Structures", *J. Vib. Acoust. Stress Rel. Des.*, ASME, **106**, p. 224 (1984).
51. Swaminadham, M., Soni, M.L., Stange, W.A. and Reed, J.D., "On Model Generation and Nodal Analysis of Flexible Bladed-Disc Assemblies", *Bladed-Disc Assemblies*, **6**, p. 49, ASME Vibration Conf., Cambridge, MA (September 27–30, 1987).
52. Afolabi, D., "The Frequency Response of Mistuned Bladed-Disc Assemblies", *Vibrations of Blades and Bladed-Disc Assemblies*, ASME, p. 15 (1985).
53. Afolabi, D., "The Eigenvalue Spectrum of a Mistuned Bladed-Disc", *Vibration of Blades and Bladed-Disc Assemblies*, ASME, p. 23 (1985).
54. Ewins, D.J. and Han, Z.S., "Resonant Vibration Levels of a Mistuned Bladed-Disc", *J. Vib. Acoust. Stress Rel. Des.*, ASME, **106**, p. 211 (1984).
55. Kubota, Y., Suzuki, T., Tomita, H., Nagafugi, T. and Okamura, C., "Vibration of Rotating Blade-Disc Excited by Stationary Distributed Forces", *Bull. JSME*, **26**, p. 1952 (1983).
56. MacBain, J.C. and Whaley, P.W., "Maximum Resonant Response of Mistuned Bladed-Discs", *J. Vib. Acoust. Stress Rel. Des.*, ASME, **106**, p. 218 (1984).
57. Haupt, U. and Rautenberg, M., "Blade Vibration Measurements on Centrifugal Compressors by Means of Telemetry and Holographic Interferometry", *J. Engrng. Gas Turbines Power*, ASME, **106**, p. 70 (1984).
58. Jay, R.L. and Burns, D.W., "Characteristics of the Diametral Resonant Response of a Shrouded Fan under a Prescribed Distortion", *Vibrations of Blades and Bladed-disc Assemblies*, ASME, p. 49 (1985).
59. Jay, R.L., MacBain, J.C. and Burns, D.W., "Structural Response Due to Blade Vane Interaction", *J. Engrg. Gas Turbines Power*, ASME, **106**, p. 50 (1984).
60. Michimura, S. et al, "Vibrations of Impellers (5th Report on Measurement of Resonant Vibratory Stresses of an Impeller and Pressure Distribution due to Aerodynamic Excitations)", *Bull. JSME*, **27**, p. 534 (1984).
61. Srinivasan, A.V. and Cutts, D.G., "Measurement of Relative Vibratory Motion at the Shroud Interfaces of a Fan", *J. Vib. Acoust. Stress Rel. Des.*, ASME, **106**, p. 189 (1984).
62. Srinivasan, A.V. and Cutts, D.G., "Aerodynamically Excited Vibrations of A Part-Span Shrouded Fan", *J. Engrg. Gas Turbines Power*, ASME, **107**, p. 399 (1985).
63. Whaley, P.W. and MacBain, J.C., "Effects of Mistuning on the Forced Vibration of Bladed-Discs in Subsonic Flow", *Proc. Struct. Dynam. Matls. Conf.*, Pt. 2, p. 490 (1985).

64. Kaza, K.R.V. and Kielb, R.E., "Flutter of Turbofan Rotors with Mistuned Blades", *AIAA J.*, **22**, p. 1618 (1984).
65. Kielb, R.E. and Kaza, K.R.V., "Effects of Structural Coupling on Mistuned Cascade Flutter Response", *J. Engrg. Gas Turbines Power*, ASME, **106**, p. 17 (1984).
66. Bendicksen, O.O., "Flutter of Mistuned Turbomachinery Rotors", *J. Engrg. Gas Turbines Power*, ASME, **106**, p. 255 (1984).
67. Griffin, J.H. and Sinha, A., "The Interaction Between Mistuning and Friction in the Forced Response of Bladed-Disc Assemblies", *J. Engrg. Gas Turbines Power*, ASME, **107**, p. 205 (1985).
68. Sinha, A., "Calculating the Statistics of Forced Response of A Mistuned Bladed-Disc Assembly", *AIAA J.*, **24**, p. 1797 (1986).
69. Muszynska, A. and Jones, D.I.G., "A Parametric Study of Dynamic Response of A Discrete Model of Turbomachinery Bladed-Disc", *J. Vib. Acoust. Stress Rel. Des.*, ASME, **105**, p. 434 (1983).
70. Muszynska, A. and Jones, D.I.G., "On Tuned Bladed-Disc Dynamics — Some Aspects of Friction Related Mistuning", *J. Sound Vib.*, **86**, p. 107 (1983).
71. Sinha, A., Griffin, J.H. and Kielb, R.E., "Influence of Friction Dampers on Torsional Blade Flutter", *J. Engrg. Gas Turbines Power*, ASME, **108**, p. 313 (1986).
72. Sinha, A. and Griffin, J.H., "Effects of Friction Dampers on Aerodynamically Unstable Rotor Stages", *AIAA J.*, **23**, p. 262 (1985).
73. Sinha, A. and Griffin, J.H., "Stability of Limit Cycles in Frictionally Damped and Aerodynamically Unstable Rotor Stages", *J. Sound Vib.*, **103**, p. 341 (1985).
74. Menq, C.H., Griffin, J.H. and Bielak, J., "The Forced Response of Shrouded Fan Stages", *J. Vib. Acoust. Stress Rel. Des*, ASME, **108**, p. 50 (1986).
75. McLeod, A.J. and Bishop, R.E.D., "The Forced Vibration of Circular Flat Plates", *Mech. Engrg. Sci. Monograph*, No. 1.
76. Cottney, D.J., *Natural Frequency Changes Due to Shrouding of a Bladed-Disc*, M.Sc. Thesis, Imperial College, London (1971).
77. Ewins, D.J., Imperial College Progress Account No. 7 (June–September 1973).
78. *CDC PATRAN-G Reference Manual*, Control Data Corp. Publication, 60459330 and 60459340 (1980).

Chapter 11

Some Thoughts on Fatigue Life Estimation

Though design procedures for preventing fatigue induced failures of turbine blading are practised, there is no indication that a satisfactory situation exists with reference to basic data that leads to the establishment of sound procedures in this regard. This is because the blade fatigue is a multi-technology problem which is undergoing continuous investigations. In this book, we discussed various aspects of this multi-faceted technology, viz., blade excitation, free vibration analysis, damping mechanisms and response under steady and transient conditions. There are several other factors involved in the technology development, e.g., material behaviour under high temperature and loading conditions, fatigue failure theories, crack initiation and propagation, creep under high temperature conditions pertinent to metallurgical studies, besides other problems in aeroelastic and thermoelastic subjects which are of significance in the turbine blade design.

Papers by Rieger [1, 2], and Rao and Vyas [3] discussed various aspects involved in the life analysis of turbomachine blading. Blade fatigue behaviour is known to be mainly influenced by the static and dynamic stress fields on the blade, fatigue properties of the blade material, loading history and the environment of blade operation. Fatigue crack usually initiates in a zone of high stress at some metallurgical or structural discontinuity and if critical conditions of blade operations sustain, the crack may grow to lead to failure. The factors cited above may affect the blade behaviour in several different ways to make a fatigue problem very specific.

The major hurdles faced in approaches to blade fatigue analysis have been, see Rao [4] and Rieger [5], precise evaluation of excitation forces in terms of the magnitude and phasing of harmonics and their relation to stage flow and geometry; obtaining valid damping coefficients as functions of rotor speed, amplitude and blade material; estimation of blade dynamic stresses under forced vibration and resonant conditions and their dependence on excitation harmonics, vibrational modes, damping; analysis of blade dynamic behaviour during transients, intermittent loading, partial admission; evaluation of fatigue strength parameters under broad conditions of operating environment.

In the previous chapters we considered certain basic aspects, viz., free vibration analysis to determine the natural frequencies and mode shapes, stage flow interference to determine the blade excitation forces at the nozzle passing frequency, blade damping due to various sources as a function of rotational speed and tip displacement for various modes and steady as well as transient stress analysis of turbine blades. In this chapter, some basic principles of fatigue damage criteria are discussed and the stress analysis results previously given will be used to work out life estimates for a test blade. Fatigue strength parameters for laboratory conditions of operation of the test blade are assumed in this evaluation, and Bagci's fatigue failure surface line and the concept of cumulative fatigue damage are employed to work out numerical examples on life estimation [6]. For a comprehensive treatment on fatigue design reference may be made to Heywood [7] and Juvenal [8].

11.1 FATIGUE DATA

A number of different materials are used for turbine blades depending on particular applications, such as:

1. Type of machine, steam or gas turbine or compressor;
2. High temperature or low temperature stage;
3. Corrosion environment (dry vapour, Wilson line stage, moisture extraction and condenser proximity);
4. Erosion requirements (foreign object ingestion, operating wetness);
5. Static strength properties (UTS, ductility, transition temperature);
6. Dynamic strength properties (endurance limit, S-N data, notch sensitivity, etc.);
7. Metallurgical factors (weldability, formability, machinability, etc.).

Some of the most commonly used materials are:

Steam Turbines

AISI 403, 410, 410-Cb, 422 series
17-4 Precipitation-Hardening steel
Ti-6Al-4V Titanium alloy

Gas Turbines

AISI 4140
Directionally solidified single crystal

AISI 403 and related high chromium alloy steels are martensitic stainless steels which are widely used in blading applications because of their good corrosion resistance, good tensile strength, good fatigue properties, good machinability amongst other factors. 17Cr-4Ni PH alloy steel is another martensitic steel, with precipitation through-hardening properties achieved by heat treatment during manufacture. 17-4 PH alloy has a higher tensile strength and endurance limit than AISI 403 and it has greater corrosion resistance and greater corrosion fatigue resistance due to its higher alloy content, see [9].

Ti-6Al-4V material is an alternate design material which is suitable for use in specific stages in the region of the Wilson line. Titanium alloy blades possess good corrosion resistance properties, and have favourable strength/mass properties for withstanding centrifugal forces. Heyman et al [10] showed superior high-cycle fatigue properties in a 22% NaCl solution. The most important shortcoming, however, of all titanium alloys is in their low damping properties. The titanium alloys may have an order of magnitude or less material damping than AISI or 17-4 steel alloys. This deficit would lead to larger resonant stresses, and a greater number of high-stress cycles following transient excitation.

It is customary design practice to use a modified Goodman diagram or Soderberg diagram of the form shown in Fig. 11.1 to present test data on the effects of mean stress on material endurance strength. Such charts have been extensively discussed in the fatigue literature, for example, see Heywood [7], and are very useful for blade design where nominal stress (steady plus dynamic) data can be compared with material fatigue strength data. Where triaxial stress conditions exist, i.e. at most locations on a turbine or compressor blade, questions may arise concerning which equivalent stresses should be used. Juvenal [8] recommends using a Goodman diagram based on bending fatigue data, which is appropriate for the particular material, size, surface treatment, and fatigue life involved. To represent the alternating stress, an equivalent bending stress should be developed, based on distortion-energy theory and including effects from all alternating load components. The mean stress should likewise be represented as an equivalent mean bending stress, based on the highest algebraic principal stress caused by the mean load components. This method allows each alternating load component to be multiplied by its own stress concentration factors. The resulting maximized stress components can then be combined into the maximum equivalent bending stress in the most suitable proportions, rather than using an overall stress concentration factor for the combined equivalent nominal stress.

Recently proposed fatigue failure surface line by Bagci [11] updates the modified Goodman, Gerber and Kececioglu mean stress diagrams for fatigue design of machine elements subjected to cyclic combined stresses having non-vanishing mean stress. It is reported to fit the most recent fatigue data well and defines the two design zones, namely, the zone of fatigue failure experiencing brittle fracture and the zone of failure due to yielding in one, providing a unified fatigue design equation for machine members.

Fig. 11.1 illustrates the Bagci line along with modified Goodman, Gerber and Kececioglu lines on the mean stress diagram. Fig. 11.2 shows the Bagci line coupled with S-N diagram defining the fatigue failure surface. Bagci's fatigue failure surface line is of the fourth order and is given by

$$\sigma_{af} = \sigma_e \left(1 - \left\{ \frac{\sigma_{mf}}{S_y} \right\}^4 \right) \tag{11.1}$$

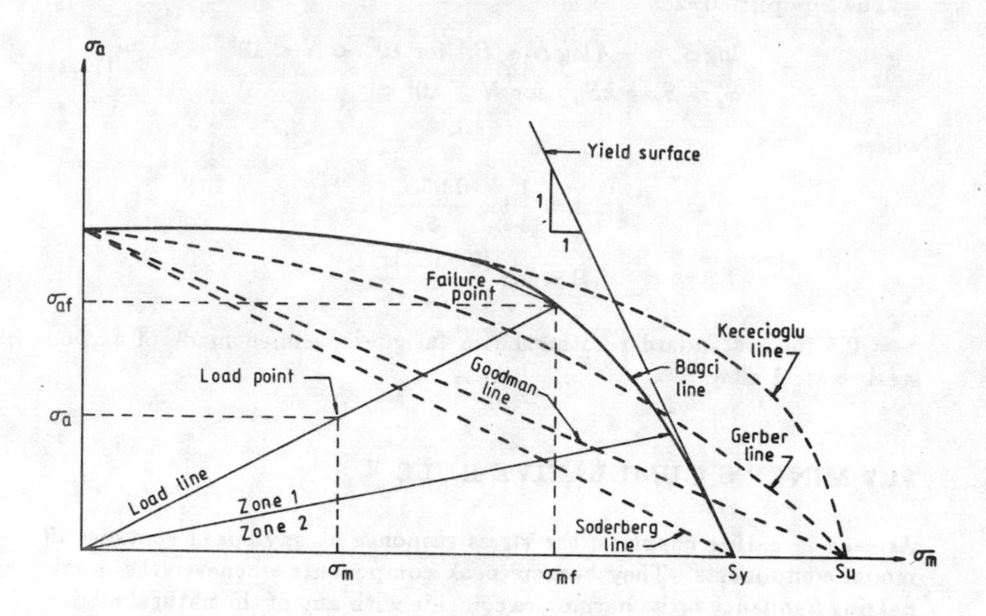

Fig. 11.1 Mean stress diagram

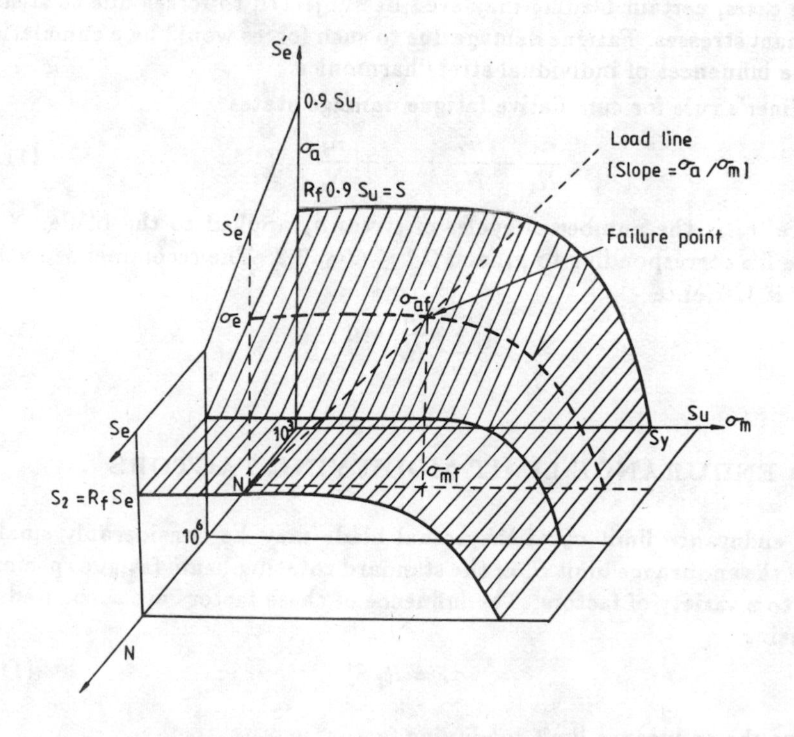

Fig. 11.2 Fatigue failure surface defined by Bagci line

The S-N plane reads

$$\begin{aligned} \log S'_e &= -A \log N + B \quad \text{for } 10^3 < N < 10^6 \\ S'_e &= S_e = kS_u \quad \text{for } N \geq 10^6 \end{aligned} \tag{11.2}$$

where,

$$A = \frac{1}{3} \log \frac{0.95 S_u}{S_e}$$

$$B = \log \frac{(0.95 S_u)^2}{S_e}$$

$k = 0.5$ for a standard rotating beam fatigue specimen made of carbon steel or steel alloy.

11.2 MINER'S CUMULATIVE RULE

As seen in earlier chapters, the stress response at any speed contains all modal components. They become peak components whenever the nozzle passing frequency or its harmonics coincide with any of the natural modes. Each blade experiences transient resonant stress amplitudes for a short period of time while transiting through such a critical rotor speed. In some cases, certain blading may even be subjected to forces due to steady resonant stresses. Fatigue damage due to such forces would be a cumulation of the influences of individual stress harmonics.

Miner's rule for cumulative fatigue damage states

$$\frac{n_1}{N_1} + \frac{n_2}{N_2} + \cdots + \frac{n_i}{N_i} = C \tag{11.3}$$

where, n_i is the number of cycles of stress σ_i applied to the blade, N_i is blade life corresponding to σ_i, and $0.7 \leq C \leq 2.2$. The recommended value of C is 1. Hence

$$\sum_i \frac{n_i}{N_i} = 1 \tag{11.3a}$$

11.3 ENDURANCE LIMIT MODIFYING FACTORS

The endurance limit σ_e of the actual blade may be considerably smaller than the endurance limit S'_e of the standard rotating beam fatigue specimen due to a variety of factors. The influence of these factors are accounted for by using

$$\sigma_e = R_f S'_e \tag{11.4}$$

where the endurance limit modifying factor

$$R_f = k_a k_b k_c k_d k_e k_f \tag{11.5}$$

with

$$
\begin{aligned}
k_a &= \text{surface factor} \\
k_b &= \text{size factor} \\
k_c &= \text{reliability factor} \\
k_d &= \text{temperature factor} \\
k_e &= \text{stress concentration factor} \\
k_f &= \text{miscellaneous factor}
\end{aligned}
$$

For the blade material used in the examples to follow, 40 Ni 3 steel, we have

$$S_u = 900 \text{ M Pa}; \quad S_y = 650 \text{ M Pa}; \quad S_e = 450 \text{ M Pa} \tag{11.6}$$

The endurance limit modifying factors can be chosen, Shigley [12], as follows.

Surface factor

For a forged blade with tensile strength of 900 M Pa, $k_a = 0.35$.

Size factor

Recommended values for bending and torsion are

$$k_b = \begin{cases} 1.00 & \text{for } d \leq 7.6 \text{ mm} \\ 0.85 & \text{for } 7.6 < d \leq 50 \text{ mm} \\ 0.75 & \text{for } d > 50 \text{ mm} \end{cases}$$

where, d corresponds to the thickness dimension in the case of non-circular sections. For the root-section of the blade in Fig. 7.18, $k_b = 1.0$.

Reliability factor

For a reliability of 95% $k_c = 0.868$, by statistical considerations.

Temperature factor

This factor for steels is

$$k_d = \begin{cases} 1 & \text{for } T \leq 70° \text{ C} \\ \dfrac{3100}{2400 + 9T} & \text{for } T > 70° \text{ C} \end{cases}$$

This factor is assumed 1.0 in the examples later for the operation of a blade in spin rig at room temparature.

Stress Concentration factor

This is given by

$$k_e = \frac{1}{1 + q(k_t - 1)}$$

where, q is the notch sensitivity factor, k_t is a factor dependent on geometric values of the member. A sharp notch of 0.2 cm radius exists at the root of the blade considered later. For this notch, $q = 0.85$ and $k_t = 2.6$, which gives $k_e = 0.424$.

Miscellaneous Effects factor

In absence of any other information, assuming that the blade is undamaged, k_f can be considered to be 1.0. The endurance limit modifying factor can therefore be

$$R_f = 0.35 \times 1.0 \times 0.868 \times 1.0 \times 0.424 \times 1.0 = 0.129$$

Therefore in Fig. 11.2 on the S-N plane, we get [see equation (11.2)],

$$\begin{aligned} \log \sigma_e &= -A \log N + B \quad \text{for } 10^3 < N < 10^6 \\ \sigma_e &= S_e = kS_u \quad \text{for } N \geq 10^6 \end{aligned} \tag{11.2}$$

where,

$$\begin{aligned} A &= \frac{1}{3} \log \frac{S_1}{S_2} \\ B &= \log \frac{S_1^2}{S_2} \\ S_1 &= 0.9 \times 0.129 \times 900 = 104.5 \text{ M Pa} \\ S_2 &= 0.129 \times 450 = 58.05 \text{ M Pa} \end{aligned} \tag{11.7}$$

11.4 MEAN STRESS DUE TO BLADE ROTATION

In addition to the vibratory stresses comprising of mean and alternative stresses dicussed earlier, the blade experiences centrifugal stresses due to its rotation which would contribute to the mean stress value influencing fatigue life. These stresses at any blade cross-section can be readily computed using

$$\sigma_{zc} = (R + z) \left(\frac{2\pi N}{60} \right)^2 \frac{1}{A_z} \int_z^L \rho A_z \, dz \tag{11.8}$$

where, A_z = blade cross-sectional area at z, and N is rotor speed in RPM.

For the blade given in Table 8.4, the centrifugal stress so computed at the blade root is plotted in Fig. 11.3 as a function of rotor speed.

11.5 ILLUSTRATIONS

Vyas [6] developed a computer program based on equations (11.1 to 11.7) to generate the fatigue failure surface, and locate the failure point of a blade subjected to mean and alternating forces, at a desired rotor speed. Using Maximum Distortion Energy Theory, the basic mean and alternating stresses are obtained for the blade given in Fig. 7.18, vibrating under simulated nozzle excitation. They are shown for different harmonics in

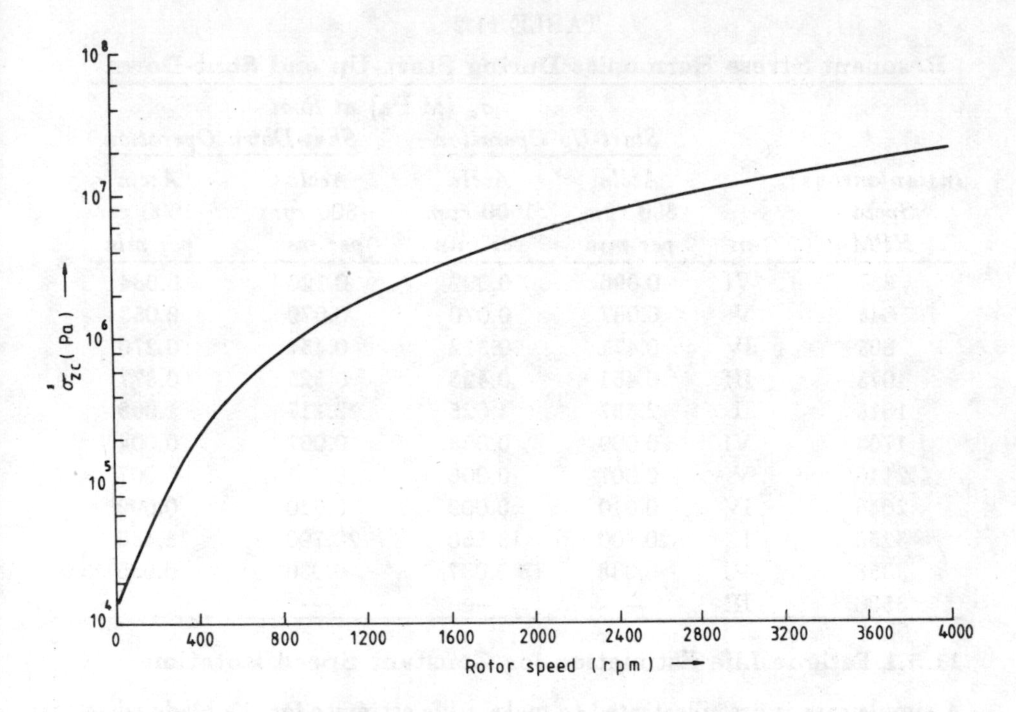

Fig. 11.3 Centrifugal root stress vs rotor speed

Figs. 8.10 to 8.15 and Table 8.8. The stress levels, during step-up and step-down operations, around resonant rotor speeds have been given earlier in Figs. 9.4 and 9.5. Tables 11.1 and 11.2 give the above stress components in tabular form for constant speed rotation and start-up, shut-down operations respectively. This data is used in illustrating blade life estimation procedures.

TABLE 11.1

Stress Harmonics at Resonant Rotor Speeds for Constant Speed Operation

Rotor Speed	σ_{mv}	σ_a (M Pa) *at Root for Harmonics I to VI*					
RPM	*M Pa*	*I*	*II*	*III*	*IV*	*V*	*VI*
537	**.2907**	**.355**	**.082**	**.026**	**.027**	**.012**	**.177**
644	**.2906**	**.358**	**.085**	**.030**	**.039**	**.110**	**.016**
805	**.2904**	**.364**	**.093**	**.041**	**.762**	**.007**	**.007**
1075	**.2897**	**.378**	**.120**	**.647**	**.020**	**.003**	**.004**
1615	**.2881**	**.425**	**3.851**	**.018**	**.009**	**.002**	**.004**
1763	**.2875**	**.451**	**.301**	**.015**	**.008**	**.002**	**.160**
2116	**.2859**	**.530**	**.096**	**.011**	**.006**	**.119**	**.002**
2646	**.2832**	**.895**	**.051**	**.008**	**.137**	**.001**	**.001**
3250	**.2795**	**23.589**	**.037**	**.007**	**.005**	**—**	**.007**
3358	**.2788**	**4.270**	**.035**	**.007**	**.004**	**—**	**.040**
3530	**.2754**	**1.561**	**.033**	**.049**	**.004**	**—**	**.003**

TABLE 11.2

Resonant Stress Harmonics During Start-Up and Shut-Down

		σ_a (M Pa) *at Root*			
		Start-Up Operation		*Shut-Down Operation*	
Instantaneous Speed RPM	*m*	*Accln* 800 *rpm per min*	*Accln* 1600 *rpm per min*	*Accln* −800 *rpm per min*	*Accln* −1600 *rpm per min*
537	VI	0.096	0.092	0.126	0.084
644	V	0.087	0.070	0.079	0.053
805	IV	0.471	0.312	0.454	0.276
1075	III	0.451	0.423	0.525	0.377
1615	II	2.337	1.625	2.117	1.668
1763	VI	0.009	0.008	0.007	0.004
2116	V	0.007	0.006	0.008	0.007
2646	IV	0.010	0.009	0.010	0.099
3250	I	20.800	18.550	20.790	18.480
3358	VI	0.038	0.037	0.036	0.035
3530	III	—	—	—	—

11.5.1 Fatigue Life Estimation for Constant Speed Rotation

A simple case is first illustrated to make a life estimate for the blade when it is rotating at a constant speed of 3250 rpm, where maximum resonant stress would be experienced by the blade (in the range 0 to 4000 rpm) due to the interaction between the first harmonic of excitation and the first blade natural mode.

For 3250 rpm from Table 11.1 and Fig. 11.3, we have

$$\begin{aligned}
&\sigma_a && = 23.5889 \text{ MPa} \\
&\sigma_{mv} && = 0.2795 \text{ MPa} \\
&\sigma_{mc} = \sigma_{zc} && = 12.913 \text{ MPa} \\
&\sigma_m = \sigma_{mv} + \sigma_{mc} && = 13.1925 \text{ MPa}
\end{aligned}$$

It can be readily seen that the load point due to σ_a and σ_m falls well below the failure surface, indicating that there is no influence of resonant stresses on the fatigue life of the blade, and working at 3250 rpm can be considered safe.

As stated earlier, resonance at 3250 rpm is caused by the interaction of the first coupled flapwise (y-y) mode and the first nozzle passing harmonic. Such a resonance should generally be the most dangerous operating point. But the example worked out predicts safe working. The reason is evident from the values of the forcing function employed (Fig. 8.8). The stagger angle of the blade being 90 deg, the component F_y of the magnetic attractive force in the flapwise direction is small, reaching a maximum value of only 0.0940 N/cm for blade traverse across an electromagnet. In practice, the F_y component of nozzle excitation would be large enough to drive the rotor at desired speeds for power generation. In further illustrations, it is assumed that similar but stronger electromagnets are employed which give similar but magnified stress fields for both constant speed rotation and start-up, shut-down operations.

11.5.2 Constant Speed 3250 rpm with 3 Times Excitation Force

Considering operation at a constant speed of 3250 rpm with excitation three times stronger than the example of 11.5.1, the stress harmonics are

$$\begin{aligned}
\sigma_{aI} &= 70.7667 \text{ MPa} \\
\sigma_{aII} &= 0.111 \text{ MPa} \\
\sigma_{aIII} &= 0.021 \text{ MPa} \\
\sigma_{aIV} &= 0.0015 \text{ MPa} \\
\sigma_{aV} &= 0 \\
\sigma_{aVI} &= 0.021 \text{ MPa} \\
\sigma_{mv} &= 0.8385 \text{ MPa}
\end{aligned}$$

and from Fig. 11.3

$$\sigma_{mc} = 12.913 \text{ MPa}$$

Locating the load point on the fatigue failure surface, with the above stresses, it can be found that

$$N_I = 97294 \text{ cycles}$$

$$N_{II}, N_{III}, N_{IV}, N_V, N_{VI} = \infty$$

Since the blade is in resonance with the 1st harmonic excitation, II to VI stress harmonics do not have any influence on the blade fatigue behaviour. The blade fatigue life under the influence of the first stress harmonic (frequency = 650 Hz) can be readily calculated from N_I, to be 149.683 secs = 2.5 mins. Similar life estimation can be carried out for the rest of the resonant speeds. It can be seen that apart from 3250 rpm, other resonant speeds in 0 to 4000 rpm range are safe. It can also be computed that blade fatigue failure would occur only between the speed range 3237 to 3264 rpm (ref. stresses in Fig. 8.10).

11.5.3 Constant Speed 3250 rpm with 5 Times Excitation Force

If the excitation is five times stronger than the case 11.5.1, we have the stress harmonics for constant speed at 3250 rpm

$$\begin{aligned}
\sigma_{aI} &= 117.94 \text{ MPa} \\
\sigma_{aII} &= 0.185 \text{ MPa} \\
\sigma_{aIII} &= 0.035 \text{ MPa} \\
\sigma_{aIV} &= 0.025 \text{ MPa} \\
\sigma_{aV} &= 0 \\
\sigma_{aVI} &= 0.035 \text{ MPa} \\
\sigma_{mv} &= 1.3975 \text{ MPa} \\
\sigma_{mc} &= 12.913 \text{ MPa}
\end{aligned}$$

The stress due to the first harmonic, $\sigma_{aI} = 117.94$ MPa is higher than the endurance limit $S_1 = 104.5$ MPa at $N = 1000$ cycles. The failure will be in the brittle region and almost sudden and therefore need not be classified as a fatigue failure. The other stress harmonics do not register any effect. The blade for this excitation is safe at all other resonant speeds, the failure zone being limited to the range 3227 to 3274 rpm.

11.5.4 Cumulative Damage

Case (a): *Acceleration Value* ±800 *rpm/min*

The blades above are assumed to be in a machine operating at 4000 rpm. The machine is assumed to have uniform start-up and shut-down rates of ±800 rpm/min and a working run of 8 hours at 4000 rpm. The blade loading pattern in terms of rotor speed and time is shown in Fig. 11.4

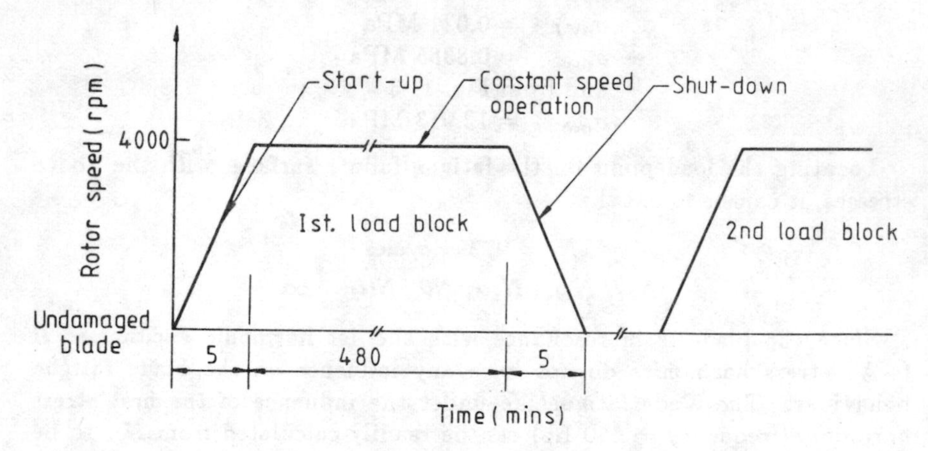

Fig. 11.4 Blade loading pattern

The excitation is assumed to be that of case 11.5.2. From Figs. 8.10 to 8.15, after multiplying the values by 3 times, we have

$$\sigma_{aI} = 1.7625 \text{ MPa}$$
$$\sigma_{aII} = 0.0868 \text{ MPa}$$
$$\sigma_{aIII} = 0.0168 \text{ MPa}$$
$$\sigma_{aIV} = 0.0090 \text{ MPa}$$
$$\sigma_{aV} = 0.0471 \text{ MPa}$$
$$\sigma_{aVI} = 0.0051 \text{ MPa}$$
$$\sigma_{mv} = 0.8223 \text{ MPa}$$
and
$$\sigma_{mc} = 19.5608 \text{ MPa}$$

It can be shown that if the rotor were to operate indefinitely at 4000 rpm without being ever shut down, the blade will have infinite life. However, with the loading pattern shown in Fig. 11.4, the blade is excessively stressed while passing through resonant rotor speeds, during start-up and shut-down in each load block.

From Table 11.2, the maximum cyclic stresses experienced by the blade are at the instantaneous resonance at 3250 rpm. These stresses magnified three times are

$$\sigma_{aI\text{-start}} = 20.80 \times 3 = 62.40 \text{ MPa}$$
$$\sigma_{aI\text{-shut}} = 20.79 \times 3 = 62.37 \text{ MPa}$$
$$\sigma_{mv} = 0.2795 \times 3 = 0.8385 \text{ MPa}$$
$$\sigma_{mc} = 12.913 \text{ MPa}$$
$$\sigma_m = \sigma_{mv} + \sigma_{mc} = 13.7525 \text{ MPa}$$

The stress levels experienced at other resonant instantaneous rotor speeds can be seen to be negligible. The alternating stress levels are a combined result of the free and forced vibration response of the blade (as seen in Chapter 9). Since the instantaneous resonance at 3250 rpm occurs due to interaction of the first natural frequency (650 Hz at 3250 rpm rotor speed) and the first nozzle passing frequency (3250 rpm × 12 nozzles = 650 Hz) the cyclic frequency of these stress levels is 650 Hz.

The resonant stress level σ_{aI} is experienced by the blade momentarily during its passage through 3250 rpm. To estimate the fatigue damage, if any, during this transit, the failure value of σ_{af} of the alternating stress is fixed using equation (11.1), with $\sigma_{mf} = \sigma_m = 13.7525$ MPa and $\sigma_a = S_2 = 58.05$ MPa. The resonant stress peak and the failure value of σ_{af} of the alternating stress for the blade passing through 3250 rpm are shown on an expanded scale in Fig. 11.5. Since the resonant stress peak lies above the failure value, fatigue damage would be caused to the blade. The peak is segmented into a number of stress blocks as shown for step by step

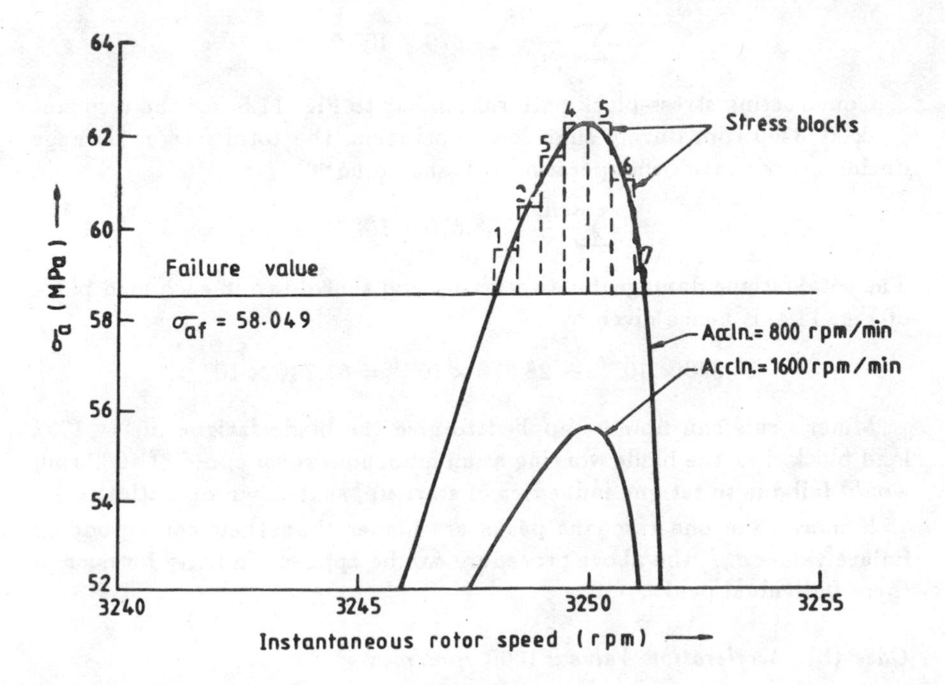

Fig. 11.5 Magnified stress levels at 3250 RPM during start-up

estimation of fatigue damage. The stress levels for each block and the mean rotor speed are listed in Table 11.3. The number of cycles N to failure is obtained by locating the load point on the failure surface and the number of stress cycles experienced, n, is obtained from the mean nozzle passing frequency and the known acceleration rate. Fatigue damage is obtained in terms of the ratio n/N.

TABLE 11.3

Stress-Block Approximation of Fatigue Damage for Step-Up Operation

Stress Block Number	*Stress Level M Pa*	*Mean Speed RPM*	*Cycles To Failure N*	*Cycles Experienced n*	*Fatigue Damage* $n/N \times 10^5$
1	59.5	3248.25	751200	24.362	3.243
2	60.4	3248.75	627300	24.366	3.884
3	61.5	3249.25	507400	24.369	4.802
4	62.4	3249.75	427800	24.373	5.697
5	62.4	3250.25	427800	24.377	5.698
6	61.0	3250.75	558200	24.381	4.365
7	59.0	3251.10	826300	9.753	1.180

The total fatigue damage during start-up in each block of Fig. 11.4 is

$$\sum \frac{n}{N} = 28.870 \times 10^{-5}$$

Constructing stress-block patterns similar to Fig. 11.5, for the resonant peak at 3250 rpm during shut-down operation, the total fatigue damage during every shut-down operation is found to be

$$\sum \frac{n}{N} = 28.876 \times 10^{-5}$$

The total fatigue damage due to start-up and shut-down in each load block of Fig. 11.4, is hence given by

$$28.870 \times 10^{-5} + 28.876 \times 10^{-5} = 57.746 \times 10^{-5}$$

Miner's rule can now be applied to give the blade fatigue life = 1732 load block. i.e., the blade working at an innocuous rotor speed of 4000 rpm would fail due to fatigue influences of start-up, shut-down operations.

If more than one resonant peaks are higher than their corresponding failure values σ_{af}, the above procedure can be applied similarly for each of these individual peaks.

Case (b): *Acceleration Value* ± 1600 *rpm/min*

If the blade has an acceleration rate of ± 1600 rpm/min during start-up and shut-down, the relevant stress levels experienced by the blade at the instantaneous rotor speed of 3250 rpm are

$$\sigma_{aI\text{-start}} = 18.550 \times 3 = 55.65 \text{ MPa}$$
$$\sigma_{aI\text{-shut}} = 18.480 \times 3 = 55.44 \text{ MPa}$$
$$\sigma_{mv} = 0.2795 \times 3 = 0.8385 \text{ MPa}$$
$$\sigma_{mc} = 12.913 \text{ MPa}$$
$$\sigma_m = \sigma_{mv} + \sigma_{mc} = 13.7525 \text{ MPa}$$

The start-up resonant peak plotted on an expanded scale in Fig. 11.5, is below the failure value σ_{af}, indicating a safe operation.

REFERENCES

1. Rieger, N.F., "Finite Element Analysis of Turbomachine Blade Problems", *ASME Vibrations Conf.*, p. 93 (1977).
2. Rieger, N.F. , "Blade Fatigue", *Proc. Tech. Comm. Rotor Dynamics, Sixth IFToMM Congress*, New Delhi, p. 66 (1983).
3. Rao, J.S. and Vyas N.S., "On Life Estimation of Turbine Blading", *Proc. Rotor Dynamics Tech. Committee, Seventh IFToMM Cong.*, Sevilla, Spain (1987).
4. Rao, J.S., "Single Blade Dynamics", *Proc. Tech. Comm. Rotor Dynamics, Sixth IFToMM Cong.*, New Delhi, p. 41 (1983).
5. Rieger, N.F., "Flow Path Excitation Mechanisms for Turbine Blades", *Seventh Cong. IFToMM, Proc. Rotor Dynamics Technical Committee*, Sevilla, Spain (1987).
6. Vyas, N.S., *Vibratory Stress Analysis and Fatigue Life Estimation of Turbine Blade*, Ph. D. Thesis, I.I.T., New Delhi (1986).
7. Heywood, R.B., *Designing Against Fatigue*, Chapman and Hall (1962).
8. Juvenal, R.C., *Engineering Considerations of Stress, Strain and Strength*, McGraw-Hill (1967).
9. Rust, T. and Swaminadhan, V.P., "Corrosion Fatigue Testing of Steam Turbine Blading Alloys", *Proc. EPRI Workshop on Steam Turbine Reliability*, Boston, MA (1982).
10. Heyman, F.J., Swaminadhan, V.P. and Cunningham, J.W., "Steam Turbine Blades: Considerations in Design and a Survey of Blade Failures", EPRI Topical Report CS-1967, Project 912-1 (1981).
11. Bagci, C., "Fatigue Design of Machine Elements Using Bagci Line Defining the Fatigue Failure Surface Line (Mean Stress Diagram)", *Mechanism and Machine Theory*, **16**, No. 4, p. 339 (1981).
12. Shigley, J.E., *Mechanical Engineering Design*, McGraw-Hill (1977).
13. Collins, J.A., *Failure of Materials in Mechanical Design*, John Wiley (1981).

Chapter 12

Some Reflections

In this book, certain basic ideas that would lead to sound design principles of turbine blades are outlined. Briefly, they are

- Determination of natural frequencies and mode shapes of single free standing beam type of blades, free standing small aspect ratio blades, packeted blades and flexible bladed-discs accomplished by various methods, such as energy principles, discretized models, finite element models and receptance techniques. Each method has its own merit and can be adopted depending on the circumstances.
- Determination of nonsteady forces due to stage flow interaction, accomplished by Kemp and Sears type models, which are valid for two dimensional thin airfoil stage in potential flow. The effects of compressibility and viscous wakes shed by the stator are also considered. The theories presented are useful for subsonic flows upto Mach numbers 0.6. The method of hydraulic analogy to simulate a gas turbine stage flow is also described. The modified analogy is shown to give accurate modeling techniques and predicting capabilities in determining the nonsteady forces.
- This book describes in a comprehensive manner, methods to evaluate the damping of a turbine blade in actual practice and modeling techniques for using such data in evaluating forced response. A finite element formulation for determining free vibration characteristics of blade root junctions with Coulomb damping is also presented.
- Modal analysis techniques to determine the dynamic stresses under steady state as well as transient conditions of excitation are presented. The importance of proper damping models in this process is outlined through analytical and experimental results.
- Finally, the information presented in various chapters in the above topics is used in outlining cumulative fatigue damage models and life estimation methods. It is shown here that the acceleration rate of the machine to transit the blades through critical speeds significantly influences its dynamic response.

In the exercise of a good mechanical design of a turbine blade, some other pieces in the puzzle require additional work. Briefly they are as follows.

- Unsteady aerodynamic analysis of a stage in transonic and supersonic flow is required to determine the nonsteady excitation forces in this regime.
- In this book, the airfoil is treated as rigid while estimating the nonsteady excitation forces, and they are in turn used to determine the vibratory characteristics of the blades, i.e., the aeroelastic coupling is neglected. There exist good aeroelastic models, with reference to aerodynamic analysis, but when the question of the complexity of the geometry of the blade and its rotational characteristics are considered, these models in general are poor. Some kind of nonlinear models are to be developed, by which the blade response can be determined from the blade excitation forces which are given as functions of angle of attack, blade displacement and rotation.
- It is significant to consider the thermo-mechanical coupling in the dynamic behaviour of turbine blades under steady as well as transient conditions of operation. Not much work is available in this direction. The response under transient conditions obtained from thermo-mechanical analysis could lead to useful information on the understanding of low cycle fatigue failures of turbine blades.
- It is demonstrated in this book that the overall damping of a turbine blade is complex in nature and is a function of rotational speed, vibratory amplitudes and the mode of vibration. It is also shown that there is a definite threshold speed, beyond which the blade gets interlocked and that there is no contribution of damping from the root junction to dissipate the vibratory energy. A lot of damping data is perhaps generated by steam and gas turbine manufacturers, a small portion of which only is available in open literature. It is generally believed that most of this data is generated in pull tests and may not be reliable. Also, this data is not directly in a useful form to conduct modal analysis; it is only used as a guide to compare different blade-root junctions. A comprehensive test program can be undertaken to determine the overall damping envelopes as a function of rotational speed and tip displacement for each of the significant modal components. This study can perhaps lead to better junctions providing damping for a larger speed range when the machine is started. It will be also useful to conduct finite element studies for evaluating the damping characteristics of turbine blades, with friction at the root and material hysteresis.
- In this book, procedures to determine the dynamic stresses of a turbine blade subjected to steady and transient conditions of excitation in a fairly accurate manner are presented. Programs can be generalized considering different types of partial admission conditions that may exist in practice. Finite element methods are versatile, but are always suspect to the accuracy of the results under dynamic analysis, particularly with the case of turbine blades. A more comprehensive study can be undertaken to evaluate the best elements that can be used in turbine blade analysis.

- **Considering various aspects, it is possible to lay down a good fatigue damage design practice. In this book, some simple cases are illustrated. A particular case of interest could be the study of crack initiation and crack propagation for turbine blades subjected to a typical load history in its life time. A good number of machines in the world are now aging and life extension studies could be useful to extract as much work as possible from the plants, before they are discarded.**

Appendix 1

Heat Transfer and Thermal Stresses of Turbine Blades

1. INTRODUCTION

The efficiency of a turbine is greatly enhanced by using higher inlet pressures and temperatures. Although higher inlet temperatures are desirable from the thermodynamics point of view, they severely increase the thermal loading and consequently cause high thermal stresses in the blades. These thermal stresses are known to peak during the acceleration or deceleration of the gas turbine engine. This is because in the transient state, there is a large difference of temperature between various points of the blade. As the demand on the gas turbine engine for more power output became the primary factor in its design, the heat transfer and thermal stress analysis problem thus became more important. Some of the recent studies include [1 to 4]. In these works, the heat exchange process at the surface by convection was only considered. The effect of temperature dependent mechanical properties on thermal stresses in cooled turbine blades was studied in [5]. The influence of turbine blade airfoil and root thickness variation along the blade height on its thermal stress state was studied in [6].

The radiative heat exchange mechanism is quite significant at the operating temperatures of the modern gas turbines. Thus it is important to account for this effect in prediction of the thermal stresses and temperature distributions of modern gas turbine blades. Bahree, Sharan and Rao [7] included the radiative heat transfer mechanism by using a finite element method and the analysis given below follows this closely.

2. HEAT TRANSFER PROBLEM

The governing differential equation for heat flow in any solid and the corresponding boundary condition can be written as [8]

$$k_x \frac{\partial^2 T}{\partial x^2} + k_y \frac{\partial^2 T}{\partial y^2} + k_z \frac{\partial^2 T}{\partial z^2} + Q = \rho c \frac{\partial T}{\partial t} \tag{1}$$

$$k_x \frac{\partial T}{\partial x} l_x + k_y \frac{\partial T}{\partial y} l_y + k_z \frac{\partial T}{\partial z} l_z + q + h(T - T_\infty) + \sigma\epsilon(T^4 - T_\infty^4) = 0 \tag{2}$$

In the above

k thermal conductivity W/m °K
Q specific heat within the body W/m^3
ρ mass density kg/m^3
c specific heat J/kg °K Ws/kg °K
T temperature
t time
q specific heat flux W/m^2
h convective heat transfer coefficient W/m^2 °K
σ Stefan-Boltzmann constant 5.669×10^{-8} W/m^2 °K^4
ϵ emissivity of the body (0.7)

For developing a finite element method, we first derive the corresponding variational problem of the above differential equation and the boundary condition. Equation (1) can be rewritten as

$$-k_x \frac{\partial^2 T}{\partial x^2} - k_y \frac{\partial^2 T}{\partial y^2} - k_z \frac{\partial^2 T}{\partial z^2} - \left\{ Q - \rho c \frac{\partial T}{\partial t} \right\} = 0 \tag{3}$$

The variational functional can be set up by multiplying equations (2) and (3) by the first variation of T and integrating over the whole domain as

$$\begin{aligned} \delta\chi = &\int_V \left[-k_x \frac{\partial^2 T}{\partial x^2} - k_y \frac{\partial^2 T}{\partial y^2} - k_z \frac{\partial^2 T}{\partial z^2} - \left\{ Q - \rho c \frac{\partial T}{\partial t} \right\} \right] \delta T \, dV \\ &+ \int_S \left\{ k_x \frac{\partial T}{\partial x} l_x + k_y \frac{\partial T}{\partial y} l_y + k_z \frac{\partial T}{\partial z} l_z + q \right. \\ &\left. + h(T - T_\infty) + \sigma\epsilon(T^4 - T_\infty^4) \right\} \delta T \, dS \end{aligned} \tag{4}$$

i.e.,

$$\begin{aligned} \delta\chi = &\int_S k_x \frac{\partial T}{\partial x} l_x \delta T \, dS - \int_V k_x \frac{\partial^2 T}{\partial x^2} \delta T \, dV \\ &+ \int_S k_y \frac{\partial T}{\partial y} l_y \delta T \, dS - \int_V k_y \frac{\partial^2 T}{\partial y^2} \delta T \, dV \\ &+ \int_S k_z \frac{\partial T}{\partial z} l_z \delta T \, dS - \int_V k_z \frac{\partial^2 T}{\partial z^2} \delta T \, dV \\ &+ \int_S q \, \delta T \, dS - \int_V \left(Q - \rho c \frac{\partial T}{\partial t} \right) \delta T \, dV \\ &+ \int_S h(T - T_\infty) \delta T \, dS + \int_S \sigma\epsilon(T^4 - T_\infty^4) \delta T \, dS \end{aligned} \tag{5}$$

We can use Green's theorem to transform surface integrals in the above to volume integrals, i.e.,

$$\int_S k_x \frac{\partial T}{\partial x} l_x \, \delta T \, dS = \int_V k_x \frac{\partial^2 T}{\partial x^2} \delta T \, dV + \int_V \frac{k_x}{2} \delta \left\{ \frac{\partial T}{\partial x} \right\}^2 dV \tag{6}$$

$$\int_S k_y \frac{\partial T}{\partial y} l_y\, \delta T\, dS = \int_V k_y \frac{\partial^2 T}{\partial y^2} \delta T\, dV + \int_V \frac{k_y}{2} \delta \left\{ \frac{\partial T}{\partial y} \right\}^2 dV \tag{7}$$

$$\int_S k_z \frac{\partial T}{\partial z} l_z\, \delta T\, dS = \int_V k_z \frac{\partial^2 T}{\partial z^2} \delta T\, dV + \int_V \frac{k_z}{2} \delta \left\{ \frac{\partial T}{\partial z} \right\}^2 dV \tag{8}$$

Using the above in equation (5) we get

$$\begin{aligned}\delta\chi = &\int_V \frac{k_x}{2} \delta \left\{ \frac{\partial T}{\partial x} \right\}^2 dV + \int_V \frac{k_y}{2} \delta \left\{ \frac{\partial T}{\partial y} \right\}^2 dV + \int_V \frac{k_z}{2} \delta \left\{ \frac{\partial T}{\partial z} \right\}^2 dV \\ &- \int_V Q\, \delta T\, dV + \int_V \rho c \frac{\partial T}{\partial t} \delta T\, dV + \int_S q\, \delta T\, dS \\ &+ \int_S \frac{h}{2} \delta (T - T_\infty)^2 dS + \int_S \sigma \epsilon \delta \left\{ \frac{T^5}{5} - T_\infty^4 T \right\} dS \end{aligned} \tag{9}$$

We can rewrite the above as

$$\begin{aligned}\chi = &\int_V \frac{1}{2} \left[k_x \left\{ \frac{\partial T}{\partial x} \right\}^2 + k_y \left\{ \frac{\partial T}{\partial y} \right\}^2 + k_z \left\{ \frac{\partial T}{\partial z} \right\}^2 \right. \\ &\left. - 2QT + 2\rho c \frac{\partial T}{\partial t} T \right] dV + \int_{S_1} qT\, dS \\ &+ \int_{S_2} \left[\frac{h}{2} \delta (T - T_\infty)^2 + \sigma \epsilon \left\{ \frac{T^5}{5} - T_\infty^4 T \right\} \right] dS \end{aligned} \tag{10}$$

where S_1 is the surface experiencing heat flux and S_2 is the surface experiencing convection and radiation.

To solve the heat transfer problem of the blade, we assume that the temperature of the blade is not continuous over the whole domain but defined over an individual element. As it has been shown in [2], that the thermal gradient along the blade height z of the turbine blade is negligible, let us consider the simplified two-dimensional problem with triangular finite elements. Let T_i, T_j, T_k be the nodal temperatures of the element taken in the counter-clockwise sense from node i, then

$$T^e = [N_i^e\ N_j^e\ N_k^e] \begin{Bmatrix} T_i \\ T_j \\ T_k \end{Bmatrix} \tag{11a}$$

$$[D^e] = \begin{bmatrix} k_x^e & 0 \\ 0 & k_y^e \end{bmatrix} \tag{11b}$$

$$[B^e] = \begin{bmatrix} \dfrac{\partial N_i^e}{\partial x} & \dfrac{\partial N_j^e}{\partial x} & \dfrac{\partial N_k^e}{\partial x} \\ \dfrac{\partial N_i^e}{\partial y} & \dfrac{\partial N_j^e}{\partial y} & \dfrac{\partial N_k^e}{\partial y} \end{bmatrix} \tag{11c}$$

$$\{g^e\} = \begin{Bmatrix} \dfrac{\partial T}{\partial x} \\ \dfrac{\partial T}{\partial y} \end{Bmatrix} = [B^e] \begin{Bmatrix} T_i \\ T_j \\ T_k \end{Bmatrix} \tag{11d}$$

$$\frac{\partial T^e}{\partial t} = [N_i^e \, N_j^e \, N_k^e] \begin{Bmatrix} \dfrac{\partial T_i}{\partial t} \\ \dfrac{\partial T_j}{\partial t} \\ \dfrac{\partial T_k}{\partial t} \end{Bmatrix} \tag{11e}$$

For the turbine blade problem, there is no heat generation and there are no specified heat fluxes, therefore Q and q are both zero in the functional (10), hence for the two dimensional problem this becomes

$$\begin{aligned} \chi^e = & \int_{V^e} \frac{1}{2}\{T^e\}^T [B^e]^T [D^e][B^e]\{T^e\} dV \\ & + \int_{V^e} \rho c^e [N^e] \frac{\partial T^e}{\partial t} [N^e]\{T^e\} dV \\ & + \int_{S^e} \frac{h^e}{2} \left([N^e]\{T^e\}\right)^2 dS \\ & - \int_{S^e} h^e T_\infty [N^e]\{T^e\} dS + \int_{S^e} \frac{h^e}{2} T_\infty^2 \, dS \\ & + \int_{S^e} \frac{1}{5} \sigma\epsilon \left([N^e]\{T^e\}\right)^5 dS - \int_{S^e} \sigma\epsilon T_\infty^4 [N^e]\{T^e\} dS \end{aligned} \tag{12}$$

We can minimize the functional χ with respect to the nodal temperature vector to give

$$\frac{\partial \chi}{\partial \{T^e\}} = \frac{\partial \chi^1}{\partial \{T^e\}} + \frac{\partial \chi^2}{\partial \{T^e\}} + \cdots + \frac{\partial \chi^n}{\partial \{T^e\}} = 0 \tag{13}$$

where n is the total number of elements. To differentiate equation (12), we make use of the following relations [8]

$$\frac{\partial}{\partial \{T^e\}} \int_{V^e} \frac{1}{2}\{T^e\}^T [B^e]^T [D^e][B^e]\{T^e\} dV = \int_{V^e} [B^e]^T [D^e][B^e]\{T^e\} dV \tag{14}$$

etc.

Equation (12) then gives

$$\begin{aligned} \frac{\partial \chi}{\partial \{T^e\}} = \sum_{e=1}^{n} \Bigg[& \int_{V^e} [B^e]^T [D^e][B^e]\{T^e\} dV \\ & + \int_{V^e} \rho c^e [N^e] \frac{\partial \{T^e\}}{\partial t} [N^e]^T dV + \int_{S^e} h^e [N^e]^T [N^e]\{T^e\} dS \\ & - \int_{S^e} h^e T_\infty [N^e]^T dS + \int_{S^e} \sigma\epsilon [N^e]^T \left([N^e]\{T^e\}\right)^4 dS \\ & - \int_{S^e} \sigma\epsilon T_\infty^4 [N^e]^T dS \Bigg] = 0 \end{aligned} \tag{15}$$

We can write the elemental capacitance and conduction matrices and the force vectors due to convection and radiation respectively as follows:

$$[CP^e] = \int_{V^e} \rho c^e [N^e]^T [N^e] dV \tag{16a}$$

$$[K^e] = \int_{V^e} [B^e]^T [D^e][B^e] dV + \int_{S^e} h^e [N^e]^T [N^e] dS \tag{16b}$$

$$\{F_c^e\} = \int_{S^e} h^e T_\infty [N^e]^T dS \tag{16c}$$

$$\{F_r^e\} = \int_{S^e} \sigma \epsilon T_\infty^4 [N^e]^T dS - \int_{S^e} \sigma \epsilon [N^e]^T \left([N^e]\{T^e\}\right)^4 dS \tag{16d}$$

Equations (15) can now be summed up for all the elements to give the following equation for the global matrices:

$$[CP^G] \frac{\partial \{T^G\}}{\partial t} + [K^G]\{T^G\} = \{F_c^G\} + \{F_r^G\} \tag{17}$$

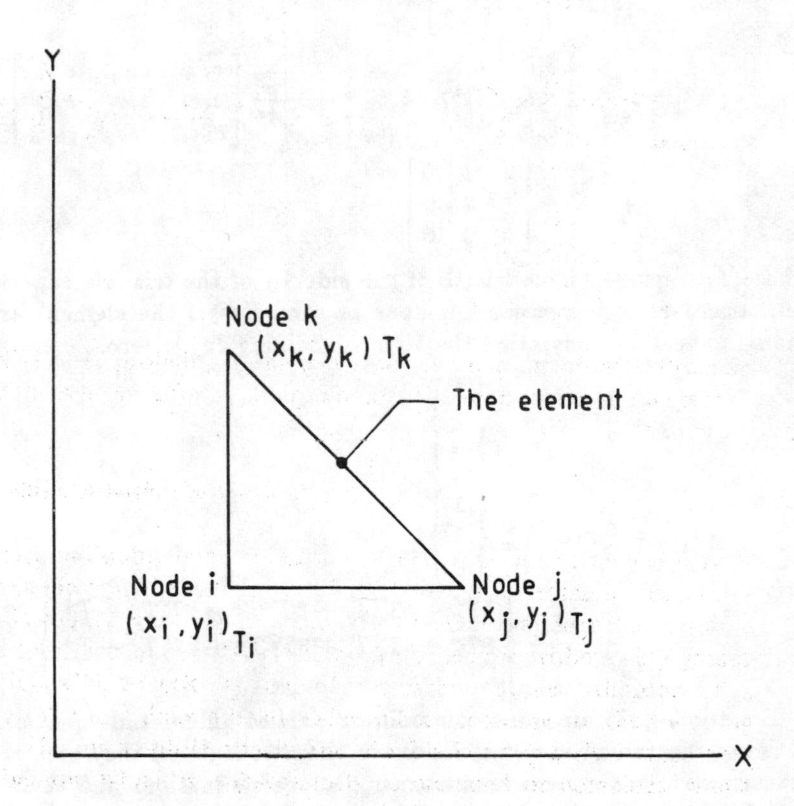

Fig. 1 The details of the triangular element

The shape functions for the linear triangular element, see Fig. 1, are

$$N_\beta = \frac{1}{2A^e}(a_\beta + b_\beta x + c_\beta y), \quad \beta = i, j, k \tag{a}$$

where

$$a_i = x_j y_k - x_k y_j$$
$$b_i = y_j - y_k$$
$$c_i = x_k - x_j \tag{b}$$
$$a_j = x_k y_i - x_i y_k$$
$$b_j = y_k - y_i$$
$$c_j = x_i - x_k \tag{c}$$
$$a_k = x_i y_j - x_j y_i$$
$$b_k = y_i - y_j$$
$$c_k = x_j - x_i \tag{d}$$

Consequently equations (16) become

$$[CP^e] = \frac{\rho c^e A^e}{12} \begin{bmatrix} 2 & 1 & 1 \\ 1 & 2 & 1 \\ 1 & 1 & 2 \end{bmatrix} \tag{16A}$$

$$[K^e] = \frac{k_x^e}{4A^e} \begin{bmatrix} b_i b_i & b_i b_j & b_i b_k \\ b_j b_i & b_j b_j & b_j b_k \\ b_k b_i & b_k b_j & b_k b_k \end{bmatrix} + \frac{k_y^e}{4A^e} \begin{bmatrix} c_i c_i & c_i c_j & c_i c_k \\ c_j c_i & c_j c_j & c_j c_k \\ c_k c_i & c_k c_j & c_k c_k \end{bmatrix} + \frac{h^e L_{ij}}{6} \begin{bmatrix} 2 & 1 & 0 \\ 1 & 2 & 0 \\ 0 & 0 & 0 \end{bmatrix} \tag{16B}$$

where L_{ij} represents the length of the side i-j of the triangle experiences heat transfer by convection. If none of the sides of the element are exchanging heat by convection the term containing L_{ij} is zero.

$$\{F_c^e\} = \frac{1}{2} h^e T_\infty L_{ij} \begin{Bmatrix} 1 \\ 1 \\ 0 \end{Bmatrix} \tag{16C}$$

$$\{F_r^e\} = \frac{1}{2} \sigma \epsilon L_{ij} T_\infty^4 \begin{Bmatrix} 1 \\ 1 \\ 0 \end{Bmatrix} - \frac{\sigma \epsilon L_{ij}}{30} \begin{Bmatrix} 5T_i^4 + 4T_i^3 T_j + 3T_i^2 T_j^2 + 2T_i T_j^3 + T_j^4 \\ 5T_j^4 + 4T_j^3 T_i + 3T_j^2 T_i^2 + 2T_j T_i^3 + T_i^4 \\ 0 \end{Bmatrix} \tag{16D}$$

Equations (17) are a set of nonlinear partial differential equations because of the radiation term. To solve these equations, they can be first transformed to a system of nonlinear algebraic equations and then use a suitable method such as Crank-Nicolson finite difference method which is unconditionally stable. The first derivative of the nodal temperature vector can be expressed between two points separated by Δt in the time domain as

$$\frac{d\{T^G\}_t}{dt} = \frac{\{T^G\}_{t+\frac{\Delta t}{2}} - \{T^G\}_{t-\frac{\Delta t}{2}}}{\Delta t} \tag{18}$$

Also

$$\{T^G\}_t = \frac{1}{2}\left(\{T^G\}_{t+\frac{\Delta t}{2}} + \{T^G\}_{t-\frac{\Delta t}{2}}\right) \tag{19}$$

$$\{F^G\}_t = \frac{1}{2}\left(\{F^G\}_{t+\frac{\Delta t}{2}} + \{F^G\}_{t-\frac{\Delta t}{2}}\right) \tag{20}$$

Using equations (18 to 20) in (17) we get

$$\frac{1}{\Delta t}[CP^G]\{T^G\}_{t+\frac{\Delta t}{2}} - \frac{1}{\Delta t}[CP^G]\{T^G\}_{t-\frac{\Delta t}{2}}$$
$$+\frac{1}{2}[K^G]\{T^G\}_{t+\frac{\Delta t}{2}} + \frac{1}{2}[K^G]\{T^G\}_{t-\frac{\Delta t}{2}} = \frac{1}{2}[F_c^G]_{t+\frac{\Delta t}{2}}$$
$$+\frac{1}{2}[F_c^G]_{t-\frac{\Delta t}{2}} + \frac{1}{2}[F_r^G]_{t+\frac{\Delta t}{2}} + \frac{1}{2}[F_r^G]_{t-\frac{\Delta t}{2}} \tag{21}$$

In the above, $\{T^G\}$, $\{F_c^G\}$ and $\{F_r^G\}$ are unknowns at time $t + \Delta t/2$. Therefore we can rewrite the above equations as

$$\left([K^G] + \frac{2}{\Delta t}[CP^G]\right)\{T^G\}_{t+\frac{\Delta t}{2}} = \left(\frac{2}{\Delta t}[CP^G] - [K^G]\right)\{T^G\}_{t-\frac{\Delta t}{2}}$$
$$+ [F_c^G]_{t+\frac{\Delta t}{2}} + [F_c^G]_{t-\frac{\Delta t}{2}} + [F_r^G]_{t+\frac{\Delta t}{2}} + [F_r^G]_{t-\frac{\Delta t}{2}} \tag{22}$$

Since the nodal temperatures and the force vectors are known at $t-\Delta t/2$, the above can be written as

$$\left([K^G] + \frac{2}{\Delta t}[CP^G]\right)\{T^G\}_{t+\frac{\Delta t}{2}} = \{A_1\} + [F_c^G]_{t+\frac{\Delta t}{2}} + [F_r^G]_{t+\frac{\Delta t}{2}} \tag{23}$$

where $\{A_1\}$ containing $t - \Delta t/2$ terms is known at this stage. We can make use of an iteration scheme to determine the unknown temperatures at $t+\Delta t/2$, by initially assuming the nodal temperature vector at $t+\Delta t/2$ to be the same as at $t-\Delta t/2$. This results in the evaluation of the force terms on the right hand side of equation (23) and thus determining the nodal termperature vector at $t+\Delta t/2$. There will be of course a difference in the newly calculated value from the assumed value and if they do not meet the convergency criteria, then the calculated nodal temperature vector becomes the assumed nodal temperature vector for the next iteration. In this way the transient temperatures in the blade cross-section can be determined. The thermal gradients $\{g^e\}$ can be then determined by making use of equation (11d).

3. TRANSIENT THERMAL STRESSES

The variation of temperature across the cross-section of a blade causes thermal stresses along the x, y and z directions. Stress components other than σ_z are found to be negligible [2]. Let T_{av} be the average initial temperature under no stress condition of the three nodes of the element and ΔT_{av} is the difference in T_{av} at any instant of time, then the stress σ_z induced in the material due to rise in temperature from the stress free state, at the center of the element x_c, y_c is given by [9]

$$\sigma_z = E\left[x_c \frac{\int_{A^e} E\alpha x_c \Delta T_{av} dA^e}{\int_{A^e} E x_c^2 dA^e} + y_c \frac{\int_{A^e} E\alpha y_c \Delta T_{av} dA^e}{\int_{A^e} E y_c^2 dA^e} + \frac{\int_{A^e} E\alpha \Delta T_{av} dA^e}{\int_{A^e} E\, dA^e} - \alpha\, \Delta T_{av}\right] \tag{24}$$

where α is the coefficient of thermal expansion and E is the Young's modulus, which are functions of temperature T and evaluated at the average temperature of the element.

The above can be written as

$$\sigma_z = E(T)\left[x_c \frac{\sum_{e=1}^{n} A^e E(T)\alpha(T) x_c^e \Delta T_{av}}{\sum_{e=1}^{n} A^e E(T)(x_c^e)^2} + y_c \frac{\sum_{e=1}^{n} A^e E(T)\alpha(T) y_c^e \Delta T_{av}}{\sum_{e=1}^{n} A^e E(T)(y_c^e)^2} + \frac{\sum_{e=1}^{n} A^e E(T)\alpha(T)\Delta T_{av}}{\sum_{e=1}^{n} A^e E(T)} - \alpha(T)\Delta T_{av}\right] \tag{25}$$

4. NUMERICAL EXAMPLE

The heat transfer process within a turbine blade made of MAR-M200 superalloy nickel, is studied by using a computer program according to the theory outlined in Section 2. A cross-section of this blade is shown in Fig. 2.

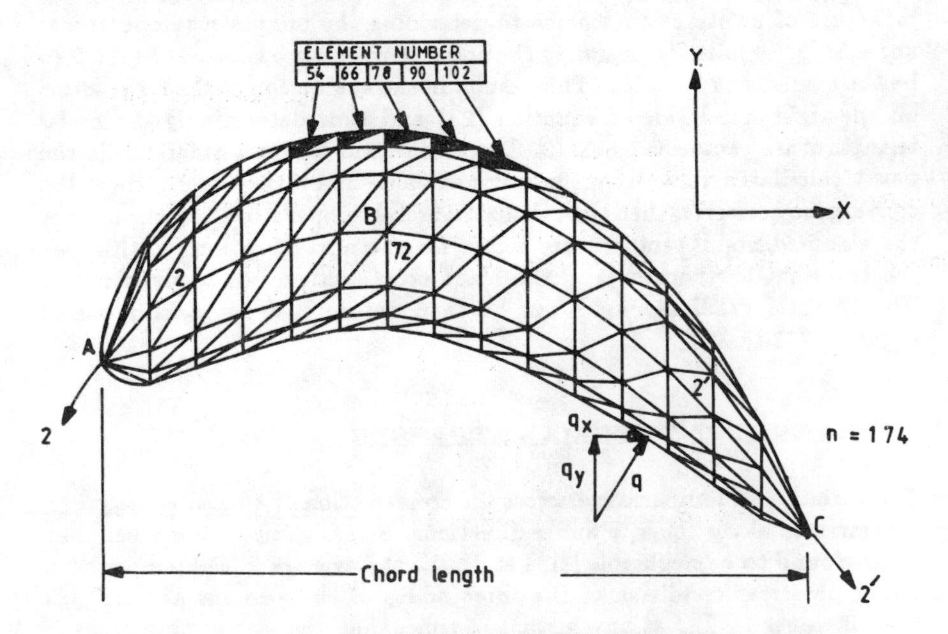

Fig. 2 The finite element discretization of the turbine blade cross-section

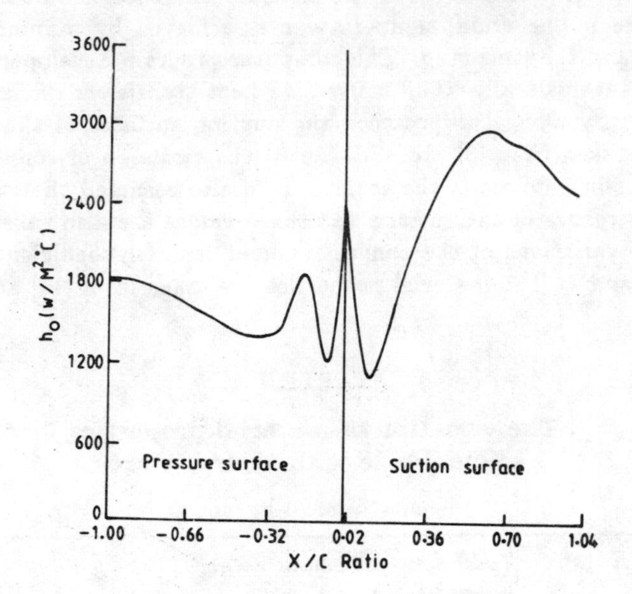

Fig. 3 Heat transfer distribution around the blade surface at full load operation

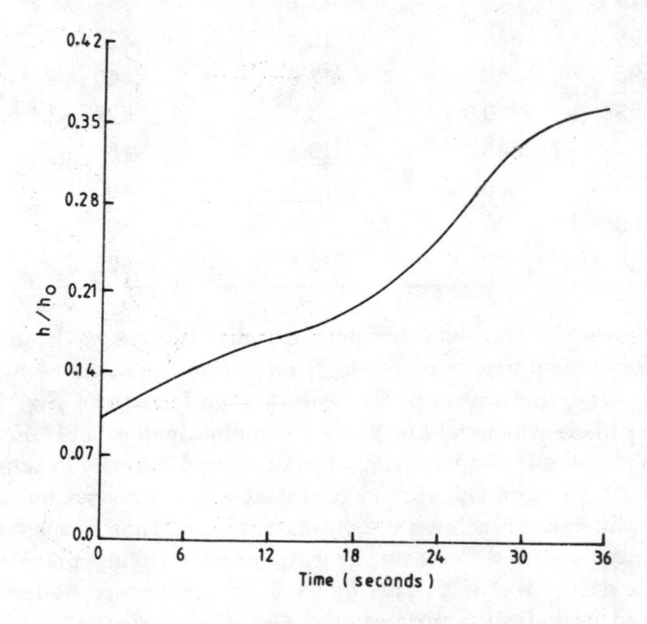

Fig. 4 Heat transfer coefficient during quick start as a function of time

The airfoil cross-section is divided into 174 linear triangular elements. The discretization is finer near both the pressure and suction surfaces and the convergence in the nodal temperatures is achieved by refining both the mesh size and the time step. The computer program developed generates the mesh automatically. The convective heat transfer coefficients are assumed to vary along the pressure and suction surfaces of the blade and they were taken from [3]. In addition to the variation of convective heat transfer coefficients along the surface, it is also assumed that they varied with temperature of the surface and these values are also taken from [3]. These two variations of the convective heat transfer coefficient are given in Figs. 3 and 4. The material properties are taken from [10] and given in Table 1.

TABLE 1

The variation of material properties of the blade with temperature

density = 8526 kg/cu. m

T °C	E GPa	Yield Stress MPa	Thermal Exp. Coeff. micron/m °K	c J/kg °K	k W/m °K
21	220	840	11.9	400	12.7
93	215	842	12.07	400	13.0
205	215	844	12.07	395	13.5
315	195	846.3	12.4	420	13.8
425	190	848.4	12.8	440	15.1
540	185	850	13.1	420	15.2
650	175	855	13.5	460	17.3
760	170	840	14.0	480	14.0
870	160	760	14.8	500	21.6
980	145	470	15.8	525	24.9

Fig. 5 shows the transient temperature distribution with and without radiation for three different nodes A, B and C on the mean camber line at the leading edge, midway and the trailing edge (shown in Fig. 2) respectively of the blade when the hot gases are maintained at 1143 K (870 °C). This figure also shows the variation of maximum difference in temperature, denoted as Δt between the various nodes at a given instant of time. This temperatre difference reaches a maximum value and then drops down which is an indicator of the differential heating process taking place at various nodes. It is quite clear from this figure that the surface nodes are much more effected by radiation process than the inner nodes.

The temperature gradients in the x direction along the pressure surface at six instants of time are shown in Fig. 6. This gradient changes sign twice

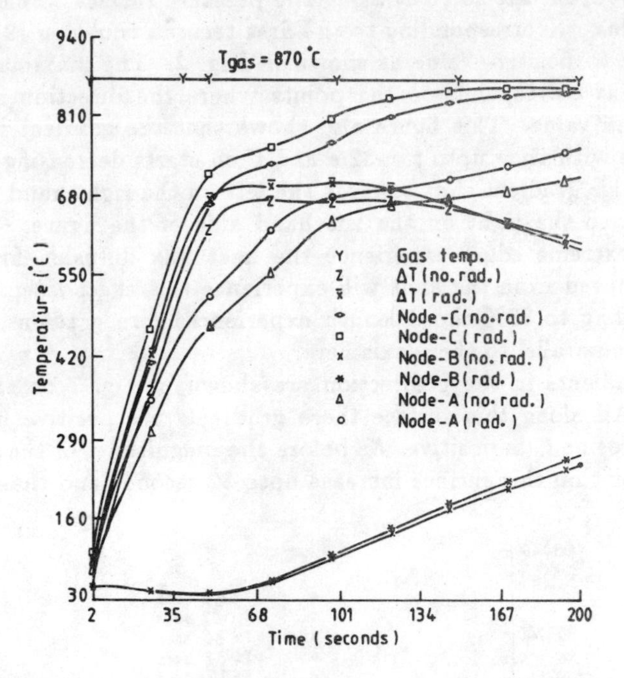

Fig. 5 The transient temperature distribution with and without radiation ($T_g = 870\,^{\circ}$C)

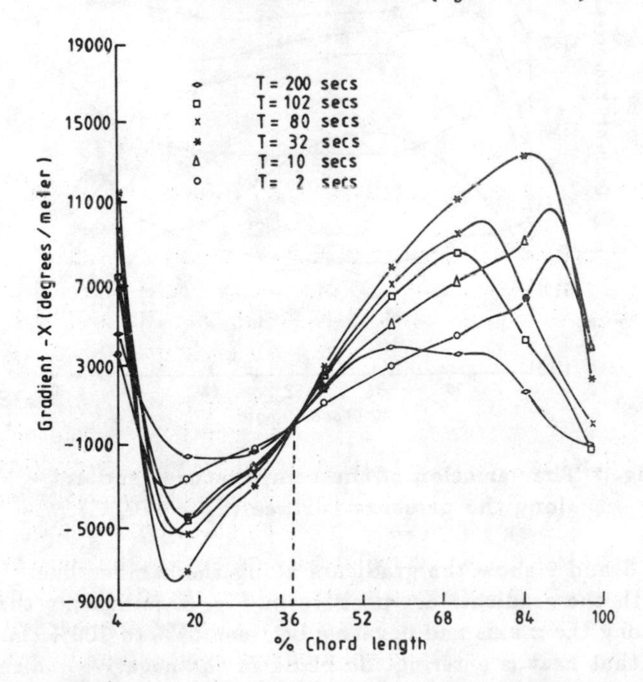

Fig. 6 The variation of the temperature gradient – X along the pressure surface ($T_g = 870\,^{\circ}$C)

as we move from left to right along the pressure surface. This is because the heat flux q_x corresponding to the first term in equation (2) goes from a negative to positive value as shown in Fig. 2. The maximum value of the gradient corresponds to the points where the direction cosine l_x is a maximum value. This figure also shows that the gradient increases in magnitude with time upto $t = 32$ s and then starts decreasing. The peak values of this gradient shift towards the left on the right hand side of this figure and to the right on the left hand side of the figure. The points near the extreme edges experience the heat flux diffusion first and the points removed from the ends will experience this effect later. The point corresponding to 38% chord length experiences zero gradient, where the heat flux is parallel to the y-axis here.

The gradients in the y direction are shown in Fig. 7 on the pressure surface. All along this surface these gradients are positive because the direction cosine l_y is positive. As before the magnitudes of the gradient at a given point on this surface increase upto 32 seconds and then decrease.

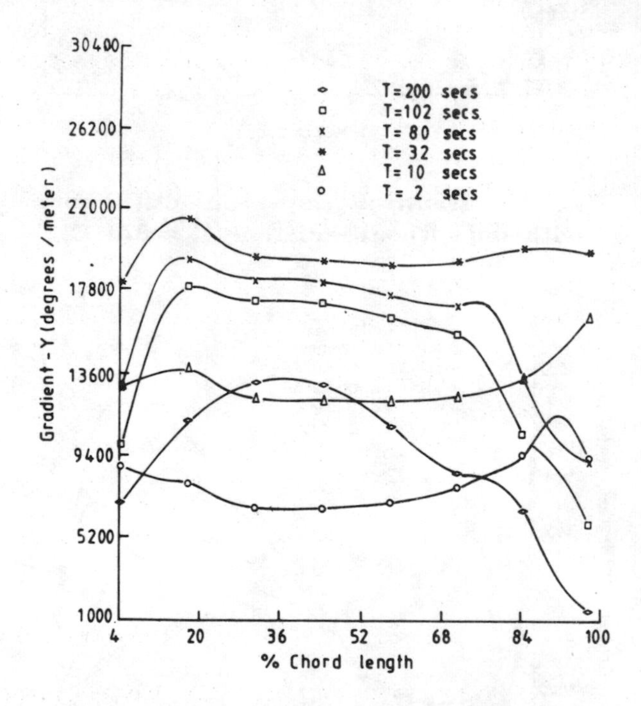

Fig. 7 The variation of the temperature gradient – Y along the pressure surface ($T_g = 870\,°C$)

Figures 8 and 9 show the gradients along the camber line. Upto 25% chord length the gradients are positive in Fig. 8, indicating that heat is entering along the x axis and negative between 55% to 100% chord length indicating that heat is entering the blade in the negative x direction. In Fig. 9 at 2 seconds, the y gradients are negative for all elements indicating that the heat flux q_y is entering from the suction surface, but with passage

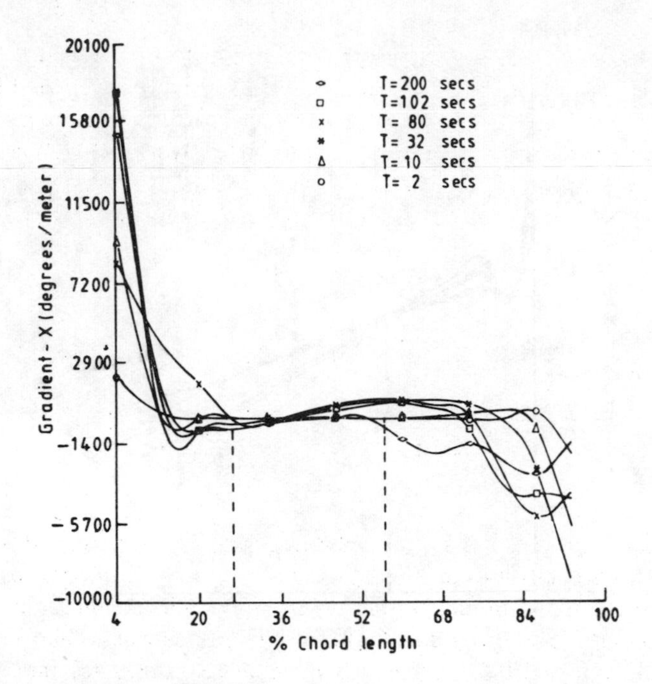

Fig. 8 The variation of the temperature gradient – X along the mid-section (2-2′) at $T_g = 870\,°C$

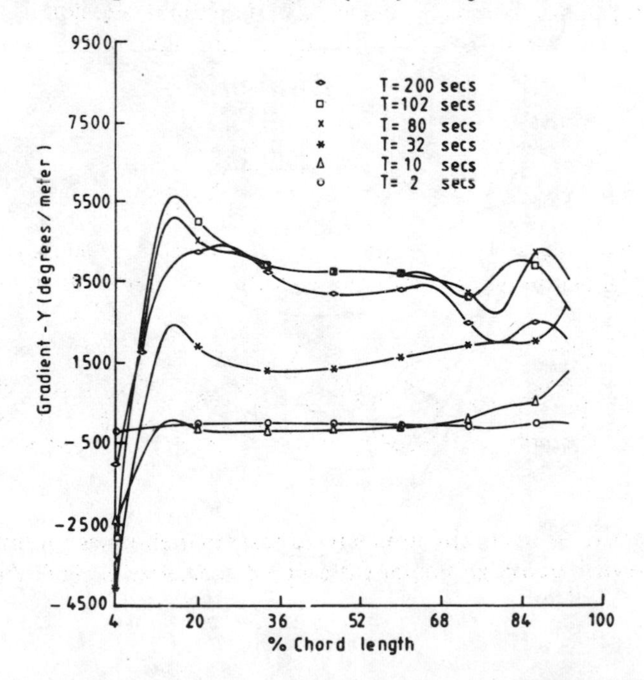

Fig. 9 The variation of the temperature gradient – Y along the mid-section (2-2′) at $T_g = 870\,°C$

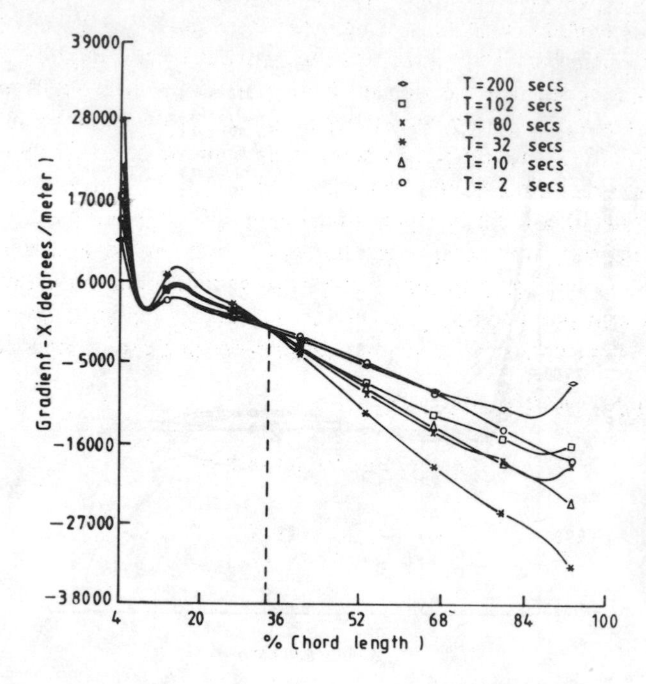

Fig. 10 The variation of the temperature gradient – X along the section surface ($T_g = 870\,^{\circ}C$)

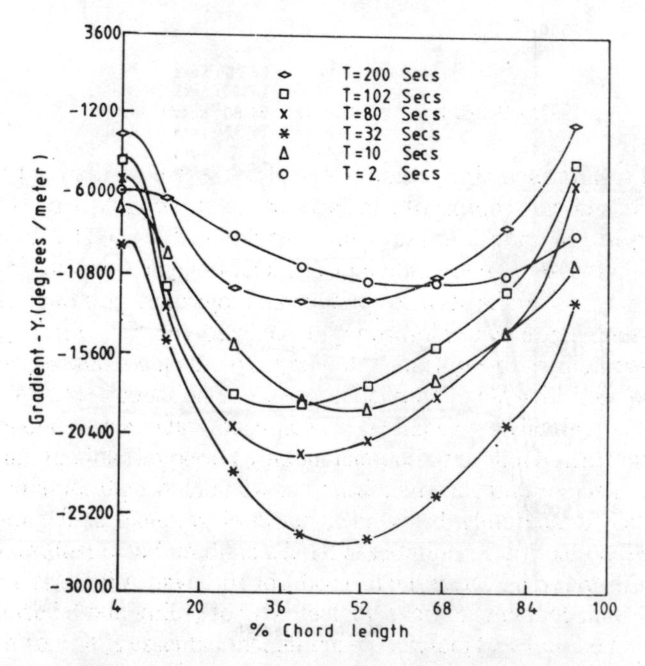

Fig. 11 The variation of the temperature gradient – Y along the section surface ($T_g = 870\,^{\circ}C$)

of time at elements away from the leading edge this gradient changes its sign. The global maximum value of the gradient occurs at $t = 102$ s at 15% chord length. This indicates that there is a delayed response in the occurrence of the peak gradients between surface nodes and the internal nodes of the blade due to the thermal diffusivity. Figs. 10 and 11 show the gradients in the x and y directions respectively at the suction surface.

The variation of normal stress with time at the leading edge, the trailing edge and at the internal node B are shown in Fig. 12. These stresses have been normalized with respect to the yield stress at the corresponding temperature. (Therefore the normalizing values for different elements are not same). It is found that the core of the section was in tension and the outside was under compression throughout the heating period. Thus the stress at the location B was tensile and the other two were compresive in nature. It is clear from Fig. 12 that the peak value of stress at the trailing edge occurs at a different instant of time than that of the leading edge. The absolute magnitude of the peak stress is larger for the leading edge as compared to the trailing edge. This figure also shows the maximum stress searched amongst all the elements, the peak value of which is more than unity, suggesting that the heating path would be infeasible.

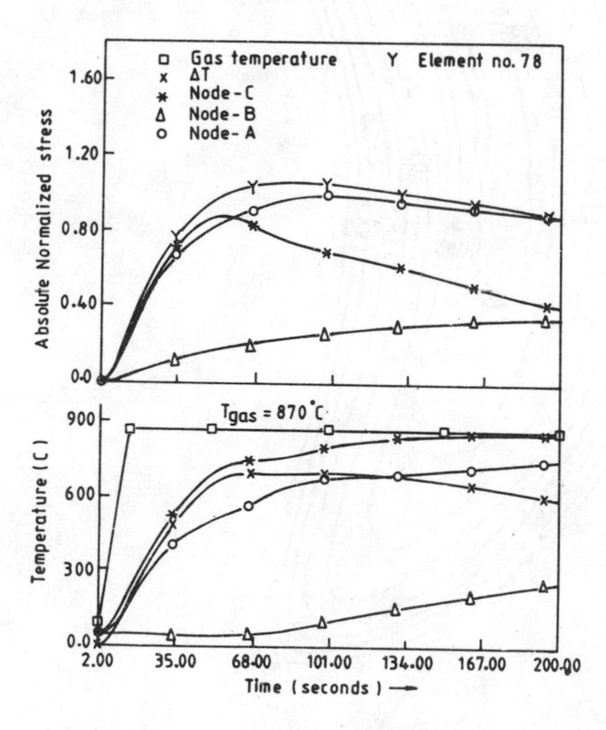

Fig. 12 The variation of the normalized stresses and ΔT as a function of time corresponding to the turbine heating rate of 56 °C/sec & final $T_g = 870$ °C

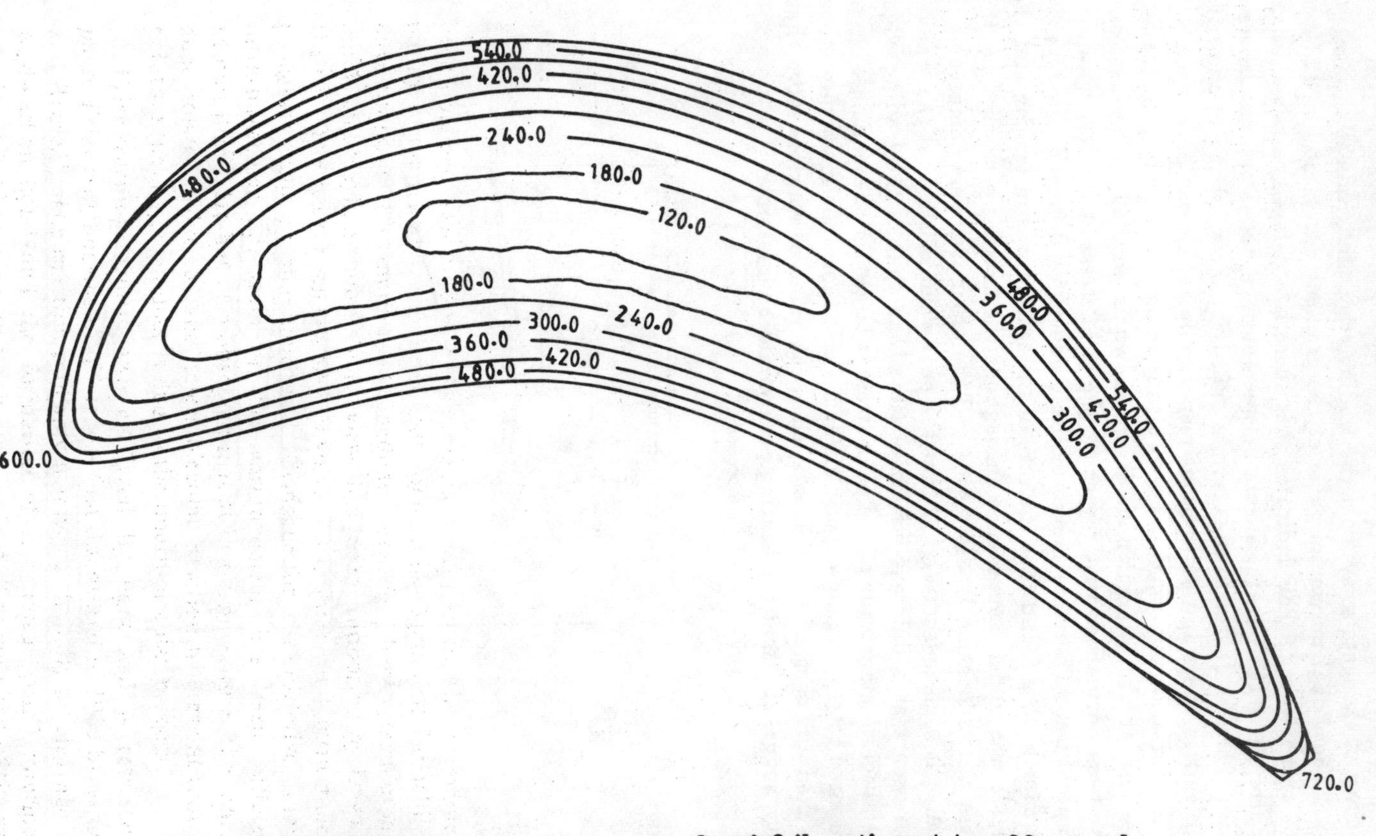

Fig. 13 The temperature contours across the airfoil section at t = 88 seconds

Fig. 12 also shows the variation of ΔT with time. At $t = 88$ s, when the peak stress occurs, element 72 is found to have minimum temperature. At this time, elements 54, 66, 78, 90 and 102 have high values of σ_z. In Table 2, the parameter ΔT^* which is the difference of temperature between an element having high stress value and the element 72 is given along with the distance, S, between the corresponding centroids of the elements. Obviously $\Delta T^*/S$ is the temperature gradient between the centroids. The maximum value of the gradient occurred between the elements 78 and 72, so also the maximum stress at this instant of time within the body at element 78. This shows that it is possible to find the location of the element having the maximum stress just from transient temperature analysis.

TABLE 2

Overall temperature gradient of elements having high stress state at t = 88 s

Element No.	ΔT^*	$S(m)$	$\Delta T^*/S$
54	406.83	0.005867	68990
66	431.28	0.005434	73960
78	445.81	0.004955	89960
90	454.95	0.005507	89700
102	462.36	0.005776	80050

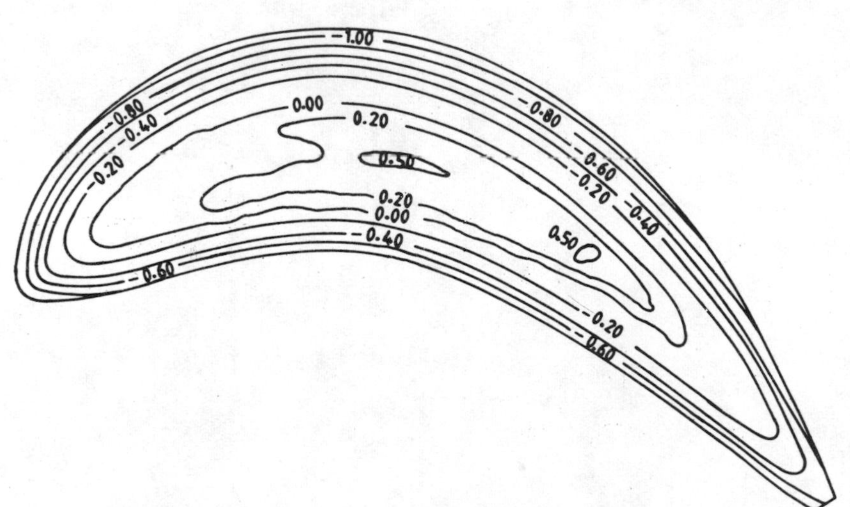

Fig. 14 The stress contours across the airfoil section at T = 88 seconds

The temperature and stress distributions at $t = 88$ s are shown in Figs. 13 and 14. In Fig. 14, the stresses are high at the leading and trailing edges, but the highest stresses occur at the suction surface on the elements shown in Fig. 2.

REFERENCES

1. Mohanty, A.K., Raghavachar, T.S. and Nanda, R.S., "Heat Transfer from Rotating Short Radial Blades", *Intl. J. Heat and Mass Transfer*, V. 20, 1977, p. 1417.
2. Maya, T., Katsumata, I. and Itoh, M., "A Study of Thermal Fatigue Life Prediction of Air Cooled Turbine Blades", ASME 78-GT-63.
3. Mukherjee, D.K., "Stresses in Turbine Blades due to Temperature Load Variation", ASME 78-GT-158.
4. Daniels, L.D. and Browns, W.B., "Calculation of Heat Transfer Rates to Gas Turbine Blades", *Intl. J. Heat and Mass Transfer*, V. 24, 1981, p. 871.
5. Allen, J.M., "Effect of Temperature Dependent Mechanical Properties on Thermal Stresses in Cooled Turbine Blades", *J. of Engg. Power*, ASME, V. 104, 1982, p. 349.
6. Bogov, I.A., "Influence of Turbine Blade Airfoil and Root Thickness Variation along the Blade on its Thermal Stress State", *Energomashiostroenie*, 1978, No. 8, p. 10.
7. Bahree, R., Sharan, A.M. and Rao, J.S., "The Design of Rotor Blades due to the Combined Effects of Vibratory and Thermal Loads", *J. Engg. Power and Gas Turbines*, ASME, V. 111, 1989, p. 610.
8. Segerlind, L.J., *Applied Finite Element Analysis*, John Wiley and Sons, 1976.
9. Littler, D.J., "Thermal Stresses and Thermal Fatigue", Proc. Intl. Conf. held at Berkeley Castle, Gloucestershire, England, September 22-26, 1969, p. 374.
10. Cubberly, W.H., *ASM Metals Reference Book*, Amer. Soc. for Metals, Metals Park, Ohio, 1980, Vol. 3, p. 248-49.

Appendix 2

SI Units

length	m
mass	kg
time	s
temperature	K
area	m^2
volume	m^3
frequency	Hz
density	kg/m^3
velocity	m/s
angular velocity	rad/s
acceleration	m/s^2
angular acceleration	rad/s^2
force	N, kgm/s^2
pressure	Pa, N/m^2, kg/ms^2
viscosity	$Pa \cdot s$, Ns/m^2, kg/ms
work, torque, energy	J, Nm, Ws, kgm^2/s^2
power, heat flux	W, J/s, kgm^2/s^3, Nm/s
heat flux density	W/m^2, kg/s^3
heat transfer coefficient	W/m^2 K, kg/s^3 K
specific heat capacity	J/kg K, m^2/s^2 K
volumetric heat release rate	W/m^3, kg/ms^3, J/sm^3
thermal conductivity	W/m K, Jm/sm^2 K, kgm/s^3 K
Boltzman constant	W/m^2 K^4 (5.669×10^{-8})

Conversion Tables (SI — Gravitational Units)

Length Units

	cm	m	in	ft
cm	1	0.01	0.3937008	0.0328084
m	100	1	39.37008	3.28084
in	2.54	0.0254	1	0.083333
ft	30.48	0.3048	12	1

Area Units

	cm^2	m^2	in^2
cm^2	1	0.0001	0.1550003
m^2	10000	1	1550.003
in^2	6.4516	0.00064516	1

Second moment of area Units

	cm^4	in^4
cm^4	1	0.0240251
in^4	41.62314256	1

Mass Units

	kg	lb	kgf s^2/cm	lbf s^2/in
kg	1	2.204623	0.001019716	0.005710147
lb	0.4535924	1	0.000462535	0.002590079
kgf s^2/cm	980.665	2161.996614	1	5.599742241
lbf s^2/in	175.1268108	386.088595	0.178579611	1

Mass moment of inertia Units

	kg m^2	kg cm^2	kgf cm s^2	lbf in s^2
kg m^2	1	10000	10.1971621	8.85074
kg cm^2	0.0001	1	0.001019716	0.000885074
kgf cm s^2	0.0980665	980.665	1	0.867961805
lbf in s^2	0.1129848293	1129.848293	1.152124429	1

Density Units

	kg/m^3	kg/cm^3	lb/in^3
kg/m^3	1	0.000001	0.00003612728
kg/cm^3	1000000	1	36.12728
lb/in^3	27679.91	0.02767991	1

Force Units

	N	kgf	lbf
N	1	0.101971621	0.224808981
kgf	9.80665	1	2.204623
lbf	4.448220852	0.4535924	1

Torque Units

	N m	N cm	kgf cm	lbf in
N m	1	100	10.1971621	8.8507473
N cm	0.01	1	0.101971621	0.088507473
kgf cm	0.0980665	9.80665	1	0.867961811
lbf in	0.1129848096	11.29848096	1.152124422	1

Pressure Units

	M Pa	kgf/cm^2	lbf/in^2
M Pa	1	10.19716	145.0377
kgf/cm^2	0.0980665	1	14.22334
lbf/in^2	0.006894757	0.07030696	1

1 mm Hg (torr) = 133.3224 Pa

1 Pa = 0.007500617 torr

1 atm = 0.101325 MPa = 1.033227 kgf/cm^2 = 14.69595 lbf/in^2

Linear stiffness

	N/m	N/cm	kgf/cm	lbf/in
N/m	1	0.01	0.00101971621	0.00571014795
N/cm	100	1	0.101971621	0.571014795
kgf/cm	980.665	9.80665	1	5.599742241
lbf/in	175.1268108	1.751268108	0.178579648	1

Linear damping

	Ns/m	Ns/cm	kgf s/cm	lbf s/in
Ns/m	1	0.01	0.00101971621	0.00571014795
Ns/cm	100	1	0.101971621	0.571014795
kgf s/cm	980.665	9.80665	1	5.599742241
lbf s/in	175.1268108	1.751268108	0.178579648	1

Torsional stiffness

	N m/rad	N cm/rad	kgf cm/rad	lbf in/rad
N m/rad	1	100	10.1971621	8.8507473
N cm/rad	0.01	1	0.101971621	0.088507473
kgf cm/rad	0.0980665	9.80665	1	0.867961811
lbf in/rad	0.1129848096	11.29848096	1.152124422	1

Torsional damping

	N m s/rad	N cm s/rad	kgf cm s/rad	lbf in s/rad
N m s/rad	1	100	10.1971621	8.8507473
N cm s/rad	0.01	1	0.101971621	0.088507473
kgf cm s/rad	0.0980665	9.80665	1	0.867961811
lbf in s/rad	0.1129848096	11.29848096	1.152124422	1

Mass moment of inertia per unit length

	kg m^2/m	kg cm^2/cm	kgf s^2	lbf s^2
kg m	1	100	0.1019716	0.2248089
kg cm	0.01	1	0.001019716	0.002248089
kgf s^2	9.80665	980.665	1	2.204623
lbf s^2	4.448220852	444.8220852	0.453592292	1

Torsional rigidity (stiffness per unit length)

	N m^2/rad	N cm^2/rad	kgf cm^2/rad	lbf in^2/rad
N m^2/rad	1	10000	1019.71621	348.454595
N cm^2/rad	0.0001	1	0.101971621	0.0348454595
kgf cm^2/rad	0.000980665	9.80665	1	0.341717226
lbf in^2/rad	0.002869814	28.69814	2.926396221	1

Temperature

$$°\mathrm{R} = \frac{5}{9}\mathrm{K}$$

$$t°(\mathrm{C}) = T(\mathrm{K}) - 273.15$$

$$t°(\mathrm{F}) = T°(\mathrm{R}) - 459.67$$

Appendix 3

Nomenclature

A	Area of cross-section of the blade Also, area of cross-section of the nozzle
$[A(r)]_j$	Matrix of coefficients, see equation (10.132)
A_0	Amplitude of force distribution
$A_0 \cdots A_n$	Glauert constants
A^*	Critical area of cross-section of the nozzle
$A_i \cdots F_i$	Arbitrary parameters in shape functions
A_{in}, B_{in}	Arbitrary constants, see equation (10.4)
A_t, A_a	Flexibility factors of the hinge in tangential and axial directions respectively
A_z	Area of cross-section at distance z
A, A'	Polynomial coefficients, see equation (3.16)
A, B	Coefficients of first quadratic form Also, S-N plane coefficients, see equation (11.2)
a	Constant related to Theodorsen's function Also, radius of the disc
a_r	Response coefficients in rth mode, see equation (10.66)
a_j	Mistuning components
a_{ns}	Disc inertia coefficient
a, b	Forcing function coefficients, see equation (8.3)
a, b, c	Arbitrary parameters in assumed solutions
a, b, c, d	Blade flexure coefficients, see equation (5.32)
$\{a_i\}$	Displacement vector of ith node
$\{a_i^0\}$	Vector of clearances between bodies A and B
a_i	Flexibility coefficient, see equation (5.37)
a_x, a_z	Lacing wire coefficients, see equation (5.78)
a_0	Sound velocity
$[a]$, $[b]$	Transfer matrix relations, see equation (5.80)
$\{a\}$	Displacement vector for complete solid element

B	Breadth of the beam
	Also, defines the camber line of an airfoil
B_i	Body force distribution
$[B]$	Strain displacement relation matrix
	Also, eigen matrix of a multi-stage disc
B, B'	Polynomial coefficients, see equation (3.16)
B_m	Bending moment of the disc annulus
b	Breadth of the cross-section
	Also, lacing wire coefficient, see equation (5.78)
	Also, axial distance between stator and rotor cascade centers
	Also, βa
b_{nz}	Non-dimensional frequency of the disc
b^*	Constant related to Theodorsen's function
C	Torsional stiffness
	Also, couple
	Also, viscous damping coefficient
C, C'	Polynomial coefficients, see equation (3.16)
$[C]$	Damping matrix
$[C_{ij}]$	Matrix of flexibility coefficients at node i due to forces at j
$\{C\}$	Vector of coefficients of fundamental solutions
C_D	Profile drag coefficient
C_M	Correction factor for Froude number
C_v	Proportional damping coefficient
$C_{\overline{M}}$	Correction factor for hypothetical Mach number
C_P	Correction factor for pressure distribution
C_T	Correction factor for temperature distribution
$C(\omega)$	Theodorsen's function
$C_a(\omega)$	Lift response function in convecting gust (parabolic blade)
$C'_a(\omega)$	Lift response function in convecting gust (skewed blade)
C_t	Torsional stiffness
C_a	Flexibility coefficient, see equation (5.54)
c	Blade semichord
$\{c\}$	Nodal local coordinates vector
c_3	See equation (2.39)
c_γ	Additional torsional stiffness due to pretwist
D	Thickness of the beam
	Also, platform height, see Fig. 5.4
	$\dfrac{2Eh^3}{3(1-\nu^2)}$
	Also, functions in disc receptances, see equation (10.14)

$[D]$	Material property matrix
D, D'	Polynomial coefficients, see equation (3.16)
d	Blade pitch in the direction of rotation Also, depth of cross-section of the blade
$\{d\}$	Nodal displacement vector
$[d]$	Transfer matrix relations, see equation (5.80)
dF, dM	Extensional force and moment at the blade shroud junction on the shroud Also, deflections due to unit force and moment
dZ	Length of element
dx	Distance between two stations of the blade
d_a	Flexibility coefficient, see equation (5.56)
E	Young's modulus Also, polynomial coefficients, see equation (3.16)
$\overline{E}$	Complex elastic modulus
E_N	Energy loss per unit volume
E_d	Energy dissipated by packet
E_s	Energy supplied to blade packet
e	Blade twisting coefficient, see equation (5.34) Also, twist coefficient, see equation (5.58)
e_i	Experimental strain
e_{zz}	Longitudinal strain
e_{xz}, e_{yz}	Shear strains
$\overline{e}$	Rotating unit vector, thus, $\overline{e}_{x1}$
F	Coefficient of quasi steady circulation Also, Froude number Also, point force
F^*	Maximum deviation of mean line from the chord line
F_i	Functions of frequency in blade receptances
F_m	Froude number based on modified velocity
F_x, F_y	Shear forces, see Fig. 3.4 Also, forcing functions
F_i, f_i	Shape functions
F, f	Centrifugal terms, see equation (5.70)
$\{F\}$	State vector
$\{F_j\}$	Force vector at jth location of the disc
$\{F\}_k$	Six-component force vector at kth station on the disc
f	Ratio of maximum camber over half-chord length
f_i, g_i, h_i	Shape functions

f_N	Force acting on Nth blade of the packet
$f_{ns}^{\vert r\vert}$	Radial mode shape function of the disc
$\{f_j\}$	Force vector at jth blade root
$\{f\}_j$	Six component force vector at the root of jth blade
$[f]$	Matrix of fundamental solution quantities
$\{f\}$	Force vector
G	Modulus of rigidity
$[G]$	Transformation matrices, see equation (10.173)
$\{G_j\}$	Force vector at jth position of the shroud
$\overline{G}_m$	Coefficient of velocity induced at a blade (upwash component)
g	Blade flexure coefficient, see equation (5.32) Also, lacing wire coefficient, see equation (5.78)
$\{g_j\}$	Force vector at the blade tip
H	Weight coefficients of Gaussian quadrature Also, blade profile parameter
H_{mk}	mth Harmonic response in kth mode
$H^*(\omega)$	Lift response function for convecting stream-wise gust, general camber airfoil
$H^*(\omega, \lambda)$	Lift response function for non-convecting stream-wise gust, general camber airfoil
$H_L(\omega)$	Lift response function due to the angle of incidence for convecting gust, general camber airfoil
$H_L(\omega, \lambda)$	Lift response function due to the angle of incidence for non-convecting gust, general camber airfoil
$\overline{H}_m$	Steady bound vortex distribution function
$H_k^{\vert 2\vert}$	Hankel function of the second kind and kth order
h	Thickness of the plate Also, flexibility coefficient, see equation (5.39) Also, lacing wire coefficient, see equation (5.78) Also, height of the flow on the water table Also, half thickness of disc Also, hysteretic damping coefficient
h^*	Orientation parameter
$\overline{h}$	h/b
I	Second moment of area
I, J, K, N	Bessel functions of order n
I_A	Second moment of area for axial bending
I_{xx}, I_{yy}	Second moments of area about centroidal axes
I_{XX}, I_{YY}	Principal second moments of area

I_{xy}	Product moment of area about centroidal axes
I_{cf}	Mass moment of inertia per unit length about c.f.
I_{cg}	Mass moment of inertia per unit length about c.g.
I_p	Polar moment of area
I_R	Reissner's functional
I_1, I_2	See equation (2.22a) Also, time integrals, see equations (9.6a) and (9.15)
$\widehat{I}$, $\widehat{J}$, $\widehat{K}$	Unit vectors in ξ, η, Z coordinates
$\widehat{i}, \widehat{j}, \widehat{k}$	Unit vectors in x, y, z directions
J	Lumped inertia Also, constant of proportionality Also, number of annuli in the disc
J_n	Mass moment of inertia at nth station
$J(n)$	$J_0(n) - iJ_1(n)$
$[J]$	Jacobian matrix
j	Blade number Also, annulus number
K	Gaussian curvature of the shell surface
K_k	Generalized stiffness in kth mode
$[K]$	Stiffness matrix
$[\overline{K}]$	Reduced stiffness matrix
$[\widehat{K}]$	Effective stiffness matrix
$[\widetilde{K}]$	Stiffness matrix in upper triangular form
K_x, K_z	Support stiffnesses in bending and torsion
k	Shear correction factor Also, summation index
k_a, k_b k_f	Endurance limit modifying factors
k_t	Factor depending on the geometry of member
L	Blade length Also, elemental length Also, Lagrangian function Also, lift Also, Reissner's dynamic functional
$[L]$	Strain operator matrix
L, M, N	Coefficients of second quadratic form
l	Length of the plate Also, lacing wire length Also, gust wave length Also, axial blade coordinate

M	Bending or twisting moment Also, number of blades in a laced packet Also, Mach number in the flow Also, moment on airfoil
$\overline{M}$	Radius vector of undeformed middle surface Also, hypothetical Mach number in the gas flow
$M_a(\omega)$	Moment response function for convecting stream-wise gust, flat plate
M_c	Corrected value of Froude number
M_D	Mass of the disc
M_k	Generalized mass in kth mode
$M_i(\omega, \lambda)$	Moment response function due to the angle of incidence for non-convecting gust, general camber airfoil
M_s	Steam bending moment
M_4	Polar moment of inertia, see equation (10.138)
$M^*(\omega, \lambda)$	Moment response function for non-convecting stream-wise gust, general camber airfoil
$\overline{M}_{,\alpha}/A$	Unit vectors in α direction
$\overline{M}_{,\beta}/B$	Unit vectors in β direction
$[M]$	Mass matrix
$[\overline{M}]$	Reduced mass matrix
m	Mass per unit length Also, lumped mass Also, number of elements Also, master degrees of freedom Also, frequency parameter, see equation (5.23) Also, harmonic number
m_n	Mass at nth station
m_s	Mass per pitch length of the shroud
N	Number of blades Also, force Also, number of segments in the bladed-disc Also, number of stress cycles to failure Also, rotor speed in rpm
N, N', N''	Functions in disc receptances, see equation (10.14)
N^*	Blade number adjacent to band middle location Also, the middle blade number
$[N]$	Shape function matrix
$[\overline{N}]$	Global and nodal local coordinates relation matrix
N_i	Shape function for the ith mode solid element
NG	Number of Gaussian points

n	Number of terms in assumed solution
	Also, number of spans in each blade
	Also, summation index
	Also, shape factor
	Also, damping exponent
	Also, number of nodal diameters
	Also, rotor speed in rpm
n_s	Number of nozzles
$\hat{n}$	Unit vector normal to blade surface
P	Pitch length
	Also, pressure
	Also, normal force
P_1	Real part of moment response function
P_{ns}	(n, s)th principal coordinate loading on the annulus
$[P]$	Strain and displacement derivatives relation matrix
$[P^B][Q^B]$	Matrices, see equation (7.27)
$\{P\}_j$	Excitation force on the blade
p	Frequency
$\{p^j\}$	Nodal contact force vector on body A
p_i	Generalized coordinates
p_n	nth natural frequency
p_{ns}	Deflection of the annulus
$\overline{p}$	Non-dimensional frequency
Q	Amplitude of harmonic force per unit radial distance
Q_a	Driving force
Q_1	Imaginary part of moment response function
$\overline{Q}_m$	Coefficient of velocity induced at a blade (downwash)
Q_{ns}	Generalized force, see equation (10.27)
$Q_1(n)$	Axial force
$Q_2(n)$	Tangential torque
$Q_5(n)$	Radial torque
$[Q]$	Matrix relating derivatives of displacements with local and global coordinates
$\{Q\}$	Forcing vector, see equation (8.8)
$\{Q\}_i$	Force vector at the blade end point i
q	Time function
	Also, notch sensitivity factor
$\{q\}_i$	Response vector at the blade end point i
q_j	Deflection of the jth annulus
q_{ns}	Principal coordinate, see equation (10.28)

R	Disc radius Also, distance from the center of the disc Also, rotational speed in Hz
$[R]$	Matrix relating vectors of displacement derivatives and nodal displacements Also, vector of reaction forces
$\overline{R}$	R/L
R_1, R_2	Radii of curvature in α, β directions
R_{12}	Radius of torsion
R_f	Endurance limit modifying factor
$R(\alpha)$	Assumed α mode shapes
r	Aspect ratio, l/b Also, distance of an element from centroid in a symmetrical cross-section Also, number of blade segments Also, radial coordinate
$\overline{r}$	I_{YY}/I_{XX}
$\overline{r}_g$	Non-dimensional radius of gyration
r_1	Distance of an element from centroid
r_2	Distance of an element from center of torsion
r_s	Radius of the shroud
r_x, r_y	Coordinates of center of flexure from centroid
S	Horizontal distance between center of flexure and c.g. Also, shear force Also, number of blades in a packet Also, stimulus coefficient
S_1	Blade endurance limit for 1000 cycles
S_2	Blade endurance limit for 10^6 stress cycles
S_e	Endurance limit for 10^6 cycles
S'_e	Endurance limit for N stress cycles
S_f	Shear force of the disc annulus
$S_L(\omega)$	Lift response function for convecting transverse gust
$S_L(\omega, \lambda)$	Lift response function for non-convecting transverse gust
$S_M(\omega, \lambda)$	Moment response function for non-convecting transverse gust
S_u	Ultimate tensile strength
S_y	Yield strength
S	Slave degrees of freedom Also, surface area Also, number of blades in packet Also, number of nodal circles

T	Kinetic energy Also, twisting moment Also, temperature
T_i	Traction force
T_θ	Twisting moment
$T(\omega)$	Horlock's function
$T(\omega, \lambda)$	Lift response function for non-convecting stream-wise gust, flat plate
$[T]$	Transfer matrix, see equation (5.95)
t	Time Also, thickness of the cross-section
U	Extensional displacement of the band Also, velocity of the rotor cascade
U_1	Downwash component at stator or rotor blade
U_F	Gain in potential energy due to rotation
$\overline{U}$	Displacement vector of M surface Also, position vector
U^*	Complementary energy density function
$[U]$	Modal matrix
$\{U\}$	Nodal displacement vector
$\{U\}^n$	Nodal displacement vector of the nth segment
$\{\overline{U}\}^K$	Vector of all interior grid point displacements
u	Longitudinal displacement Also, velocity in chord-wise gust or in the x direction
u_0	Amplitude of velocity perturbation in the chord-wise gust
u_s	Perturbation velocity at the center of the wake
u_i	Displacement in ith direction
u_T	Perturbation velocity due to all viscous wakes
u, v, w	Displacements of a point in an element Also, deflections of the middle surface in α, β and negative n directions respectively
V	Potential energy Also, strain energy Also, shear force Also, speed of the flow relative to the blade Also, velocity on the water table
$\{V\}$	State vector at jth annulus of the disc
$\overline{V}$	Velocity of propagation of non-convecting gust
V_1	Upwash component at stator or rotor blade Also, extensional strain energy
V_2	Shearing strain energy

V_{up}	Induced velocity perpendicular to a given line
V_{uq}	Induced velocity parallel to a given line
$\tilde{V}_{1i}$, $\tilde{V}_{2i}$	Orthogonal vectors
$\tilde{V}_{3i}$	Vector along thickness direction at ith node
v	Volume Also, velocity of particle Also, velocity in the transverse gust or in y direction
v_0	Amplitude of velocity perturbation in the transverse gust
vF, vM	Slopes due to unit force and moment, see Fig. 3.3
vT	Torsional deflection due to unit moment
W	Potential energy due to external loads Also, complex velocity Also, shape function
W_d	Energy dissipated
w	Weights used in Gaussian quadrature Also, transverse displacement
X	Coefficient of first quadratic form Also, displacement response
$\overline{X}$	Displacement vector
$\{X_j\}$	Displacement vector of the disc
$\{X\}_j$	Displacement vector at jth location of the disc
X, Y	Coordinates in the incompressible plane
XX, YY	Principal axes
X_1, Y_1	Real and imaginary parts of lift response function
x, y	Bending displacements in x, y directions
x'', y''	Oblique coordinates
x^*	Coordinate, see equation (6.162)
x_j	Distance from disc center to jth span
x_1, x_2	Bending displacements in x_1, x_2 directions
x', y'	Slopes
xx, yy	Coordinate axes through center of flexure
x_1x_1, y_1y_1	Coordinate axes through center of gravity
x, y, z	Global coordinates
x', y', z'	Local orthogonal Cartesian coordinates
$\{x_j\}$	Displacement vector at the blade root
$\{x\}_j$	Six-component displacement vector at jth blade root
Y	y/L
Y_m	Maximum value of the mode shape

Y_n	Bessel function of the second kind
	Also, blade mode shape in a packet
Y_v	Value of Y_m
Y_w	Half width of wake behind an airfoil
$\{Y_j\}$	Displacement vector of the shroud
y, z	Displacements
z	Distance along the length of the blade
Z	Normal mode functions
	Also, z/L
	Also, p^2
Z_1	ω^2
z	Distance between middle and M^* surfaces
α	Non-dimensional rotational parameter
	Also, non-dimensional support stiffness in torsion
	Also, slope
	Also, stagger angle of blades in a cascade
	Also, phase lag of force between two successive blades
	Also, angular acceleration
	Also, disc receptance
α, α', α''	Disc diametral receptances
α_{jk}	Point receptances of the disc
α_{r1}, α_{r2}	Angles, see Fig. 6.8
α_{s1}, α_{s2}	Angles, see Fig. 6.8
α, β	Curvilinear coordinates
α, β, γ	Moment coefficients due to pretwist, see equation (5.68)
$\overline{\alpha}$, $\overline{\beta}$	α/b, β/b
α_i, β_i	Rotations of normals at ith node
α^{*0}, β^{*0}	Orthogonal set of coordinates
β	Non-dimensional stiffness parameter, see equation (2.49)
	Also, frequency parameter, see equation (5.26)
	Also, angle between stator and rotor blades, $\alpha_s + \alpha_r$
	Also, arbitrarily fixed phase of first blade force
	Also, blade receptance
β^4	$\omega^2\mu/D$
β^*	$\sqrt{1-M^2}$
Γ	Unit pretwist angle, γ/l
	Also, circulation
γ	Pretwist angle
	Also, shear strains
	Also, strength of bound vortex distribution
	Also, ratio of specific heats
	Also, shroud receptance

γ_z	Pretwist angle at distance z from the root
Δ	Inward displacement Also, differential operator, see equation (10.3)
$[\Delta]$	Frequency determinant
$\{\Delta\}$	Rigid body displacement vector containing U, V, θ components
δ	Variational parameter Also, angle of incidence Also, log decrement
ϵ	Normal strains Also, strength of the wake vortex distribution Also, half angular width of the blade
ϵ_i	Length of small sections of the blade
$\{\epsilon\}$	Strain vector
ϵ_α, ϵ_β	Longitudinal strains of middle surface
ζ	Modal damping ratio
η	Generalized coordinate Also, loss factor
η, ξ, z	Local curvilinear coordinates
η_D, ξ_Z, z_D	Deflections along η, ξ, z directions respectively
$\eta\eta$, $\xi\xi$	Coordinate axes, see Fig. 2.1
θ	Torsional deflection Also, circulation parameter
$\overline{\theta}$	Rotational displacement vector at interface node
θ, μ	Slopes
$\theta(\beta)$, $\pi(\beta)$	Assumed β mode shapes
λ	Non-dimensional frequency Also, radius of gyration of lacing wire lumped mass Also, reduced non-convecting frequency, $\mu c/V$
λ^4	$\omega^2 \rho A/EI$
λ_1	Wave length
μ	Complex frequency Also, coefficient of dynamic friction Also, mass per unit area of the disc
μ_{ns}	See equations (10.10) and (10.11)
ν	Convective frequency of motion in time, $2\pi u/d$ Also, Poisson's ratio Also, nozzle passing frequency
ν_α, ν_β, ν_n	Rotations about α, β and normal axes
π	Total potential energy
ξ	Wake coordinate

ξ, η, ς	Local coordinate directions
ρ	Mass density Also, density of the flow medium
$\sum_n$	Summation from $n = 1$ to ∞
$\sum_n{}'$	Summation from $n = -\infty$ to $+\infty$ except $n = 0$
σ	Solidity ratio, $2c/d$ Also, local stress
$\sigma_1, \sigma_2, \sigma_3$	Principal stresses
σ_a	Basic alternating stress at the root
σ_{af}	Failure value of alternating stress
σ_{alt}	Basic alternating stress
σ_s	Blade endurance limit for N stress cycles
σ_m	Total basic mean stress
σ_{mc}	Mean stress due to blade rotation only
σ_{mf}	Failure value of mean stress
σ_{mv}	Basic mean stress due to 0th value NPF component
σ_z	Bending stress
$[\sigma_i]$	Kinematic transformation matrix
$\{\sigma\}$	Stress vector
τ_{xz}, τ_{yz}	Shear stresses
τ_α, τ_β	Additional bending strains of middle surface
$\tau_{\alpha\beta}$	Additional shear strain due to bending
Φ	Velocity potential
ϕ	Stagger angle Also, torsional deflection Also, angle between x direction and tangent at ith node
ϕ'	Setting angle
ϕ_c	Warping function
ϕ_m	Torsional mode shape
ϕ_{ns}	(n, s) disc mode shape
ϕ_r	rth blade mode shape
χ_α, χ_β	Longitudinal bending strains in α, β directions
$[\psi]$	Transformation matrix from global to master degrees of freedom
Ω	Rotational speed
$\overline{\Omega}$	Vector of elastic rotation
ω	Angular frequency Also, rotational speed Also, reduced convecting frequency, $\nu c/V$

ω_{ns} — (n, s) mode natural frequency

ω_0 — Initial angular frequency

ω_α, ω_β — Additional longitudinal strains of middle surface

$\omega_{\alpha\beta}$ — Shear strain of middle surface

Superscripts

A, B — Refer to the quantities belonging to bodies A and B

c — Refers to the quantities at contact points

f — Refers to the points where external forces are applied

k — Refers to kth blade quantities

R — Refers to the quantities at reaction points

r — Rotor

$r1$ — Effects at rotor due to motion of steady stator circulation

$r2$ — Effects at rotor due to stator vortex wakes

rv — Effects at rotor due to stator viscous wakes

s — Effects at stator due to steady rotor circulation

t, n — Tangenetial and normal components of contact forces

0 — Zeroth blade

$'$ — Denotes quantities relating to body B

$'$ — Some specified quantity in the incompressible plane or quantity at a reference station

$'$ — Denotes quantities of 1st and 2nd quadratic forms of the deformed middle surface

$'$ — Refers to the calculations with a unit value of V or T at the base

$''$ — Refers to the calculations with a unit value of M at the base

$'$ — Refers to values at free end due to unit root load F_x

$''$ — Refers to values at free end due to unit root load F_y

$'''$ — Refers to values at free end due to unit root load M_x

$''''$ — Refers to values at free end due to unit root load M_y

$'$ — Denotes differentiation with respect to z or r, thus, x'

$\cdot$ — Denotes differentiation with respect to time, thus, $\dot{x}$

$*$ — Refers to surface at distance z from middle surface

$-$ — Represents virtual quantities, thus, $\bar{\epsilon}$

I, II, etc. — Refer to fundamental solution quantities, see equation (5.73)

Subscripts

a — Refers to the axial vibration of the packet. Also, refers to axial component

B, C — Refers to bodies B and C

b	Denotes blade quantities
c	Quantity associated with compressibility taken into account or a corrected value
i	Quantity associated with incidence taken into account
inc	Incompressible flow quantity
j	Refers to jth span quantities Also, refers to jth blade or jth annulus
j, m	Constants for blade profile
l	Refers to lacing wire quantities
N	Refers to Nth blade quantities
n	Denotes the quantities of station n or number indicating order Also, denotes number of nodal diameters
nl	Denotes quantities to the left of station n
nr	Denotes quantities to the right of station n
r	Refers to the blade tip quantities or rotor Also, refers to radial component
s	Denotes shroud quantities or stator Also, denotes number of nodal circles
t	Stagnation condition Also, denotes quantities at time t Also, refers to tangential vibration of the packet Also, refers to tangential component
u	Quantity associated with chord-wise velocity perturbation
v	Quantity associated with transverse velocity perturbation
W	Denotes receptance due to displacement
λ	Refers to values at station λ
ϕ	Denotes receptance due to slope
$,x$	$\dfrac{\partial}{\partial x}$ Denotes coordinate quantity, thus, x

Pre-subscripts

0	Steady quantity
C	Denotes disc receptance due to applied couple
N	Denotes disc receptance due to applied force
q	Quasi-steady quantity
t	Time dependant quantity
1	Apparent mass contribution
2	Wake contribution

Index

Accelerating turbine blade 329
Ahmad's element 133
Alternating stress component forced vibration 308
Armstrong's analysis 347
Aspect ratio effect on plate frequencies 123, 125
Aspect ratio effect torsion modes 125
Asymmetric cross-section blades 42
Average Modal damping ratios 282
Axial and torsion modes packeted blades 161
Axial displacement due to bending 20

Bagci failure line 389
Bahree results beam type blades 91
Banerji coupled bending torsion modes 78
Basic stress component forced vibration 308
Basic vectors 102
Bending frequencies of rotating plates 120
Blade attachments 9
Blade critical speeds 298
Blade damping mechanisms 252
Blade flexure relations packeted blades 157
Blade groups (see packeted blades)
Blade profile parameters parabolic camber 229
Blade profile parameters, H 197
Blade receptances 343
Blade torsion relations packeted blades 157
Blade under acceleration 329
Bladed disc system results 356
Bladed disc tuned system 347
Bladed discs finite element analysis 376
Blading nomenclature 6

Camber general case 197
Cambered airfoil 194
Campbell diagram 3
Cantilever modes packeted blades 153
Carnegie's airfoil blade values 37
Carnegie's coupled modes 55
Carnegie's pre-twisted blade torsional modes 40
Carnegie's pre-twisted blade values 36
Carnegie-Dawson-Thomas tapered beam values 29
Centrifugal forces beam type blades 46
Centrifugal stress 392
Circular disc 339
Classical hydraulic analogy 232
Complementary energy 306
Convecting gust 191
Correction factors hydraulic analogy 235
Coupled axial torsion modes packeted blades 161
Coupled bending-bending beam type blades 34
Coupled bending-torsion beam type blades 43
Critical speeds 298
Cumulative damage 390

Damping mechanisms 252
Damping ratio dependent on rotor speed 286
Damping ratio dependent on strain 282
Dawson-Carnegie coupled bending modes 37
Deak and Baird analysis 167
Differential geometry 100
Disc 339
Disc blade shroud analysis 357
Disc blade shroud interaction 369
Disc radius effect bending modes 47
Disc radius effect plate modes 130
Disc radius effect pre-twisted plates 122
Disc receptances 338
Discrete system analysis 69
Displacement field beam type blade 19
Displacement field for shells 135
Downwash on rotor potential interaction 211
Downwash on rotor wake shed by stator 215
Downwash on stator blade 207, 210
Duhamel's integral 326
Dynamic contact problem finite element method 267

Elastic coefficients 72
Elasticity matrix for shells 136
Element mass matrix beam type blades 90
Element stiffness matrix beam type blades 89
Elemental mass matrix for shells 139
Elemental stiffness matrix for shells 139
Endurance limit 390
Energy dissipated in rotating blade 291
Ewins' analysis 350
Excitation with varying frequency 325
Extended Holzer method 78

Failure surface 389
Fatigue 387
Fiber bending effect 42
Finite element analysis bladed discs 376
Finite element method dynamic contact problem 267
Finite element method static contact problem 258
Finite element model beam type blades 83
Finite element model shells 132
First quadratic form coefficients 102
Fixed-supported modes packeted blades 154
Flat plate airfoil results 222
Forced vibration 291
Forced vibration bladed discs 352
Forced vibration response nonlinear damping 318
Fourier harmonic components of excitation 292

Gas dynamic damping 254
General camber mean line coefficients 197
Gerber failure line 389
Glauert expansion 205
Goodman failure line 389
Gupta plate values 116, 119
Gust parallel to flow 193
Gust transverse to flow 192

Hamilton's principle 140, 152
Horlock's function 193
Hydraulic analogy 232
Hydraulic analogy correction factors 235
Hydraulic analogy limitations 233
Hydraulic analogy modified 234

Incidence effect 195
Inextensional solution Levy method 116
Interface damping 257
Interference of rotor on stator 201
Interference of stator on rotor 210
Interlocking shroud 367
Interpolation function beam type blades 86
Inward displacement due to rotation 22

Isolated airfoil 191
Isolated airfoil compressible flow 199
Isoparametric element 85

Jacobian matrix 87, 137

Karman and Sears theory 191
Kececioglu failure line 389
Kinetic energy beam type blade 20
Kinetic energy packeted blades 152
Kinetic energy shell 108

Laced blade packet 167
Lacing wire packet 167
Lagrange's equations forced vibration 293
Lagrangian equation beam type blade 27
Lagrangian expression pre-twisted plate 113
Lazan stress damping law 253
Levy solution 116
Life estimation 394
Life estimation cumulative damage 396
Life estimation constant speed operation 394
Lift and moment potential interaction 213, 214
Lift and moment viscous wake shed by stator 220
Lift and moment wake shed by stator 216
Lift response 191 to 197

MacBain pre-twisted plate bending modes 116
Mach number correction 235
Material damping 253
Material property matrix 88
Mean stress component forced vibration 308
Miner's rule 390
Mistuned system 351
Mistuned system roots 354
Modal analysis 294
Modal analysis transient response 330
Modal damping envelopes 276
Modal damping ratios 282
Modal response 295
Modal response Reissner's analysis 307
Moment response 191 to 198
Multi-stage bladed discs 357
Multi-stage blades receptances 365
Myklestad-Prohl method 69

NASTRAN analysis bladed discs 378
Natural frequencies rotating plates 124
Naumann and Yeh's analysis 195
Non-convecting gust 192
Non-convecting streamwise gust 193
Nonlinear damping forced vibration response 318
Nonlinear damping values 282
Nozzle passing excitation 3, 292
Nozzle passing frequency 292

Orpheus engine bladed disk 372

Packeted blade continuous system analysis 151
Packeted blade group with lacing wire 167
Packeted blades coupled modes 161
Packeted blades even number 160, 164
Packeted blades even symmetry 160, 164
Packeted blades odd number 160, 164
Packeted blades odd symmetry 160, 164
Parabolic camber 194
Parabolic camber airfoil results 228
Petricone-Sisto plate bending modes 116
Plate modes 115, 127, 142
Point receptance disc 343
Poisson summation 219
Polynomial coefficients 74
Polynomial frequency equation 77
Potential energy shell 107
Potential energy gain due to rotation 109
Potential interaction of stator on rotor 211

Prandtl-Glauert equation 201
Pre-twisted beam type blade torsion modes 39
Pre-twisted beam type blades bending modes 34
Pre-twisted plate 100
Pre-twisted plate bending modes 116
Pre-twisted plate effect of rotation 121
Pre-twisted plate torsion modes 123
Pressure ratio correction 236
Principal stresses forced vibration response 308
Prohl method for forced vibration 302
Prohl's analysis packeted blades 156
Proportional damping 291

Radii of curvature 101
Rawtani-Dokainish plate values 116
Receptances blade 343
Receptances blade closed form 344
Receptances blade series form 345
Receptances damped blades 353
Receptances disc 338
Receptances disc closed form 340
Receptances disc series form 341
Receptances of multi-stage blades 365
Receptances of multi-stage disc 358
Receptances of multi-stage disc closed form 361
Receptances of multi-stage disc series form 363
Receptances of shroud 367
Reissner's functional 306
Reissner's variational principle 305
Reliability factor 391
Resonant response factors coupled motion 304
Resonant response factors packeted blades 167, 302
Resonant response nonlinear damping 318
Resonant stress components 310
Resonant stresses transient response 333
Response of accelerating blade 325
Response of Bladed discs 352
Rotary inertia effect in shells 119
Rotary inertia effect plate modes 130
Rotary inertia torsion modes of plates 126
Rotating pre-twisted plates 120
Rotating water table 238
Rotation effect beam type of blades 46
Rotation effect plate modes 130
Rotation effect torsion modes of plates 127
Rotor speeds for resonance of blades 300

S-N diagram 389
Sear's function 191
Setting angle effect 49
Shape functions pre-twisted plates 114
Shape functions shells 134
Shape functions beam type blade 25, 307
Shear and rotary inertia beam type blades 39
Shell modes 142
Shroud receptances 367
Shroud bladed discs 357
Shroud blades (see packeted blades)
Size factor 391
Skewed mean line camber 194
Stage interference 201
Stagger angle effect bending modes 49
Stagger angle effect plate modes 131
Stagger angle effect pre-twisted plates 123
Static contact finite element problem 258
Strain components in shell 104
Strain displacement relations 86, 136
Strain energy beam type blade 24
Strain energy packeted blades 152
Streamwise gust 193
Stress blocks cumulative damage 398
Stress concentration factor 391
Stress harmonics 312
Stress response 308

Stress response accelerating blade 332
Stress strain relations for shells 136
Superparametric element 133
Support stiffness in bending 57
Support stiffness in torsion 59
Surface factor 391
Swaminadham's pre-twisted blade values 120

Tangential modes packeted blades 151, 157
Tangential vibration resonant response 302
Tapered blade bending modes 29
Temperature effect 92
Temperature ratio correction 237
Theodorsen's function 191
Tie wire and cover 7
Torsion modes packeted blades 161
Torsion modes pre-twisted plates 126
Torsion modes of plates effect of rotation 127
Torsion modes of plates effect of rotary inertia 126
Torsional stiffness due to pre-twist 39
Transient response cantilever beam 325
Transient response of blades 329
Transient stress response 332
Transverse gust 191
Tuned systems 347
Turbomachine stage 202

Upwash 192
Upwash due to viscous make shed by stator 220
Upwash on rotor due to potential interaction 211
Upwash on rotor due to wake shed by stator 215
Upwash on stator blade 207, 210

Virtual work principle 139
Viscous wake interaction 216
Viscous wake of an airfoil 217
Vortex wake effect on rotor 214
Vyas' theoretical values for beam type blades 91

Wake strength 192, 215
Warburton value for pre-twisted plates 116
Warping function 42
Water table 238
Wilson-θ method 271
Work of external forces 292